Technological Breakthroughs in Modern Wireless Sensor Applications

Hamid Sharif
University of Nebraska - Lincoln, USA

Yousef S. Kavian
Shahid Chamran University of Ahvaz, Iran

A volume in the Advances in Wireless Technologies and Telecommunication (AWTT) Book Series

Information Science REFERENCE
An Imprint of IGI Global

Managing Director: Lindsay Johnston
Managing Editor: Austin DeMarco
Director of Intellectual Property & Contracts: Jan Travers
Acquisitions Editor: Kayla Wolfe
Production Editor: Christina Henning
Development Editor: Caitlyn Martin
Typesetter: Cody Page
Cover Design: Jason Mull

Published in the United States of America by
Information Science Reference (an imprint of IGI Global)
701 E. Chocolate Avenue
Hershey PA, USA 17033
Tel: 717-533-8845
Fax: 717-533-8661
E-mail: cust@igi-global.com
Web site: http://www.igi-global.com

Library of Congress Cataloging-in-Publication Data

Technological breakthroughs in modern wireless sensor applications / Hamid Sharif and Yousef S. Kavian, editors.
pages cm
Includes bibliographical references and index.
ISBN 978-1-4666-8251-1 (hardcover) -- ISBN 9781466682528 (ebook) 1. Wireless sensor networks. 2. Wireless communication systems. I. Sharif, Hamid R. (Hamid Reza), 1958- editor. II. Kavian, Yousef S., 1978- editor.
TK7872.D48T426 2015
681'.2--dc23

2015003300

This book is published in the IGI Global book series Advances in Wireless Technologies and Telecommunication (AWTT) (ISSN: 2327-3305; eISSN: 2327-3313)

British Cataloguing in Publication Data
A Cataloguing in Publication record for this book is available from the British Library.

All work contributed to this book is new, previously-unpublished material. The views expressed in this book are those of the authors, but not necessarily of the publisher.

For electronic access to this publication, please contact: eresources@igi-global.com.

Advances in Wireless Technologies and Telecommunication (AWTT) Book Series

Xiaoge Xu
The University of Nottingham Ningbo China

ISSN: 2327-3305
EISSN: 2327-3313

Mission

The wireless computing industry is constantly evolving, redesigning the ways in which individuals share information. Wireless technology and telecommunication remain one of the most important technologies in business organizations. The utilization of these technologies has enhanced business efficiency by enabling dynamic resources in all aspects of society.

The **Advances in Wireless Technologies and Telecommunication Book Series** aims to provide researchers and academic communities with quality research on the concepts and developments in the wireless technology fields. Developers, engineers, students, research strategists, and IT managers will find this series useful to gain insight into next generation wireless technologies and telecommunication.

Coverage

- Radio Communication
- Wireless Technologies
- Mobile Web Services
- Network Management
- Grid Communications
- Telecommunications
- Mobile Technology
- Mobile Communications
- Digital Communication
- Virtual Network Operations

IGI Global is currently accepting manuscripts for publication within this series. To submit a proposal for a volume in this series, please contact our Acquisition Editors at Acquisitions@igi-global.com or visit: http://www.igi-global.com/publish/.

The Advances in Wireless Technologies and Telecommunication (AWTT) Book Series (ISSN 2327-3305) is published by IGI Global, 701 E. Chocolate Avenue, Hershey, PA 17033-1240, USA, www.igi-global.com. This series is composed of titles available for purchase individually; each title is edited to be contextually exclusive from any other title within the series. For pricing and ordering information please visit http://www.igi-global.com/book-series/advances-wireless-technologies-telecommunication-awtt/73684. Postmaster: Send all address changes to above address.

Titles in this Series

For a list of additional titles in this series, please visit: www.igi-global.com

Handbook of Research on Software-Defined and Cognitive Radio Technologies for Dynamic Spectrum Management
Naima Kaabouch (University of North Dakota, USA) and Wen-Chen Hu (University of North Dakota, USA)
Information Science Reference • copyright 2015 • 927pp • H/C (ISBN: 9781466665712) • US $505.00 (our price)

Interdisciplinary Mobile Media and Communications Social, Political, and Economic Implications
Xiaoge Xu (The University of Nottingham Ningbo China, China)
Information Science Reference • copyright 2014 • 409pp • H/C (ISBN: 9781466661660) • US $205.00 (our price)

Cognitive Radio Sensor Networks Applications, Architectures, and Challenges
Mubashir Husain Rehmani (Department of Electrical Engineering, COMSATS Institute of Information Technology, Pakistan) and Yasir Faheem (Department of Computer Science, COMSATS Institute of Information Technology, Pakistan)
Information Science Reference • copyright 2014 • 313pp • H/C (ISBN: 9781466662124) • US $235.00 (our price)

Game Theory Applications in Network Design
Sungwook Kim (Sogang University, South Korea)
Information Science Reference • copyright 2014 • 500pp • H/C (ISBN: 9781466660502) • US $225.00 (our price)

Convergence of Broadband, Broadcast, and Cellular Network Technologies
Ramona Trestian (Middlesex University, UK) and Gabriel-Miro Muntean (Dublin City University, Ireland)
Information Science Reference • copyright 2014 • 333pp • H/C (ISBN: 9781466659780) • US $235.00 (our price)

Handbook of Research on Progressive Trends in Wireless Communications and Networking
M.A. Matin (Institut Teknologi Brunei, Brunei Darussalam)
Information Science Reference • copyright 2014 • 592pp • H/C (ISBN: 9781466651708) • US $380.00 (our price)

Broadband Wireless Access Networks for 4G Theory, Application, and Experimentation
Raul Aquino Santos (University of Colima, Mexico) Victor Rangel Licea (National Autonomous University of Mexico, Mexico) and Arthur Edwards-Block (University of Colima, Mexico)
Information Science Reference • copyright 2014 • 452pp • H/C (ISBN: 9781466648883) • US $235.00 (our price)

Multidisciplinary Perspectives on Telecommunications, Wireless Systems, and Mobile Computing
Wen-Chen Hu (University of North Dakota, USA)
Information Science Reference • copyright 2014 • 305pp • H/C (ISBN: 9781466647152) • US $175.00 (our price)

IGI GLOBAL
DISSEMINATOR OF KNOWLEDGE
www.igi-global.com

701 E. Chocolate Ave., Hershey, PA 17033
Order online at www.igi-global.com or call 717-533-8845 x100
To place a standing order for titles released in this series, contact: cust@igi-global.com
Mon-Fri 8:00 am - 5:00 pm (est) or fax 24 hours a day 717-533-8661

Table of Contents

Detailed Table of Contents

The safety of rail transportation has always been the top priority for the Federal Railroad Administration (FRA). Legacy technology is still largely relied upon for detection of faults. Modern technology is mostly used to detect a particular railcar rather than to monitor it for problems. Wireless Sensor Network (WSN) technology is being evaluated by the railroads for real-time or near real-time monitoring of the railcar status for timely response to problems and for trend analysis. In this chapter the authors first highlight the importance of freight rail transportation, followed by briefly discussing specific wireless technologies of interest. In particular, the authors present the shortcomings of the ZigBee protocol in the application domain of railcar monitoring. The authors then introduce hybrid technology protocol. Finally, the authors discuss HTNMote, a hardware platform to implement hybrid technology network and present the results of tests showing the benefits of the new protocol and hardware.

Underwater Wireless Sensor Networks (UWSNs) are used in applications such as mineral exploration and environmental monitoring, and must offer reliability and energy efficiency. These are related to each other in the sense that the former requires error-correction which in turn requires energy, consuming battery life in an environment where battery replacement and recharging are difficult. This chapter thus addresses the energy efficiency of three suitable error correction methods for UWSNs, namely Automatic Repeat Request (ARQ), Forward Error Correction (FEC) and Network Coding (NC). The performance of the schemes as a function of transmission distance is determined for various packet sizes by using models of attenuation and noise that represent the underwater environment. ARQ offers the lowest efficiency and NC the highest but there is a distance at which FEC becomes the best option rather than NC suggesting a hybrid FEC/NC method.

This chapter explores the design of wireless sensor networks for applications in precision agriculture. A short review of developments in precision agriculture and recent applications of wireless sensor networks in the area is presented. The authors present their design of medium access control and network layer protocols exploring the challenges and opportunities associated with the design of such a networked system. The physical layer in their network allows multiple power modes in both receive and transmit operations. The MAC layer employs these multiple power modes to implement a novel wake-up synchronization mechanism to reduce the energy overhead. The network layer ensures reliable collection of data while balancing the energy consumption among the nodes. Finally, the authors present an analytical approach to model the behavior of the MAC protocol developed and compare it against the duty-cycle based S-MAC protocol. The results are also confirmed using simulations.

Reducing power consumption and per-hop latency are major challenges in Wireless Sensor Networks (WSNs), when network nodes or "tags" must be battery-powered. On the other hand, reducing transition overhead in RFID technology is crucial for efficient data collection and quick recognition of multiple tags. This chapter presents the method of analysis performed and explains the calculations used to evaluate the various performance behaviors based on the following mechanisms: (1) a duty cycle complying with the IEEE802.15.4 standard when assuming a fixed set of tags and free operating conditions. The system can be regarded as Reader Talk First (RTF), which is implemented during a tag collection process executed at the beginning of the system activation. (2) Wake-Up Mechanism, the tag transmits an identification packet to the reader and then waits for an Acknowledgment signal from the reader.

The energy efficiency is a main challenging issue for employing wireless sensor networks (WSNs) in extreme environments where the media access progress consumes the main part of network energy. The IEEE 802.15.4 is adopted in low complexity, ultra-low power and low data rate wireless sensor applications where the energy consumption of nodes should be managed carefully in harsh and inaccessible environments. The beacon-enabled mode of the IEEE 802.15.4 provides a power management scheme. When the network traffic is variable, this mode does not work as well and the coordinator is not capable for estimating the network traffic and adjusting proper duty cycle dynamically. In this chapter an approach for estimating network traffic in star topology is proposed to overcome this issue where the coordinator could estimate the network traffic and dynamically adjusts duty cycle proportion to the variation of network traffic. The simulation results demonstrate the superiority of proposed approach for improving the energy consumption, throughput and delay in comparison with the IEEE 802.15.4 under different traffic conditions.

Sensor network deployments have enabled automated data collection at a finer granularity compared to human-centric sparse deployments of traditional telemetric data loggers. However, this fine granularity cannot be achieved in the presence of failing nodes, frequent network failures, and/or potentially malicious network behaviors. Moreover, given a perfect and secure network with long-lived sensor nodes, the collected sensor data should be fault-free to guarantee quality of reasoning. Unlike traditional devices, wireless sensor motes are resource-constrained battery-powered devices equipped with few on-board sensors which are susceptible to failures, and batteries cannot always be reliably recharged. Over the years, sensor network deployments have materialized a large number of monitoring and event-detection applications. Researchers and decision makers have observed several anomalous patterns ranging from functional failures to pervasive sensing faults in most of these deployments. In this chapter, we discuss a few of these deployments to emphasize the need for an anomaly detection mechanism.

Wireless sensor networks (WSNs) are deployed for various applications such as military applications, environmental monitoring, security and surveillance, health care applications and so on. In most of these applications, reliable data transport is of great importance and also a facet of quality of service (QoS). This chapter discusses reliable data transport approaches and protocols. The presented protocols should be energy efficient besides they guarantee reliability in data transmission. The authors review protocols mostly based on packet and event reliability taking advantage of retransmission and redundancy mechanisms. In addition, network component failures and fault tolerant techniques should be also considered in reliable WSN designs. Since different reliability evaluation modeling approaches is presented. Finally the authors highlight the challenges to WSNs reliability enhancement and some future research directions are addressed in particular.

Over the course of operation, sensor network deployments usually exhibit numerous types of anomalies that can result in missing, erroneous, malicious, or rather meaningful data points indicating an important event. Sensor data points can go missing due to network performance degradation as a result of a failing node, a network failure, or a malicious network behavior. Erroneous points are faulty measurements potentially generated by malfunctioning sensors. Malicious data points are usually injected into the network, replayed, or altered by an adversary. Finally, events manifest themselves as abnormal data points reported by sensor nodes expectedly. The significance of such anomalies calls for detection mechanisms that can identify an abnormal behavior in a sensor network deployment. In this chapter, the authors systematically analyze the different types of anomalies prevalent in a sensor network, construct a taxonomy of such anomalies, and discuss potential detection mechanisms necessary to reveal the root causes behind each anomaly.

This chapter addresses fault diagnosis agreement problem in a network with malicious members. The authors provide a new algorithm to reach an agreement among fault-free members about the faulty ones. The algorithm is designed for fully-connected networks and also can be easily adapted to partially connected networks. The authors contribution is to reduce bit complexity of Byzantine agreement process by detecting the same list of faulty units in all fault-free members. Therefore, the malicious units can be removed from other consensus processes. Also, each healthy unit detects a list of malicious units locally which results in lower packet transmission in the network. The authors provided algorithm solves fault diagnosis agreement problem in 2t+1 rounds and O(nt+1) bit complexity for each member.

Spectrum analysis is one of the momentous fields in signal processing. It has a large variety of applications in radar, sonar, speech and image processing. Parametric methods have been proposed and employed for spectrum analysis including power spectral density (PSD) estimation. These methods estimate the parameters of a statistical model and compute the PSD, afterwards. In some circumstances one is obliged to deal with observations of numerous geographically dispersed sensors, to either increase the precision or based on application demands. Having a set of sensors linked together to take the advantages of cooperation and network topology, one obtains a more comprehensive estimation. In this chapter, the authors propose and study four different algorithms capable of facing spatio-temporal variations for parametric modeling and PSD estimation using wireless sensor networks (WSNs). For this purpose, the authors first validate the proposed algorithms using theoretical and mathematical formulations. Thereafter, performing simulation tasks demonstrates and supports the theoretical achievements. The next section of the chapter illustrates the concepts to a greater degree, the authors analyze and compare the performance of these algorithms with each other, as well as with the simple PSD estimation using individual sensors, wherein there is no cooperation among the nodes.

In this chapter, the authors introduce a threat in Wireless Sensor Network (WSN) E-Health applications that can be used for leaking patients' medical information, identity, personal data, and medical history. This threat is called covert communication, which establishes a secret path hidden within legitimate network traffic to transfer information. Based on the exploit used for this type of communication, covert channels have different categories and functionalities. The authors introduce various covert communication methods and their possible applications in transmitting e-health data secretly in WSNs. In addition, the authors discuss different methods and approaches to detect this type of communication in WSNs to prevent this form of malicious e-health information transfer.

Preface

The fast growing trend of sensing and measuring the every parameter that is directly related in our lives has led the sensors technology for producing sensors with effective information gathering, efficient energy consumption, resource sensitive communications, miniaturized size, and other key operational factors. With this focus, sensors have been designed and introduced to support in most domains including transportation, communications, security, health care, education, manufacturing, data processing, agriculture, food science, etc.

As the number of the sensors increases in an application, the challenge of communications among the sensors becomes significant. The approaches for communications to create a network of sensors include either wired or wireless. With the advances in wireless technology, Wireless Sensor Networks (WSN) have become the focal point of sensing and monitoring technology. A WSN can provide an effective communication network for transmitting sensed information with effective energy consumption, low cost of infrastructure, and reasonable geographical coverage.

The design of a WSN depends significantly on an application. The variety of possible applications of WSNs to the real world is practically unlimited. It is evident that each application strongly affects the choice of the WSN elements including the wireless technology.

Several classifications of WSN applications have been proposed to distinguish applications according to the type of data that must be gathered in the network. In general, the WSN applications could be classified into two categories of "Event Detection" and "Spatial Process Estimation". The class of Event Detection includes detection of a specific event which a sensor has to compare the measured quantity with a given threshold to detect an event. The density of nodes must ensure that the event is detected and forwarded with a suitable probability of success while maintaining a low probability of a false alarm.

In Spatial Processing Estimation, the focus is on estimating a given physical phenomenon to which can be modelled as a random process typically. In this case, the objective is to obtain the entire behavior of the spatial process estimation based on the samples taken by sensors that are typically placed in random positions. It is clear that higher nodes density lead to a more accurate scalar field reconstruction at the expense of a larger network. Additionally, the estimation error is strictly related to nodes density as well as on the spatial variability of the process.

The system requirements for WSNs are also application dependent. Because of the variety of possible applications for WSNs, the system requirements could change significantly. The dominant system requirement could include energy efficiency, data rate, and wireless links.

WSN has been classified by MIT as one of the ten merging technologies that will change the word due to its flexibility in support of unlimited applications including the Internet of Things.

Recent advances in semiconductor, networking and material science technologies are driving the ubiquitous deployment of new generation of WSNs. Together, these technologies enable a new generation of WSNs which differ greatly from wireless networks developed and deployed as recently as 5 to 10 years ago. Today's state-of-the-art WSNs enables applications to support lower deployment and maintenance costs, lasting longer and more rugged.

WSN deployment requires a broad range of specialties including energy storage, wireless networking protocols, semiconductor design and sensor/actuator technology. The complexity of WSN design represents one of the most significant barriers to the widespread adoption of modern WSN applications but also provides an opportunity for hardware and software technology research.

This book provides a series of chapters outlining the latest research findings of well-known scientists around the world to understanding the complex tradeoffs among many modern WSN application variables including the challenges of wireless communications, energy management system, deployment costs, hardware and software, system reliability, security and performance.

The safety of rail transport has always been the top priority for the Federal Railroad Administration (FRA). Legacy technology, like wayside monitoring, is still in place and is largely relied upon for detection of faults. Modern technology like Radio Frequency Identification (RFID) has been introduced recently. However, this is largely used to detect a particular railcar rather than to monitor it for problems. Wireless Sensor Network (WSN) technology is being evaluated by the railroads for real-time or near real-time monitoring of the status of railcars for timely response to problems and also for trend analysis.

ZigBee has been the networking protocol of choice for the railroads for its low power consumption and cost of implementation. The railroad scenario presents a long linear-chain like network topology which ZigBee was not designed to handle. It has been found that a ZigBee-only network in the railroad environment suffers from drawbacks like long synchronization delays, severe problems with route discovery and maintenance, aggregation of data errors leading to unacceptable packet loss rates, lack of a mechanism to decide traffic priority for critical packets, like alarm, so that they can reliably traverse the network to the collecting node in the locomotive etc. Hybrid Technology Networking (HTN) protocol has been suggested which addresses the shortcomings of ZigBee in the railroad scenario. It proposes a standards-based multi-protocol approach that is well-suited for the railroad scenario. The current crop of sensor platforms does not provide an integrated environment for the implementation of HTN.

In *Chapter 1*, *Rakshit, Rezaei, Lal Shrestha, Hempel, and Sharif* develop a new hardware platform, HTNMote, to implement the HTN protocol in the resource constrained, high stress railroad environment to enable fine-grained and near real-time monitoring of railcar parameters.

Underwater Wireless Sensor Networks (UWSNs) are used in applications such as mineral exploration and environmental monitoring, and must offer reliability and energy efficiency. These are related to each other in the sense that the former requires error-correction which in turn requires energy, consuming battery life in an environment where battery replacement and recharging are difficult. In *Chapter 2*, *Leeson and Patel* address the energy efficiency of three suitable error correction methods for UWSNs, namely Automatic Repeat Request (ARQ), Forward Error Correction (FEC) and Network Coding (NC). The performance of the schemes as a function of transmission distance is determined for various packet sizes by using models of attenuation and noise that represent the underwater environment. ARQ offers the lowest efficiency and NC the highest but there is a distance at which FEC becomes the best option rather than NC suggesting a hybrid FEC/NC method.

Chapter 3, by *Sahota, Kumar, and Kamal,* explores the design of wireless sensor networks for the implementation of information and control technologies in precision agriculture. A short review of the historical developments in precision agriculture and the recent applications of wireless sensor networks in the area is presented. A new design of medium access control (MAC) and network layer protocols that are suited to the application requirements is presented and the challenges and the opportunities associated with the design of such a networked system is explored. The physical layer in presented network allows multiple power modes in both receive and transmit operations. The MAC layer employs these multiple power modes to implement a novel wake-up synchronization mechanism to reduce the overhead involved in the process. The network layer ensures reliable collection of data while balancing the energy consumption among the various nodes. Finally, an analytical approach to model the behavior of the MAC protocol developed and compare it against the well-known duty-cycle based S-MAC protocol is considered. The savings achieved in energy and delay compared to the latter are also confirmed using simulations.

Radio Frequency Identification (RFID) is a type of automatic identification using radio waves to automatically identify people or objects. The focus of Wireless Sensor Network (WSN) protocols is usually on long-range and real-time applications. Reducing power consumption and per-hop latency are major challenges in WSNs, when network nodes or "tags" must be battery-powered. On the other hand, reducing transition overhead in RFID technology is crucial for efficient data collection and quick recognition of multiple tags. *Chapter 4,* by *Abdulla and Ismail*, presents the method of analysis performed and explains the calculations used to evaluate the various performance behaviors based on the following mechanisms: (1) a duty cycle complying with the IEEE802.15.4 standard when assuming a fixed set of tags and free operating conditions. The system can be regarded as Reader Talk First (RTF), which is implemented during a tag collection process executed at the beginning of the system activation. (2) Wake-Up Mechanism, the tag transmits an identification packet to the reader and then waits for an Acknowledgment signal from the reader. This process, known as Tag Talk First (TTF), is implemented during the interrupt step. This research also discusses the effects of changing parameters in conventional IEEE802.15.4 protocol (RTF) and in the RFID system (TTF), and predicts the behavior of sensor network applications using comparisons between them, including the effect of TTF and RTF raw and effective data transmission rates.

The energy efficiency is a main challenging issue for employing wireless sensor networks in extreme environments where the media access progress consumes the main part of network energy. The IEEE 802.15.4 is adopted in low complexity, ultra-low power and low data rate wireless sensor applications where the energy consumption of nodes should be managed carefully in harsh and inaccessible environments. The beacon-enabled mode of the IEEE 802.15.4 provides a power management scheme. When the network traffic is variable, this mode does not work as well and the coordinator is not capable for estimating the network traffic and adjusting proper duty cycle dynamically. In *Chapter 5,* by *Kavian and Rasouli*, an approach for estimating network traffic in star topology is proposed to overcome this issue where the coordinator could estimate the network traffic and dynamically adjusts duty cycle proportion to the variation of network traffic. The simulation results demonstrate the superiority of proposed approach for improving the energy consumption, throughput and delay in comparison with the IEEE 802.15.4 standard under different traffic conditions.

Sensor network deployments have enabled automated data collection at a finer granularity compared to human-centric sparse deployments of traditional telemetric data loggers, driving scientists or application administrators to draw better conclusions from the collected sensor data. However, this fine granularity cannot be achieved in the presence of failing nodes, frequent network failures, and/or potentially malicious network behaviors. Moreover, given a perfect and secure network with long-lived sensor nodes, finer-grained data samples may still not guarantee the quality of reasoning during the decision making process unless the sensor dataset is fault-free. Unlike traditional devices, wireless sensor motes are resource-constrained battery-powered devices equipped with few on-board sensors which are susceptible to failures, and batteries cannot always be reliably recharged. Over the years, sensor network deployments have materialized a large number of monitoring and event-detection applications. Researchers and decision makers have observed several anomalous patterns ranging from functional failures to pervasive sensing faults in most of these deployments. In *Chapter 6*, *Abuaitah and Wang* discuss a few of these deployments to emphasize the need for an anomaly detection mechanism.

Wireless sensor networks (WSNs) are deployed for various applications such as military applications, environmental monitoring, security and surveillance, health care applications and so on. In most of these applications, reliable data transport is of great importance and also a facet of quality of service (QoS). In *Chapter 7*, *Ansari and Mahani* discuss reliable data transport approaches and protocols. The presented protocols should be energy efficient besides they guarantee reliability in data transmission. Some protocols which are mostly based on packet and event reliability taking advantage of retransmission and redundancy mechanisms are reviewed. In addition, network component failures and fault tolerant techniques should be also considered in reliable WSN designs. Since different reliability evaluation modeling approaches is presented. Finally the challenges to WSNs reliability enhancement and some future research directions are addressed in particular.

Recent sensor network deployments have enabled event detection and sensor data analysis at a finer granularity. Over the course of operation, sensor network deployments usually exhibit numerous types of anomalies that may result in missing, erroneous, malicious, or rather meaningful data points that indicate an important event. Sensor data points can go missing due to the degradation in network performance and lifetime as a result of a failing node, a network failure, or a malicious network behavior. Erroneous points are those faulty measurements potentially generated by malfunctioning sensors. Malicious data points are harmful to the decision making process and are usually injected into the network, replayed, or altered by an adversary. Finally, events manifest themselves as abnormal data points reported by sensor nodes expectedly. The significance of such anomalies calls for detection mechanisms that can identify an abnormal behavior in a sensor network deployment. In *Chapter 8*, *Abuaitah and Wang* systematically analyze the different types of anomalies prevalent in a sensor network, construct a taxonomy of such anomalies, and discuss potential detection mechanisms necessary to reveal the root causes behind each anomaly.

Chapter 9, by *Khosravi and Kavian,* addresses fault diagnosis agreement problem in ad-hoc network with malicious members. A new algorithm to reach an agreement among fault-free members about the faulty ones is presented. The algorithm is designed for fully-connected networks and also can be easily adapted to partially connected networks. The contribution is to reduce bit complexity of Byzantine agreement process by detecting the same list of faulty units in all fault-free members. Therefore, the malicious units can be removed from other consensus processes. Also, each healthy unit detects a list of malicious units locally which results in lower packet transmission in the network.

In *Chapter 10, Stoyanova, Kerasiotis, and Papadopoulos* discuss the feasibility of sensor node localization by exploiting the inherent resources of WSN technology, such as the Received Signal Strength (RSS) of the exchanged messages. A brief overview of various factors influencing the RSS, including the RF-signal propagation and other topology parameters which influence the localization process and accuracy is presented. Moreover, the RSS variability due to internal factors, related to the hardware implementation of a sensor node, is investigated in order to be considered in simulations of RSS-based outdoor localization scenarios. Localization considerations referring to techniques, topology parameters and factors influencing the localization accuracy are combined in simulation examples to evaluate their significance concerning target positioning performance. Finally, the RF propagation model and the topology parameters being identified are validated in real outdoor localization scenario.

Chapter 11, by *Zhu and Hua,* discusses Mobile Wireless Sensor Networks (MWSNs). First, the evolution of MWSNs from sensor networks to wireless sensor networks, and finally to mobile networks is considered. Second, chapter compares the peer work of MWSNs in chronological order. More typical issues will be discussed in the third section, including localization, deployment, resource/energy efficiency and coverage issues. Hence, Cross-layer design is considered as a useful way to improve MWSNs in the future.

Mobile target tracking is one of the most important applications of wireless sensor networks (WSNs). But, the use of sensor networks for object tracking faces a number of issues in which the limited energy supply is the most important. So in target tracking problem, using methods to decrease the energy consumption as well as high accuracy and quality of tracking is the main goal. Hence, reducing the number of participant nodes in tracking phase, increasing the sleep duration of noninvolved nodes and decreasing the number of transmitted packets to the sink are the most referred methods. In *Chapter 12, Mirsadeghi and Mahani* introduce the most suitable methods for energy efficient mobile object tracking.

Spectrum analysis is one of the momentous fields in signal processing. It has large variety of applications in radar, sonar, speech and image processing. Parametric methods have been proposed and employed for spectrum analysis including Power Spectral Density (PSD) estimation. These methods estimate the parameters of a statistical model and compute the PSD, afterwards. In some circumstances, it is obliged to deal with observations of numerous geographically dispersed sensors, to either increase the precision or based on application demands. Having a set of sensors linked together to take the advantages of cooperation and network topology, we will obtain a more comprehensive estimation. In *Chapter 13, Nosrati, Taheri, Shamsi, and Sedaaghi* propose and study four different algorithms capable of facing spatio-temporal variations for parametric modeling and PSD estimation using Wireless Sensor Networks (WSNs). For this purpose, they first validate the proposed algorithms using theoretical and mathematical formulations. Thereafter, performing simulation tasks demonstrate and well support the theoretical achievements. At the next compartment of the chapter in order to illustrate the concepts to a greater degree, they elaborate on analyzing and then comparing the performance of these algorithms with each other, as well as with the simple PSD estimation using individual sensors, wherein there is no cooperation among the nodes.

In *Chapter 14, Rezaei, Lal Shrestha, Hempel, Rakshit, and Sharif* introduce a threat in Wireless Sensor Network (WSN) E-Health applications that can be used for leaking patients' medical information, identity, personal data and medical history. This threat is called covert communication, which establishes a secret path hidden within legitimate network traffic to transfer information. Based on the exploit used for this type of communication, covert channels have different categories and functionalities. They introduce various covert communication methods and their possible applications in transmitting e-health data secretly in WSNs. In addition, different methods and approaches to detect this type of communication in WSNs to prevent this form of malicious e-health information transfer is discussed.

Hamid Sharif
University of Nebraska - Lincoln, USA

Yousef S. Kavian
Shahid Chamran University of Ahvaz, Iran

Acknowledgment

We would like to thank the chapter authors for providing us with the fundamental constituents of this volume and the reviewers who provided detailed and constructive feedback. Also, thanks are due to members of staff at the publisher who have guided us through the publishing process.

Finally, to my family and friends such as Yousef for being such an outstanding colleague. Hamid Sharif

Yousef dedicates this work to his parents, wife and daughter.

Hamid Sharif
University of Nebraska-Lincoln, USA

Yousef S. Kavian
Shahid Chamran University of Ahvaz, Iran

Chapter 1
HTNMote:
A Platform for On-Board Real-Time Monitoring of Railcars

Sushanta Mohan Rakshit
University of Nebraska-Lincoln, USA

Pradhumna Lal Shrestha
University of Nebraska-Lincoln, USA

Fahimeh Rezaei
University of Nebraska-Lincoln, USA

Michael Hempel
University of Nebraska-Lincoln, USA

Hamid Sharif
University of Nebraska-Lincoln, USA

ABSTRACT

The safety of rail transportation has always been the top priority for the Federal Railroad Administration (FRA). Legacy technology is still largely relied upon for detection of faults. Modern technology is mostly used to detect a particular railcar rather than to monitor it for problems. Wireless Sensor Network (WSN) technology is being evaluated by the railroads for real-time or near real-time monitoring of the railcar status for timely response to problems and for trend analysis. In this chapter the authors first highlight the importance of freight rail transportation, followed by briefly discussing specific wireless technologies of interest. In particular, the authors present the shortcomings of the ZigBee protocol in the application domain of railcar monitoring. The authors then introduce hybrid technology protocol. Finally, the authors discuss HTNMote, a hardware platform to implement hybrid technology network and present the results of tests showing the benefits of the new protocol and hardware.

INTRODUCTION

Importance of Freight Railroad

Freight railroad is very important to the economic health of a country as it moves commodities between source and the market place, between the place of origin and the place where those commodities are used as raw materials. Hence freight railroads directly help businesses and industries thrive which in turn drives the economy of a country.

In the United States of America freight railroad accounts for 40% of the total freight volume (US Department of Transportation; Federal Railroad Administration, n.d.), when measured in ton-miles.

DOI: 10.4018/978-1-4666-8251-1.ch001

Ton-mile is defined as the movement of one ton of freight over a distance of one mile. The commodities moved by freight rail vary from coal, chemicals, food and related products, automobiles and their related products, lumber and wood products, minerals, metallic ores, petroleum and coal and other miscellaneous products (Overview of Freight Railroad Industry, n.d.) (Class I Railroad Statistics). Figure 1 and 2 illustrate this distribution.

There are several advantages of using freight railroad as well. They are presented as under:

1. A major advantage of freight rail is that it is as much as three times as fuel efficient as trucks.
2. The movement of freight by rail is environment friendly. Environment Protection Agency (EPA) estimates that freight rail produces only 9 percent of total transportation related nitrous emissions and 4 percent of total transportation related particulate emissions although it accounts for nearly 40 percent of inter-city ton-miles (Overview of Freight Railroad Industry, n.d.).

Figure 1. Percent in ton-miles of freight carried by different modes of transport (Adapted from http://www.fra.dot.gov/Page/P0362)

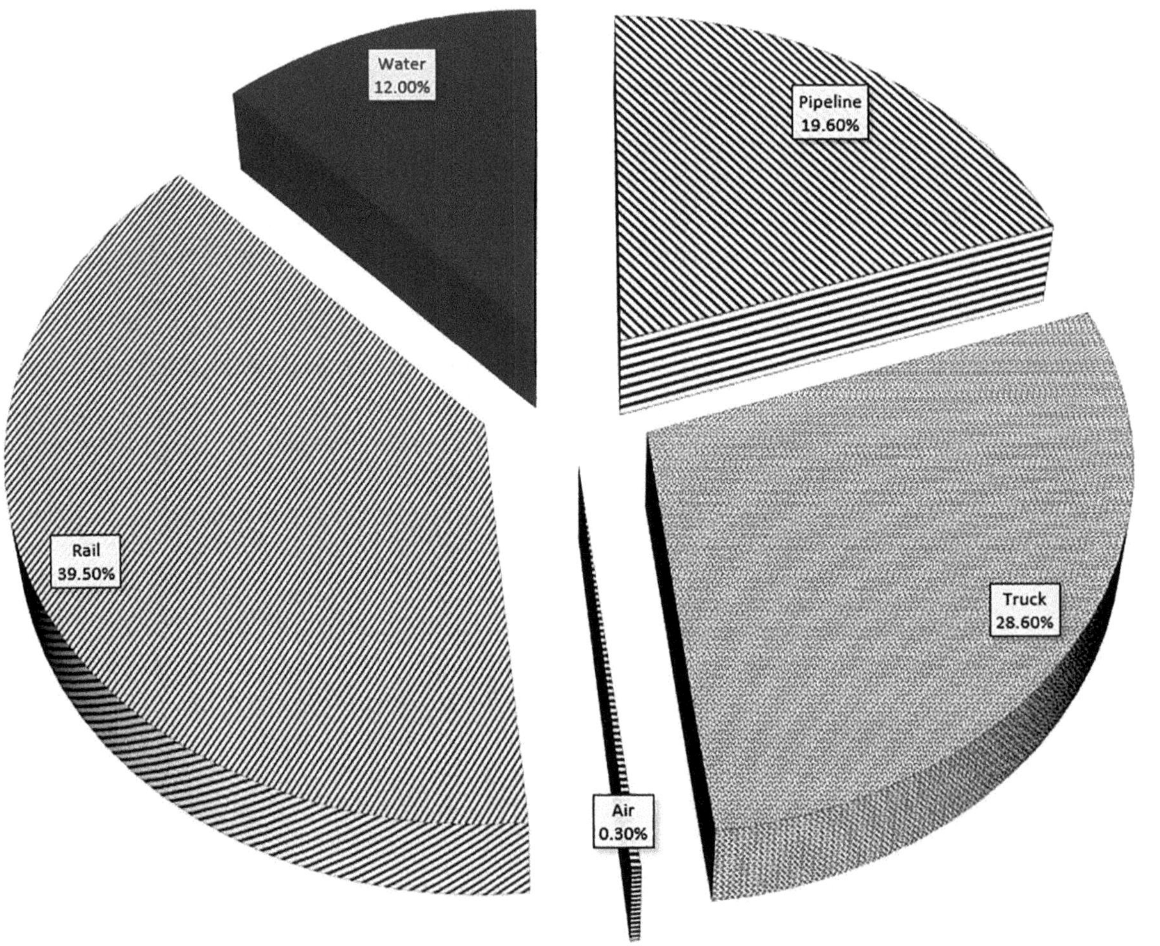

Figure 2. 2005 Class I Railroad revenues by commodity group (Adapted from Association of American Railroads, Class I Railroad Statistics. (n.d.))

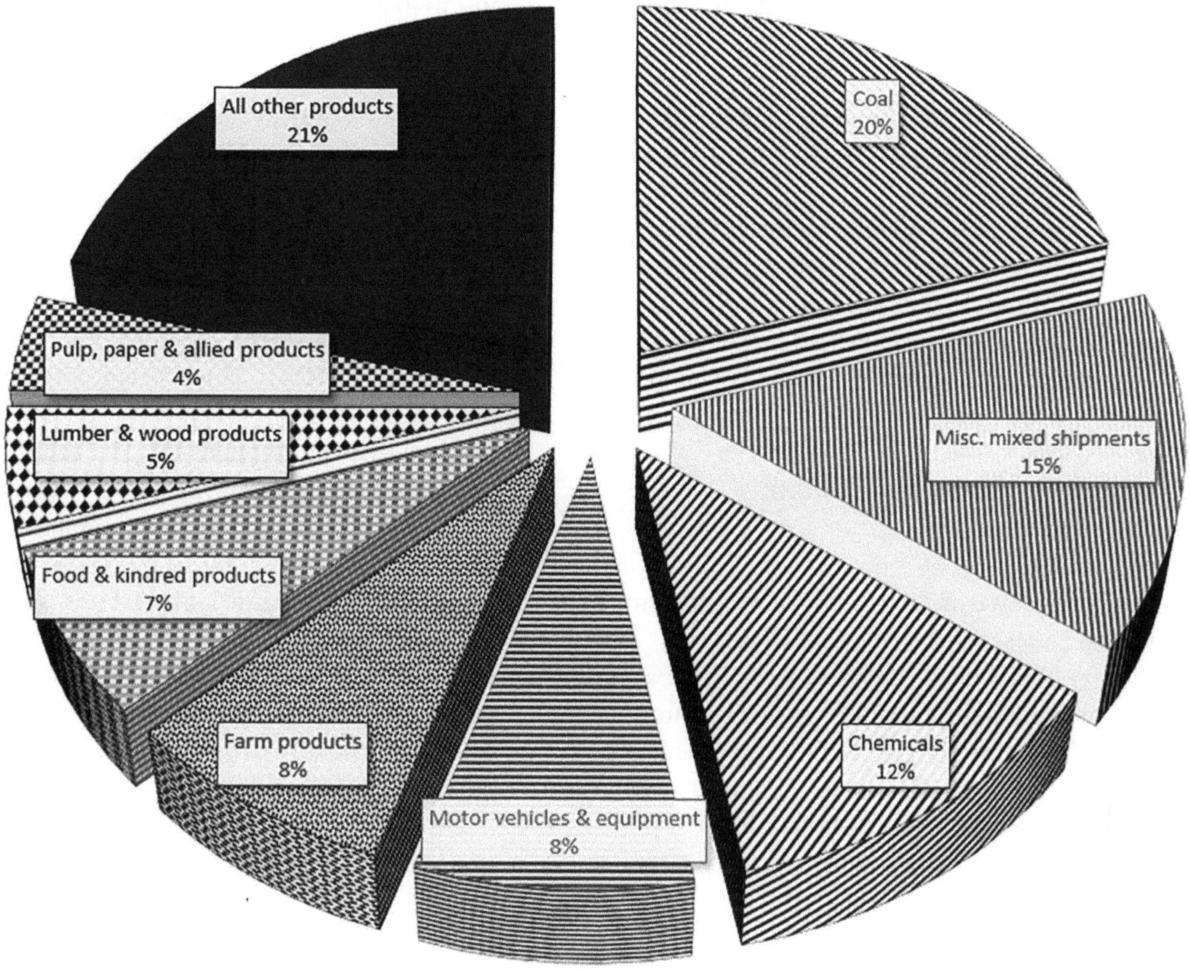

3. It is estimated that a single freight rail takes 280 – 550 trucks off the road (US Department of Transportation; Federal Railroad Administration, n.d.). This contributes significantly to the alleviation of congestion from the road network.
4. Freight railroad is the safest way to carry hazardous materials.

Reasons for Monitoring Freight Railroad

Each freight railcar consists of hundreds of moving mechanical parts, electrical wirings, goods and commodities that are prone to damage and maybe perishable.

The bearing in the wheels of a railcar may overheat and lead to a catastrophic failure and derailment. Electrical systems may fail leading to shut down of a refrigeration unit that would spoil the perishable goods that railcar was carrying. Doors of the railcar may be opened with the intention of stealing merchandise.

Any of these result in huge losses in the form of time and money for both the railroads and the businesses which depend on the goods transported by the railroads.

Hence the ability to monitor status of various components and goods in a railcar becomes of critical importance. Also such monitoring can be used to look at and evaluate trends that can be used to upgrade maintenance schedules and put best practices in place.

Existing Methods for Railcar Status Monitoring

Currently employed methods of monitoring railcar status can be classified as the following:

1. **Wayside Monitoring:** In wayside monitoring certain parameters of a railcar are logged when it passes by the detector. Figure 3 shows a wayside monitoring installation.
2. **Automatic Equipment Identification (AEI):** This form of monitoring is used just to identify the railcar or rail equipment that passed by the detector. It makes use of passive tags that are mounted on the side of railcars or on the rail equipment that is read by a reader mounted by the trackside. This technology makes use of Radio Frequency Identification technology (Automatic Equipment Identification, n.d.) (RFID). Figure 4 shows an AEI setup in the field.
3. **On-Board Monitoring and Control System (OBMCS):** This system puts the sensing equipment on-board the railcar. It has two distinct components. The sensors on-board a single railcar are interconnected by a CAN bus. While intra-train communication and control from the locomotive is achieved using IEEE 802.11(b). The system is also equipped with cellular radio and GPS for transmission of information to remote monitoring stations and location services. The system details are present in (Edwards, et al., 16-18 March 2005).

Figure 3. Wayside monitoring equipment in the field. Note: Wayside Monitoring Equipment. (n.d.). (Railway Technology) Retrieved from http://en.wikipedia.org/wiki/Hot_box licensed under CC BY-SA 3.0 via Wikimedia Commons)

Figure 4. A railcar with an AEI tag mounted on its side. Note: Automatic Equipment Identification. (n.d.). (Wikipedia) Retrieved from http://en.wikipedia.org/wiki/Automatic_equipment_identification (Attribution: AEI tag affixed to the side of a freight car (AEI), 11 January 2007, 14:11:21, Self-photographed, Harvey Henkelmann)

Drawbacks of Current Methods

The methods of railcar status monitoring mentioned briefly above suffer from some severe drawbacks. They are shortly discussed below:

1. Lack of real-time monitoring is a severe drawback plaguing the wayside monitoring technique. The wayside installations are spaced far apart from one another. The distance between two stations or yards may be in the hundreds of miles. Bearing failures are often catastrophic and happen very quickly. There is a good chance that it may take place between two stations. Also failure of the refrigeration unit, opening and closing of doors are events that cannot be effectively monitored by wayside installations.
2. The very limited number of parameters that can be monitored by a wayside installation is also a noticeable drawback. These installations can measure bearing temperature and some very limited other parameters when a railcar passes by.
3. The OBMCS addresses these problems but introduces some unique to it. The use of a wired CAN bus within a railcar fixes the location of the sensors and also makes it hard to introduce a new sensor later on. The use of cellular radio is possible only in areas covered by cellular network. There are vast stretches of land, through which trains need to pass, where there is no cellular coverage. In those areas this system becomes non real-time. Also, the use of IEEE 802.11(b), cellular radio and GPS (Global Positioning System, n.d.) make the

system power hungry. The use of dynamo to generate power to recharge the batteries needs invasive installation on the wheels.

BACKGROUND

This chapter will provide a brief overview of the involved communication protocols, namely IEEE 802.11, , Fixed.4: Displaytech SD028ATFT LCD with a picture displayedFi modulee have been provided at the end of the section.uring the IEEE 802.15.4 and ZigBee.

IEEE 802.11

IEEE 802.11 (IEEE 802.11: Wireless LAN Medium Access Control (MAC) and Physical Layer (PHY) Specifications. (2012 revision)., 5 April 2012) consists of a group of standards that are used to implement Wireless Local Area Network (WLAN) communication in the 2.4, 3.6, 5.0 and 60.0 GHz frequency bands. In 1999, the Wi-Fi Alliance (Wi-Fi Alliance, n.d.) was formed which is a body that evaluates and certifies a product as Wi-Fi compliant. Wi-Fi is the name that has come to be commonly associated with products that conform to the IEEE 802.11 group of standards.

The standard provides for two types of networks. The infrastructure mode depends on all the nodes connecting to an access point. In the ad-hoc mode an access point is not required. The individual nodes establish a network among themselves. Figure 5 is a representation of several nodes in the ad-hoc mode.

The original IEEE 802.11 standard defined communication at 1 or 2 Mbps using Frequency Hopped Spread Spectrum (FHSS) or Direct Sequence Spread Spectrum (DSSS). Later on Orthogonal Frequency Division Multiplexing (OFDM) and High Rate-Direct Sequence Spread Spectrum (HR-DSSS) were introduced to support higher data rates.

Figure 6 shows the channel assignment within the 2.4 GHz band in IEEE 802.11.

The IEEE 802.11 Data Link Layer consists of the Logical Link Control (LLC) sub-layer and the Medium Access Control (MAC) sub-layer.

The LLC is the same for IEEE 802.11 as it is for any other 802 communication protocol (IEEE 802.11 Standards Explanation, n.d.).

The MAC layer in IEEE 802.11 uses Carrier Sense Multiple Access/Collision Avoidance (CSMA/CA) to detect and avoid collisions when two nodes try to transmit at the same time over the same channel. Request-To-Send (RTS) and Clear-To-Send (CTS) constructs are used to tide over the hidden node problem. This layer is also responsible to reliable delivery of large payloads that need fragmentation and reassembly.

Transmission security is one of the foremost concerns in any deployment of the IEEE 802.11 networks. Since it is over-the-air the ease of eavesdropping is much easier than in wired networks. The IEEE 802.11 standard defines a Wired Equivalent Privacy (WEP) for the security of communication. This has been shown to be very weak in the face of sophisticated attacks (Lashkari, Danesh, & Samadi, 8-11 Aug. 2009) (Majstor, 20-24 Oct. 2003) (Sandirigama & Idamekorala, 26-27 Sept. 2009).

In summary, IEEE 802.11 provides for a high data rate long range communication protocol. The finer details of the protocol and its various implementation details can be found in (IEEE 802.11: Wireless LAN Medium Access Control (MAC) and Physical Layer (PHY) Specifications. (2012 revision)., 5 April 2012).

IEEE 802.15.4

IEEE 802.15.4 (IEEE 802.15.4, "Part 15.4: Wireless Medium Access Control (MAC) and Physical Layer (PHY) Specifications for Low Rate Wireless Personal Area Networks (LR WPANs), 2003) standard specifies the physical layer and the medium access layer for Low Rate-

Figure 5. Devices in an ad-hoc IEEE 802.11 network. Note: Devices in Ad-Hoc Mode, (n.d.) (Wikipedia). Retrieved from http://en.wikipedia.org/wiki/Peer-to-peer

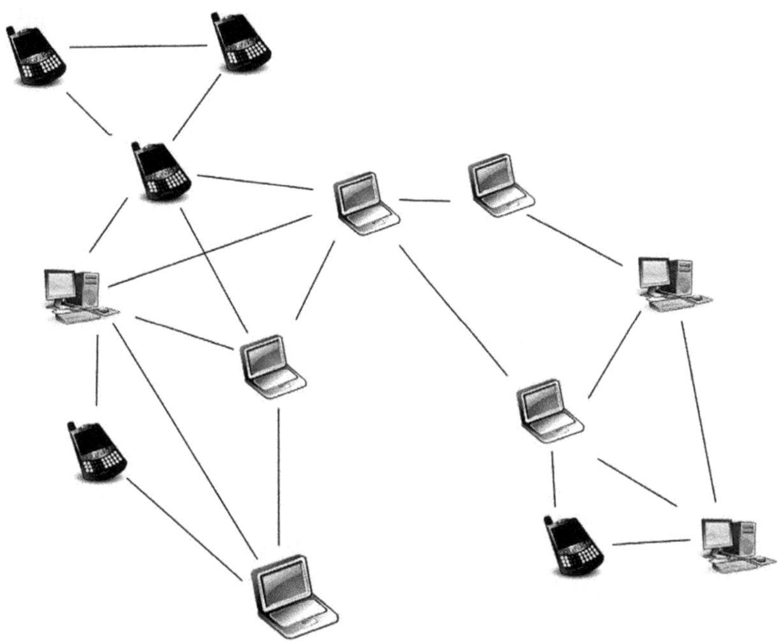

Figure 6. Channel assignment in the 2.4 GHz band for IEEE 802.11. Note: Wi-Fi Architecture, (n.d.) (Wikipedia). Retrieved from http://en.wikipedia.org/wiki/IEEE_802.11 licensed under CC BY-SA 3.0 via Wikimedia Commons)

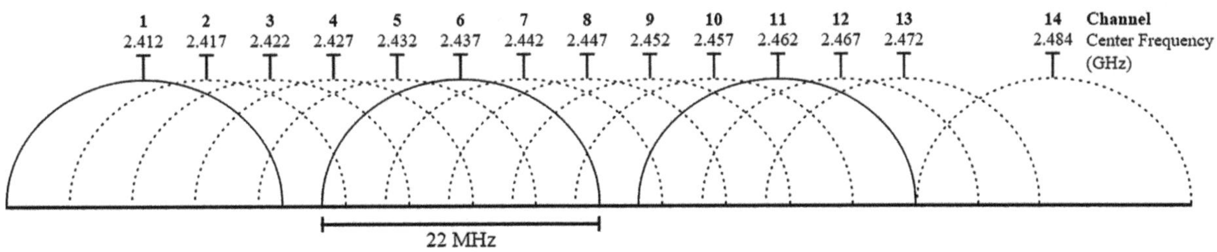

Wireless Personal Area Networks (LR-WPAN). It is designed for networks where communication is within devices close by and infrastructure for network formation and maintenance is virtually absent. It focusses on low power consumption, low data rate and small area of coverage. It is in sharp contrast to IEEE 802.11 or Wi-Fi where the networks deployed are principally infrastructure based and power consumption is not an issue but end-user experience is (Wireless Standards Technology, n.d.).

IEEE 802.15.4 was initially theorized to have a communication distance of 10 m and a maximum theoretical data rate of 250 kbps.

The two versions of IEEE 802.15.4 that are of interest in this research work and their brief introduction is provided in the Table 1 below:

Table 1. IEEE 802.15.4 versions. Note: IEEE 802.15.4 Overview, (n.d.) (Wikipedia) Retrieved from https://en.wikipedia.org/wiki/IEEE_802.15.4.

IEEE 802.15.4 Version	Brief Introduction of Features
IEEE 802.15.4 – 2003	This was the first release of the IEEE 802.15.4 standard. It provided two physical layer bands, one in the 868 and 915 MHz and the other at 2.4 GHz.
IEEE 802.15.4 – 2006	This version of the standard upgraded the data rates possible in the 868 and 915 MHz bands. It also introduced new modulation schemes – three for the 868 and 915 MHz and one for the 2.4 GHz band.

The IEEE 802.15.4 standard specifies the physical and the MAC layers only. The upper layers have to be implemented according to usage scenarios. The details of the layers and their interactions are provided in (IEEE 802.15.4, "Part 15.4: Wireless Medium Access Control (MAC) and Physical Layer (PHY) Specifications for Low Rate Wireless Personal Area Networks (LR WPANs), 2003) (IEEE 802.2 standards, n.d.).

There are some popular upper layer implementations using IEEE 802.15.4 as the base and they include, but are not limited to, ZigBee (ZigBee, n.d.), 6LowPAN (Mulligan, 2007), ISA-100.11a (ISA, n.d.) and MiWi (David Flowers).

There are two types of network topologies that can be used within the IEEE 802.15.4 standard. These are briefly defined below and shown in Figure 7 (Networking Topologies, n.d.):

1. **Star Network:** In this topology there is a central PAN coordinator node and every other node communicates with this node.
2. **Peer-to-Peer Network:** In this topology there still exists the PAN coordinator node but the other nodes can also communicate between themselves without having to go through the coordinator node.

There are three device classes that can exist within an IEEE 802.15.4 network. A Full Function Device (FFD) can send, receive and route data. They consume more power. A special type of FFD is the PAN Coordinator which creates, coordinates and manages the network. Lastly, Reduced Function Devices (RFD) can send and receive data. Since they do not route data they can sleep once they are done sending or receiving data. This makes them very power efficient and are generally used as end-nodes in a network.

ZigBee

ZigBee is a standard which defines the network and application layer over the IEEE 802.15.4 physical and MAC layers for wireless sensor networks. It is maintained by ZigBee alliance, which an open, non-profit collaboration of hundreds of companies. Before any device is declared ZigBee compliant it has to undergo the ZigBee Certification Program (ZigBee Primer). ZigBee has the following characteristics:

1. Leads to low power consumption hence longer battery life.
2. Leads to small device footprint.
3. Allows mesh networking.
4. Self-forming, self-healing network.
5. Interoperability between devices from various vendors but conforming to the ZigBee standard.
6. Security mechanisms in place with 128 bit AES encryption possible.

In addition to the standard the ZigBee Alliance has defined many application profiles as well. Every data request in ZigBee is sent and received on an application profile. Some of the commonly used application profiles are Home Automation, Industrial Plant Monitoring, Commercial Building Automation, Telecom applications, Personal Home and Hospital care etc (ZigBee Specification, ZigBee document 053474r13, 1 December, 2006).

Figure 7. IEEE 802.15.4 network topologies a) Star and b) Peer-to-Peer. Note: ZigBee Network Topologies (n.d) (Wikipedia). Retrieved from http://en.wikipedia.org/wiki/IEEE_802.15.4

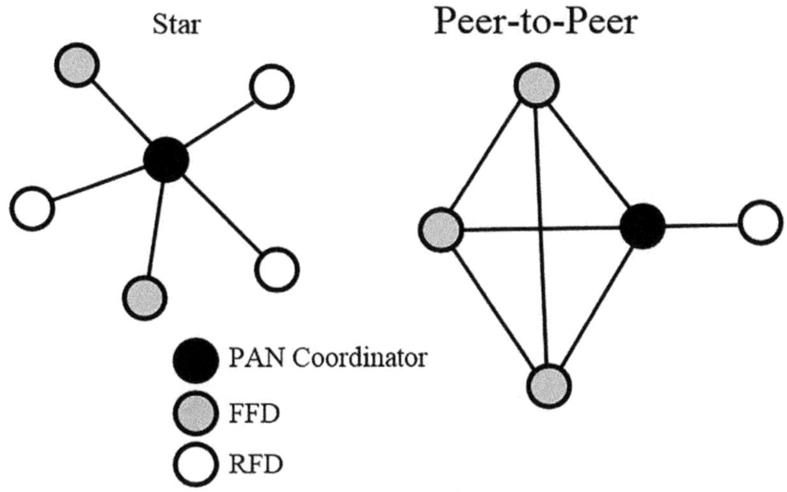

In ZigBee standard the device classes are known as Coordinator, Routers and End-Devices and correspond to PAN Coordinator, Full Function Devices and Reduced Function Devices as defined in the IEEE 802.15.4 standard.

ZigBee is capable of star and mesh topologies which are very much similar to the star and peer-to-peer topology in IEEE 802.15.4. A hierarchical tree structure is introduced in ZigBee. A tree network topology is as shown in the Figure 8 below:

Issues with ZigBee in the Railroads Environment

In the railroad environment typically there will be at least hundred railcars strung together to form a freight rail. Each of the railcars will be equipped with several sensors. Let us assume that each railcar has only one sensor node. Then the data from each of these sensor nodes will have to travel hop-by-hop to the monitoring station in the locomotive. So the network topology that is most commonly found in the railroad environment is a long linear-chain like topology (S. M. Rakshit et al., 2012).

ZigBee was primarily designed for star topology (J. Zheng, 2004). Tests carried out at the Advanced Telecommunication Laboratory at the University of Nebraska, has shown that although ZigBee supports multi-hop communication but it is unsuitable for topologies involving as many hops as in the railroads environment (P. Mahasukhon et al., 4-8 July 2011) (P. Mahasukhon et al., 2010) (P. Mahasukhon et al., 2010).

Similar research studies carried out by others have also found similar issues and performance degradations with ZigBee (F. Osterlind, 2008) and also link layer performance bottlenecks as found in (J. Misic et al., 2005).

In the following pages we briefly revisit the main issues and challenges that were found to plague a ZigBee-only network in the railroads environment.

Route Discovery: A route in a network, wired or wireless, refers to the path that packets take from their origin to the intended destination. Each link from one node to its next communication partner is referred to as a hop. If a router has more than one hop, we refer to this as multi-hop routing (S. M. Rakshit et al., 2012).

Figure 8. ZigBee Tree Topology. Note: ZigBee Network Topologies (n.d.) (Wikipedia). Retrieved from http://en.wikipedia.org/wiki/IEEE_802.15.4 licensed under CC BY-SA 2.5 es via Wikimedia Commons)

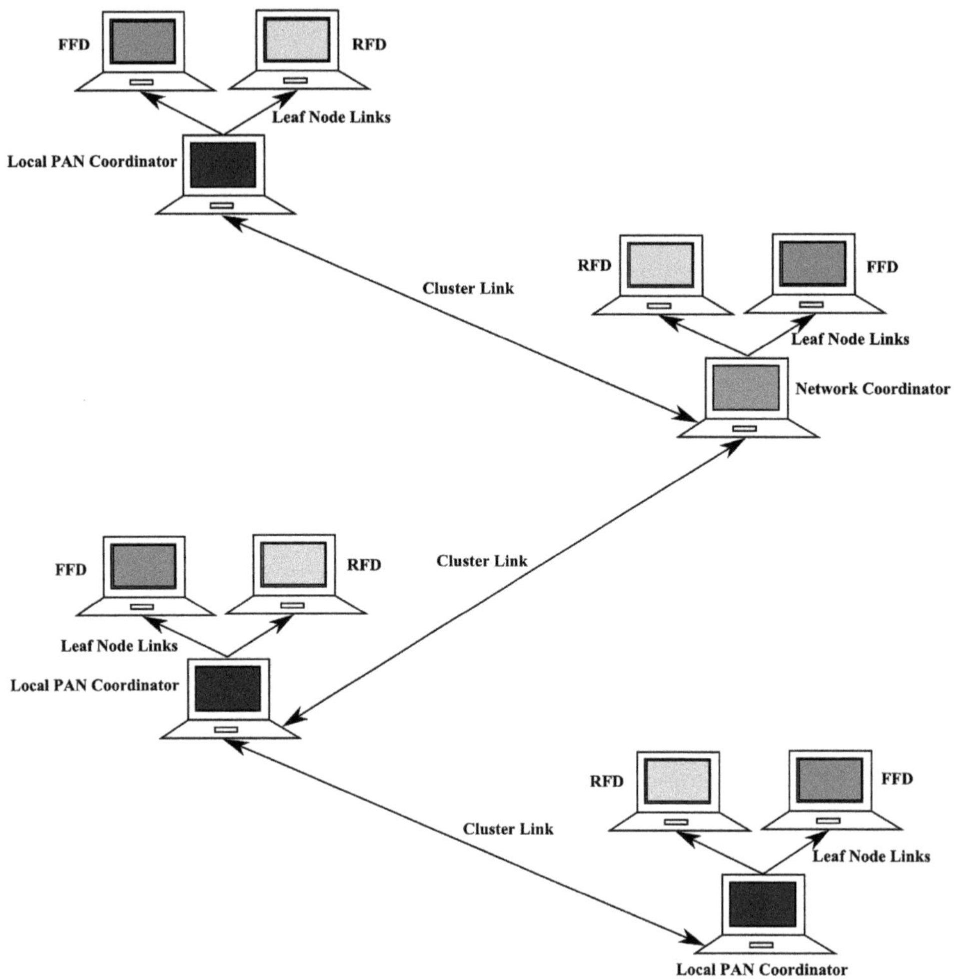

ZigBee was designed for star topologies. However, it is severely restricted in the number of hops it supports for this topology. This is due to the bandwidth limitation of ZigBee. The restricted data payload size will lead to the route request messages quickly overwhelming the available bandwidth. A hop count of 100 or more is far beyond the capability of ZigBee. In the chain-like topology we find on a freight train, ZigBee fails to discover routes after about 20 – 25 hops (S. M. Rakshit et al., 2012).

Synchronization Delay: All nodes within a single PAN network are synchronized with each other. Each time the layout of the network changes due to node failure or disconnections, a resynchronization becomes necessary. This can also be triggered by fluctuating channel conditions that prevent a node from successfully communicating with the rest of the network for a period of time. Fluctuating channel conditions are a common occurrence in wireless networks. For a 100 railcar scenario, this synchronization can stretch to several minutes. For a large number of nodes the request for synchronization messages will lead to network congestion and thus add to the delay in synchronization. During this time, no communication is possible between the unsynchronized nodes and the rest of the network (S. M. Rakshit et al., 2012).

Packet and Link Loss: If we study the behavior of wireless communication systems we can find that the probability of successful packet delivery decreases as the number of hops increases. As reported in (P. Mahasukhon H. S., 4-8 July 2011), our analysis of this problem when applied to freight train monitoring using ZigBee. (S. M. Rakshit et al., 2012).

$$PER = 1 - \left(1 - BER\right)^{\text{number of bits}} \tag{1}$$

In a multihop environment equation 1 can be represented as

$$PER_{multihop} = 1 - \left(1 - PER\right)^{\text{number of bits}} \tag{2}$$

A Bit Error Rate of 10^{-5}, which in wireless networks is a very low BER signifying good channel conditions, we still experience a packet loss of over 60% due to errors. Less than 40% of packets are being delivered successfully for a hop count of 100. Under more realistic conditions, with BERs between 10^{-3} and 10^{-4}, we can observe that none of the packets from a majority of nodes on the train will successfully be delivered to the locomotive. This means that the network will completely cease to function. We use a packet size of 127 bytes.

Lack of Quality-of-Service: Quality-of-Service refers to the ability of a network to allow transport of packets with special requirements, and appropriately allocate transmission resources to it. ZigBee does not have a way of assigning priorities to different packets.

This is extremely important in the freight train monitoring system. An important aspect of the envisioned system is the ability to generate alerts in case of equipment failure, the detection of dangerous substances or leaks, etc. But if these alert messages are treated the same way as a message containing periodic status updates from, say, a refrigeration unit, then it shares the same probability of being discarded by the network due to errors or congestion. It also will be entered into every queue at the end, like all other packets, and thus be significantly delayed. For alerting, this is simply not feasible (S. M. Rakshit et al., 2012).

Data Forwarding and Aggregation: A chain topology using ZigBee means that data from the last node needs to be relayed to the first node. In addition to relaying that information, however, each node also has its own data to report. Hence the amount of data quickly aggregates. Simulation shows that if a freight train consists of more than 25 railcars then it cannot even sustain traffic of small 50-byte message per node generated every 20 seconds (S. M. Rakshit et al., 2012). An illustration is provided in Figure 9.

Network Lifetime: The nodes closest to the data destination have a lot more data to forward than nodes towards the end of the train. Hence, their increased number of operations also means that they consume a lot more energy and deplete

Figure 9. Amount of data present in channel as a progression of hops from the end of train

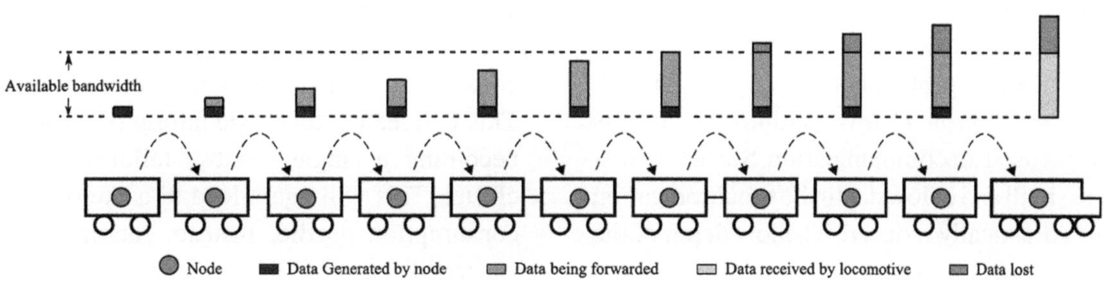

their reserves much faster. Due to the point-of-failure issue, if one of those nodes fails, no data from behind that node will be able to reach the destination anymore. At this point, the network effectively fails to operate (S. M. Rakshit et al., 2012). The graphical representation of this situation is shown in Figures 10 and 11.

The battery life profile illustrates the claim that the nodes closer to the coordinator node run out of battery faster than the ones farther away from it. We can see that, although node 0 would be operational for around 2000 hours, the network will fail after less than 200 hours (S. M. Rakshit et al., 2012).

THE HYBRID TECHNOLOGY NETWORKING PROTOCOL

The Advanced Telecommunication Engineering Laboratory at the University of Nebraska-Lincoln has come up with a solution that addresses the issues that a ZigBee-only network faces in the railroads application environment. The solution is named Hybrid Technology Networking (HTN) protocol (S. M. Rakshit et al., 2012).

The basic idea behind the protocol envisions railcar monitoring as a two-step process.

In the first step clusters of sensors are formed spanning the length of the freight rail. Each of the clusters have only one coordinator node which is called the gateway. Each sensor node communicates via single or limited number of hops to the gateway. In the next step, the gateway aggregates the sensor data and relays it to the locomotive via Wi-Fi hops over similar gateways spanning the length of the freight rail (S. M. Rakshit et al., 2012).

The high-level overview of the Hybrid Technology Networking protocol is shown in Figure 12.

Synchronization Delay: Within the clusters ZigBee is used for communication. Since the clusters are small and of low depth the synchronization delay is drastically reduced. The low depth of the network is ideal for ZigBee. It is very efficient in its utilization of power in this network topology. The network communication is also characterized by low delay, faster synchronization and lower packet losses in such a network topology. As can be seen from Figure 13, synchronization delay is reduced by 90% compared to a ZigBee-only network implementation.

Throughput: Since the ZigBee networks operate in small clusters the chance of the network getting overwhelmed due to data aggregation is minuscule. The aggregated data at the gateway in each cluster is then transmitted over Wi-Fi to the locomotive over multiple Wi-Fi hops. This reduces the overall number of hops that it takes for data from the end of the train to reach the locomotive. This results from Wi-Fi having a larger coverage distance and higher bandwidth as opposed to a ZigBee-only network.

As can be seen in Figure 14 the per node throughput using ZigBee-only network is shown with the red bar, the improvement in throughput with HTN is clearly visible.

We can clearly see that a ZigBee-only network, with its low bandwidth and higher number of required hops, gets overwhelmed very quickly. Whereas with a ZigBee and Wi-Fi hybrid approach the combined bandwidth does not saturate even at 100 hops from the end of the railcar to the locomotive.

Energy Consumption Balance: The protocol also suggests rotation of the gateway role among all the capable nodes in a cluster hence balancing the power consumption profile. The gateway will utilize Wi-Fi which consumes more power. Rotating this responsibility among other nodes in the cluster averages the energy consumption. This rotation ensures that there is not just a single node in a cluster acting as the gateway and hence consuming higher energy than the other nodes. This will lead to this node failing first and hence becoming a single point-of-failure within the cluster. This will again lead to a skewed energy consumption profile. Instead, rotating the role ensures that there will be a more gradual decrease

Figure 10. Energy consumption per node for a 100-railcar train

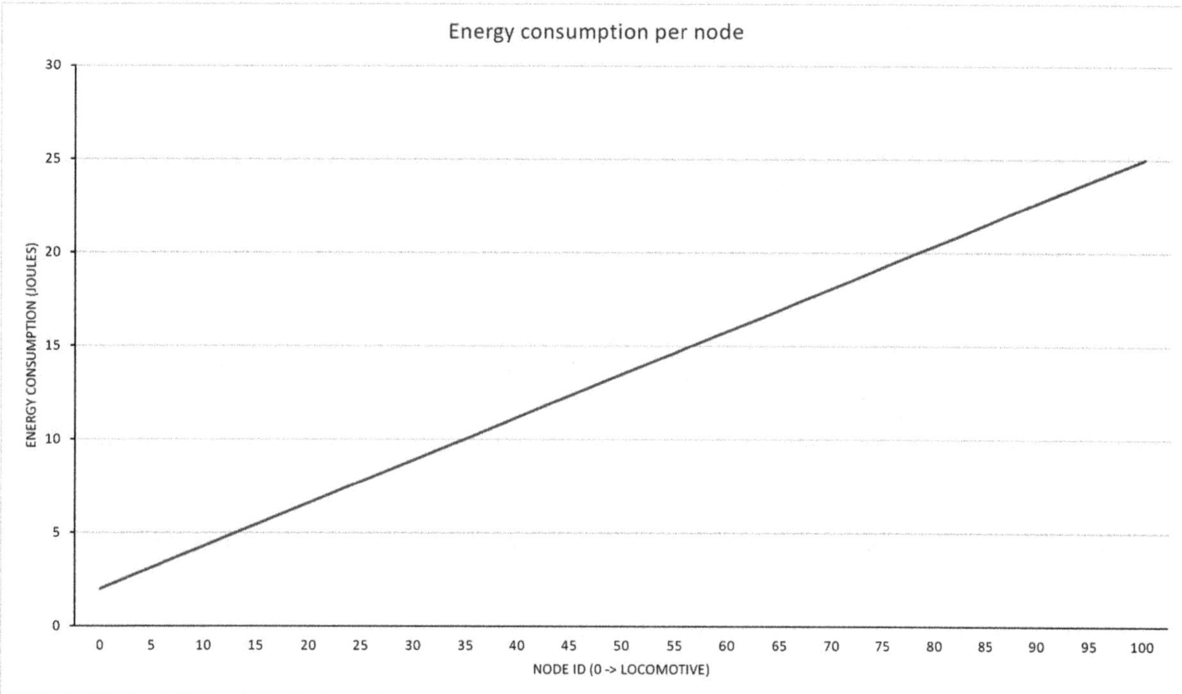

Figure 11. Battery life in hours for each node in a 100-railcar train

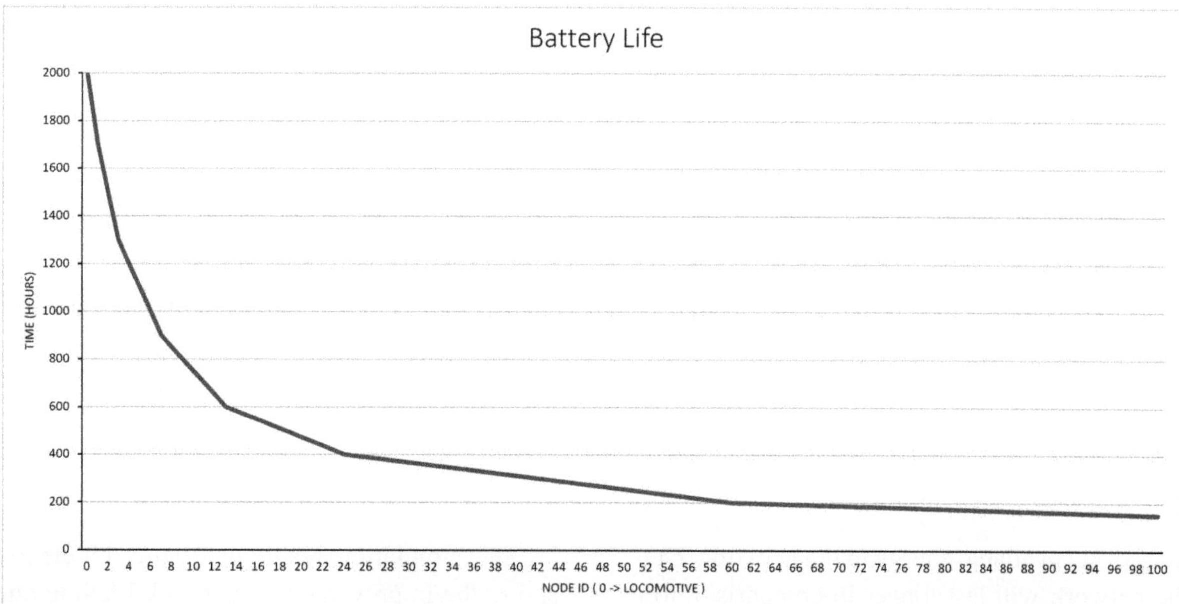

Figure 12. High-level representation of hybrid technology networking

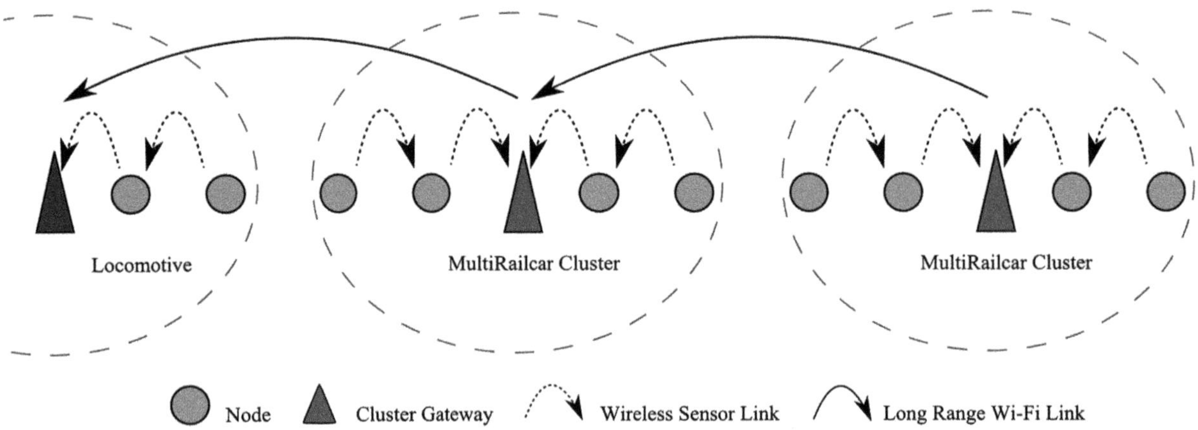

Figure 13. Time required for synchronization vs. Total number of hops in network in HTN

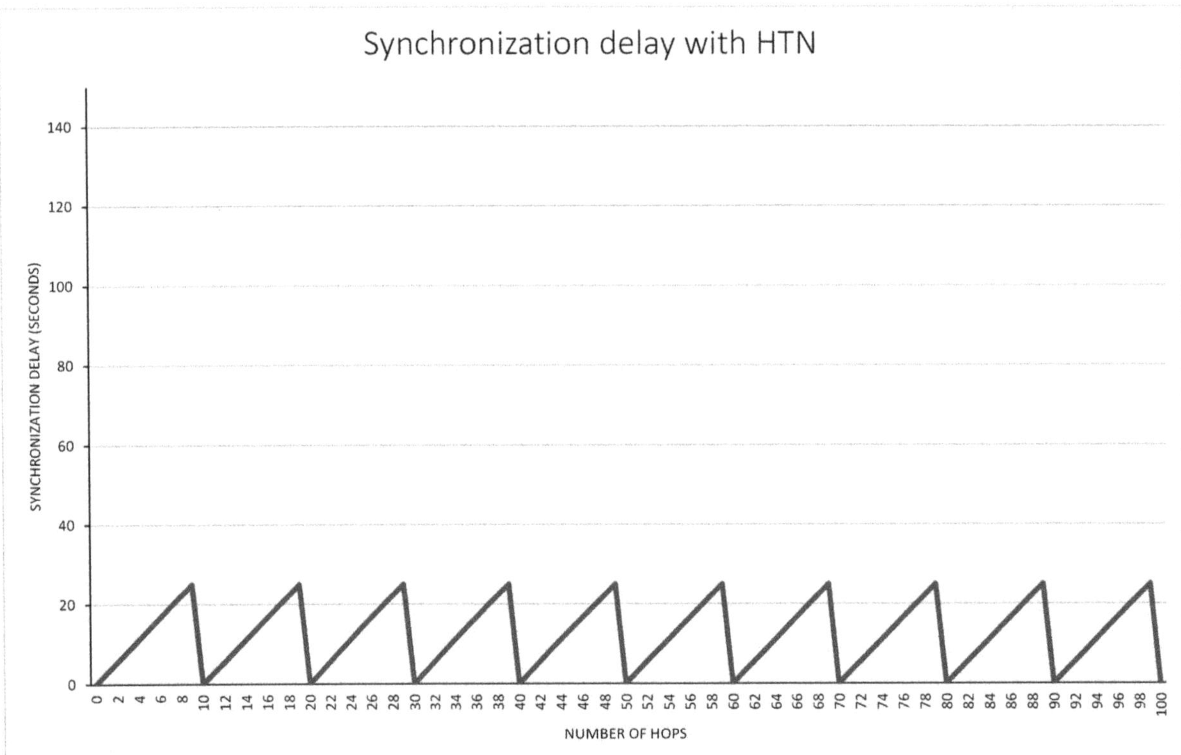

in the total available energy in the network. Thus the network will last longer in comparison to a network where the gateway role is not rotated.

The simulation result in Figure 15 clearly shows the improvement in network lifetime due to a balanced power consumption profile.

Figure 14. Per node throughput vs. Total number of hops in HTN

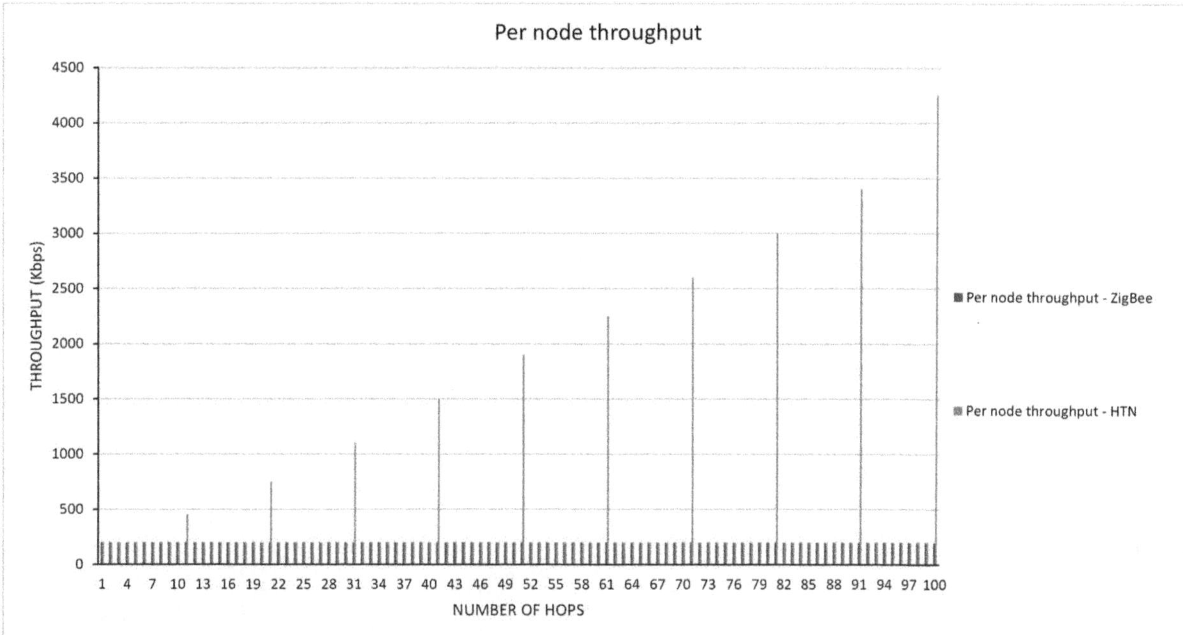

Quality-of-Service: Quality-of-Service mechanism is a very important aspect of a railcar status monitoring network. We envision several kinds of messages to be in such a network. There will be regular reports from nodes like status of equipment, temperature etc. as well as alarm messages when an equipment exceeds an established maximum allowed value. In a ZigBee-only network all of these messages will be treated as similar from the network perspective. Both will be treated similarly in face of congestion and network losses. But alarm messages should be accorded higher priority so that they reliably arrive at the locomotive within an acceptable frame of time after being generated. We will need to assign priorities to network messages. This is done by the Quality-of-Service mechanism implemented in the HTN protocol. There will also be actuating messages from the locomotive that will need to be reliably delivered to the intended node. This mandates that the nodes acknowledge receipt of this message to the locomotive. The HTN protocol has end-to-end acknowledgement mechanism that ensures this.

As mentioned in (S. M. Rakshit et al., 2012), simulation results with HTN shows that there are the significant advantages over a ZigBee-only network in the railroads application environment with 90% reduction is synchronization delay, 40 times improvement in system throughput, 250% increase in network lifetime and a customizable Quality-of-Service mechanism.

Need for a Custom Design

The available sensor platforms like MICAz (MICAz Datasheet at MEMSIC), Stargate (Stargate, n.d.) and recent development Libelium Waspmote (Libelium Waspmote, n.d.) are all capable of providing the ZigBee communication part as well as the Wi-Fi communication part but not in an integrated solution. The current available sensor platforms will have to be plugged into one another to realize both the communication radios together.

Several technical and logistical problems arise from this approach. The data rate of plug-in boards will always be limited by the bus that is used to

Figure 15. Node lifetime vs. Total number of hops in HTN

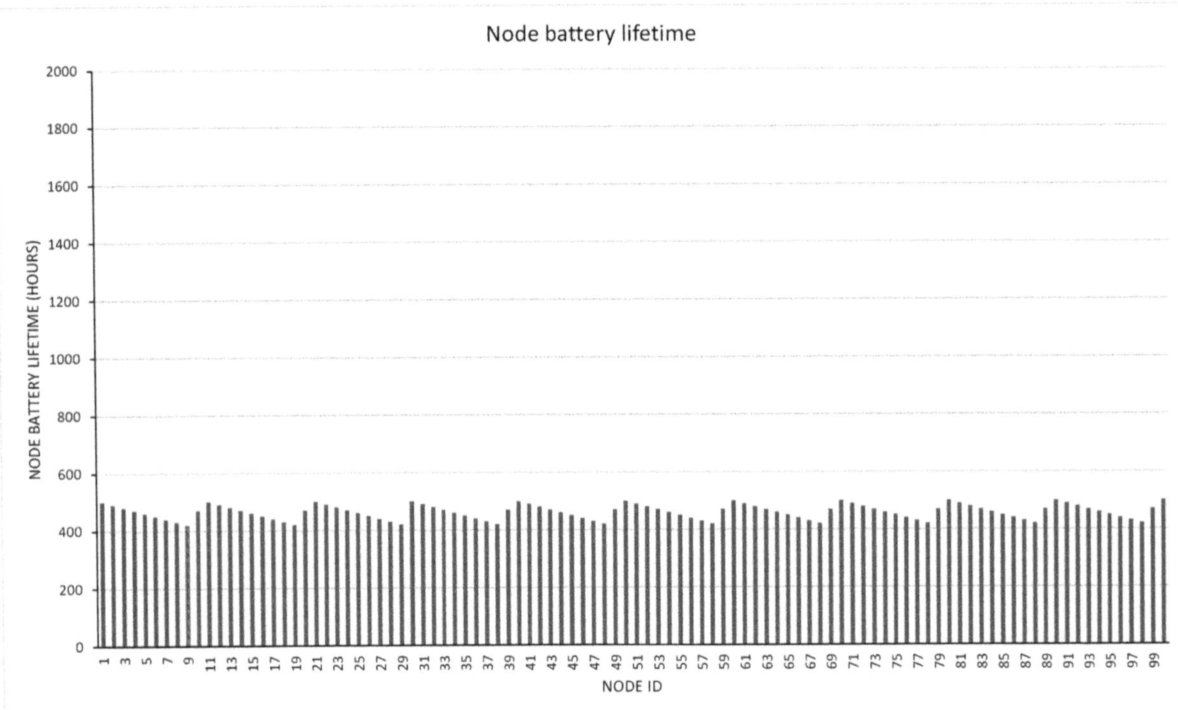

connect them. In all of the above cases, except the Stargate, the only available interfacing bus is the UART. This is slow speed and hence will cause a bottleneck for Wi-Fi traffic. In Stargate, the PCMCIA interface is used which removes the speed bottleneck but is an order of magnitude more power hungry than the UART.

The cost of obtaining various pieces of hardware from different vendors, assembling them and coding for the required firmware is high and is a logistical nightmare when the numbers of such platforms required are in millions. The railroads are looking for standards compliant solution so that vendor independence is maintained. Some of the off-shelf solutions may not be standards compliant.

A custom hardware can be made with specific design goals in mind. The commercially available components were designed with some design goals and when they are brought together they may not meet the goals of the railroads.

As can be understood from the above discussion there are major advantages in designing a custom integrated hardware platform that will be standards compliant and also will meet and exceed any design goals that may be set by the railroads.

We designed a hardware platform, that we call HTNMote, to mitigate the above issues. Our hardware integrates two radio communication technologies on a single board that is controlled by a single microcontroller. The interfacing bus between the components are high speed, thereby eliminating the speed bottlenecks. There are several power levels available on the board to power components and external sensors which require different power levels. There are also several interfacing modes available, like serial port, USB and LCD, which allow us to interface with the board. A view of the assembled hardware is provided in Figure 16.

We ruggedized this board by putting it in a NEMA enclosure so that it can withstand the harsh conditions of the railroad environment. We took this board to the Transportation Technology Center (TTCI) in Pueblo, Colorado to test it on an actual freight train with 36 railcars. The results bore out our claims that the network of HTNMotes implementing the HTN protocol had superior performance over a ZigBee-only network.

Performance Tests

In the intended usage scenario of the HTNMote the most critical components on the board, apart from the microcontroller, are the Wi-Fi and the ZigBee modules. If these two modules do not perform above a certain level in comparison with other off-the-shelf devices then the HTNMote solution will be rendered infeasible. The following bunch of performance tests helps to show the capabilities of the communication hardware on the HTNMote.

Current Consumption Test Results

The current consumption observed for various scenarios of interest are shown below:

When looking at the current consumption figures present in Table 2 above it must be noted that all of these figures are not stand-alone consumption values for a corresponding component. Each of these values includes consumption due to all peripherals active as well as the code running in the microcontroller. This holistic view is necessary as in the field it is this overall current consumption which is of importance rather than individual component consumption.

Wi-Fi Test Results and Analysis

The test is conducted with constant channel conditions. The transmitted packet payload size is varied from 50 bytes to a Wi-Fi maximum packet payload size of 1400 bytes.

It is observed from Figure 17 that the lower packet payload sizes do not make the device Wi-Fi operate at the saturation level of throughput and hence the throughput gradually increases. As the packet payload size increases the throughput starts to level off. It can be seen that using packet payload sizes of 800 bytes and above will guarantee that the device is operating close to the maximum throughput possible given all the other conditions. It can also be seen that the Wi-Fi module allows nearly 2.2 Mbps as the maximum throughput.

Figure 18 shows that the overall packet loss for the Wi-Fi module is in the vicinity of a maximum of 32 packets per second when the payload size is 1400 bytes, the throughput is 2.2 Mbps and the path loss is 113 dB.

Figure 19 shows the overall percentage packet loss observed for a given path loss for different payload sizes at 2.412 GHz center frequency of operation. It is important to note that the above result will be affected by change in channel conditions and should be seen as the worst-case packet loss given similar conditions. It is observed that the overall packet loss increases dramatically as the channel deteriorates. For smaller payload sizes we still are able to receive about 50% of the packets when there is greater than 100 dB path loss. But as the payload size increases this amount becomes less and less and worst-case we observe near 100% packet loss for 1400 bytes payload size at a path loss of 100 dB and above.

We use the following formula convert the path loss value into a distance value:

$$\text{Path Loss} = 10n\log_{10}\left(\frac{4\pi d}{\lambda}\right)$$

In the above equation the path loss exponent (n) is assumed to be 2. The distance (d) between the sender and the receiver is in unit of meters. The wavelength (λ) is for the center frequency of 2.412 GHz. The path loss in the equation is in the unit of decibels (dB).

Using the above formula we can see that for a distance of separation of roughly 1 Km between the

Figure 16. HTNMote - assembled

Table 2. HTNMote current consumption

Scenario	Current Consumption
HTNMote switched on: voltage regulators, power-on LED, RS-232, SD-Card, USB, clock circuitry, microcontroller active. No LCD.	48.4 mA
HTNMote switched on and a single indicator LED glowing	49.3 mA
HTNMote switched on and RS-232 transaction on-going	51.5 mA
HTNMote switched on and and SD Card transfer on-going	50.2 mA
HTNMote switched on and connected as USB mass storage device	59.3 mA
HTNMote switched on and LCD displaying graphic	77.5 mA
HTNMote switched on and ZigBee sending data to coordinator	80.1 mA
HTNMote switched on and ZigBee receiving data from end-device	91.1 mA
HTNMote switched on and WiFi module in idle but full power mode	163.4 mA
HTNMote switched on and WiFi module sending data	197.1 mA
HTNMote switched on and WiFi module receiving data	190.1 mA

sender and the receiver the packet loss is a meager 10 - 12 packets per second when the transmission is very fast with 2.2 Mbps throughput using 1400 bytes packet size.

Comparable hardware platforms include the Stargate gateway node (Stargate, n.d.). It uses PCMCIA slot to provide Wi-Fi connectivity. The Wi-Fi transfer speeds on the Stargate range from 1.5 Mbps to 3.75 Mbps depending on the Wi-Fi settings on both the sender and receiver sides (Stargate Wi-Fi, n.d.). But this comes at a significant increase in power consumption as

Figure 17. Throughput vs. packet size test result

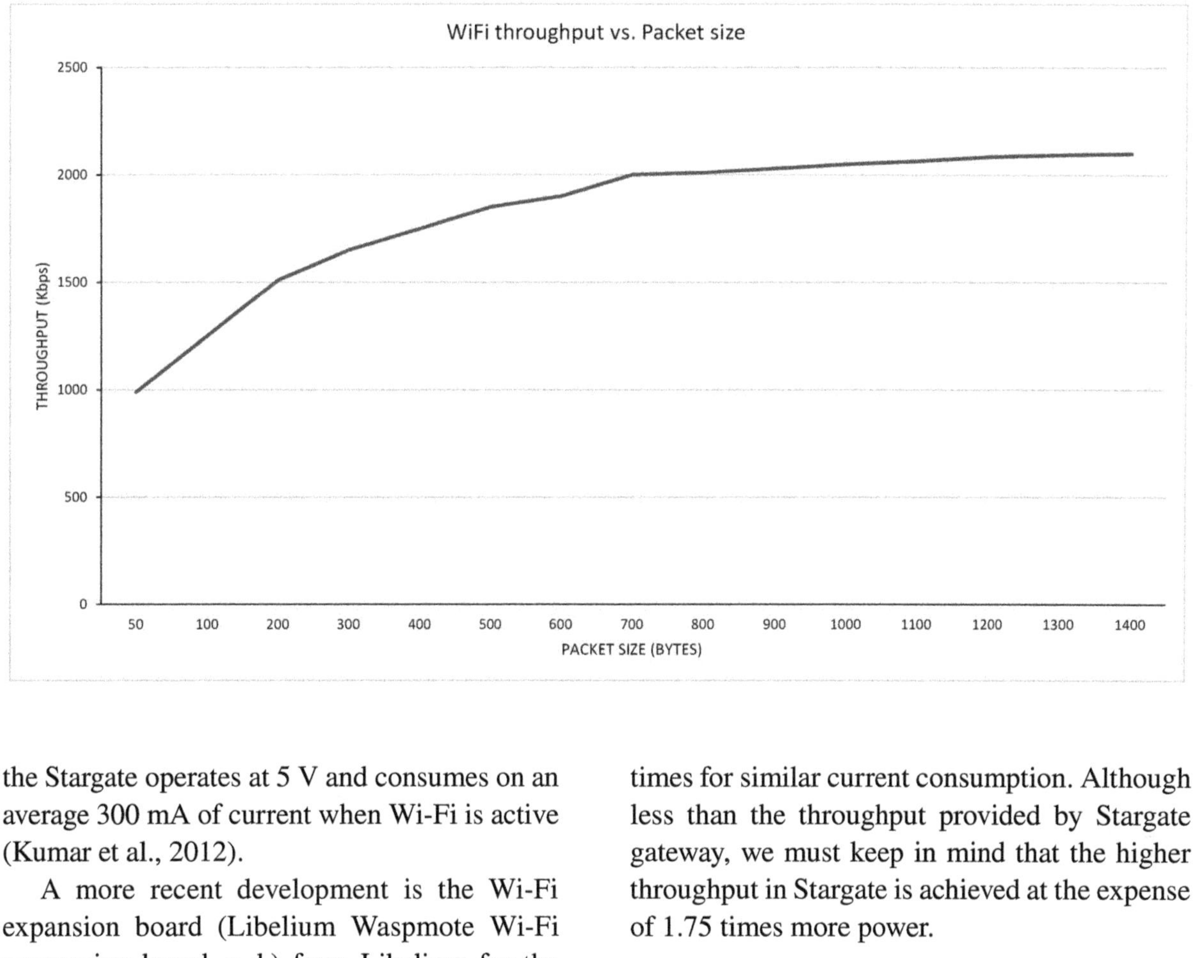

the Stargate operates at 5 V and consumes on an average 300 mA of current when Wi-Fi is active (Kumar et al., 2012).

A more recent development is the Wi-Fi expansion board (Libelium Waspmote Wi-Fi exapansion board, n.d.) from Libelium for the Waspmote sensor platform. This board interfaces to the microcontroller core of the Waspmote using UART interface. This is a bottleneck for data transfer between the microcontroller and the Wi-Fi radio. In a practical scenario, the current consumed by the Waspmote hardware and the Wi-Fi board working together is close to 200 mA (Libelium Waspmote Wi-Fi exapansion board, n.d.). The maximum transmission throughput achievable when the UART baud rate is set to 57600 is 15.68 kbps.

The Wi-Fi results show us that there is a huge performance gain with HTNMote when the parameter under consideration is throughput. The HTNMote provides a maximum throughput of close to 2.2 Mbps. This is an improvement of 140 times for similar current consumption. Although less than the throughput provided by Stargate gateway, we must keep in mind that the higher throughput in Stargate is achieved at the expense of 1.75 times more power.

ZigBee Test Results and Analysis

Figure 20 shows that given the software interface provided by the Texas Instruments Z-Stack ZigBee Network Processor API, the maximum throughput achievable is roughly 52 kbps.

This is observed towards the end of the graph where the channel is substantially deteriorated. The increase in throughput comes from re-transmissions in the channel.

The term goodput denotes the number of useful information bits that were transmitted per unit time from a source to the destination. As we can see from Figure 21 the goodput falls as the channel deteriorates. This stems from increased

Figure 18. Packet loss vs. path loss result

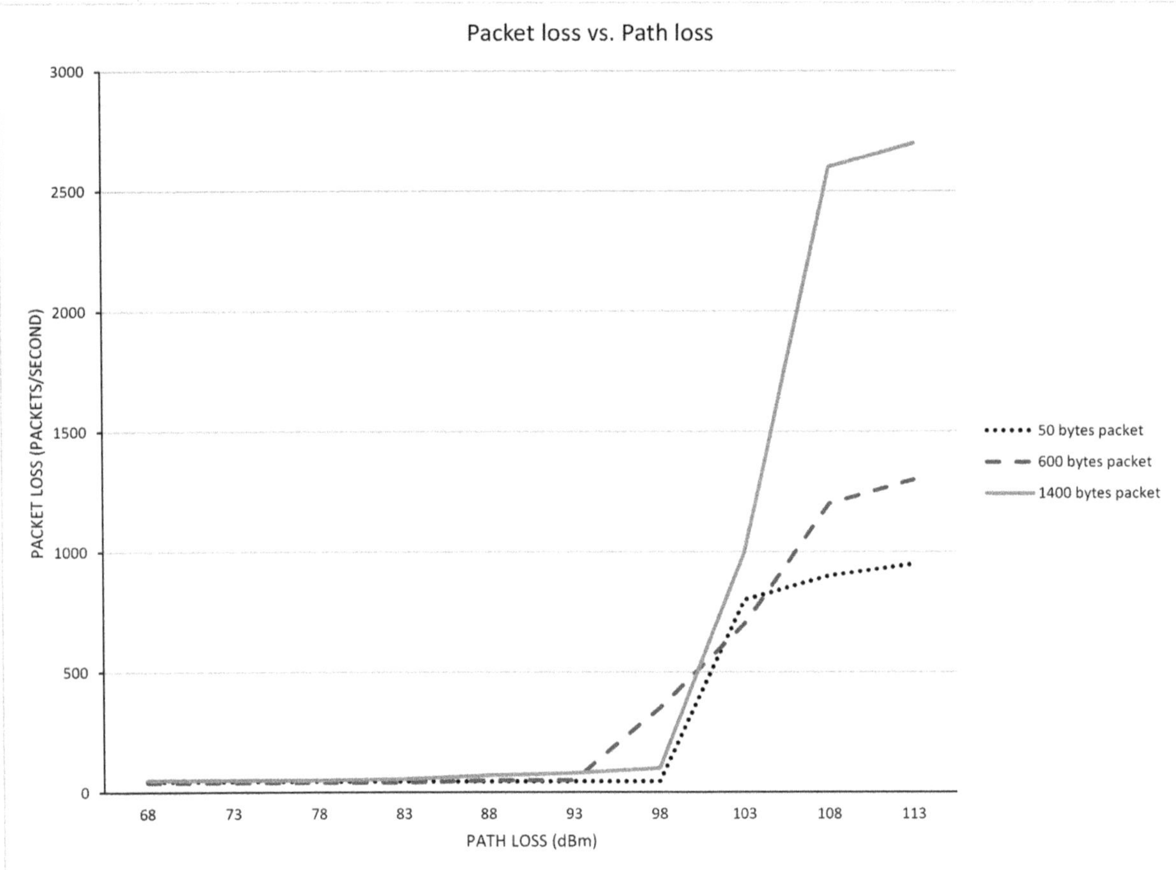

packet retransmissions and packet losses when the channel deteriorates.

Figure 22 shows us that the current hardware will be able to receive transmissions reliably down to 85 dB of path loss.

The above results confirm what the results for throughput and goodput show us. As the channel conditions deteriorate the number of retransmissions and packet loss increases. This means there is more traffic being generated and transmitted into the channel by the sender, this results in an increase in throughput. But the number of transmissions actually reaching the intended recipient goes down due to deteriorated and congested channel and thus the goodput decreases.

The above results show that the HTNMote ZigBee hardware is comparable to the performance of MICAz and Waspmotes. This is to be expected since the radio chip used on the MICAz and HTNMote share a similar architecture. The XBee ZigBee module (Xbee Module Datasheet, n.d.) used with the Waspmote is theoretically capable of delivering better performance but is severely restricted by the UART interface between the microcontroller and the XBee module.

FUTURE RESEARCH DIRECTIONS

The initial version of HTNMote has been shown to be quite capable of handling the demands of the Hybrid Technology Networking protocol. It is already better than comparable products available in the market.

HTNMote

Figure 19. Packet loss percentage vs. path Loss result

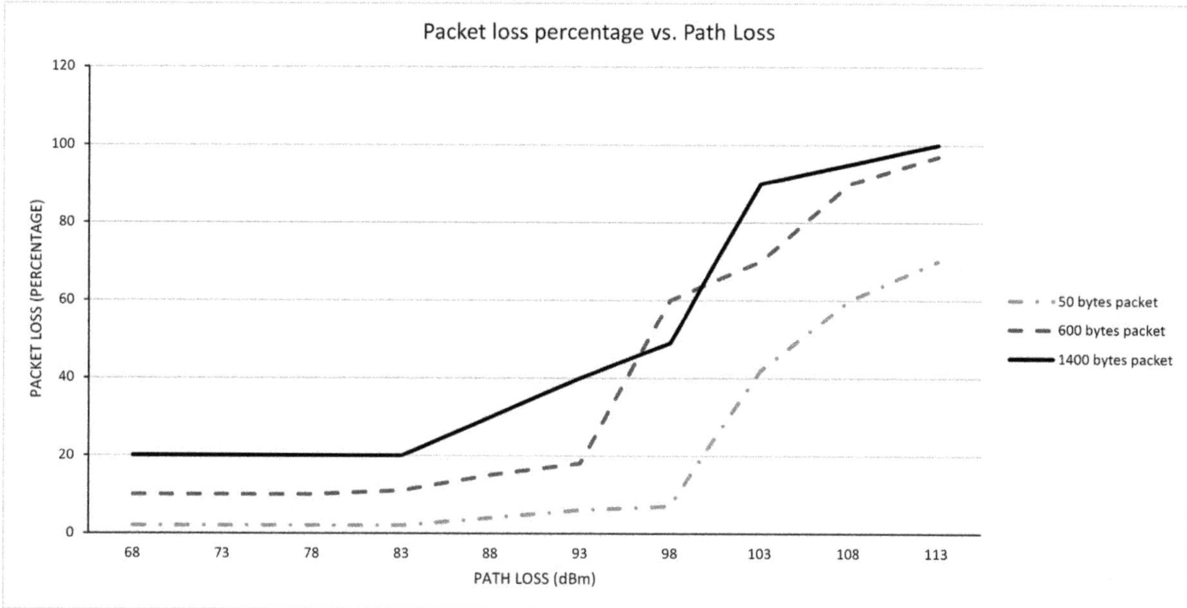

Figure 20. ZigBee throughput vs. path loss test result

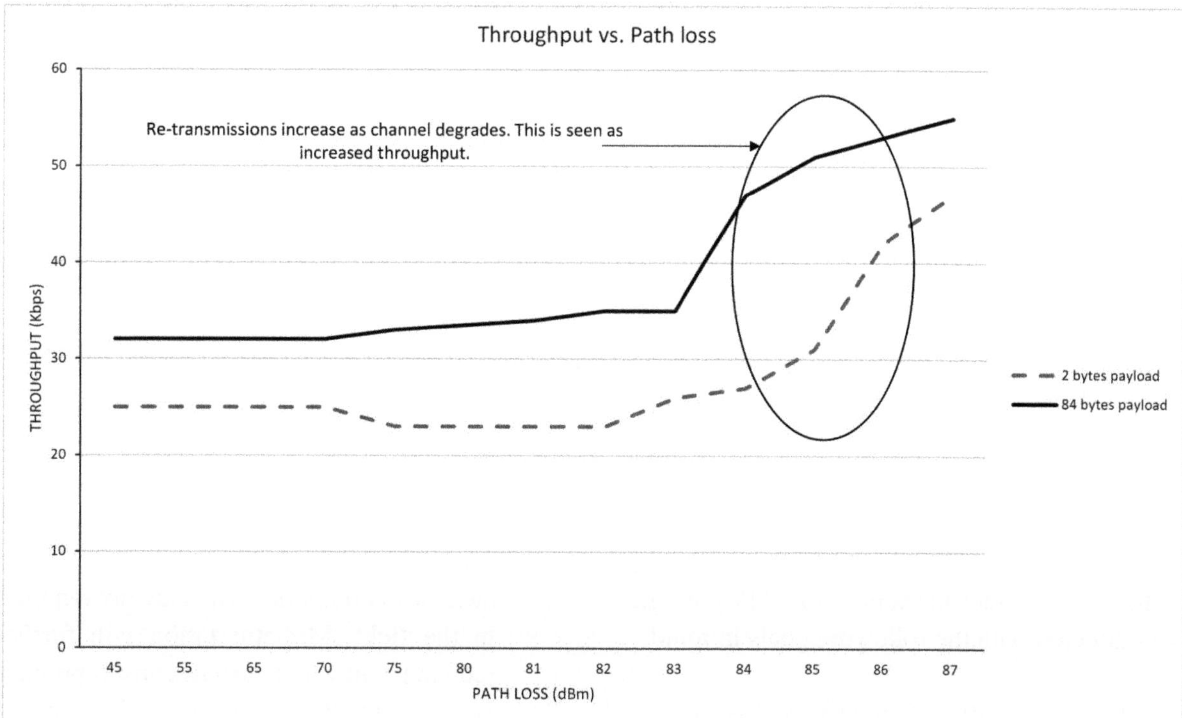

Figure 21. ZigBee goodput vs. path loss test result

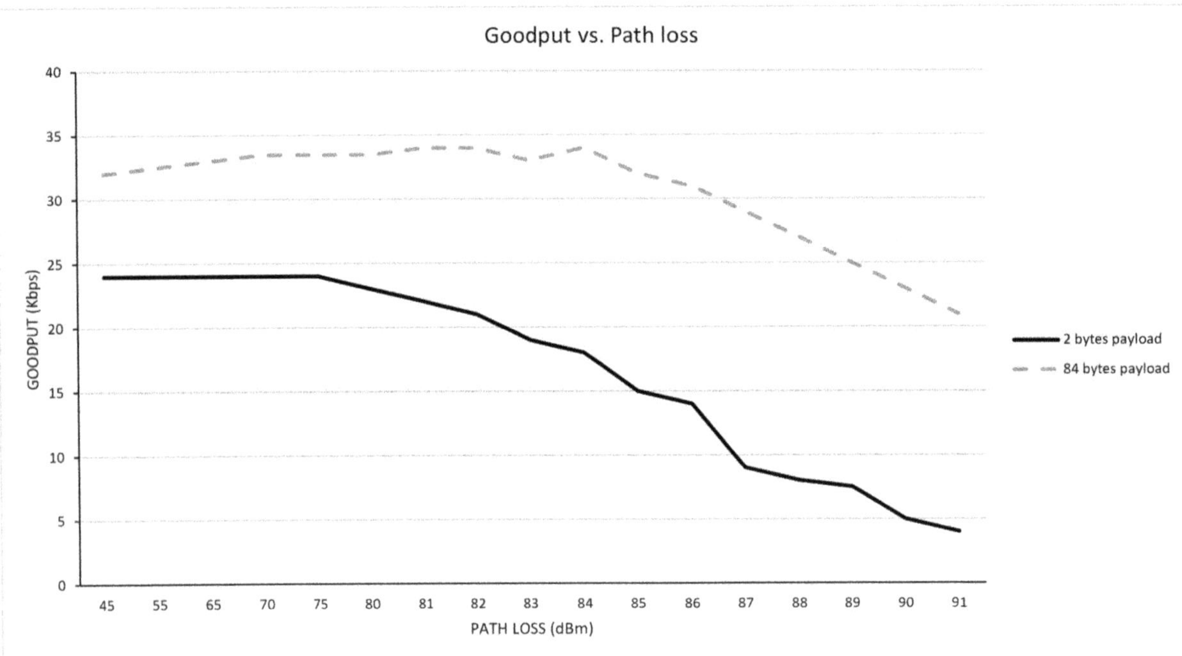

Figure 22. ZigBee packet loss rate vs. path loss test results

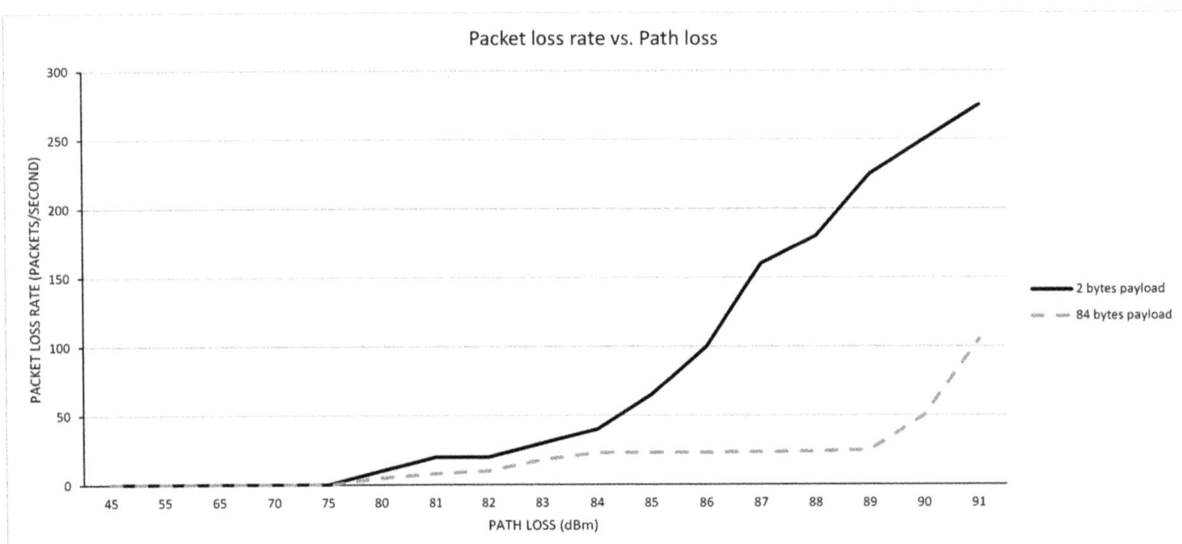

In future further research on HTNMote will be conducted with the following goals in mind:

1. **Miniaturization:** The next goal is to miniaturize the HTNMote hardware by doing away with components that are not required in the field. Miniaturization will further reduce latency and current consumption.
2. **Improving Performance:** The current implementation has shown that even though

capable hardware and a fast interface to that hardware exists but still performance bottlenecks come from the software implementation. In the next revision, software optimization with an eye to throughput and other performance improvement will be done.

3. **Large Deployment Tests:** In future, large deployments of HTNMotes will be carried out to test how the HTNMotes perform when they are in a network with several other HTNMotes.

CONCLUSION

The Federal Railroad Administration is looking at WSN to monitor freight rail in real-time or near real-time. The freight rail is an important contributor to the health of the United States economy. Freight rail is also a greener mode of transport. The current methods of monitoring freight rail, like wayside monitoring etc., suffer from the drawbacks of not being able to monitor a large parameter set and the monitoring is not in real-time.

WSN is a natural choice in achieving the real-time monitoring goals. ZigBee is the protocol of choice for such deployments owing to its low cost and low power consumption.

However, the railroads deployment scenario is very unique when it comes to the topology in which the network exists. The long linear chain-like topology is one for which ZigBee was not designed. Hence, the use of ZigBee in such scenarios present significant problems.

These problems include synchronization delay, route discovery issues, packet and link loss, lack of Quality-of-Service, data forwarding and aggregation errors and skewed network lifetime issues.

Hybrid Technology Networking protocol has been suggested by the Advanced Telecommunication Engineering laboratory at the University of Nebraska-Lincoln which addresses the above problems. This protocol envisions ZigBee to operate in small clusters and the data forwarding from one cluster to the next to happen over Wi-Fi. This has been shown to improve end-to-end throughput, balance power consumption and also is standards based.

The current available sensor hardware platform does not have an integrated solution for the application of Hybrid Technology Networking. Hence, the need to design an integrated standards-based sensor hardware platform that will at least be equivalent in performance.

HTNMote is designed to address the drawbacks of available hardware. It combines a Cortex M-4 microcontroller with both Wi-Fi and ZigBee radio capabilities residing on the same board. SD Card storage facility is present. Communication with the external world is carried out using RS-232 and USB apart from Wi-Fi and ZigBee.

The test results of the HTNMote hardware show that the Wi-Fi capability is far better than any solution available in the market now. The ZigBee performance is comparable to any other hardware currently available. Coupled with the possibilities of implementing extremely low power consuming code, as all the peripherals support power saving features, the HTNMote is an ideal platform to implement the Hybrid Technology Networking protocol and provide a reliable way to enable on-board real-time monitoring of freight railroad.

REFERENCES

Association of American Railroads. (n.d.). *Class I Railroad Statistics*. Retrieved from http://www.aar.org/PubCommon/Documents/AboutTheIndustry/Statistics.pdf

Digi International Inc. (n.d.). *Xbee Module Datasheet*. Retrieved from www.digi.com/pdf/ds_xbeezbmodules.pdf

Edwards, M., Donelson, J. I., Zavis, W., Prabhakaran, A., Brabb, D., & Jackson, A. (2005, March 16-18). Improving freight rail safety with on-board monitoring and control systems. In *Proceedings of the 2005 ASME/IEEE Joint Rail Conference* (pp. 117-122).

Flowers, D., & Yang, Y. (n.d.). *The MiWi Wireless Networking Protocol is a simple protocol designedfor low data rate, short distance, low-cost networks.* Retrieved from http://ww1.microchip.com/downloads/en/AppNotes/AN1066%20-%20MiWi%20App%20Note.pdf

Global Positioning System. (n.d.). *GPS.gov Web site.* Retrieved from http://www.gps.gov/

IEEE-SA. (2003). *IEEE 802.15.4, Part 15.4: Wireless Medium Access Control (MAC) and Physical Layer (PHY) Specifications for Low Rate Wireless Personal Area Networks (LR WPANs).* IEEE SA Standards Board.

IEEE-SA. (2012, April 5). *IEEE 802.11: Wireless LAN Medium Access Control (MAC) and Physical Layer (PHY) Specifications* (2012 revision).

IEEE-SA. (n.d.). *IEEE 802.2 standards.* Retrieved from http://standards.ieee.org/about/get/802/802.2.html

ISA. (n.d.). *ISA.org Web site.* Retrieved from http://www.isa.org/

Kumar, A., Namboothiri, P., Deshpande, S., Vidhyadharan, S., Sivalingam, K., & Murty, S. (3-5 Feb. 2012). Testbed based throughput analysis in a Wireless Sensor Network. In *National Conference on Communications (NCC)* (pp. 1-5). doi:10.1109/NCC.2012.6176804

Lashkari, A., Danesh, M., & Samadi, B. (2009 Aug. 8-11). A survey on wireless security protocols (WEP, WPA and WPA2/802.11i). In *2nd IEEE International Conference on Computer Science and Information Technology* (pp. 48-52).

Libelium. (n.d.). *Libelium Waspmote.* Retrieved from http://www.libelium.com/products/waspmote

Libelium. (n.d.). *Libelium Waspmote Wi-Fi exapansion board.* Retrieved from http://www.libelium.com/expansion_radio_board/

Mahasukhon, P., Sharif, H., Hempel, M., Zhou, T., Ma, T., & Shrestha, P. L. (4-8 July 2011). A study on energy efficient multi-tier multi-hop wireless sensor networks for freight-train monitoring. In *7th International Wireless Communications and Mobile Computing Conference (IWCMC)*, (pp. 297-301). doi:10.1109/IWCMC.2011.5982549

Mahasukhon, P., Sharif, H., Hempel, M., Zhou, T., Wang, W., & Ma, T. (2010). Multi-Tier Multi-Hop Routing in Large-Scale Wireless Sensor Networks for Freight Train Monitoring. In *ITS World Congress.*

Mahasukhon, P., Sharif, H., Hempel, M., Zhou, T., Wang, W., & Ma, T. (2010). *Multi-Tier Multi-Hop Routing in Large-Scale Wireless Sensor Networks for Real-Time Monitoring.* IEEE Sensors. doi:10.1109/ICSENS.2010.5690077

Majstor, F. (20-24 Oct. 2003). WLAN security threats & solutions. In *28th Annual IEEE International Conference on Local Computer Networks*, (p. 650).

MEMSIC. (n.d.). *MICAz Datasheet.* Retrieved from http://www.memsic.com/support/documentation/wireless-sensor-networks/category/7-datasheets.html?download=148%3Amicaz

Misic, J., Shafi, S., & Misic, V. B. (2005). Avoiding the Bottlenecks in the MAC Layer in 802.15.4 Low-Rate WPAN. In *Proceedings of 11th International Conference on Parallel and Distributed Systems Workshops* (pp. 363-367).

Mulligan, G. (2007). The 6LoWPAN architecture. In *Proceedings of the 4th workshop on Embedded networked sensors*, ACM. doi:10.1145/1278972.1278992

nationalatlas.gov. (n.d.). *Overview of Freight Railroad Industry*. Retrieved from http://nationalatlas.gov/articles/transportation/a_freightrr.html

Osterlind, F., & Dunkels, A. (2008). *Approaching the Maximum 802.15.4 Multi-hop Throughput* (SICS Tech. Rep. T2008:05).

Radio-Electronics.com. (n.d.). *Networking Topologies*. Retrieved from http://www.radio-electronics.com/info/wireless/ieee-802-15-4/mesh-networking-topology-topologies.php

Radio-Electronics.com. (n.d.). *Wireless Standards Technology*. Retrieved from http://www.radio-electronics.com/info/wireless/ieee-802-15-4/wireless-standard-technology.php

Rakshit, S. M. (2013). *Design and Implementation of a Hybrid Technology Networking System*. Dissertations & Student Research in Computer Electronics & Engineering.

Rakshit, S. M., Hempel, M., Sharif, H., Punwani, J., & Stewart, M. (2012). Hybrid Technology Networking: A Novel Wireless Networking Approach for Real-Time Railcar Status Monitoring. In *ASME Rail Transportation Division Fall Technical Conference*, (pp. 1-7). doi:10.1115/RTDF2012-9402

Rakshit, S. M., Hempel, M., Sharif, H., Punwani, J., Stewart, M., & Mehrvarzi, S. (2012). Challenges in current Wireless Sensor Technology for Railcar Status Monitoring for North America's Freight Railroad Industry. In *ASME/ASCE/IEEE Joint Rail Conference*, (pp. 1-9). doi:10.1115/JRC2012-74114

Sandirigama, M., & Idamekorala, R. (2009, Sept. 26-27). Security weaknesses of WEP protocol IEEE 802.11b and enhancing the security with dynamic keys. In *2009 IEEE Toronto International Conference-Science and Technology for Humanity* (pp. 433-438).

SD Card Association. (n.d.). *SD Association Web site*. Retrieved from https://www.sdcard.org/

USB Standard. (n.d.). *USB 3.1 Specification*. Retrieved from http://www.usb.org/developers/docs/

Stargate. (n.d.). *Stargate Gateway*. Retrieved from www.willow.co.uk/html/spb400-_stargate_gateway.html

Stargate Wi-Fi. (n.d.). *WiFi FAQ*. Retrieved from http://platformx.sourceforge.net/Documents/nuts/WiFiFAQ.html

US Department of Transportation, & the Federal Railroad Administration. (n.d.). *Freight Rail Today*. Retrieved from http://www.fra.dot.gov/Page/P0362

US Department of Transportation; Federal Railroad Administration. (n.d.). *Freight Rail Overview*. Retrieved from http://www.fra.dot.gov/Page/P0362

Wi-Fi Alliance. (n.d.). *Wi Fi Alliance Web site*. Retrieved from http://www.wi-fi.org

Wikipedia. (n.d.). *Automatic Equipment Identification*. Retrieved from http://en.wikipedia.org/wiki/Automatic_equipment_identification</ eref>

Wikipedia. (n.d.). *Devices in Ad-Hoc Mode*. Retrieved from http://en.wikipedia.org/wiki/Peer-to-peer

Wikipedia. (n.d.). *IEEE 802.11 Standard*. Retrieved from http://en.wikipedia.org/wiki/IEEE_802.11

Wikipedia. (n.d.). *IEEE 802.15.4*. Retrieved from https://en.wikipedia.org/wiki/IEEE_802.15.4

Wikipedia. (n.d.). *Wayside Monitoring Equipment*. Retrieved from http://en.wikipedia.org/wiki/Hot_box

Wikipedia. (n.d.). *Wi-Fi Architecture*. Retrieved from http://en.wikipedia.org/wiki/IEEE_802.11

Wikipedia. (n.d.). *ZigBee Network Topologies*. Retrieved from http://en.wikipedia.org/wiki/IEEE_802.15.4

Zheng, J., & Lee, M. J. (2004). Will IEEE 802.15.4 make ubiquitous networking a reality? A discussion on a potential low power low bit rate standard. *IEEE Communications Magazine*, *42*(6), 140–146. doi:10.1109/MCOM.2004.1304251

ZigBee. (n.d.). *ZigBee*. Retrieved from http://www.sensor-networks.org/index.php?page=0903503549

ZigBee. (n.d.). *ZigBee Primer*. Retrieved from http://www.daintree.net/downloads/whitepapers/ZigBee_primer.pdf

ZigBee Alliance. (2006, Dec 1). *ZigBee Specification* (ZigBee document 053474r13). Retrieved from https://openwsn.googlecode.com/files/ZigBee-Specification-2006.pdf

KEY TERMS AND DEFINITIONS

Freight Railroad: Trains carrying goods from origin to destination.

Monitoring System: The sensors and the network they form together constitute a monitoring system for critical parameters of a rail car.

North American Railroad: The rail industry in North America.

On-board: The sensors and communication systems will be on-board the railcar.

Power Optimized: These sensors and the communication protocol are highly power optimized so as to keep the sensors operating for a long time.

Real-Time: The sensors monitor the parameters in real-time.

Self-Healing: The network can recover from local failures automatically.

Wi-Fi: A communication protocol specified in IEEE 802.11. A high range high bandwidth protocol.

Wireless Sensor Networks: A network of sensors communicating between themselves wirelessly.

ZigBee: A communication protocol based on the IEEE 802.15.4 specification suited for small low data rate networks.

Chapter 2
Energy Efficiency of Coding Schemes for Underwater Wireless Sensor Networks

Mark S. Leeson
University of Warwick, UK

Sahil Patel
University of Warwick, UK

ABSTRACT

Underwater Wireless Sensor Networks (UWSNs) are used in applications such as mineral exploration and environmental monitoring, and must offer reliability and energy efficiency. These are related to each other in the sense that the former requires error-correction which in turn requires energy, consuming battery life in an environment where battery replacement and recharging are difficult. This chapter thus addresses the energy efficiency of three suitable error correction methods for UWSNs, namely Automatic Repeat Request (ARQ), Forward Error Correction (FEC) and Network Coding (NC). The performance of the schemes as a function of transmission distance is determined for various packet sizes by using models of attenuation and noise that represent the underwater environment. ARQ offers the lowest efficiency and NC the highest but there is a distance at which FEC becomes the best option rather than NC suggesting a hybrid FEC/NC method.

INTRODUCTION

The majority of the earth's surface is covered with water and underwater communication applications such as oceanographic data collection, pollution detection and offshore exploration require monitoring of the different parts of the ocean (Akyildiz et al., 2005). One common method for ocean monitoring is to use oceanographic sensors, record the information and then recover the instruments (Yick et al., 2008). However this method is not efficient because it creates long lags in the recorded data, and also if by chance any problem occurs while recovering the data, they all tend to get lost (Sozer et al., 2000). The best solution to this problem is to establish a network configuration which helps in the real-time communication between the underwater devices

DOI: 10.4018/978-1-4666-8251-1.ch002

and a control center. Two way acoustic links are established between different underwater devices such as Autonomous Underwater Vehicles (AUVs) and sensors, thus giving rise to basic Underwater Acoustic (UWA) Networks (Sozer et al., 2000). Such networks are connected to a surface station which in turn is connected to the Internet through a radio frequency (RF) link. Using this network configuration, scientists can easily extract real-time data from a number of distant underwater devices. After examining the received data, control information is sent to individual instruments and the network adapts itself to the changing situations. Thus, significant data loss does not occur because the data are transferred as soon as they become available.

UWA network design involves major challenges such as limited available bandwidth, the nature of the underwater channel and limited battery power, making them prone to failure (Akyildiz et al., 2005; Heidemann et al., 2006). The most suitable option is underwater sensor networks (UWSNs) (Cui et al., 2006) consisting of a number of sensors and vehicles which interact with each other to carry out some collaborative tasks such as detection of the target and tactical surveillance. Wireless acoustic communications have to be established in these applications so that a considerable area can be monitored with a small number of nodes. Such communications coupled with appropriate networking protocols help to overcome the limitations of underwater channels to some extent (Cherkaoui et al., 2012) but UWSNs and their design challenges are different from terrestrial sensor networks (Partan et al., 2006). One major challenge that is considered in this project is the energy efficiency in the underwater acoustic channels. The wide variety of applications of UWA communications in UWSNs and multi-AUV cooperative missions (Akyildiz et al., 2004; Akyildiz et al., 2005) ranging from simple monitoring and data collection to exploration, deployment and rescue work, highlight the significance of UWA networks. However, there has historically been little progress in developing networking topologies for the underwater acoustic medium (Harris and Zorzi, 2007). Although wireless connectivity can be achieved underwater, when using the acoustic medium for inter-device networking, the acoustic channel differs greatly from the RF channel (Stojanovic, 2006). In the underwater environment, bandwidth varies from a few kHz in a long range (typically 1000 km) system to hundreds of kHz in short-range (less than 100m) systems (Akyildiz et al., 2004). We will now briefly summarize the factors which characterize underwater acoustic propagation, indicating which ones are not included in our analysis and why. Factors that make major contributions to the transmission performance of the systems considered in this chapter are analyzed in greater depth within subsequent sections.

Path Loss

The overall signal strength of the acoustic signal transmitted between the sensor nodes of the network decreases as the distance increases because of a number of factors that govern the sound propagation factors in the ocean. This decrease in the acoustic intensity between the source and the receiver is termed the propagation loss or the path loss and comprises three main sources namely attenuation, geometrical spreading and propagation anomaly which are now briefly introduced.

Attenuation is the signal loss that occurs due to the conversion of acoustic energy within the sea-water into heat, known as absorption loss. This loss exhibits dependence on many factors particularly the frequency of the propagating wave. The classic model to determine the signal attenuation in the ocean water is due to (Thorp, 1965; Thorp, 1967). This has been extended to accommodate temperature and pressure (Fisher & Simmons, 1977), and salinity and acidity (Ainslie & McColm, 1998).

Geometrical spreading loss refers to the signal level reduction seen at the receiver due to the fact that the acoustic wave spreads out as it propagates (Urban, 2002). This spreads the energy over a surface area that increases with distance reducing the collected power. Spherical spreading of the signal occurs when the underwater transmission distance is large (over several km), while cylindrical spreading takes place for smaller distances. The exact point at which the transition from cylindrical to spherical spreading takes place depends on the scenario (Urick, 1983; Urban, 2002) and here we consider the distances such that spreading remains cylindrical.

The *propagation anomaly* includes all the losses that might occur due to leaky communication ducts, scattering and diffraction and which have not been associated so far with attenuation and geometrical spreading. It thus requires knowledge of the operation environment to attempt its estimation. The effects are particularly pronounced in thermocline and halocline areas (Urick, 1983; Urban, 2002) and less significant in the deep-sea scenario that we consider here. Hence, we do not include a loss for propagation anomaly meaning that our results may in some case be slightly optimistic but since the loss is linear, the relative performance of the error correction schemes will remain the same.

Noise

In common with many communication systems, there are two kinds of noise that effect underwater acoustic communications (Preisig, 2007). First, *man-made noise*: this is caused by machinery such as pumps, reduction gears and the like, and also by shipping activities. Second, *environmental noise*: this is caused by natural phenomena such as tides, current, storms, wind and rain and also by biological phenomena (Preisig, 2007). The noise which occurs while choosing the appropriate frequency band in underwater communications systems is the ambient noise (Preisig, 2007). Later, details of the method employed to include noise in our analysis are given along with quantitative exposition (see Figure 3).

Multipath Effects

In some environments, the ocean is considered as a waveguide for communication signals (Liu & Huang, 1999) that comprises reflecting surfaces and an ocean floor. The speed of sound varies within this waveguide and reflection, refraction and diffraction of the acoustic signals results in the formation of multiple propagation paths from the source to the receiver. Here, we consider the near-vertical deep-water channel in which multipath effects are small or can be ameliorated (Crocker, 1998), and thus our model does not include a multipath term. We also consider that the nodes are nominally stationary so will suffer minimal effects from expansion or compression of the emitted signal at the receiver, known as Doppler shift (Stojanovic, 2008). Moreover, compensation methods are available for drift (Sharif et al., 2000) that will make the effect of Doppler shift negligible in comparison with the other effects considered here.

Chapter Scope

Previous work by (Babiker & Zakaria, 2011) has compared the energy efficiency of two major established underwater error correction schemes namely Automatic Repeat Request (ARQ) and Forward Error Correction (FEC) in terms of parameters such as communication distance, packet size, shipping factor and wind speed. Here we include Network Coding (NC) (Médard and Sprintson, 2012) as a newer scheme in terms of underwater communications. We present the mathematical energy efficiency analysis of ARQ, FEC and NC, and follow this with a comparison of the three approaches in terms of the energy efficiency by varying the communication distance and packet size. The effects of shipping and wind speed are also studied.

BACKGROUND

As a result of the frequent use of underwater sensors for the monitoring and exploitation of underwater resources, underwater wireless sensor networks (UWSNs), based largely on acoustic transmission technologies, have been a subject of considerable interest in recent decades (Proakis et al., 2001; Akyildiz et al., 2004; Heidemann et al., 2006; Cui, et al. 2006). Although a number of data transport protocols have been developed for UWSNs (Sozer et al., 2000; Chitre et al., 2008), there is still no single internationally accepted standard for underwater communication. This is mainly as a result of a number of technical challenges encountered while trying to establish efficient underwater wireless data communication, not least of which is the poor propagation of electromagnetic signals in both the standard radio frequency and optical ranges. Although the latter has been the subject some success (Farr et al., 2010), there is still significant scattering and absorption from the water, and the attenuation suffered by the former rules it out except for very short links (Cella et al., 2009). Thus, the field to date has been dominated by acoustic transmissions that propagate for relatively long distances but at reduced bandwidths (in the kHz range at most) (Akyildiz et al., 2004). Some of the factors that disrupt smooth underwater acoustic transmission are: the path loss caused by spreading and absorption; noise arising from diverse sources such as the water itself, seismic activity, biological phenomena; reflection and refraction of the signals ocean surfaces (Ainslie, 2010).

Design of a good data transport protocol for the underwater environment requires consideration of two main issues, viz. reliability and energy efficiency. Both these issues are related to each other in the sense that the former requires error-correction which in turn requires energy. This energy cost for reliability causes detrimental effects on those applications that require nodes to be operated underwater for long periods of time without battery recharging and in aquatic environments where recharging or replacing batteries is difficult (Chan and Motani, 2007; Xie and Cui, 2007). Now, because of the severe energy constraints on UWSNs, an appropriate strategy has to be adopted which not only ensures reliable data transmission but also conserves the energy at the same time. The most established error correction techniques to ensure reliable data transmission in underwater acoustic links are Automatic Repeat ReQuest (ARQ) and Forward Error Correction (FEC) (Proakis et al., 2001; Akyildiz et al., 2007; Gao et al., 2009). More recently NC has been applied to underwater acoustic links (Guo et al., 2009). This chapter presents a mathematical energy efficiency analysis the three methods, ARQ, FEC and NC but we first briefly review some of the most relevant past work in the area.

Previous Work

A large amount of research has been carried out concerning the usage of the error control schemes in underwater sensor networks (UWSNs) to increase their robustness. There have also been studies related to the energy efficiency of error correction techniques in other sensor networks. A suitable propagation model for the underwater acoustic channel to calculate the signal to noise ratio at the receiver was designed and implemented by (Harris & Zorzi, 2007). An optimization metric for energy efficiency was first introduced by (Sankarasubramaniam et al., 2003), who also examined the energy efficiency of ARQ and came up with the conclusion that its retransmission strategy fails to improve sensor network energy efficiency. (Tian et.al, 2008) used the same optimization metric to perform energy efficiency calculations for the ARQ and FEC schemes in terrestrial wireless sensor networks. They compared the two techniques based on the communication distance and packet size, concluding that the energy efficiency of ARQ remains unchanged with the number of retransmissions. The model introduced by (Sankarasubramaniam et al., 2003) has also been employed by (Karvonen et al., 2004; Shelby et al., 2004; Kashani & Shiva, 2006) in their further investigations of wireless sensor network energy efficiency. (Sankarasubramaniam et al., 2003) and (Balakrishnan et al., 2007) performed energy efficiency analyses of FEC, showing that using Bose, Chaudhuri and Hocquenghem (BCH) codes was superior to employing convolutional coding because of their low encoding and decoding energy consumption. Modern signal processing methods mean that phase-coherent 8-Phase Shift Keying (8PSK) is suitable for underwater acoustic communications (PSK), offering good signal to noise performance coupled with manageable complexity (Stojanovic et al., 1991).

Apart from ARQ and FEC-based techniques, another scheme which has been proposed to make UWSNs more robust is the multi-path forwarding scheme (Seah & Tan, 2006; Yang et al., 2010; Xu et al., 2011). This utilizes the transmission of redundant packets through a number of different paths to enhance the packet delivery ratio. However, multi-path forwarding itself is not enough as the duplicated packets, which are eventually removed, use up a lot of energy without providing any useful information. In addition, it is possible to employ multiple virtual sinks for improving robustness and eliminating errors (Seah & Tan, 2006). In this approach, the source uses acoustic communications to send the packets to different virtual sinks, which in turn forward the packets to the final receiver using wireless radio communication with a large-bandwidth. NC was first shown to achieve broadcast capacity in multicast tree networks by (Ahlswede et.al, 2000). Subsequently, (Li et al., 2003) investigated how linear network codes could be used in multicast networks, and (Koetter & Médard, 2003) showed how the coefficients of linear coding and decoding functions could be found. Application of NC to UWSNs with multipath instead of single path routing was demonstrated by (Guo et al., 2006) showing benefits from the broadcast property of the underwater acoustic channel to increase the collaboration among the nodes. More recently, the idea of Vector-Based Forwarding (VBF) (Guo et al., 2009) in which offers energy efficiency and the ability to deal with node mobility has been proposed. In VBF, a "routing pipe" is formed by the routers so that any arriving packets may be deemed either eligible for forwarding or discarded. The combination of VBF and NC (VBF-NC) has then delivered a reliable transport protocol for UWSNs. A natural extension to three disjoint paths produces Multiple Paths and Network Coding (MPNC) (Cai et al., 2013), which has been shown to both higher delivery ratios and lower energy consumption than VBF-NC. As is apparent later in this chapter, the frequency-dependent nature of the underwater acoustic channel is highly significant in network design. Recent work on energy efficiency has included optimization of node placements for its improvement (Kam et al., 2014).

The applications of UWSNs fall into similar categories to those applicable to terrestrial sensor networks and it is the nature of the underwater environment that introduces new challenges as described later in this chapter. The broad categories are summarized as follows by (Heidemann et al., 2012): scientific applications concern the observation of the environment in terms for example of geology, water characteristics or animal life; industrial applications encompass the monitoring and control of mineral extraction activities; military applications include communicating with submarines and divers. The motivations for the deployment of UWSNs are similar to those for terrestrial wireless sensor networking, namely reduced deployment costs and the opportunity to adjust sensor operations in real time. Some more specific examples include (Heidemann et al., 2006; Akyildiz & Vuran, 2011):

- **Seismic Monitoring:** Frequent seismic monitoring is important in ensuring the safety and effective operation of oil extraction from underwater fields. This task has traditionally involved the expensive process of a ship towing an array of hydrophones so the prospect of using UWSNs instead is highly appealing (Parameshwaraiah et al., 2013). UWSN seismic monitoring also offers the prospect of tsunami warnings to coastal areas (Comfort et al., 2012).
- **Environmental Monitoring:** UWSNs may be utilized for example to detect chemicals in water (Yang et al., 2002) or to ascertain the presence of temperature gradients thought to promote the breeding of marine microorganisms (Zhang et al., 2004).
- **Equipment Monitoring and Control:** Long-term communication with undersea equipment is generally by means of wired connections and also requires development of the appropriate control strategies (Altamiranda et al., 2009). Nevertheless, UWSNs present opportunities to replace cabled connections for temporary installations and in a more general context (Proctor et al., 2010). The challenges are similar to those for other applications with the main difference being that stable and frequent sensing is needed for equipment monitoring in contrast to the infrequent bursts of data seen in seismic monitoring.
- **Coordination of Groups of Underwater Robots:** For purposes such as sensing of chemical leaks or biological phenomena and equipment monitoring similar to that described above. Historically, underwater robots have been either autonomous but with limited networking abilities or tethered with the resultant good communication abilities but restricted maneuverability. The provision of wireless networking capabilities (Zhao & Guo, 2010) opens up the prospect devices that are both well-connected and flexible in location. This application leads naturally into the area of *ocean sampling* (Curtin & Bellingham, 2009) which enables a much better picture of physical characteristics to be obtained than is possible with fixed point or satellite observations. As a result of being able to capture the stochastic nature of currents and other ocean properties, improved modeling and prediction of ecosystem impacts from climate change is possible (Duffy et al., 2013).
- **Assisted Navigation:** UWSNs can be used to identify submerged hazards such as mines (Elbergui et al., 2012), dangerous currents (Huang et al., 2012) and shipwrecks (Cui et al., 2006).
- **Distributed Tactical Surveillance:** An UWSN can be configured to identify submarines, divers and so forth by processing data collected from a range of microsensors (Petrut et al., 2014). This fusion of data and the dynamic, flexible nature of UWSNs deliver increased accuracy, coverage and robustness in comparison with traditional sonar or radar techniques (Li, 2012). Another interesting recent development that has arisen from progress in identification is *forensic detection* by law enforcement authorities to reduce the need for expensive and dangerous diving operations (Schultz et al., 2013).

These applications continue to be of increasing relevance today with particular interest as one would expect in environmental monitoring (Suzuki et al., 2014) and oilfield applications (Jawhar et al., 2013). Moreover, developments in protocols (Chen et al., 2014) and localization (Krishna & Yadav, 2014) are having a particular impact on the implementation of underwater monitoring and robotics.

General Framework of the Study

A scenario is considered within an UWSN with a single source and sink, connected by a single path. Three error-correction schemes, namely ARQ, FEC and NC are used and these are analyzed for a single hop from source to sink. Later, we will describe the underwater propagation model employed and the determination of the receiver Signal to Noise ratio (SNR) leading to the bit error rate (BER), P_b for each scheme. We begin though with the generality of the calculation of the Packet Error Rate (PER) from the BER for a packet of length L thus:

$$PER = 1 - \left(1 - P_b\right)^L \tag{1}$$

The energy efficiency (η) is a suitable optimization metric in this work, and is generally defined as the useful fraction of the total energy expenditure in a communication link between sensors (Al-Suhail et al., 2012). The energy efficiency is the product of two factors (Sankarasubramaniam et al., 2003; Tian et.al, 2008), the energy throughput (η_e) and the packet acceptance rate (r). Now r helps to determine data reliability and is given by:

$$r = \left(1 - PER\right) \tag{2}$$

The energy throughput is the ratio of the energy consumed by successfully delivered packets to the total energy consumed by all packets from the source, meaning that:

$$\eta = \frac{E_{Payload}}{E_{total}} r = \frac{E_{Payload}}{E_{total}} \left(1 - PER\right) \tag{3}$$

The calculation of η for ARQ and FEC depends on the packet size but also on the size of the overheads used. For NC, this is not the case and the energy throughput calculation depends on two metrics, namely the successful delivery ratio (R) and the normalized energy consumption (T). The successful delivery ratio (R) is defined as the ratio of the number of successfully delivered packets to the total number of packets from the source. T is actually the total number of transmissions across a link but this approximately corresponds to the energy normalized by R (Guo et al., 2009). For a single hop, we can consider an arbitrary packet (whether transmitted successfully or not) that is transmitted on average T_1 times, then:

$$T = \frac{T_1}{R} \tag{4}$$

The work of (Babiker & Zakaria, 2011) encompasses ARQ and FEC, so here we introduce NC for an extended comparison of the options when transmitting data via an underwater acoustic modem. Details are now provided of the three schemes followed by the underwater propagation model employed. Ocean background noise arises from a number of sources, and varies as the frequency and location changes (Wenz, 1962). Here, the total ambient noise in the ocean is calculated by taking into consideration the noise from the four most prominent sources namely turbulence, shipping, wind driven waves and thermal noise (Coates, 1989).

DETAILS OF THE ERROR CORRECTION TECHNIQUES EMPLOYED

To realize the potential of UWSNs, appropriate methods are needed to ameliorate the effects of the underwater propagation environment, which is hostile to long distance communications. We thus present a brief overview of the major approaches that have been used to improve UWSN operation, and which subsequently form the basis of our numerical study.

Automatic Repeat Request (ARQ)

This method, also known as Automatic Repeat Query is an error-control method for data transmission that uses error-detection codes, positive and/or negative acknowledgement messages and timeouts to achieve reliable data transmission over an unreliable service (Costello et al., 1998). Acknowledgements are messages sent by the receiver to indicate that it has successfully received a data packet, whereas timeouts are specified periods of time for which the sender waits to receive an acknowledgement. The basic idea behind ARQ is that the sender waits for a specified interval of time after sending the packet (the timeout) to receive an acknowledgement. If a positive acknowledgement (ACK) is received, then it at once transmits the next packet. However, if negative acknowledgement (NAK) is received or if the sender does not receive any acknowledgement before timeout, then it usually retransmits the same packet. This process continues until the sender receives an ACK from the receiver or until a specific number of retransmissions have been reached. Three different types of ARQ protocols are used in data communications, namely Stop-and-Wait (S&W) ARQ, Go-Back-N (GBN) ARQ and Selective Repeat (SR) ARQ (Costello et al., 1998). These operate at either the Data Link or Transport layers of the OSI model and each of them use some kind of sliding window protocol to ask the transmitter to determine which (if any) packets need to be retransmitted (Costello et al., 1998).

In this work, the simple Stop-and-wait (S&W) ARQ protocol is employed to carry out the energy efficiency analysis; this is the main choice for underwater acoustic communications because of the half-duplex property of the acoustic modem (Gao et al., 2009). (Tan et al., 2007) proposed that in a multi-hop acoustic channel, the acknowledgement can be achieved either explicitly by sending an acknowledgement packet for each successfully received packet or implicitly by using the broadcast nature of the medium. An efficient ARQ technique was proposed by (Lee et al., 2008) by making use of the channel sharing property embedded inside the underwater environment. The packet size is varied in such a way that the propagation delay becomes larger than the transmission time and the packets are scheduled between multiple nodes so as to achieve collision free transmission. In a multiple-hop scenario, overhearing is used as an acknowledgement method, in which a packet transmitted from the next hop is overheard and this overheard packet replaces the acknowledgement packet. This method saves energy and also reduces the overhead and the transmission latency. Evaluation of this scheme involves comparing it with the existing S&W ARQ in terms of the latency, which is reduced. (Lee et al., 2008) also implemented a continuous ARQ scheme to transmit and receive by making use of the long propagation delay in the underwater environment, and without taking into account the half-duplex property of the acoustic modem. Since this scheme made use of more than one packet in the channel between the transmitter and the receiver, the propagation time was reduced and the throughput was increased compared to other ARQ schemes. However, it was found that the scheme was not appropriate when the channel conditions were poor and when the distance ranges were large.

Forward Error Correction (FEC)

In telecommunications, information theory and coding theory, Forward Error Correction (FEC) or channel coding (Costello et al., 1998) is a technique used for controlling errors and achieving reliable data transmission in a communication system over unreliable or noisy communication channels (Moon, 2005). The basic idea behind FEC is that the message is encoded by the sender in a redundant way by making use of an error-correcting code (ECC). In FEC, the errors are corrected at the receiver side, hence the name "forward", and the concept has become an essential part of modern communications since it was pioneered by Hamming in the middle of the twentieth century (Hamming, 1950).

FEC can operate at either at the bit, the byte or the packet level (Tsai et al., 2010) and at each of these appropriate redundant bits are added to the data by the sender to recover either bits, bytes or whole packets efficiently. Redundant bits are formed using functions of the original information bits, which may be present in an unmodified form in the encoded output (systematic codes) or not (non-systematic codes) (Moon, 2005). The addition of redundancy enables the receiver to detect and correct errors without retransmission. Thus FEC enables the receiver to correct errors without requiring a reverse channel to request data retransmission but at the expense of a fixed and a larger forward channel bandwidth. It is therefore applied in cases where retransmissions are either very expensive or almost impossible, such as one-way communication links and when broadcasting to different receivers in multicast scenarios (Park et al., 2003). The FEC information is sparsely used in modems and generally stored in receiver mass storage devices so as to be able to recover the corrupted data. The code rate and the type of coding used determine the number of errors that can be corrected using FEC meaning that different FEC codes are chosen for different conditions.

FEC methods are categorized as block codes or convolutional codes (Morelos-Zarogaoz, 2006). The former operate on fixed-size blocks (packets) and their decoding generally performed in polynomial time. There are many different types of block codes (Morelos-Zarogoza, 2006) such as Reed-Solomon, Golay, and the BCH codes that are employed here based on the results of (Sankarasubramaniam et al., 2003; Balakrishnan et al., 2007). Convolutional codes operate on bit or symbol streams of arbitrary length and are usually decoded using the Viterbi algorithm (Morelos-Zarogoza, 2006) but their decoding may become extremely complex.

It should also be noted that FEC has been combined with ARQ in the segmented data reliable transport (SDRT) protocol, used to send data block by block and hop by hop (Xie and Cui, 2007). The transmitter uses erasure codes to encode the packet and then in turn send it to an intermediate node, which reconstructs the packet, encodes it and then forwards it to the next hop. SDRT helps in improving the channel utilization and simplifying protocol management but continuously sends data until an acknowledgement is received making it not particularly energy efficient. The concept has also been developed further in the Adaptive Redundancy Reliable Transport Protocol (ARRTP) (Liu et al., 2008) which makes use of distance to adapt and the Adaptive Hybrid Error Correction Technique (AHECT) (Babiker et al., 2011) which utilizes BER as its adaptation metric.

Network Coding (NC)

In contrast to the previous methods, NC is a relatively new area in information theory that began in 2000 (Ahlswede et.al, 2000) and has many interesting applications in practical networking systems. In the seminal work, the broadcast capacity of a multicast tree could not be achieved by the traditional store-and-forward approach to packet transmission but only by the application of NC (Ho et al., 2003b). The main idea of NC is that store-and-forward at intermediate nodes is replaced by the transmission of packets that are linear combinations of the previously received data. In other words, the nodes recombine a number of input packets into one or more output packets (Fragouli et.al, 2006). Two advantages of this approach in the context of the work here are that there is an increase in the throughput and robustness may be increased by using diverse paths for different packet combinations. The advantages of NC can be demonstrated by means of the example shown in Figure 1 (Guo et al., 2009). When NC is employed, as shown in Figure 1(a) the source produces three packets A, B and C, encodes them into X_1, X_2 and X_3 and then delivers them to the sink. The broadcast property of the acoustic channel means that these packets pass through the three relays R_1, R_2 and R_3 simultaneously. Out of the three packets, relay R_1 only receives X_1 and X_3 successfully and encodes them into packets Y_{11} and Y_{12}. Similarly, relay R_2 receives X_1 and X_2 successfully and encodes them into Y_{21} and Y_{22} whereas relay R_3 receives all three packets and encodes them into Y_{31}, Y_{32} and Y_{33} respectively. These encoded packets are then forwarded by the relays to the sink but this only receives Y_{11}, Y_{21} and Y_{32}. The use of an appropriate NC scheme such as random linear coding (Ho et al., 2003b, Li et al., 2003) ensures that the sink can recover all the three original packets using Gaussian elimination. However, the situation shown in Figure 1(b) where the packets are merely forwarded means that packet C is lost.

NC is well-matched to UWSNs because in general underwater sensor nodes are generally bigger than land-based sensors (Ayaz et al., 2011), with greater computational capabilities (Nicolaou et al., 2007; Xie et al., 2006). Moreover, the broadcast nature of acoustic results in a large number of highly interleaved paths from the source to the sink which may be employed as in Figure 1 above.

A typical NC scheme takes a generation of packets into consideration instead of a single packet. Ideally, packets which are generated by the source are first divided into generations each containing K packets, which are linearly combined by the source using randomly generated coefficients. Mathematically speaking, if $X_1, X_2, \ldots, X_K$ represent the K packets in a generation, then these are linearly combined by the source to produce K' outgoing packets:

$$Y_i = \sum_{j=1}^{K} g_{ij} X_j \qquad i = 1, 2, \ldots, K \tag{5}$$

The coefficient g_{ij} is randomly selected from a finite field F_2G. The complete set of coefficients $\left(g_{i1}, g_{i2}, \ldots, g_{iK}\right)$ is called the encoding vector for the encoded packet Y_i (Fragouli et.al, 2006) and is present in the packet as overhead. The value of K' is chosen in such a way that $K' \geq K$ to introduce redundancy, reducing the effect of lost packets. It is to be noted that the network setting considered here has only one source, hence the nodes in the first hop receive packets from the source only, while the nodes in the other relay sets receive packets from various nodes in the previous hop. It is for this reason that some redundancy is purposely introduced at the source.

In general, packets from multiple paths within the network are stored in a buffer at a relay node and combined with others from the same generation. Since the focus here is on a comparison of single hop transmission, the reader is referred to (Guo et al., 2009) for full details of NC within an underwater network. The sink on receiving all the K packets with linearly independent encoding vectors, recovers all the original packets by the process of matrix inversion (Fragouli et.al, 2006). The computational power of the underwater sensor nodes, coupled with multiple paths in the acoustic medium, makes NC a viable choice for UWSNs.

UNDERWATER PROPAGATION MODEL

Work related to underwater sensor networks is based upon an accurate understanding and modelling of the underwater acoustic channel. The varying channel conditions and communication distances between the sensor nodes, which occur due to water currents define the underwater acoustic propagation channel (Sozer at al., 2000). Different models exist to predict and calculate the attenuation, and this in turn affects all the other aspects of the underwater channel model. Parameters such as frequency, distance, depth, acidity, salinity and temperature of the underwater environment help in determining how a channel behaves potentially causing network performance

Figure 1. An example which demonstrates the advantages of using NC in underwater sensor networks; struck out packets are lost (Guo et al., 2009)

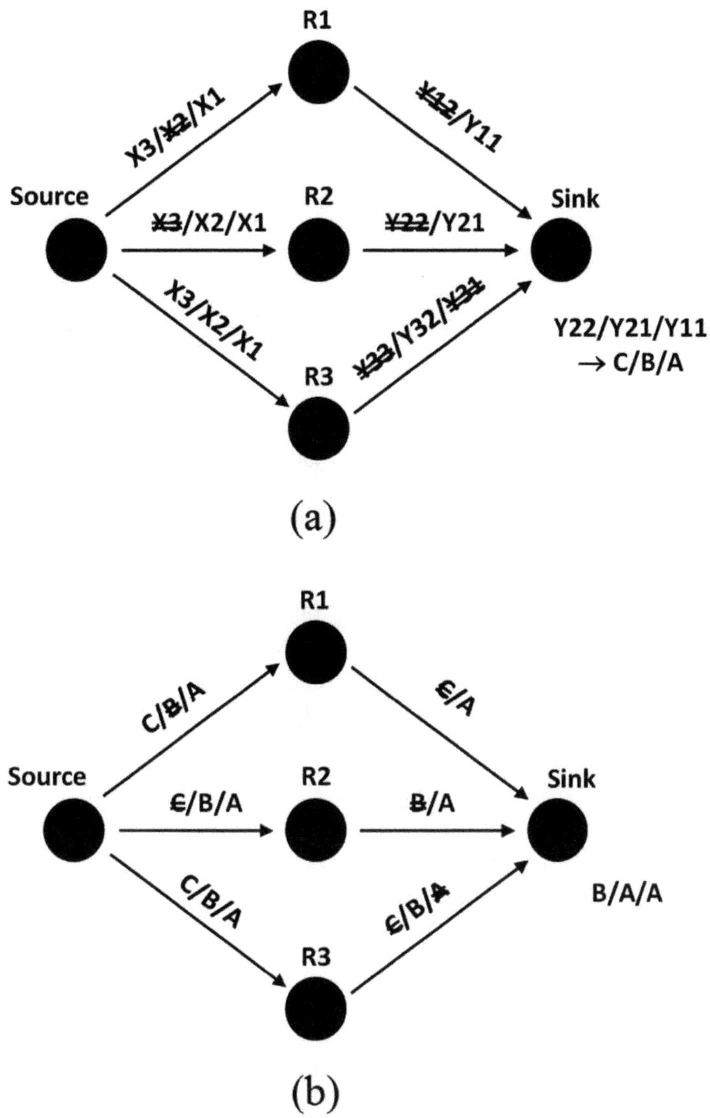

variations. Thus, understanding the relationship between the various parameters and how they affect network performance is vital for the construction of an efficient underwater channel model. The underwater propagation channel model considered here is utilized to compute the receiver Signal to Noise ratio (SNR) by taking both the attenuation of the acoustic signal in water and the ambient noise into consideration. The total attenuation is calculated by taking into account both the absorption of water and he fact that the signal spreads out as it travels.

Attenuation

Acoustic propagation underwater involves many factors and attenuation may occur as a result localized (e.g. obstructions) and more general (e.g.

water temperature) effects (Lurton, 2010). Here we take a view based on empirical expressions that have been developed over many years for underwater communication systems calculations. The path loss comprises two different elements, the first being the absorption loss and the second being the spreading loss. To determine the former denoted by $a(f)$ at a particular frequency we employ Thorp's approximation function (Brekhovskikh & Lysanov, 1982). This function gives an empirical fit to the loss in dB per km as a function of the frequency in kHz thus (See Box 1).

The absorption coefficient increases rapidly, limiting the transmission frequency that may be employed even in the absence of any other losses or noise, as may be seen in the plot of $a(f)$ versus frequency shown in Figure 2.

To account for the spreading loss as the signal propagates, a path loss exponent k is utilized to produce a combination of absorption and the spreading loss over a distance l in km (Brekhovskikh & Lysanov, 1982; Urick, 1983):

$$A\{l,f\} = l^k \{a(f)\}^l \tag{7}$$

It should be noted that this is a numerical loss per km so the value of a(f) must be extracted from Thorp's formula after the calculation of the dB value. The value of k depends on the environment in which the propagation takes place. For shallow water it is 1, representing cylindrical spreading, for deep water is it 2 (spherical spreading) and for a practical compromise k is taken as 1.5 (Urick, 1983). Here, propagation is assumed to take place in deep water where the effects of multipath reflections are not significant, thus spherical spreading occurs and $k = 2$.

Noise

Background noise in the ocean is due to a number of factors that vary with frequency and location (Wenz, 1962). The four major contributors to the total ambient noise in the underwater environment are turbulence, shipping, wind and heat. The power spectral densities (PSDs) for the four respective noise components, $N_t(f)$, $N_s(f)$, $N_w(f)$ and $N_{th}(f)$ in dB re μPa per Hz are given by the formulas in Box 2 when f is in kHz (Coates, 1989).

In (8b), s is a shipping variable which has values ranging from 0 to 1 and in (8c), w is a wind variable denoting the wind speed in ms^{-1}. For a particular frequency f (in kHz), the overall noise power spectral density in dB/kHz is given by:

$$N(f) = N_t(f) + N_s(f) + N_w(f) + N_{th}(f) \tag{9}$$

To illustrate the nature of the ambient noise, Figure 3 shows a plot of the total noise PSD in the absence of shipping (s = 0) and wind (w = 0). It may be observed that over the range of frequencies employed underwater i.e. a few tens of kHz to avoid overly large attenuation losses, the noise PSD reduces at an approximate rate of 18 dB per decade. This combination of variable attenuation and noise means that the signal experiences a signal to noise ratio (SNR) that varies over its bandwidth.

Box 1.

$$10\log_{10}\{a(f)\} = \begin{cases} 0.002 + 0.11\dfrac{f^2}{1+f^2} + 0.001f^2, f \le 400Hz \\ 0.11\dfrac{f^2}{1+f^2} + 44\dfrac{f^2}{4100+f^2} + 2.75\times10^{-4} + 3\times10^{-3}, f \le 400Hz \end{cases} \tag{6}$$

Figure 2. Underwater absorption coefficient in dB per km as a function of frequency

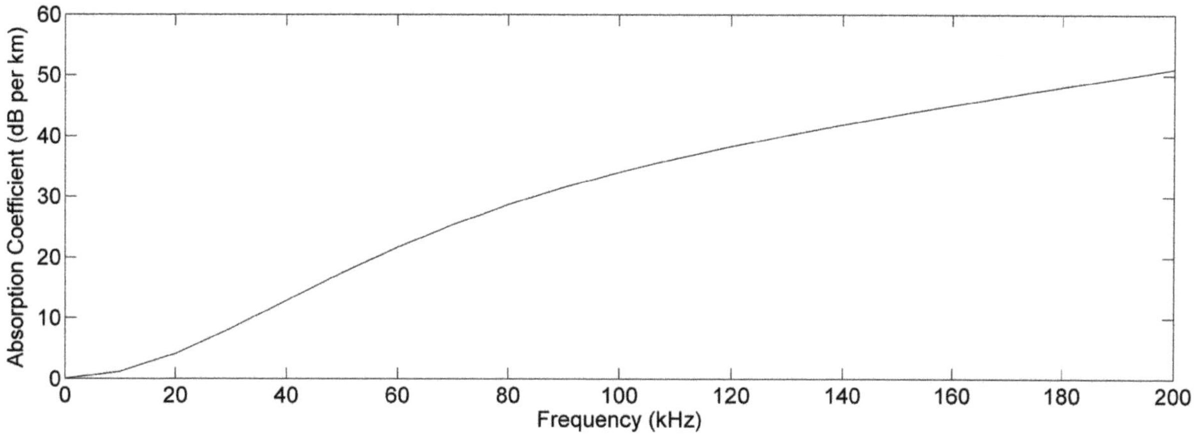

Box 2.

$$10\log_{10}\left[N_t\left(f\right)\right] = 17 - 30\log\left(f\right) \quad (8a)$$

$$10\log_{10}\left[N_s\left(f\right)\right] = 40 + 20\left(s - 0.5\right) + 26\log\left(f\right) - 60\log\left(f + 0.03\right) \quad (8b)$$

$$10\log_{10}\left[N_w\left(f\right)\right] = 50 + 7.5w^{0.5} + 20\log\left(f\right) - 40\log\left(f + 0.4\right) \quad (8c)$$

$$10\log_{10}\left[N_{th}\left(f\right)\right] = -15 + 20\log\left(f\right) \quad (8d)$$

Figure 3. Power spectral density of the ambient noise, N(f) in dB re μ Pa

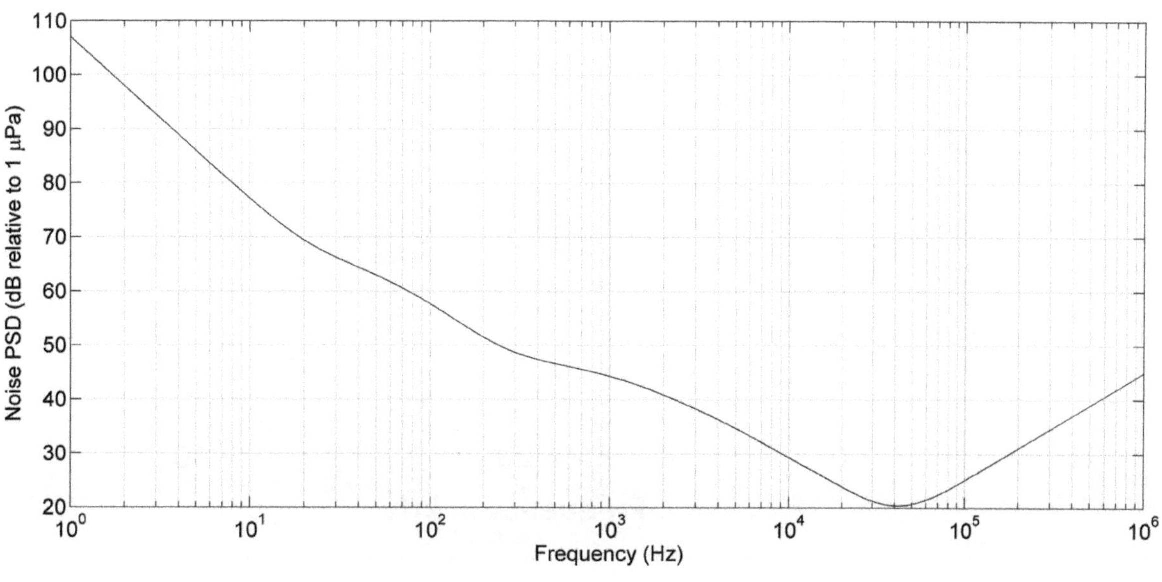

SNR

To obtain the SNR, it is conventional in underwater communications to consider a narrow band of frequencies Δf about some center frequency fc. The SNR is then obtained using the expressions above and the power of the omnidirectional source, S(f), thus:

$$SNR = \frac{S(f)}{A\{l,f\}N(f)\Delta f} \tag{10}$$

The quantity $A\{l,f\}N(f)$ is known as the attenuation noise (AN) factor and we may use it to show that the narrowband SNR in (10) is a function of frequency for a given length as illustrated in Figure 4 for distances of between 0.5 km and 3 km. The available bandwidth becomes increasingly limited as the distance increases. In this work, the transmission distances covered are up to a few km and so in subsequent calculations a value of 15 kHz is used because it is unlikely that sensors will be able to dynamically retune their frequencies or different distances, and this value represents a good compromise in terms of acceptable loss over the relevant distance range.

Once the attenuation and noise are calculated, it is usual to express the SNR of the emitted underwater omnidirectional signal at the receiver in terms of the source level SL in dB re μPa:

$$SNR = SL - A\{l,f\}\Big|_{dB} - \int_{\Delta f} N(f)\,df\Big|_{dB} \tag{11}$$

This is the passive sonar equation where, $A\{l,f\}\big|_{dB}$ is the A(1,f)=total attenuation in dB given from equation (7), $N(f)$ is the total noise power spectral density given from equation (9), Δf is the modem bandwidth and SL is the source level expressed in dB re μPa. In typical seawater, the source level for a 1 Watt output is given by 170.9 dB re μPa (Burrowes & Khan, 2011). For the calculations that follow, a 1 W source was assumed and the modem bandwidth taken to be 10 kHz centered at the 15 kHz operating frequency (Melodia et al., 2013). The noise value in dB was found

Figure 4. Frequency dependent attenuation noise factor in the absence of shipping and wind

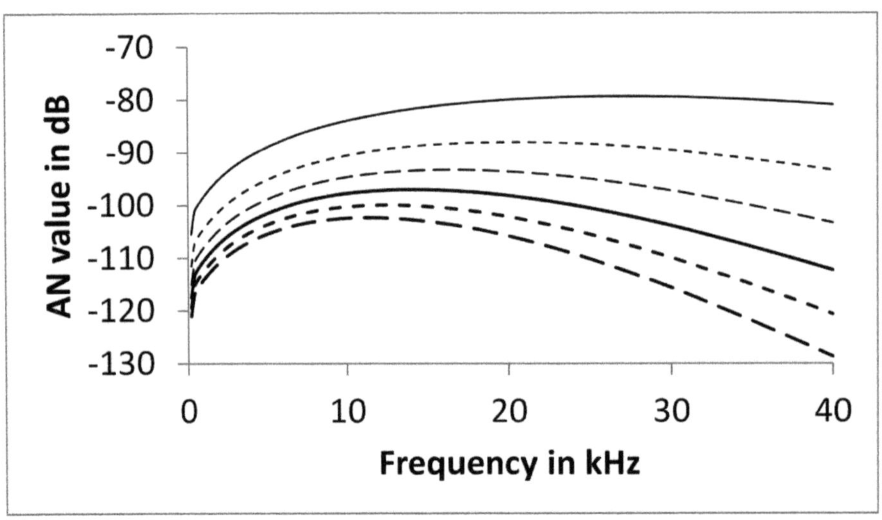

by numerically integrating $N(f)$ over the bandwidth to give a value of 66.6 dB, which was used in the obtaining of all quantitative results.

MATHEMATICAL ANALYSIS OF SCHEME ENERGY EFFICENCIES

To compute the energy efficiency for each of the three error correction schemes: ARQ, FEC and Network Coding, first the BER and then the PER needs to be calculated. For ARQ and NC, 8-PSK is used for BER calculations, whilst convolutional coding is utilized for FEC.

ARQ

The analysis is based on the Stop and Wait (S&W) ARQ protocol since as discussed previously it is the only viable option for underwater communications. Each data packet of ARQ consists of a header, a payload and a Frame Check Sequence (FCS). The header field is of size α bits, the payload is of size *n* bits and the FCS is of size τ bits (Al-Suhail, et al., 2012); the length of the acknowledgement packet is taken as l_{ack}.

The energy efficiency of ARQ is independent of the retransmission attempts and is invariant with respect to the number of retransmissions (Tian et.al, 2008). For communication in a single hop, the energy consumption of a sensor node is the sum of the energy consumed by the sender in sending the data and receiving the acknowledgement (E_{ARQ}^{TX}) and the energy consumed by the receiver in receiving the data and sending back the acknowledgement (E_{ARQ}^{RX}), thus (Babiker & Zakaria, 2011):

$$E_{ARQ} = E_{ARQ}^{TX} + E_{ARQ}^{RX} \quad (12)$$

So explicitly,

$$E_{ARQ}^{TX} = E_{DATA}^{TX} + E_{ACK}^{TX} = \left(P_{TX} l_{data} + P_{RX} l_{ack}\right) T_b \quad (13)$$

$$E_{ARQ}^{RX} = E_{DATA}^{RX} + E_{ACK}^{RX} = \left(P_{RX} l_{data} + P_{TX} l_{ack}\right) T_b \quad (14)$$

where, P_{TX} is the power consumed in transmitting, P_{RX} is the power consumed in receiving, l_{data} is the length of the ARQ data packet in bits (α+τ+n) and T_b is the bit time.

Now, the bit error rate for the ARQ scheme is calculated by employing the 8-PSK modulation technique. Using an M-PSK scheme, the symbol error probability for ARQ is given by (Labrador et al., 2009):

$$P_s \approx \operatorname{erfc}\left[\sqrt{\gamma_s} \sin\left(\frac{\pi}{M}\right)\right] \quad (15)$$

where erfc(•) denotes the complementary error function and γ_s is the ratio of the energy per *symbol* to the noise PSD. Now, the number of bits per symbol is the logarithm to the base 2 of M (denoted by lgM). So the BER may be approximately related to the SER and the ratio of the energy per *bit* to the noise PSD (E_b/N_0) via:

$$P_b \approx \frac{1}{\lg M} \operatorname{erfc}\left[\sqrt{\lg M \frac{E_b}{N_0}} \sin\left(\frac{\pi}{M}\right)\right] \quad (16)$$

Using 8-PSK, means:

$$P_b \approx \frac{1}{3} \operatorname{erfc}\left[\sqrt{\frac{3E_b}{N_0}} \sin\left(\frac{\pi}{8}\right)\right] \quad (17)$$

Given that the signal and the noise have the same bandwidth (that of the modem), we can use the SNR as equivalent to E_b/N_0. Thus it is possible to obtain the ratio needed in (17) above using the SNR expression (11). Now, assuming independent bit errors, the PER for ARQ may be found from (1) with L = α+τ+n:

$$PER_{ARQ} = 1 - \left(1 - P_b\right)^{\alpha+\tau+n} \tag{18}$$

Moreover, the energy efficiency of ARQ is given by:

$$\eta_{ARQ} = \frac{Payload\ Energy}{Total\ Energy}\left(1 - PER_{ARQ}\right)$$

$$\frac{\left(P_{TX} + P_{RX}\right) nT_{TX}}{\left(P_{TX} + P_{RX}\right)\left(\alpha + \tau + n + l_{ack}\right) T_{TX}}\left(1 - PER_{ARQ}\right)$$

$$\eta_{ARQ} = \frac{n}{\alpha + \tau + n + l_{ack}}\left(1 - PER_{ARQ}\right) \tag{19}$$

It should be noted that the ARQ energy efficiency expression for any number of packets remains the same since each transmission attempt is nominally identical so the numerator and denominator in (19) grow at the same rate.

FEC

In FEC, each data packet consists of a header of size α bits, a parity check (τ bits) and a payload of *n* bits. For communication in a single hop, the energy consumption of a sensor node in FEC is given by (Sankarasubramaniam et al., 2003):

$$E_{FEC} = E_{FEC}^{TX} + E_{FEC}^{RX} + E_{enc} + E_{dec} \tag{20}$$

where, E_{FEC}^{TX} is the transmitter energy consumed in sending the data, E_{FEC}^{RX} is the receiver energy consumed in receiving the data, E_{enc} is the encoding energy and E_{dec} is the decoding energy. Here, the coding and encoding energies are much smaller than the transmission energies, assuming CMOS implementations (Sankarasubramaniam et al., 2003; Tian et.al, 2008) and so are not included from this point onwards. For a fair comparison with ARQ, the same order of PSK is assumed for the underlying transmission. Thus the underlying BER is given for a code rate R_c, by

$$P_b \approx \frac{1}{\lg M}\operatorname{erfc}\left[\sqrt{R_c \lg M \frac{E_b}{N_0}} \sin\left(\frac{\pi}{M}\right)\right] \tag{21}$$

The rate is for the BCH code is given by:

$$R_c = \frac{n + \alpha}{n + \alpha + \tau} \tag{22}$$

The number of parity bits may be upper bounded by $\tau = mt$ (Sankarasubramaniam et al., 2003), where $m = \left\lfloor \lg\left(n\right) + 1 \right\rfloor$ is the size of the field of the code and *t* is its error correcting capability. Then (Sankarasubramaniam et al., 2003):

$$1 - PER_{FEC} = \sum_{j=0}^{t} \binom{n + \alpha + \tau}{t} P_b^j \left(1 - P_b\right)^{n+\alpha+\tau-j} \tag{23}$$

$$\eta_{FEC} = \frac{Payload\ Energy}{Total\ Energy}\left(1 - PER_{FEC}\right)$$

$$\frac{\left(P_{TX} + P_{RX}\right) nT_{TX}}{\left(P_{TX} + P_{RX}\right)\left(n + \alpha + \tau\right) T_{TX}}\left(1 - PER_{FEC}\right)$$

$$\eta_{FEC} = \frac{n}{n + \alpha + \tau}\left(1 - PER_{FEC}\right) \tag{24}$$

NC

We consider an arbitrary generation of K packets. Using linear random coding, if a sink receives at least K packets in the generation, then there is a very high probability that it will recover all the *original* packets for a large finite field (Ho et al., 2003a). Hence, for the sake of simplicity this is assumed to be true so we just need to receive at least K packets for completely successful transmission, The source linearly combines the K packets in the generation to form K' outgoing packets, where $K' \geq K$. The addition of a small amount of redundancy via K' reduces the effects of packet loss and here we follow (Guo et al., 2009) and set $K' = K + 2$. Then we let $\beta_{1,k}$ be the probability that the sink receives k packets (when $0 \leq k < K$) or at least k packets (when $k = K$ from the source in a single hop. Then (Guo et al., 2009):

$$\beta_{1,k} = \begin{cases} K'k\left(1 - PER_{NC}\right)^{k} PER_{NC}^{(K-k)} & 0 \leq k < K \\ 1 - \sum_{j=0}^{K-1} \beta_{1,j} & k=K \end{cases} \tag{25}$$

Assuming independent bit errors, the PER for NC (PER_{NC}) for an NC packet length of L and BER of P_b can be expressed as:

$$PER_{NC} = 1 - \left(1 - P_b\right)^{L} \tag{26}$$

As in the previous cases 8-PSK is assumed and thus P_b is given by (ARQ) above. Since the communication takes place in a single hop and the generation of packets considered is arbitrary, the successful delivery ratio (R) is just $\beta_{1,k}$. Moreover, the average number of times that a packet is transmitted from the source to the sink for a single hop (T_1) is given by K'/K (Guo et al., 2006), and this also represents the average energy consumed by each packet from the source to the sink. The energy throughput (η_e), for the network coding scheme is thus:

$$\eta_e = \frac{T_1 \times \#\text{ of successfully delivered packets}}{T_1 \times \text{Total }\#\text{ of source packets}} = \frac{T_1 RK}{T_1 K} = R \tag{27}$$

Therefore, the NC energy efficiency can be expressed as:

$$\eta_{NC} = R\left(1 - PER_{NC}\right) \tag{28}$$

RESULTS AND DISCUSSION

We now turn to the results obtained for the coding schemes discussed. The general range of the parameters chosen is in the range of typical acoustic modems (LinkQuest, 2014; DSPComm, 2014). We thus consider a combined header plus FCS length of 11 bytes and acknowledgment length of 7 bytes for ARQ operating within the 10 kHz bandwidth discussed previously at a data rate of 10 kbps. For FEC, the overhead α is not directly related to the coding scheme and so we follow (Sankarasubramaniam et al., 2003) assuming that it is of size 16 bits. In the case of NC, we take the parameter settings $K = 3$ and $K' = 5$ from (Guo et al., 2009), and the same 11 byte length overhead used for ARQ for an equitable comparison. The numerical results were obtained using the MATLAB® 2013b software package.

Comparison with No Shipping Noise or Wind as a Function of Packet Size

The first comparison is between the three schemes as a function of packet size in the absence of shipping noise and wind. Figure 5 below shows plots of the three schemes against the transmission

Figure 5. Energy efficiency of ARQ, FEC and NC in the absence of shipping and wind for packet sizes of (a) 64 bits, (b) 256 bits, (c) 1024 bits and (d) 4096 bits

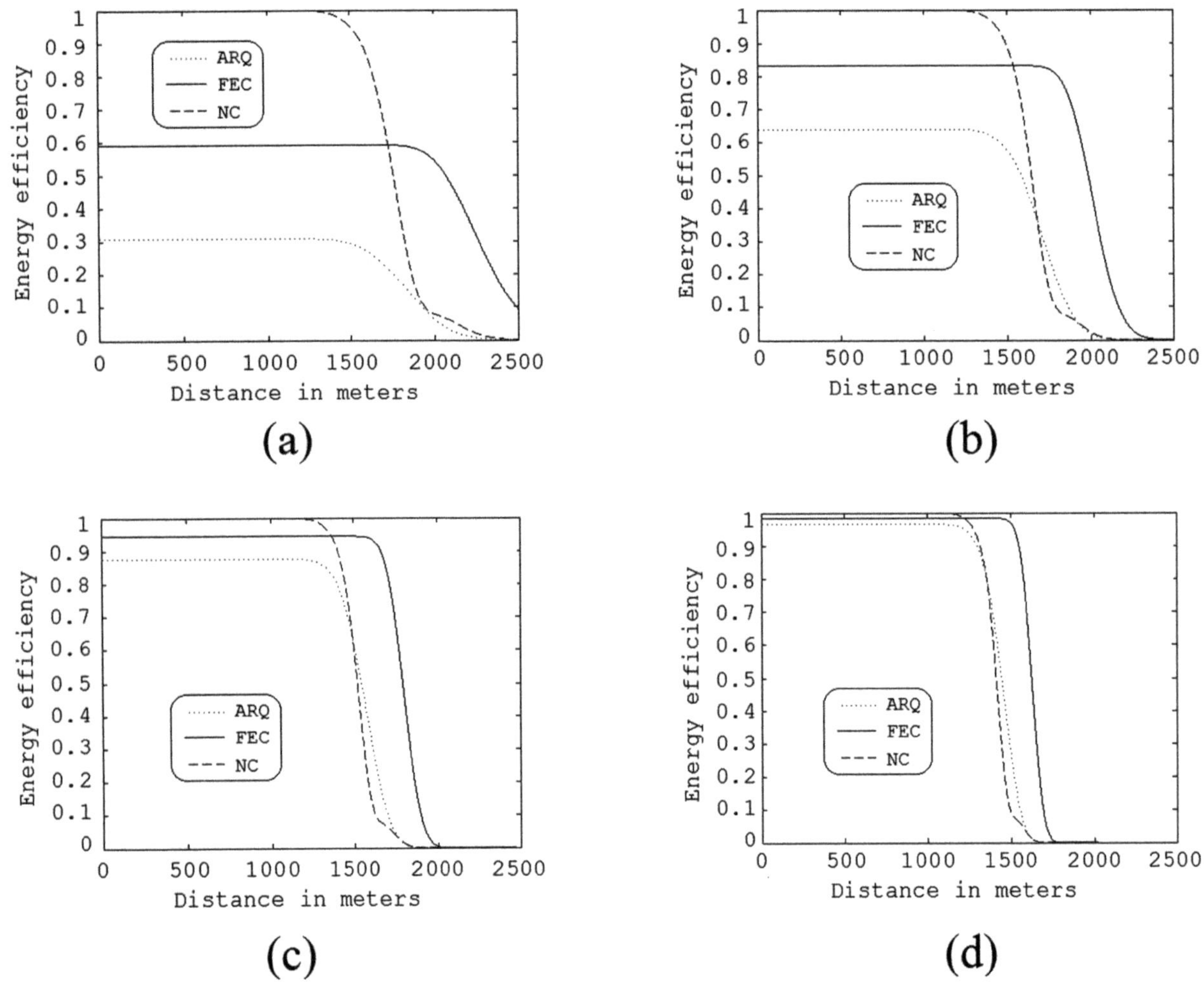

Figure 6. Distance at which FEC becomes the method of choice as a function of base 2 logarithm of the packet size (diamonds) and a quadratic fit to the data (solid line)

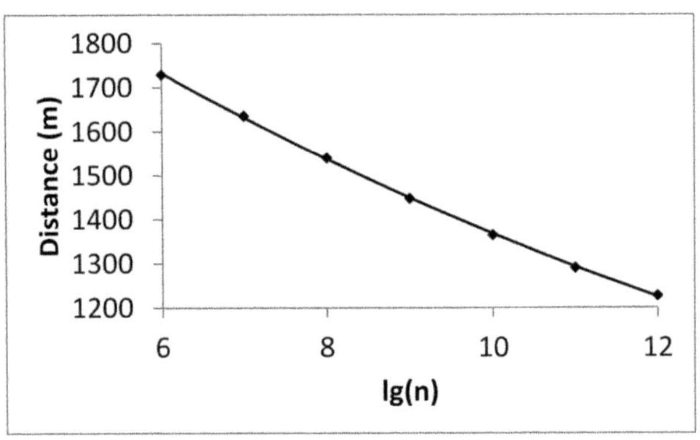

distance for packet sizes of 64 bits, 256 bits, 1024 bits and 4096 bits. For all three schemes we see the same general trends in that energy efficiency is highest over the shortest distances because the number of errors is small. As the BER and hence the PER increases as the transmission distance goes up, the efficiency reaches a critical point when the error rate increases dramatically. The distance at which this occurs varies to a great degree between the schemes as does the maximum efficiency achieved.

As would be expected, all the schemes deliver much better efficiencies for larger packet sizes since the overheads represent a smaller fraction of the total bits transmitted. However, the relative positions of the three schemes remain relative to each other with ARQ always being the least efficient and NC the most efficient with FEC in between. Since ARQ and NC have the same underlying BER, their efficiencies begin to decline at similar distances. The interaction of the probabilities for NC produces a small kink in the efficiency curve in the region where NC is less efficient that ARQ. In general, NC is the most efficient for short distances but FEC takes over this role as the distance increases. The distance at which FEC becomes the method of choice varies with the logarithm to the base 2 of the packet length denoted by lg(n) as shown in Figure 6.

The dependence may be fitted extremely well ($R^2 > 0.99$) by a quadratic expression for the distance:

$$d = 3.3\left[\lg\left(n\right)\right]^2 - 144\left[\lg\left(n\right)\right] + 2478 \qquad (29)$$

This expression is of practical use when choosing the method of transmission in underwater acoustic communications. Clearly in cases with large packets and short distances, ARQ may be chosen for simplicity but in other cases it is likely that FEC or NC will be appropriate. Using (29) for a given distance and desired packet size facilities the choice between the two in terms of energy efficiency.

Effect of Shipping Factor

To ascertain the impact of the shipping factor on the performance of the coding schemes, a moderate sized packet of 256 bits was chosen. Figure 7 shows the performance of ARQ, FEC and NC with s = 0 (white marker fill) and s = 1 (black marker fill).

The scale of the figure is expanded relative to most on the others so that the modest effect of the shipping factor may be seen. The transmission distance before the efficiency declines is reduced by a few tens of meters at most and thus the transmission scheme it relatively immune to the effects of shipping noise.

Effect of Wind Speed

To investigate the impact of wind speed on the performance of the coding schemes, the same moderate packet size of 256 bits was also used. Figure 8 shows the performance of ARQ, FEC and NC without (white marker fill) and with (white marker fill) a wind speed of 1ms^{-1}.

It is immediately apparent that the wind has a very great effect on the transmission distance but not the efficiency values. This is because it increases the noise level but all the schemes operate at negligible BER values over short distances allowing them to deliver their maximum efficiency values. Once the SNR begins to fall, the efficiency decline is rapid given the nature of the complementary error function. Increasing the noise level from wind naturally moves the point of rapid decline to shorter distances since the signal level stays the same.

Comparison with Other Work

There are no other studies to our knowledge that include ARQ, FEC and NC. Therefore, we sum-

Figure 7. Energy efficiency of ARQ, FEC and NC with s = 0 (while marker fill) and s = 1 (black marker fill)

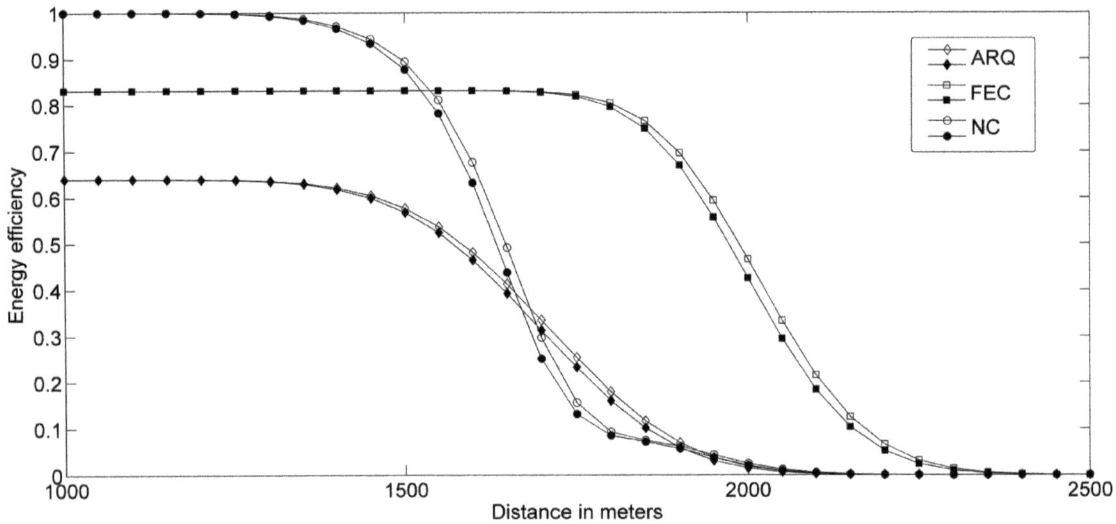

Figure 8. Energy efficiency of ARQ, FEC and NC without (white marker fill) and with (white marker fill) a wind speed of 1ms^{-1}

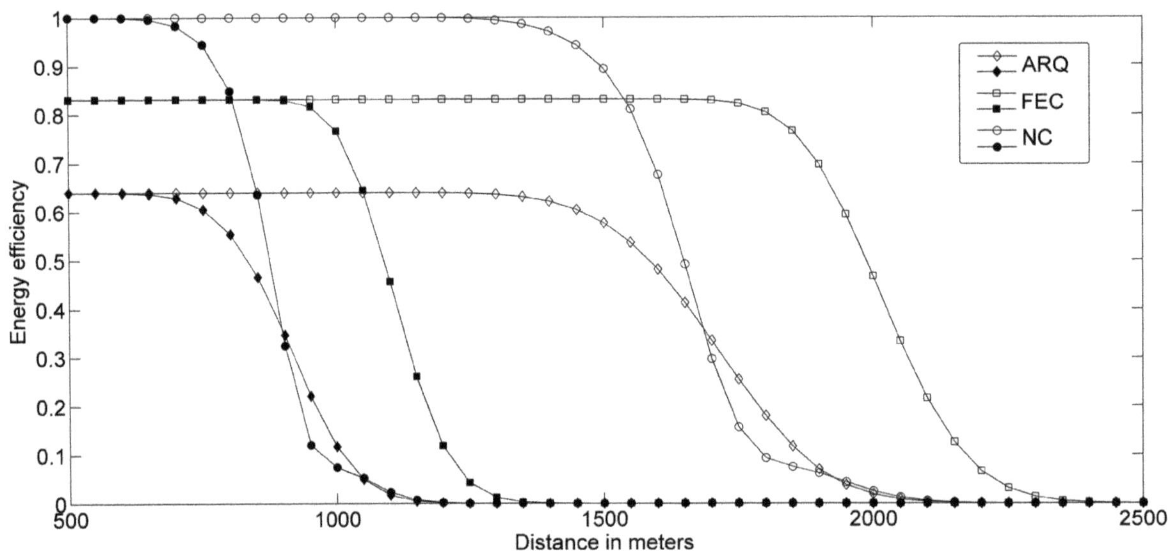

marize here some of the representative results from studies that compare two of the schemes. The work of (Babiker & Zakaria, 2009) uses convolutional turbo coding (Martins et al., 2002) as its FEC and shows that ARQ delivers efficiency of ≈0.72, whereas FEC produces a value of ≈0.65 but doubles the useable transmission distance to approximately 3 km. Moreover, the effects of the shipping factor and wind speed were as observed here in that the former has little effect but results are highly sensitive to the latter. (Guo et al., 2009) compares multipath forwarding, FEC and NC, showing that NC achieves a superior packet delivery ratio and in the majority of cases, is more energy efficient. Moreover, hop-by-hop FEC outperforms end-to-end FEC. The use of multiple paths for UWSNs is seen to be beneficial with NC delivering an extra 10% or so of packets as the BER increases without significant extra energy consumption. The work presented here extends that existing in the literature, offering the first quantification of the switchover points between different schemes – nothing like Figure 6 has been presented anywhere previously.

CONCLUSION AND FURTHER WORK

We have considered underwater communication, which has a variety of applications often requiring the accurate monitoring of different parts of the ocean. UWSNs provide a good information collection and transmission solution but not one without challenges such as the energy efficiency considered in this chapter. This aspect is of particular relevance given that the battery life of underwater sensors is limited and it is difficult to recharge or replace batteries in the aquatic medium. To investigate UWSN energy efficiency we have thus presented an overview of the existing work, and a general framework for energy efficiency. Following this, we have given brief summaries of the essential features of the error correction techniques considered relevant to energy efficiency. An underwater propagation model is developed to calculate the attenuation of the acoustic signal, the ambient noise in the ocean and the signal to noise ratio at the receiver.

We have then taken three major error-correction techniques namely ARQ, FEC and NC, and ascertained their energy efficiency in the UWSN context as a function of the transmission distance. The parameters chosen are those typical of current acoustic modems, and the calculations of attenuation and noise conform to the methods used in the existing literature. One can thus have some confidence in the results to be representative of the performance of the coding schemes in the field. As one may expect, the energy efficiency of all three schemes increases with greater packet sizes since the overheads represent a smaller percentage of the overall transmitted bits. The price paid is a reduction of several hundred meters in transmission length because the PER becomes intolerable more quickly when there are more bits that can be in error in the larger packets. For all packet sizes, ARQ is the least efficient and its performance declines at a distance of approximately 1500 meters. NC maintains and efficiency of 100% until its performance also declines at a rate greater than that of ARQ from a distance of 1500 meters until it under performs ARQ as the efficiency reaches a few tens of percent. FEC offers the greatest transmission distance before its efficiency declines, typically 2000 meters for the cases considered. Its performance is between NC and ARQ (ranging from 59% for small packets to 98% using large packets compared to ARQ, which offers 30% and 97% respectively with NC at 100%). Thus NC should be employed for short transmission distances and FEC chosen as the distance increased. To illustrate this point we have determined that a quadratic fit to the distance above which FEC should be employed as a function of the base 2 logarithm of the packet size is appropriate.

The shipping factor has a negligible effect on the energy efficiency of the schemes considered, whereas the results are highly influenced by the wind speed – a 1 ms^{-1} wind reduces the transmission distance by approximately 15%.

The results presented here suggest that a hybrid energy efficient error-correction protocol for the underwater wireless sensor networks can be designed and implemented in the near future. This hybrid protocol must adapt to the varying channel conditions and to the varying distances between the sensor nodes. An adaptation algorithm can be developed based on the packet acceptance rate and the encoding technique used. In addition, the quadratic fit to the switchover distance between NC and FEC also offers an opportunity to develop an adaptive method.

REFERENCES

Ahlswede, R., Cai, N., Li, S.-Y. R., & Yeung, R. W. (2000, July). Network Information Flow. *IEEE Transactions on Information Theory*, *46*(4), 1204–1216. doi:10.1109/18.850663

Ainslie, M. A. (2010). *Principles of Sonar Performance Modelling*. Heidelberg, Germany: Springer. doi:10.1007/978-3-540-87662-5

Ainslie, M. A., & McColm, J. G. (1998). A simplified formula for viscous and chemical absorption in seawater. *The Journal of the Acoustical Society of America*, *103*(3), 1671–1672. doi:10.1121/1.421258

Akyildiz, I. F., Pompili, D., & Melodia, T. (2004). Challenges for efficient communication in underwater acoustic sensor networks. *ACM SIGBED Review*, *1*(2), 3–8. doi:10.1145/1121776.1121779

Akyildiz, I. F., Pompili, D., & Melodia, T. (2005). Underwater Acoustic Sensor Networks: Research Challenges. *Ad Hoc Networks*, *3*(3), 257–279. doi:10.1016/j.adhoc.2005.01.004

Akyildiz, I. F., Pompili, D., & Melodia, T. (2007). State-of-the-art in protocol research for underwater acoustic sensor networks. *SIGMOBILE Mobile Computing and Communications Review*, *11*(4), 11–22. doi:10.1145/1347364.1347371

Akyildiz, I. F., & Vuran, M. C. (2011). *Wireless Sensor Networks*. New York: Wiley.

Al-Suhail, G. A., Louis, K. W., & Abdallah, T. Y. (2012). Energy Efficiency Analysis of Adaptive Error Correction in Wireless Sensor Networks. *International Journal of Computer Science Issues*, *9*(4), 79–84.

Altamiranda, E., Kiaer, L., & Hu, X. (2009). Condition monitoring and diagnosis for subsea control systems: A subsystem prototype. In *Proceedings of OCEANS 09 – Europe*.

Ayaz, M., Baig, I., Abdullah, A., & Faye, I. (2011). A survey on routing techniques in underwater wireless sensor networks. *Journal of Network and Computer Applications*, *34*(6), 1908–1927. doi:10.1016/j.jnca.2011.06.009

Babiker, A. E., & Zakaria, M. N. B. (2011). Energy efficiency analysis of error correction techniques in underwater wireless sensor networks. *Journal of Engineering Science and Technology*, *6*(1), 17–28.

Babiker, A. E., Zakaria, M. N. B., Yosif, H., & Ibrahim, S. B. (2011). An Efficient Energy Adaptive Hybrid Error Correction Technique for Underwater Wireless Sensor Networks. *World Academy of Science, Engineering and Technology*, *5*, 1108–1124.

Balakrishnan, G., Yang, M., Jiang, Y., & Kim, Y. (2007). *Performance Analysis of Error Control Codes for Wireless Sensor Networks*. Paper presented at 4th International Conference on Information Technology, Las Vegas, Nevada. doi:10.1109/ITNG.2007.149

Brekhovskikh, L. M., & Lysanov, Y. P. (1982). *Fundamentals of ocean acoustics*. Berlin, Germany: Springer-Verlag. doi:10.1007/978-3-662-02342-6

Burrowes, B., & Khan, J. Y. (2011). Short-Range Underwater Acoustic Communication Networks. In N. A. Cruz (Ed.), *Autonomous Underwater Vehicles* (pp. 174–198). Rijeka, Croatia: Intech Europe. doi:10.5772/24098

Cai, S., Gao, Z., Yang, D. S., & Yao, N. (2013). A network coding based protocol for reliable data transfer in underwater acoustic sensor. *Ad Hoc Networks*, *11*(5), 1603–1609. doi:10.1016/j.adhoc.2013.02.001

Cella, U. M., Johnstone, R., & Shuley, N. (2009). *Electromagnetic wave wireless communication in shallow water coastal environment: theoretical analysis and experimental results*. In *Proceedings of 4th ACM Int. Workshop on Underwater Networks* (pp. 9:1–9:8). doi:10.1145/1654130.1654139

Chan, C. Y. M., & Motani, M. (2007). An integrated energy efficient data retrieval protocol for underwater delay tolerant networks. In Proceedings of MTS/IEEE OCEANS 07 – Europe (pp. 1-6). doi:10.1109/OCEANSE.2007.4302341

Chen, K., Ma, M., Cheng, E., Yuan, F., & Su, W. (2014). A Survey on MAC Protocols for Underwater Wireless Sensor Networks. *IEEE Communications Surveys and Tutorials*, *16*(3), 1433–1447. doi:10.1109/SURV.2014.013014.00032

Cherkaoui, E. H. Azad S., Casari, P., Toni, L., Agoulmine, N. & Zorzi, M. (2012). Packet Error Recovery via Multipath Routing and Reed-Solomon Codes in Underwater Networks. In *Proceedings of MTS/IEEE OCEANS 12*.

Chitre, M., Shahabudeen, S., Freitag, L., & Stojanovic, M. (2008). Recent Advances in Underwater Acoustic Communications & Networking. In Proceedings of MTS/IEEE OCEANS 2008 (pp. 1-10). doi:10.1109/OCEANS.2008.5152045

Coates, R. (1989). *Underwater Acoustic Systems*. New York, NY: Wiley.

Comfort, L. K., Znati, T., & Voortman, M., Xerandy & Freitag L. E. (2012). Early Detection of Near-Field Tsunamis using Underwater Sensor Networks. *Journal of Tsunami Society International*, *31*(4), 232–243.

Costello, D. J., Hagenauer, J., Imai, H., & Wicker, S. B. (1998). Applications of Error-Control Coding. *IEEE Transactions on Information Theory*, *44*(6), 2531–2560. doi:10.1109/18.720548

Crocker, M. J. (Ed.). (1998). *Handbook of Acoustics*. New York, NY: Wiley.

Cui, J. H., Kong, J., Gerla, M., & Zhou, S. (2006). The challenges of building mobile underwater wireless networks for aquatic applications. *IEEE Network*, *20*(3), 12–18. doi:10.1109/MNET.2006.1637927

Curtin, T. B., & Bellingham, J. G. (2009). Progress toward autonomous ocean sampling networks. *Deep-sea Research. Part II, Topical Studies in Oceanography*, *56*(3-5), 62–67. doi:10.1016/j.dsr2.2008.09.005

DSPComm. (2014). *Aquacomm: Underwater Wireless Modem*. Retrieved April 30, 2014, from http://www.dspcomm.com

Duffy, J. E., Amaral-Zettler, L. A., Fautin, D. G., Paulay, G., Rynearson, T. A., Sosik, H. M., & Stachowicz, J. J. (2013). Envisioning a Marine Biodiversity Observation Network. *Bioscience*, *63*(5), 350–361. doi:10.1525/bio.2013.63.5.8

Elbergui, A., Quidu, I., Zerr, B., & Solaiman, B. (2012). *Model Based Classification of Mine-Like Objects in Sidescan Sonar Using the Highlight Information*. In *proceedings of 11th European Conference on Underwater Acoustics ECUA 2012* (pp. 1158-1165).

Farr, N., Bowen, A., Ware, J., Pontbriand, C. & Tivey, M. (2010). An integrated, underwater optical/acoustic communications system. In *Proceedings of MTS/IEEE OCEANS* 10.

Fisher, F. H., & Simmons, V. P. (1977). Sound absorption in sea water. *The Journal of the Acoustical Society of America*, *62*(3), 558–564. doi:10.1121/1.381574

Fragouli, C., Le Boudec, J.-Y., & Widmer, J. (2006). Network Coding: An instant Primer. *Computer Communication Review*, *36*(1), 63–68. doi:10.1145/1111322.1111337

Gao, M., Soh, W.-S., & Tao, M. (2009). A Transmission Scheme for Continuous ARQ Protocols over Underwater Acoustic Channels. In *Proceedings of IEEE International Conference on Communications* (pp. 1-5). doi:10.1109/ICC.2009.5198684

Guo, Z., Wang, B., Xie, P., Zeng, W., & Cui, J.-H. (2009). Efficient error recovery with network coding in underwater sensor networks. *Ad Hoc Networks*, *7*(4), 791–802. doi:10.1016/j.adhoc.2008.07.011

Guo, Z., Xie, P., Cui, J. H., & Wang, B. (2006). On applying network coding to underwater sensor networks. In *Proceedings of 1st CM International Workshop on Underwater Networks* (pp. 109-112). doi:10.1145/1161039.1161062

Hamming, R. W. (1950). Error Detecting and Error Correcting Codes. *The Bell System Technical Journal*, *29*(2), 147–160. doi:10.1002/j.1538-7305.1950.tb00463.x

Harris, A. F., & Zorzi, M. (2007). *Modeling the underwater acoustic channel in ns2*. In *Proceedings of 2nd International Conference on Performance Evaluation Methodologies and Tools*. (pp. 1-8). doi:10.4108/nstools.2007.2024

Heidemann, J. Ye., & Wills, W., J., Syed A. & Li Y. (2006). *Research Challenges and Applications for Underwater Sensor Networking*. IEEE Wireless Communications and Networking Conference, 228-235. doi:10.1109/WCNC.2006.1683469

Ho, T., Karger, D. R., Médard, M., & Koetter, R. (2003a). *Network Coding from a Network Flow Perspective*. Paper presented at IEEE International Symposium on Information Theory, 2003, Yokohama, Japan. doi:10.1109/ISIT.2003.1228458

Ho, T., Koetter, R., Médard, M., Karger, D. R., & Effros, M. (2003b). *The Benefits of Coding over Routing in a Randomized Setting*. Paper presented at IEEE International Symposium on Information Theory, Yokohama, Japan. doi:10.1109/ISIT.2003.1228459

Huang, C.-F., Yang, T. C., Kuo, N.-T., Liu, J.-Y., & Schindall, J. (2012). *Experimental Demonstration of Current Mapping Using Distributed Networked Underwater Sensors*. Presented at Seventh ACM International Conference on Underwater Networks and Systems (WUWNet '12). doi:10.1145/2398936.2398983

Jawhar, I., Mohamed, N., Al-Jaroodi, J., & Zhang, S. (2013). *An Efficient Framework for Autonomous Underwater Vehicle Extended Sensor Networks for Pipeline Monitoring*. Paper presented at 11th IEEE International Symposium on Robotic and Sensors Environments (ROSE), Washington, DC. doi:10.1109/ROSE.2013.6698430

Kam, C., Kompella, S., Nguyen, G. D., Ephremides, A., & Zaihan, J. (2014). Frequency Selection and Relay Placement for Energy Efficiency in Underwater Acoustic Networks. *IEEE Journal of Oceanic Engineering*, *39*(2), 331–342. doi:10.1109/JOE.2013.2245057

Karvonen, H., Shelby, Z., & Pomalaza-Raez, C. (2004). *Coding for energy efficient wireless embedded networks*. Paper presented at the International Workshop on Wireless Ad-hoc Networks, Oulu, Finland. doi:10.1109/IWWAN.2004.1525590

Kashani, Z. H., & Shiva, M. (2006). BCH Coding and Multi-hop Communication in Wireless Sensor Networks. Paper presented at the International Conference on Embedded and Ubiquitous Computing, Barcelona, Spain. doi:10.1109/WOCN.2006.1666619

Koetter, R., & Médard, M. (2003). An Algebraic Approach to Network Coding. *IEEE/ACM Transactions on Networking*, *11*(5), 782–795. doi:10.1109/TNET.2003.818197

Krishna, C. R., & Yadav, P. S. (2014). *A Hybrid Localization Scheme for Underwater Wireless Sensor Networks*. In *proceedings of International Conference on Issues and Challenges in Intelligent Computing Techniques (ICICT)*, (pp. 579-582). doi:10.1109/ICICICT.2014.6781345

Labrador, Y., Karimi, M., Pan, D., & Miller, J. (2009). Modulation and error correction in the underwater acoustic communication channel. *International Journal of Computer Science and Network Security*, *9*(7), 123–130.

Lee J., Kim J., Lee J., Jang, Y. & Dho, K. (2008). An improved ARQ scheme in underwater acoustic sensor networks, In *Proceedings of MTS/IEEE OCEANS 08.*

Li, Q. (2012). *Digital Sonar Design in Underwater Acoustics: Principles and Applications*. Berlin, Germany: Springer. doi:10.1007/978-3-642-18290-7

Li, S.-Y. R., Yeung, R. W., & Cai, N. (2003). Linear Network Coding. *IEEE Transactions on Information Theory*, *49*(2), 371–381. doi:10.1109/TIT.2002.807285

LinkQuest. (2014). *Underwater Acoustic Modem Models*. Retrieved April 30, 2014, from http://www.linkquest.com

Liu, B., Garcin, F., Ren, F., & Lin, C. (2008) *A study of forward error correction schemes for reliable transport in underwater sensor networks*. In *Proceedings of the 5th Annual IEEE Communications Society Conference on Sensor, Mesh and Ad Hoc Communications and Networks*, 197–205. doi:10.1109/SAHCN.2008.33

Liu, J.-Y., & Huang, C.-F. (1999). Surface-Generated Noise in an Ocean Waveguide with a Transition Layer of Continuously Varying Density and Sound Speed. *Journal of Computational Acoustics*, *7*(4), 253–268. doi:10.1142/S0218396X99000175

Lurton, X. (2010). *An Introduction to Underwater Acoustics: Principles and Applications*. Berlin, Germany: Springer. doi:10.1007/978-3-642-13835-5

Martins, J., Giulietti, A., & Strum, M. (2002). *Performance comparison of convolutional and block turbo codes for WLAN applications*. In *Proceedings of the Fourth IEEE International Caracas Conference on Devicces, Circuits and Systems*, (pp. T032-1–T032-5). doi:10.1109/ICCDCS.2002.1004115

Médard, M., & Sprintson, A. (2012). *Network Coding*. Oxford, England: Academic Press.

Moon, T. K. (2005). *Error Correction Coding*. Hoboken, NJ: Wiley. doi:10.1002/0471739219

Morelos-Zaragoza, R. H. (2006). *The Art of Error Correcting Coding*. Chichester, England: Wiley. doi:10.1002/0470035706

Nicolaou, N., See, A., & Xie, P. Cui, J.-H., & Maggiorini, D. (2007). Improving the Robustness of Location based Routing for Underwater Sensor Networks. In Proceedings of MTS/IEEE OCEANS Europe.

Parameshwaraiah, R., Ramesh, R., & Gurumurthy, N. K. (2013). Real-Time Underwater Communication Technique for Energy Efficient Ocean Monitoring. Presented at *Seventh International Conference on Sensor Technologies and Applications (SENSORCOMM 2013)*, Barcelona, Spain.

Park, J., Mackay, S., & Wright, E. (2003). *Practical Data Communications for Instrumentation and Control*. Oxford, England: Newnes.

Partan, J., Kurose, J., & Levine, B. N. (2006). A survey of practical issues in underwater networks. In *Proceedings of ACM International Conference on Underwater Sensor Networks and Systems* (pp.17-24). doi:10.1145/1161039.1161045

Petrut, T., Ioana, C., Mauuary, D., Mallet, J., & Phillipe, O. (2014). Distributed Data Classification in Underwater Acoustic Sensors Based on Local Time-Frequency Coherence Analysis. In proceedings of MTS/IEEE OCEANS 14 Taipei, 6 pages doi:10.1109/OCEANS-TAIPEI.2014.6964531

Preisig, J. (2007). Acoustic propagation considerations for underwater acoustic communications network development. *Mobile Computing and Communications Review*, *11*(44), 2–10. doi:10.1145/1347364.1347370

Proakis, J. G., Sozer, E. M., Rice, J. A., & Stojanovic, M. (2001). Shallow water acoustic networks. *IEEE Communications Magazine*, *39*(11), 114–119. doi:10.1109/35.965368

Proctor, A. A., Kennedy, J., Gamroth, E., & Bradley, C. (2010). The Ocean Technology Test Bed: From Concept to Operation. In Proceedings of OCEANS 10.

Sankarasubramaniam, Y., Akyildiz, I. F., & McLaughlin, S. W. (2003). Energy efficiency based packet size optimization in wireless sensor networks. In *Proceedings of the First IEEE International Workshop on Sensor Network Protocols and Applications*, (pp. 1-8). doi:10.1109/SNPA.2003.1203351

Schultz, J. J., Healy, C. A., Parker, K., & Lowers, B. (2013). Detecting submerged objects: The application of side scan sonar to forensic contexts. *Forensic Science International*, *231*(1–3), 306–316. doi:10.1016/j.forsciint.2013.05.032 PMID:23890654

Seah, W. K. G., & Xian-Tan, H. (2006). Multipath Virtual Sink Architecture for Underwater Sensor Networks. In *Proceedings of MTS/IEEE OCEANS Asia Pacific 2006*, (pp. 1-6). doi:10.1109/OCEANSAP.2006.4393933

Sharif, B. S., Neasham, J., Hinton, O. R., & Adams, A. E. (2000). A Computationally Efficient Doppler Compensation System for Underwater Acoustic Communications. *IEEE Journal of Oceanic Engineering*, *25*(1), 52–61. doi:10.1109/48.820736

Shelby, Z., Pomalaza-Raez, C., & Haapola, C. (2004). *Energy optimization in multihop wireless embedded and sensor networks*. Paper presented at the 15th IEEE International Symposium on Personal, Indoor and Mobile Radio, Barcelona, Spain. doi:10.1109/PIMRC.2004.1370868

Sozer, E. M., Stojanovic, M., & Proakis, J. G. (2000). Underwater Acoustic Networks. *IEEE Journal of Oceanic Engineering*, *25*(1), 72–83. doi:10.1109/48.820738

Stojanovic, M. (2006). On the relationship between capacity and distance in an underwater acoustic communication channel. In *Proceedings of ACM International Conference on Underwater Sensor Networks and Systems* (pp. 41–47). doi:10.1145/1161039.1161049

Stojanovic, M. (2008). *Underwater Acoustic Communications: Design Considerations on the Physical Layer*. Paper presented at 5th Annual Conference on Wireless on Demand Network Systems and Services, Garmisch-Partenkirchen, Germany. doi:10.1109/WONS.2008.4459349

Stojanovic, M., Catipovic, J. A., & Proakis, J. G. (1991). Phase-coherent digital communications for underwater acoustic channels. *IEEE Journal of Oceanic Engineering*, *16*(1), 12–20.

Suzuki, T., Kato, K., Makihara, E., Kobayashi, T., Kono, H., Sawai, K., . . . Yamashiro, H. (2014). Development of Underwater Monitoring Wireless Sensor Network to Support Coral Reef Observation. International Journal of Distributed Sensor Networks, Article 189643.

Tan, H., Seah, W. K. G., & Doyle, L. (2007). A multi-hop ARQ protocol for underwater acoustic networks. In Proceedings of OCEANS 07 Europe (pp. 1-6). doi:10.1109/OCEANSE.2007.4302195

Thorp, W. H. (1965). Deep Ocean Sound Attenuation in the Sub- and Low-Kilocycle-per-Second Region. *The Journal of the Acoustical Society of America*, *38*(4), 648. doi:10.1121/1.1909768

Thorp, W. H. (1967). Analytic Description of the Low-Frequency Attenuation Coefficient. *The Journal of the Acoustical Society of America*, *42*(1), 270. doi:10.1121/1.1910566

Tian, Z., Yuan, D., & Liang, Q. (2008). Energy efficiency analysis of error control schemes in wireless sensor networks. In *proceedings of the International Wireless Communications and Mobile Computing Conference*, (pp. 401-405). doi:10.1109/IWCMC.2008.70

Tsai, M. F., Shieh, C.-K., Ke, C.-H., & Deng, D.-J. (2010). Sub-packet forward error correction mechanism for video streaming over wireless networks. *Multimedia Tools and Applications*, *47*(1), 49–69. doi:10.1007/s11042-009-0406-5

Urban, H. G. (2002). *Handbook of Underwater Acoustic Engineering*. Bremen, Germany: ATLAS Elektronik GmbH.

Urick, R. J. (1983). *Principles of Underwater Sound*. Los Altos, CA: Peninsula Publishing.

Wenz, G. M. (1962). Acoustic Ambient Noise in the Ocean: Spectra and Sources. *The Journal of the Acoustical Society of America*, *34*(12), 1936–1956. doi:10.1121/1.1909155

Xie, P., & Cui, J. H. (2007). An FEC-based Reliable data transport protocol for underwater sensor networks. In *Proceedings of the 16th International Conference on Computer Communications and Networks* (pp. 747-753). doi:10.1109/ICCCN.2007.4317907

Xie, P., Cui, J. H., & Lao, L. (2006). VBF:vector-based forwarding protocol for underwater sensor networks. In *Proceedings of International Federation for Information Processing Networking Conference* (pp. 228-235). doi:10.1007/11753810_111

Xu, J., Li, K., & Min, G. (2011). Reliable and Energy-Efficient Multipath Communications in Underwater Sensor Networks. *IEEE Transactions on Parallel and Distributed Systems*, *23*(7), 1326–1335. doi:10.1109/TPDS.2011.266

Yang, X., Ong, K. G., Dreschel, W. R., Zeng, K., Mungle, C. S., & Grimes, C. A. (2002). Design of a Wireless Sensor Network for Long-term, In-Situ Monitoring of an Aqueous Environment. *Sensors (Basel, Switzerland)*, *2*(11), 455–472. doi:10.3390/s21100455

Yang, Y., Zong, C., Sun, Y., & Yang, J. (2010). Network Coding based reliable disjoint and braided multipath routing for sensor networks. *Journal of Network and Computer Applications*, *33*(4), 422–432. doi:10.1016/j.jnca.2010.02.003

Yick, J., Mukherjee, B., & Ghosal, D. (2008). Wireless sensor network survey. *Computer Networks*, *52*(12), 2292–2330. doi:10.1016/j.comnet.2008.04.002

Zhang, B., Sukhatme, G. S., & Requicha, A. A. G. (2004). Adaptive sampling for marine microorganism monitoring. In *Proceedings of IEEE/RSJ International Conference on Intelligent Robots and Systems*, 2, (pp. 1115-1122).

Zhao, Z., & Guo, S. (2010). An FHMA acoustic communication system for multiple underwater robots. In *Proceedings of IEEE International Conference on Information and Automation (ICIA)*, (pp. 1223 – 1228). doi:10.1109/ICINFA.2010.5512294

ADDITIONAL READING

Baldo, N., Casari, P., & Zorzi, M. (2008). Cognitive Spectrum Access for Underwater Acoustic Communications. In *Proceedings of International Conference on Communications* (pp. 518-523). doi:10.1109/ICCW.2008.104

Chitre, M., Shahabudeen, S., & Stojanovic, M. (2008). Underwater acoustic communications and networking: Recent advances and future challenges. *Marine Technology Society Journal*, *42*(1), 103–116. doi:10.4031/002533208786861263

Climent, S., Capella, J. V., Meratnia, N., & Serrano, J. J. (2012). Underwater Sensor Networks: A New Energy Efficient and Robust Architecture. *Sensors (Basel, Switzerland)*, *12*(1), 704–731. doi:10.3390/s120100704 PMID:22368492

De Rango, F., Veltri, F., & Fazio, P. (2012). A Multipath Fading Channel Model for Underwater Shallow Acoustic Communications. In *Proceedings of IEEE International Conference on Communications* (pp. 3811-3815).

Erol-Kantarci, M., Mouftah, H. T., & Oktug, S. (2011). A survey of architectures and localization techniques for underwater acoustic sensor networks. *IEEE Communications Surveys and Tutorials*, *13*(3), 487–502. doi:10.1109/SURV.2011.020211.00035

Heidemann, J. Stojanovic, M. & Zorzi, M. (2012). Underwater sensor networks: applications, advances and challenges. *Philosophical Transactions of the Royal Society A*, *370*(1958), 158-175.

Ho, T., & Lun, D. (2008). *Network Coding: An Introduction*. Cambridge, England: Cambridge University Press. doi:10.1017/CBO9780511754623

Jensen, F. B., Kuperman, W. A., Porter, M. B., & Schmidt, H. (2000). *Computational Ocean Acoustics*. New York, NY: Springer.

Jones, E. (2007). The Application of Software Radio Techniques to Underwater Acoustic Communications. In Proceedings of MTS/IEEE OCEANS – Europe 2007 (pp. 1-6). doi:10.1109/OCEANSE.2007.4302199

Kulhandjian, H., Melodia, T., & Koutsonikolas, D. (2014). *Securing Underwater Acoustic Communications through Analog Network Coding*. Paper presented at IEEE International Conference on Sensing, Communication, and Networking (SECON), Singapore. doi:10.1109/SAHCN.2014.6990362

Lin and Costello. (2004). *Error Control Coding*. Englewood Cliffs, NJ: Prentice-Hall.

Manville, C., Miyajan, A., Alharbi, A., Mo, H., Zuba, M & Cui, J.-H. (2013 June). Network Coding in Underwater Sensor Networks. In *Proceedings of MTS/IEEE OCEANS 13 Bergen*.

Medwin, H., & Clay, C. S. (1998). *Fundamentals of Acoustical Oceanography*. San Diego, CA: Academic Press.

Mohamed, N., Jawhar, I., Al-Jaroodi, J., & Zhang, L. (2011). Sensor Network Architectures for Monitoring Underwater Pipelines. *Sensors (Basel, Switzerland)*, *11*(11), 10738–10764. doi:10.3390/s111110738 PMID:22346669

Molins, M., & Stojanovic, M. (2006). Slotted FAMA: a MAC protocol for underwater acoustic networks. In Proceedings of MTS/IEEE OCEANS – Asia.

Park, M. K., & Rodoplu, V. (2007). UWAN-MAC: An Energy-Efficient MAC Protocol for Underwater Acoustic Wireless Sensor Networks. *IEEE Journal of Oceanic Engineering*, *32*(3), 710–720. doi:10.1109/JOE.2007.899277

Srimathi, C., Park, S.-H., & Rajesh, N. (2103). *Proposed framework for underwater sensor cloud for environmental monitoring*. In *Proceedings of 5th International Conference on Ubiquitous and Future Networks (ICUFN)*, (pp. 104-109)

Stojanovic, M. (1996). Recent advances in high-speed underwater acoustic communications. *IEEE Journal of Oceanic Engineering*, *21*(2), 125–136. doi:10.1109/48.486787

Tan, B. A., Quek, S. S., & Zou, N. (2007). Characterization of Multipath Acoustic Channels in Very Shallow Waters for Communications. In proceedings of MTS/IEEE OCEANS 07Asia Pacific (8 pages).

Wang, C., Sklar, D., & Johnson, D. (2002). Forward Error-Correction Coding. *Crosslink*, *3*(1), 26–29.

Xiao, Y. (Ed.). (2010). *Underwater Acoustic Sensor Networks*. Boca Raton, FL: Auerbach Publications. doi:10.1201/9781420067125

Xu, M., Liu, G., & Wu, H. (2014). An Energy-Efficient Routing Algorithm for Underwater Wireless Sensor Networks Inspired by Ultrasonic Frogs. International Journal of Distributed Sensor Networks, 2014, 351520, 12 pages.

Chapter 3
Network Design and Performance Evaluation of Wireless Sensor Network for Precision Agriculture

Herman S Sahota
IBM, USA

Ratnesh Kumar
Iowa State University, USA

Ahmed E Kamal
Iowa State University, USA

ABSTRACT

This chapter explores the design of wireless sensor networks for applications in precision agriculture. A short review of developments in precision agriculture and recent applications of wireless sensor networks in the area is presented. The authors present their design of medium access control and network layer protocols exploring the challenges and opportunities associated with the design of such a networked system. The physical layer in their network allows multiple power modes in both receive and transmit operations. The MAC layer employs these multiple power modes to implement a novel wake-up synchronization mechanism to reduce the energy overhead. The network layer ensures reliable collection of data while balancing the energy consumption among the nodes. Finally, the authors present an analytical approach to model the behavior of the MAC protocol developed and compare it against the duty-cycle based S-MAC protocol. The results are also confirmed using simulations.

1. INTRODUCTION

Agricultural efficiency increased in the eighteenth and nineteenth centuries with the use of new technologies developed during the industrial revolution. The seed drill, the Dutch plough with iron parts and the threshing machine were among the most important inventions for agriculture. The twentieth century saw the widespread use of self-propelled machines such as tractors, harvesters,

DOI: 10.4018/978-1-4666-8251-1.ch003

etc., and synthetic fertilizers, which allowed agricultural practices at a scale and speed, which was unheard of. Precision agriculture was developed in the mid-twentieth century to provide cost effective methods to improve the quantity and quality of food production. However, the increased crop yields due to these inventions came at a certain toll on the environment. The use of heavy machinery is suspected to have increased the risk of soil erosion (Lal et al., 2007). Also, while fertilizers improved the agricultural production their excessive application without proper knowledge of the soil conditions caused pollution of water bodies. Fertilizer uptake within a field depends on factors such as variability in plant population, nitrogen mineralization from organic matter, water stress, soil properties, pests, etc. These factors vary in space and time. Fertilizers must be applied according to the needs of the crop. If under-applied, the crop yield suffers, while if over-applied, contamination of ground and surface water resources may take place leading to issues such as aquatic hypoxia.

We must predict the quantities of agricultural inputs (such as fertilizers, lime, and other soil amendments) needed to boost the crop yield based on the current soil composition and weather conditions. Thus, a method to collect such data about soil is needed before we can predict the quantities of the agricultural inputs to be applied. Traditionally, one of the methods of soil data collection has been laboratory testing of representative soil samples. Laboratory testing methods allowed us a tool to analyze samples of soil from farmlands to determine soil composition (in terms of available phosphorus, exchangeable potassium, calcium and magnesium, cation exchange capacity, pH, etc.) and predict the optimal quantities of agricultural inputs to be applied (Adamchuk et al., 2004). However, there are many disadvantages of this approach:

1. The method is cumbersome in that it requires the agronomist to collect samples of soil from various locations in the farm-field.
2. It does not allow data collection at arbitrarily low spatial and time resolutions due to limitations on account of human labor and to avoid disturbing the growing crop.
3. It can be a time consuming operation and not suitable for near real-time decision making to apply control inputs.
4. The method is intrusive and interferes with the growth of crops.

Another alternative, remote sensing, can monitor large areas of land for long periods of time but the spatial resolution is low. The use of on-the-go sensors is an alternative to improve the spatial density of the measurements at a relatively low cost. However, in order to apply precision agriculture to production scales the spatio-temporal resolution and reliability of soil data must be improved by automating the data collection process. Recent developments in System on chip (SoC) and wireless communication technologies allow the integration of sensors with wireless transceivers. A network of such wireless sensor nodes, deployed in a farm-field can allow us to gather soil data at much greater scales and resolutions. This data can be used to make near real-time farming decisions.

Densely scattered sensor nodes equipped with wireless transceivers can sense and collect data about soil characteristics such as nitrate concentration, moisture content, etc. Data thus collected can be analyzed at a central base station to enable the farmer to control the application of fertilizers, irrigation and pest control. Sensor networks reduce manpower and heavy machinery costs. Furthermore, through connecting the private wireless sensor network with the Internet, the data can be made available to a wider range of experts for research purposes. Ma et al. (2011) explore such an idea to equip agronomists with technology to analyze volumes of data from farm-fields around the world. The idea is for agronomists to not solely have to rely on their limited test fields for data. With larger volumes of data they can come up with better crop growth models and, in turn, help the farmers better optimize their crop yield.

In Figure 1 (adapted from Ma et al., 2011), sensor nodes in a farm-field are connected, through a gateway node, with mobile communication networks, which in turn make the data available on the Internet. Agronomists can access this data on their computers, analyze it and send recommendations to the farmers on their mobile devices. The authors aim to provide agronomists with the means to gain a better understanding of plant growth models by utilizing data from farmlands and green-houses that are owned by other agronomists and farmers in order to breed and improve specific crop varieties. Similarly, farmers can gain tremendously by such an instrument by gathering data from farm-fields in the broader geographic region and be better informed about climate variability, pests and disease outbreak. A number of barriers must be addressed before the widespread use of this promising technology to solve the food production problems that are facing the world today. For example, a sensor network to collect soil data must be deployed below the ground so that it does not interfere with the above ground agricultural activities. Also, such sensor networks must reliably collect data for periods spanning multiple crop seasons with limited human intervention. We summarize other requirements later in this section.

1.1. State of the Art

Agricultural practices have evolved over thousands of years and vary greatly around the world. The practical application of a new technology, such as wireless sensor networks, in such an age-old practice requires a thorough understanding of the customs and practices employed by the users. Burrell et al. (2004) study the customs of vineyard farmers to ascertain their needs and priorities to explore the use of wireless sensor networks. The work involved the deployment of a test wireless sensor network of 18 nodes for a period of a few weeks at a vineyard site in the state of Oregon, USA. The use of static as well as mobile sensor nodes was explored. The authors consider

Figure 1. System architecture of integration of precision agriculture WSN with internet of things

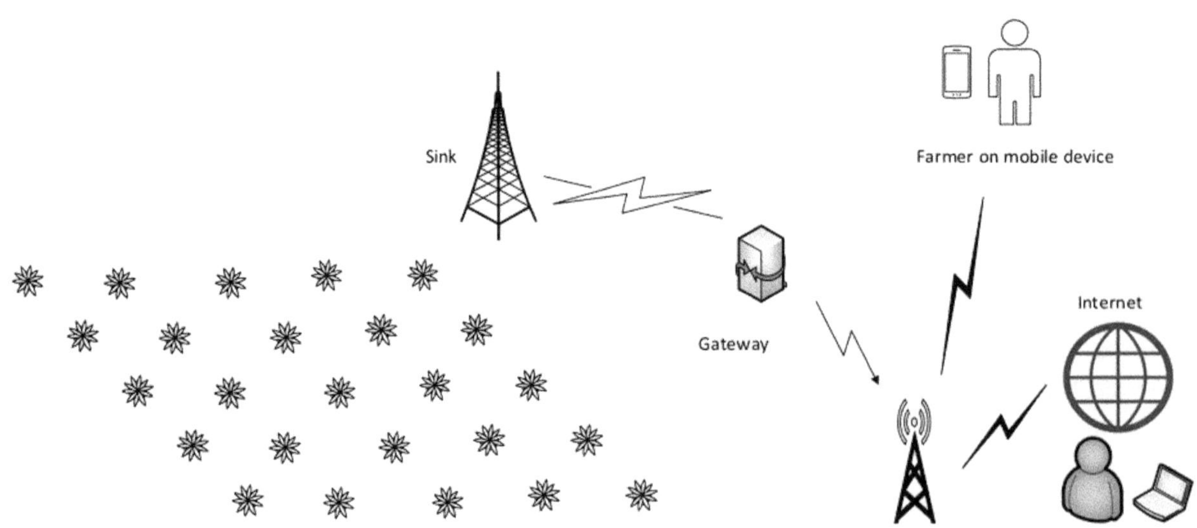

questions concerning the data to be collected, its frequency, computational analysis and presentation to the user. Beckwith et al. (2004) present an extension of the work in which they design and deploy a large-scale multi-hop wireless sensor network in a vineyard distributed across a one-hectare area. The network consists of a fixed topology of backbone nodes and leaf nodes. Duty cycling of the radio is used to conserve energy and minimize the lifetime of the network. Transmission power control is used to allow a larger transmission range for the backbone nodes. At the application level, the authors explore Nyquist frequency, in space and time, of temperature variation to maintain accurate maps of soil profile over the field. Field experiments show results on battery lifetime of the nodes and packet loss of wireless communication.

Sensor networks also open the avenue to expert help from scientists by making the data collection from farms widely available over the Internet through integration with technologies such as GPRS over GSM (Cai et al., 1997). Xia et al. (2011) present the results of such an integration effort. The authors also address several of the issues raised by Burrell et al. (2004). For example, they integrate Google Maps (Purvis et al., 2006) at the application layer to provide alarm service and sensed environment's status information in an easy-to-use interface for the farmers. The devices used by Xia et al. (2011) consist of TelosB mote hardware (Polastre et al., 2005) developed at University of California at Berkeley. Verma et al. (2010) present results from a similar deployment in an eggplant field and implementation of a fire detection system. The authors stress on the importance of energy-efficiency and recovery in the event of node power failures. In the next section, we postulate the requirements imposed on a wireless sensor network, which is to be used for precision agriculture applications, based on our literature survey.

1.2. Requirements and Challenges

Wireless sensor networks must satisfy certain practical requirements in order to be useful for precision agriculture. The following are some key requirements.

- Sensor nodes must not interfere with the agricultural field operations. The deployment of the nodes must be such that the regular agricultural practices of tillage, irrigation, fertilizer and pesticide application are not interrupted in any manner.
- The sensor network must be energy-efficient so that the battery operated nodes last for periods extending over crop season(s). It was observed by Beckwith et al. (2004) that the backbone nodes deplete their battery energy at a much faster rate than the leaf nodes, even though the radio duty cycle was not at 100%. Hence, we notice that reducing the radio duty cycle is not enough. We must design the entire protocol stack, especially the MAC and network layers, to be energy-efficient.
- Most applications of wireless sensor networks in precision agriculture are envisaged to be of a data monitoring nature in that the nodes spread over the farm-field collect spatio-temporal data periodically and send it to a central processing station. The data is analyzed and fed to crop growth models to identify the optimal values of the agricultural inputs to be fed to the control equipment. In order for the models to be accurate in prediction, the collected data must have a high spatio-temporal resolution. This is especially a challenge in an outdoor environment when channel conditions are prone to disturbances in the environment. Beckwith et al. (2004) report packet loss rates of about 77% in field conditions compared to 92% in lab performance. Thus, reliability is a key requirement.

- The application of control inputs in agriculture is time-sensitive but without a hard deadline. Hence, we envision a loose upper bound on the data propagation latency.
- Traditionally, precision agriculture farm-field shapes and sizes have been guided by factors such as the ability to differentially sense and manage different portions of the field and the patterns in which the agricultural equipment move. It is hoped that the use of wireless sensor networks will allow more flexible field sizes and shapes. For example, the sensed area can range from a small green house to a field of area of the order of several hectares. Thus, scalability of the system is a desired requirement.
- In order for commercial success, the system must be cost effective; that is, the profits obtained from the increased and improved yield must be higher than the installation costs.
- Sensor networks must be remotely manageable so that actions such as fault diagnosis, reconfiguration and software upgrade can be performed easily.
- With the advent of the Internet of Things, the data and analysis from the use of precision agriculture must be pervasively available which would require interoperability with other technologies such a Wi-Fi, cellular net works, etc.

Wireless sensor networks have been the subject of extensive research over the past decade. Since the inception of TinyOS, an operating system for embedded devices, at the University of California at Berkeley in 2000, sensor networks have attracted researchers in various sub-disciplines such as protocol design (medium access control and network layers), sensor design (micro-electro-mechanical systems), transceiver design, application design (e.g. sensor node localization), etc. This book chapter reports on our research, which is summarized in the following section.

1.3. Contribution

We study the problem of network design for a wireless sensor network used for periodically collecting soil data in a precision agriculture farm-field. We seek solutions to the design of MAC/PHY and network layers for such an application. To quantify the effectiveness of the developed protocols, we formulate analytical methods to evaluate the performance of the network in terms of different metrics of interest.

1.3.1. MAC Layer Design

MAC protocol allows the nodes to access the wireless channel to communicate with the set of nodes in its radio communication range. Sensor nodes in an application such as precision agriculture are expected to remain unattended for months, while still being required to function. Sensor nodes are battery powered and battery replacement is a costly operation. Thus, it is desirable to have a long node lifetime to minimize the frequency of such maintenance operations. In addition, due to the nature of the outdoor environment, the channel characteristics are highly unstable causing packet losses. We seek the design of a MAC layer protocol that minimizes the energy consumption in the transceiver module, while allowing reliable data communication. Section 2 addresses this topic.

1.3.2. Network Layer Design

Sensor nodes have a limited radio transmission range. In order to conserve energy, it is desirable to transmit at the lowest possible energy level to maintain communication with the nodes in the immediate geographical neighborhood only. This necessitates the use of multi-hop communication in order to allow data collection coverage over large farmlands. It is not only necessary to conserve the energy at individual nodes but also to ensure that the energy consumption is balanced throughout the entire sensor field. This is necessary so that the network does not get partitioned due to a group of nodes failing. Following a cross-layered design approach between the MAC and network layers, we compute node level transmission and reception schedules to minimize the frequency of wake-up synchronizations and data communications. We study the design of an energy-efficient and load balanced routing strategy for our sensor network application. Section 2 also addresses this topic.

1.3.3. Performance Evaluation

We require rigorous methods to test the performance of the proposed communication protocols under the given application scenario. This may be needed to calibrate the protocol parameters to optimize the performance as well as to compare different protocols on the basis of their performance. Since a wireless sensor network used in such applications may consist of hundreds of nodes, it is not readily feasible to quantify the performance metrics (energy-efficiency, throughput, latency, etc.) by actual field deployments. Mathematical modeling and simulation techniques are used in such cases. We present our analytical model for performance evaluation for our MAC and Network layer protocols along with that for a well-known duty cycle based MAC protocol, S-MAC (Ye et al., 2002), under the settings of our application in Section 3.

2. SENSOR NETWORK PROTOCOL DESIGN

2.1 Introduction

Our sensor nodes sense soil data once every hour across a farm-field and relay this information to a central base station for analysis. The nodes are placed in a rectangular grid at a spatial resolution of about one sensor for every $50m \times 50m$ of square area as required by the application. The base station is a node located at one of the four corners of the rectangular field. Note that after collection at the base station, the data may be analyzed or further routed to a wider network through gateways. Our focus is the design of the wireless sensor network in the field itself. The network stack must be energy-efficient to maximize the network lifetime. As stated earlier, reliability and latency within a certain confidence interval must also be guaranteed.

It is well established that wireless communication is one of the most energy consuming operations at a sensor node. Thus, in keeping with our efforts on designing an energy-efficient network communication stack, we introduce the design of medium access control (MAC) and network layers with the goal of minimizing the energy consumption of wireless communication.

Duty cycle based MAC protocols activate the transceiver circuitry only periodically to conserve energy in the face of intermittent network traffic. Accordingly, sensor nodes sleep a large part of the time and wake up when they have data to send to their next hop node(s). However, as the sleep durations grow longer, the clocks of the sender and receiver nodes may deviate from each other due to the inherent drifts associated with the underlying crystal oscillators and they may not awaken at the same time. Thus, sensor nodes need to follow a wake-up synchronization phase before actual data communication. Depending on the implementation of this wake-up synchronization strategy, energy is either wasted at the transmitting node in the form of over-emitting (transmission of a message when the destination is not ready to receive) and/or at the receiving node in the form of idle-listening. We consider a wake-up synchronization protocol which employs a higher power transmit signal combined with reduced sensitivity detection mode at the receiving node. The higher powered transmit mode lasts for a very short duration of time compared to the reduced sensitivity detection mode which consumes much less energy. The result is significant reduction in the energy consumption during this phase (Sahota et al., 2010). Thus, the design of a radio which operates at different power levels in the transmit mode as well as in the receive mode is motivated. The transmit mode has one higher power level in addition to the normal two levels of operation of sleep and transmit. The receive mode has one additional power level corresponding to degraded sensitivity in addition to the normal two levels of sleep and receive. Wake-up synchronization protocol between two nodes that wake up to communicate with each other is implemented at the MAC layer. Also, exploiting the knowledge of the application specific network traffic in the immediate neighborhood, namely the convergecast nature of the traffic, we optimize the MAC layer to further minimize the time spent in the higher powered transmit mode. This is achieved by making the downstream node wake up all its upstream neighbors with a common higher power transmit signal. Existing MAC protocols do not exploit such information about the network structure. We present out MAC protocol design in Section 2.4.2.

At the network layer, the routing strategy has a tremendous impact on the energy drain profile of the network. The amount of data flowing through a node in a multi-hop network directly impacts its energy consumption. Hence, the energy consumption of a node not only depends on its own data but also on the data received from all its upstream nodes in the routing tree. Therefore, the routing strategy must also incorporate energy considerations. In our network, the routes from all the sensor nodes to a sink are computed taking into consideration the remaining battery level of the nodes so as to balance the energy consumption among all nodes. Any of the four corner nodes of a field are capable of functioning as a sink node, and in order to even out the energy consumption among the sensor nodes, we rotate the role of the sink among the four corner nodes after each data collection round. For each round, the routes thus computed determine a certain order in which the nodes must forward their data (e.g. the node furthest from the sink should forward its data first). Adhering to this order, a schedule is created that determines the times at which the various nodes need to communicate with their neighbors in order to forward the data to a designated sink. This scheduling is also performed as part of the network layer (Section 2.4.2). Note that this "node scheduling" is different from packet level scheduling, which is performed at the MAC layer.

The following are the contributions of this section:

- We consider multiple power modes for the radio and show how it can help design a MAC layer that reduces the energy consumed during the wake-up synchronization (Section 2.4).
- We develop a routing protocol, and schedule the nodes according to the computed routes so as to balance out the load of communicating sensor data to a sink node, and also to keep the overheard data communications to a minimum (Section 2.4.2).
- We develop TOSSIM (Levis et al., 2003) based simulations of a wireless sensor network, modeling the PHY, MAC and network layers to study the performance of a wireless sensor network of the type described above, and show the resulting energy savings (Section 2.5).
- We present an initial implementation of our MAC protocol on CC1110 SoC with low-power RF transceiver and 8051 MCU from Texas Instruments (2014) (Section 2.5).

2.2 Related Work

Transceiver operation is one of the major sources of energy consumption in a wireless sensor node. Listening on an idle channel, reception and processing of transmissions destined for another node, retransmissions because of packet losses due to collisions, overheads involved in various channel access and communication schemes are various sources of unnecessary energy consumption. Reducing idle listening has been one of the most common objectives in the design of MAC protocols for such systems. It is achieved by duty cycling the transceiver to periodically switch between active and sleep modes. Depending on the nature of synchronization among the active schedules of neighboring nodes, the protocols can be classified as asynchronous, locally synchronous and globally synchronous.

Asynchronous protocols require a wakeup synchronization phase before data communication due to the lack of synchronization in neighboring nodes' wakeup times. The basic approach is using a preamble transmission (before actual data transmission) that is long enough to wake up the receiver node that is periodically polling the channel for activity. Protocols such as B-MAC (Polastre et al., 2004) and STEM (Schurgers et al., 2002) fall in this category. The latter employs a separate radio channel for performing pre-amble sampling to prevent the preamble transmissions from colliding with ongoing data transmissions. Improvements over the basic approach consist of reducing the collisions by employing clear channel assessment before preamble transmissions, tackling the overhearing problem by packetizing the preamble and transmitting the destination node id in the short preamble packets to prevent the overhearing problem as in ENBMAC (Lim et al., 2006) and DPS-MAC (Lim et al., 2006), reducing the length of the preamble transmission by listening for an ACK from the target receiver(s) between successive short preamble packets as in X-MAC (Buettner et al., 2006) and SpeckMAC (Wong et al., 2006), learning the schedule of neighboring nodes by overhearing or in explicit ACK messages as in WiseMAC (El-Hoyidi et al., 2004), etc. There have also been efforts to improve the throughput by employing receiver-initiated transmissions where a receiver node indicates the time for data transmission using a beacon message as in RICER (Lin et al., 2004). However, an effective contention resolution protocol must be employed among the multiple senders in such a scenario.

Asynchronous protocols are suitable for moderate data-rate sparse node density networks or applications that inherently require data flow in a broadcast manner. This is because unicast transmissions suffer under such protocols by unnecessarily waking up the non-target nodes. Additionally, channel occupation during the wakeup synchronization phase limits the available channel capacity adversely in case multiple nodes in the same neighborhood have data to transmit. Asynchronous protocols suffer from the problem of hidden terminal collisions.

Locally synchronous protocols such as S-MAC (Ye et al., 2002) employ a periodic synchronization phase to ensure that the nodes in a local neighborhood (cluster) are synchronized to the same schedule. Nodes at the boundary of two or more clusters adopt multiple schedules acting as bridges between those clusters. Multi-hop delay adversely suffers due to staggered active schedules of neighboring clusters. Consequently, research in locally synchronous MAC protocols is centered on minimizing the transmission delays. Adaptive listening to determine when a node should wake-up by anticipating the transmission from its upstream node by overhearing the latter's Clear-to-Send packets to its own upstream node as in S-MAC with adaptive listening (Ye at al., 2004) and using the Future-Request-to-Send mechanism to indicate a future time for data transmission as in T-MAC (Van Dam et al., 2003) are some approaches towards that end. T-MAC also introduces the idea of dynamically increasing or decreasing the duty cycle by listening for activation events. During the active period, if the node does not detect any activation event for a certain timeout period, it goes back to sleep reducing the energy spent in idle-listening further. The drawback of this approach is the early sleeping problem, which happens because of the broken synchronization between the nodes. Locally synchronous protocols are a natural candidate for networks with periodic data flow. These protocols eliminate the need for periodic polling and preamble transmission however hidden terminal collisions are still a potential issue.

Globally synchronous protocols employ global synchronization methodologies to assign time slots for transmission (or reception) to nodes. Such TDMA based protocols solve the hidden terminal problem by ensuring that that no two nodes in a two hop neighborhood are assigned the same time slot. Throughput is maximized under high channel contention by minimizing the chances of collisions. Such protocols are best suited for applications that already have a requirement for strict time synchronization and employ a small number of more powerful nodes to maintain synchronization in the entire network. However, global synchronization is not scalable to the network size restricting their applicability to small network sizes. Heterogeneous networks lend themselves to such protocols by allowing a small number of less resource constrained nodes to act as synchronization coordinators, thereby alleviating the need for idle listening by remaining nodes in the network. TDMA based protocols lead to low channel utilization under low contention. The mechanism of slot stealing is proposed in Z-MAC (Rhee et al., 2008) and TDMA-ASAP (Gobriel et al., 2009) to improve channel utilization whereby nodes that do not own a slot, which is idle, can contend for its use. Another mechanism to improve channel utilization includes adaptive slot assignment where only the nodes that have data to transmit are assigned transmission slots as in TRAMA (Rajendran et al., 2006).

Recently, multi-channel MAC protocols have been proposed for wireless sensor networks to achieve high channel capacity and utilization as in MuChMAC (Borms et al., 2010). However, the need for multiple transceivers makes the sensor nodes bulkier and less energy-efficient limiting their applicability to certain applications.

Since in our application sleep periods can be as long as one hour, and may be even longer, clocks may drift over such a long period, especially that no synchronization is performed between any two rounds and the drift can be significant. Hence, our MAC layer strategy is aimed at reducing the energy wastage during the wake-up synchronization phase of the transmission, as discussed in the following sections.

2.3 Application and System Model

The farmland is divided into rectangular fields. Within each field, sensors are deployed in a rectilinear grid form, with a spacing of *50m* between neighboring nodes in any of the four directions. Sensor nodes are buried in the soil at a depth between *30cm* and *50cm*. This deployment is chosen in order to ensure sensing coverage of the entire field at a resolution required by the precision agriculture application, and to also be safe from damage by field operations. Soil data samples should be taken from points in the field where the spatial lag (distance between adjacent sampling points) enables the model to interpolate the moisture and nitrogen concentration. Typically, this value varies from *10m* to *100m* for precision agriculture applications, and it was set to *50m* in our network in order accommodate a maximum transmission range of about *70m* at the chosen frequency of *27MHz* and at the chosen depth. The sensor nodes can be easily deployed according to this strategy during field operations since farm workers and equipment move in this field in regular patterns, which coincide with the deployment scheme. See Figure 2.

At the corners of each square field, satellite stations are deployed. The satellite stations are equipped with higher-powered computing equipment and are easily accessible for battery replacement. Each satellite node has a transmission range that can cover all nodes in its adjacent square field. Thus, communication between sensor nodes in a field and the satellite stations at the corners of the field is asymmetric, in the sense that a transmission from a satellite node can reach all sensor nodes in

Figure 2. Wireless sensor network, showing sensor and satellite station deployment

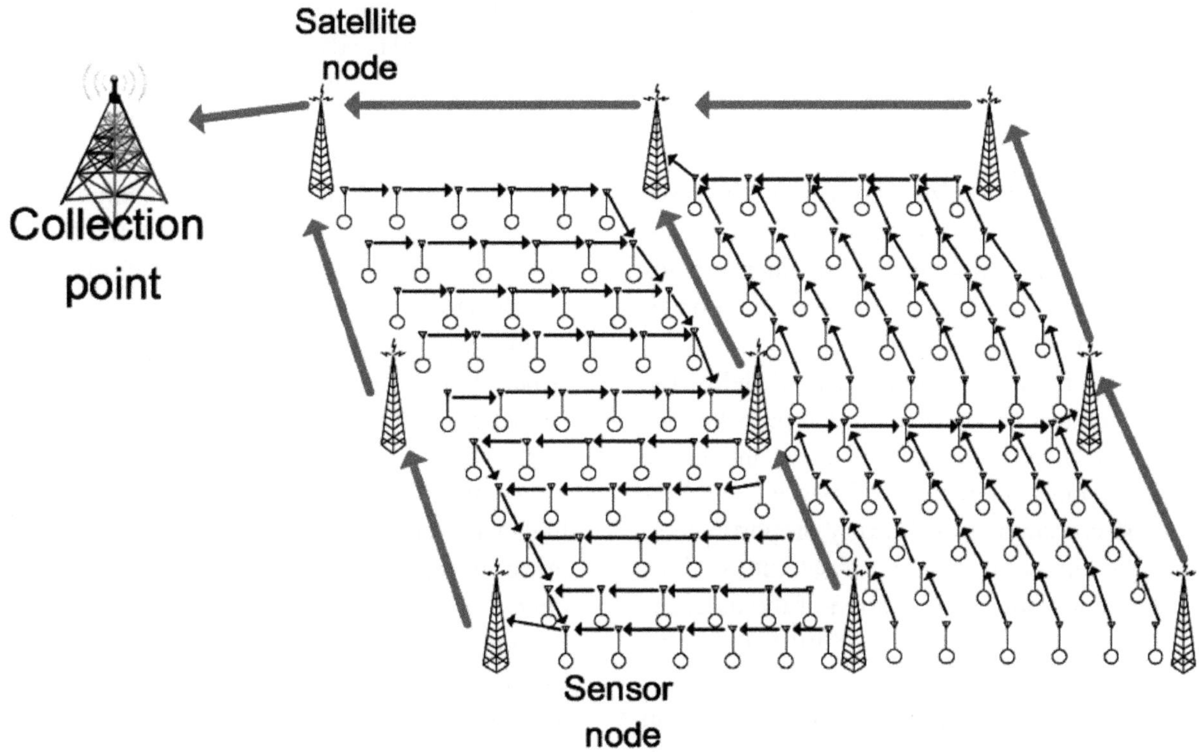

the field, but a transmission from a sensor node reaches adjacent sensor nodes only, and can reach a satellite node if it is near the corner of the field. The satellite stations do not perform any sensing operation, but they serve four purposes:

1. Collect information from the sensor nodes in a field,
2. Relay the collected information to a base station for final processing of data,
3. Dispatch routing and scheduling information to the sensors in the field, and
4. Participate in the sensor localization process.

Due to the first two points above, we propose a two-level hierarchical routing strategy, which can simplify routing and also save energy consumption of the sensor nodes by not placing the load of relaying huge volumes of data from the entire farmland on a few sensor nodes. Moreover, the third point also reduces the complexity of the sensor nodes, reduces their computational requirements, and saves their energy consumption.

The routes within a field are computed and dispatched by the satellite stations. The routing strategy computes routes over sensor nodes in the field such that satellite stations take turns in collecting data, so that energy consumption is balanced. While a static routing strategy can be devised such that the average energy consumption is the same, and is therefore simpler, the alternate routing strategy has the advantage of withstanding failures and achieves a certain level of resilience, without adding any extra measures for the purpose. That is, if a part of the network fails, including any one sensor node, under a static routing strategy, certain data measurements which traverse the failed part will never be delivered. However, with the alternating routing strategy, such a failure can only affect at most 25% of the measurements leaving 75% deliverable with integrity.

The above network model is scalable due to the following reasons:

- Since the field size is fixed and the sensor nodes communicate with adjacent sensor nodes only, increasing the number of fields does not increase energy consumption of sensor nodes.
- With increasing the number of fields, the satellite stations closer to the base stations will have to carry more data, and hence consume more energy. However, the energy consumption is linearly proportional to the number of sensors. If satellite stations perform some aggregation of the data, energy consumption can be further reduced.

It is to be noted that this network is neither an ad hoc network, nor an infrastructured network. We view this network as a hybrid network, since it combines features from both types.

2.4 Energy-Efficient Wake-Up and Data Collection

Since data sampling is periodic in our application, and the period is long, viz., one hour or longer, sensor nodes sleep most of the time and wake up at a preprogrammed time to communicate data. The relative synchronization error depends on the sensors' clocks, which are controlled by on-board crystal oscillators. Since inexpensive sensor nodes are implemented using inexpensive crystals, the drift in the sensors' clocks can be significant (typically of the order of 30 parts per million). This problem is indeed exacerbated by the fact that the periods over which clocks are synchronized are long (since nodes' sleep periods are long). In our wake-up synchronization strategy, a node is awakened by its downstream neighbor using a wake-up signal when it is ready to collect data from the former. In the worst case, the receiver of a wake-up signal needs to wake up for a time that is of the order of 4Δ, where Δ is the maximum clock drift during the period since last synchronization (this will be explained below), in order to make sure it can capture the wake-up synchronization signal. Therefore, the energy consumed by the downstream node can be significant if Δ is large, which is the case for long sleep periods.

Our approach to reduce energy consumption is based on a trade-off between the energy consumed at the transmitter and that at the receiver of the wake-up synchronization signal: the node transmitting the wake-up signal transmits at a higher power level, while the receivers use less power in the detection of the wake-up signal by degrading their sensitivity. This can be achieved by simplifying the circuitry of the receiver. For example, we can by-pass some amplifier stages, and save on the energy they consume for their operation. The wake-up signal is a signal of very short duration and carries no data. Its sole purpose is to generate an interrupt in the receiver circuit to indicate that it needs to wake up.

We term the transmission of the wake-up synchronization signal, which consumes a relatively large amount of power for a very short duration of time, the ping signal. Likewise, the mode of operation of the receiver circuit in the degraded sensitivity is termed as drowsy. These two modes of the radio are used in the wake-up synchronization strategy employed by our MAC protocol. We call our MAC protocol PD-MAC (letters `P' and `D' stand for ping and drowsy, respectively). Owing to the convergecast nature of data communication, it makes sense for a receiver node to send a ping pulse to wake up all its upstream sender nodes. The sender nodes must transition to drowsy mode before the receiver node sends out the ping. Once the receiver is awake, it sends a ping pulse to awaken and synchronize all its upstream senders, which then transmit their data one-by-one to the receiver. Figure 3 shows the time-line of events for our MAC protocol for one receiver and two senders. More details can be found in Sahota et al. (2010).

At a higher level of the protocol stack, namely network layer, we aim at further reducing the energy consumption by seeking load balancing in the computation of routes from the sensor nodes to a sink node. The routes are computed by taking into consideration the remaining battery level of the nodes, and a downstream node is chosen to be a neighbor that has the maximum remaining energy level. Also, the role of the sink is rotated among the four corner nodes after each round of data collection. The schedule for the transmission/reception of messages is computed in accordance with the order of communication as determined by the routing tree to ensure that the number of sender-receiver wake-up synchronizations and internode communications are kept to their minimum.

2.4.1 PHY Layer

We require the radio to be able to operate in a low power receive mode called drowsy, and at a high power mode called ping. The ping mode lasts for a short time only, and wakes up a neighboring node, which is in the drowsy mode. This trade-off of energy can be effective in saving energy because of the relatively short amount of time that the sender's radio has to spend in the ping mode. An RFID based system can be used to implement such a strategy (Ruzzelli et al., 2007). Each sensor node is equipped with a tag and a reader. The reader at one node is used to send the ping to the tag of a neighboring node to awaken the latter. The tag, on receiving the signal, generates an interrupt, which activates the antenna system for regular data communication. Here, the drowsy mode is realized by drawing energy from the received signal itself. Another strategy to implement the drowsy mode is to downgrade some parts of the receiver circuitry.

Figure 3. Implementation of wake-up synchronization and data phase communication in PD-MAC

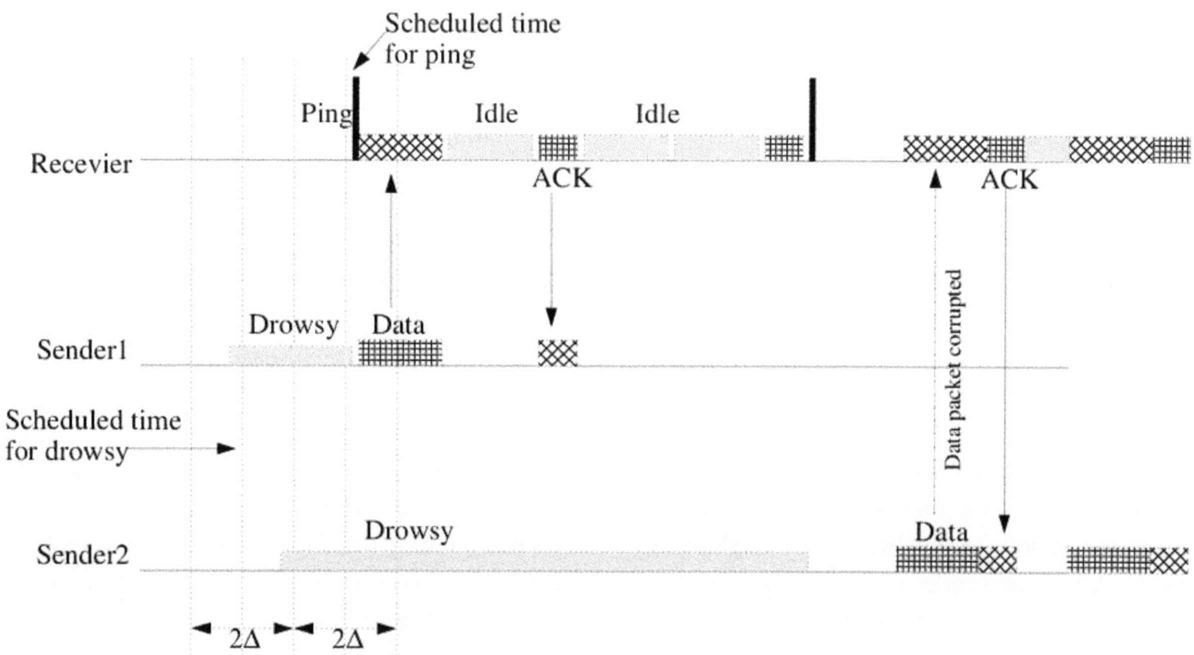

Figure 4 illustrates the energy saving achieved by using the drowsy and ping modes as opposed to the regular transmit and receive levels during the wake-up synchronization phase. Letting Δ denote the duration of clock drift during a sleep-period, w the duration of a ping, P_t and P_r the power levels of normal transmit and receive modes, and P_p and P_d the power levels of ping and drowsy modes, the energy saved during a wake-up synchronization can be up to $(4\Delta + w)(Pr - Pd) - w(P_p - P_t)$ (the first term denotes the energy saved at the receiver, and the second term subtracts the extra energy needed for ping). The appearance of the term "4Δ" can be understood as follows: The sender may wake up Δ unit earlier than scheduled, and in order to not miss the ping, the receiver must target waking up 2Δ units earlier than scheduled (so that it will wake up latest by Δ unit prior to sender's scheduled wake-up time). But then the receiver may wake up as early as 3Δ units prior to the sender's scheduled wake-up time whereas the sender may wake up as late as Δ unit after its own scheduled wake-up time, and as a result the receiver may remain drowsy for at most 4Δ units before it starts to witness the ping signal. In order that Δ does not grow unbounded, the clocks need to be synchronized periodically, and in our application this is done at the beginning of each round.

2.4.2 MAC and Network Layers

The traffic pattern in the wireless sensor network described above is of the convergecast nature, wherein many sender nodes send their data to a single receiver node to be relayed towards a sink node. Owing to this convergecast nature of data communication, it is prudent for the receiver node to broadcast the ping pulse to all its upstream sender nodes in the wake-up synchronization phase. This removes the need for multiple ping transmissions by any given node to synchronize its neighboring nodes. All of the sender nodes are programmed to wake up at time 2Δ units before the receiver node and transition to the drowsy mode. Once the receiver is awake, it sends a ping pulse to awaken and synchronize all its upstream senders, which then transmit their data one-by-one to the receiver. The receiver node acknowledges the receipt of data from each sender in a collective acknowledgement packet broadcasted to all the sender nodes. This mechanism of wake-up synchronization and the ensuing data/ack transmission is implemented at the MAC layer. Figure 3 shows the time-line of events for our MAC protocol for one receiver versus two upstream senders. In the example, the first transmission from one of the nodes succeeds. The second node misses the first ping signal but then synchronizes with the second ping and successfully transmits its data in the first attempt.

We have described the basic idea behind our MAC protocol above. The behaviors of the receiver and sender nodes are precisely described with the help of finite state machines shown in Figures 5 and 6.

Owing to the limited transmission range, the sensor nodes use a multi-hop routing strategy to forward the collected data to one of the satellite nodes at the corners of the field. The routes are computed for each data collection round by the satellite nodes and broadcasted to the sensor nodes. The gap between two data collection rounds can be of the order of hours during which the sensor nodes do not communicate with each other and their clocks may drift relative to each other. The routing information broadcast from the satellite nodes serves the additional purpose of synchronizing the clocks of the nodes so that the nodes do not go out of sync unboundedly between any two data collection rounds.

In our application, the sensor field is laid out in the form of a square with the nodes placed at the various grid points. In each round, the nodes forward their data to a sink located at one of the four corners of the field. The role of the sink is rotated among the four corners of the field to further balance the energy consumption in the field. Within a round, a node forwards its data to a neighboring node that is closer to the designated

Figure 4. Illustration of energy saving at receiver by our scheme

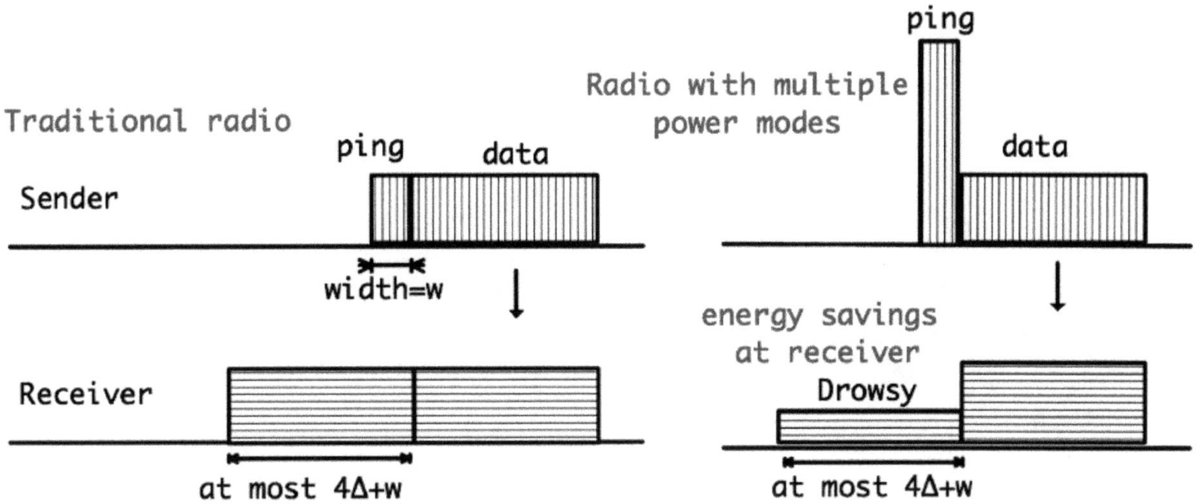

Figure 5. Finite state machine for receive protocol

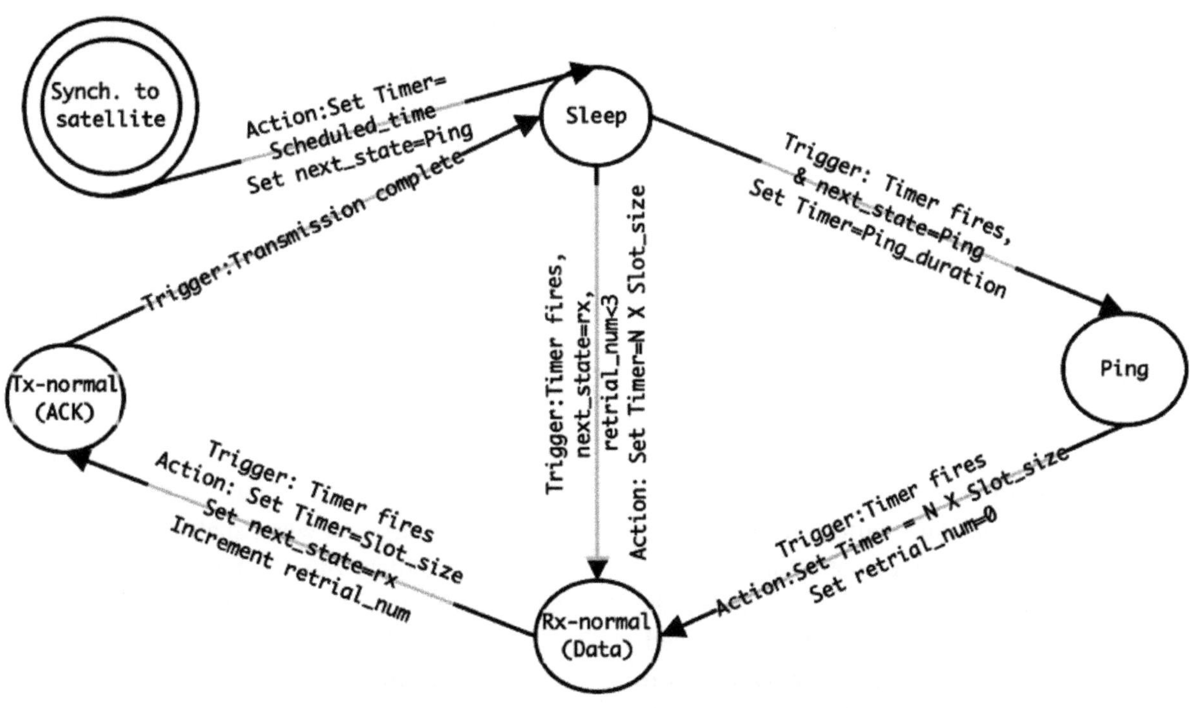

sink node. In case of multiple choices for such a neighbor, the data is forwarded to the one with the maximum remaining energy level. This routing strategy helps balance out the energy consumption across the entire sensor field. A sequence of 4 rounds, in which each corner node has once acted as a sink, is called a *macro-round*.

Figure 6. Finite state machine for send protocol

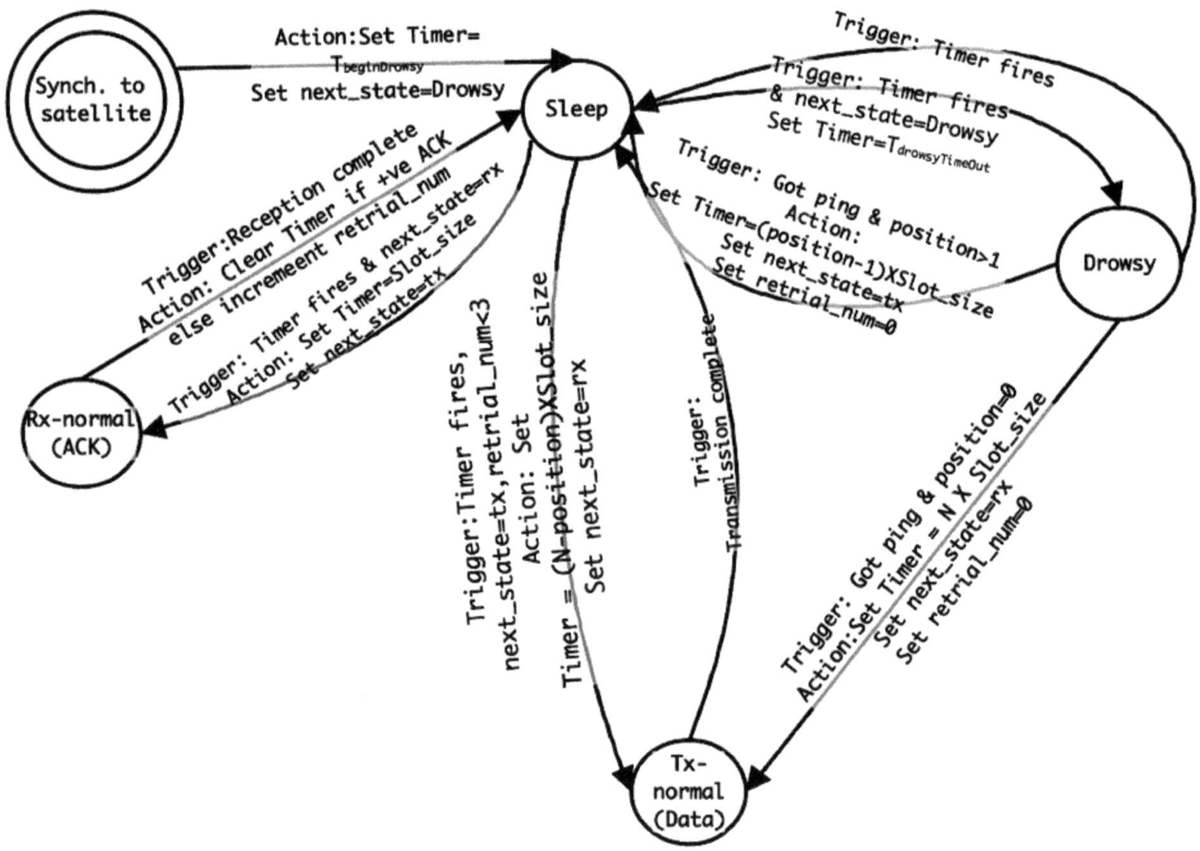

The routing strategy aims at maximizing the network lifetime by way of minimizing the energy consumption at each node and load balancing. Each data transmission incurs the additional overhead of the energy consumed in the wake-up synchronization phase in the drowsy mode and ping transmissions. Our routing strategy aims at minimizing the number of data transmissions, which, in turn, minimizes the overhead of the wake-up synchronization phase. A simple scheduling of the data transmissions across the entire routing tree can help ensure this: a node should forward to a downstream node only when it has received the data from all of its upstream nodes to avoid having to relay the data from the upstream routing tree in multiple transmissions.

The routes determine the order in which nodes must forward their data. The scheduling of the nodes' transmissions must be performed to determine the exact timings of events, such as when a node shall initiate wake-up (by generating ping), when shall a node go into drowsy (to capture a ping), and when shall a node transmit its data together with the data it has collected from its upstream nodes. All such scheduling is also performed within the network layer. A goal of this scheduling is to ensure that the data collection takes place with as few pings as possible. For this, all the senders to a receiver are grouped and awakened together by a single ping from the receiver, and further only those senders whose communications failed are allowed retransmission. We allow up to two retransmissions in our simulation studies but this value can be adjusted to meet the requirements of the packet success rate, and taking into account the dynamics of the wireless channel conditions.

2.5 Simulation Results and Initial Implementation

We simulate our network under the TOSSIM (Levis et al., 2003) framework, an event-based simulator for TinyOS (Levis et al., 2005) wireless sensor networks. MAC and PHY layers were implemented in nesC while route and schedule computations were implemented in Python.

The crystal oscillator drift is modeled as a uniform random variable, the range *(Δ)* of which depends on the duration *(Γ)* between two data collection rounds during which nodes are in sleep mode, and the drift in parts-per-million (ppm) of the crystal oscillator δ_{ppm} (manufacturer specified). Using a standard value for drift of 30 ppm for an oscillator of frequency 1 MHz: $\Delta = \delta_{ppm}\Gamma$.

The receiver wakes up at time t_{rx} and transmits a short ping to wake up its upstream sender nodes. A sender node wakes up at $t_{rx} - 2\Delta$ and goes into the drowsy mode and remains so until it receives a ping from its downstream receiver. In the event that the synchronization fails; that is, the upstream nodes do not receive a ping from the downstream node, they transition back to sleep mode after a timeout of

$4\Delta + w$, where w is the duration of the transmitted ping.

When a sender receives the ping pulse, it gets synchronized with the receiver, and transmits all its accumulated data in its assigned slot. Receiver node transmits an acknowledgement packet after all upstream senders have had a chance to transmit their data packets.

We simulate our network with wake-up synchronization strategy implemented using a traditional radio and a radio with multiple power modes for receive and transmit. Table 1 presents the values of the current drawn in various radio modes employed in our system. In Figure 7, we show the energy savings achievable by using a radio with multiple transmit and receive power modes versus using a traditional radio with only a single transmit and receive power mode. The figure also presents the results of load balancing achieved when rotating the role of the sink among the four corner nodes. It can be seen that for the case of no-rotation there is a large variation in the energy consumed by the nodes because the nodes that are closer to the sink node (shown as the lower id numbers) get more drained than the others. The most even energy consumption among the three cases is observed in the case of a radio with multiple power modes, with effective implementation of load balancing and scheduling at the network layer.

As a proof of concept, we have implemented the protocol for the case of two nodes - one uplink node and one downlink node. The hardware of the nodes is based on a CC1110Fx SoC with a low-power RF transceiver and 8051 MCU from Texas Instruments (2014). Not that the hardware platform used for our initial implementation only supports the 315/433/868/915 MHz bands that have higher attenuation for underground communication required in our application. The design of an antenna that supports the 27 MHz frequency band is the subject of future research.

We presented the protocol design of our sensor network for automated collection of soil data from a farm-field in Sahota et al. (2010).

3. NETWORK PERFORMANCE MODELING AND EVALUATION

In this section, we present an analytical approach to model the performance of PD-MAC and duty-cycle based S-MAC. We study the throughput, latency and energy consumption of each node in the network under the two MAC protocols. We

Table 1. Current drawn as per CC1110 datasheet for various power modes

Radio Mode	Power Level (dBm)	Current Drawn (mA)
Transmit (Regular)	-30	15
Transmit (Ping)	10	33.5
Receive (Regular)	-110	19.8
Receive (Drowsy)	-107	17.8

Figure 7. Energy consumption of 100 nodes over 10 macro-rounds

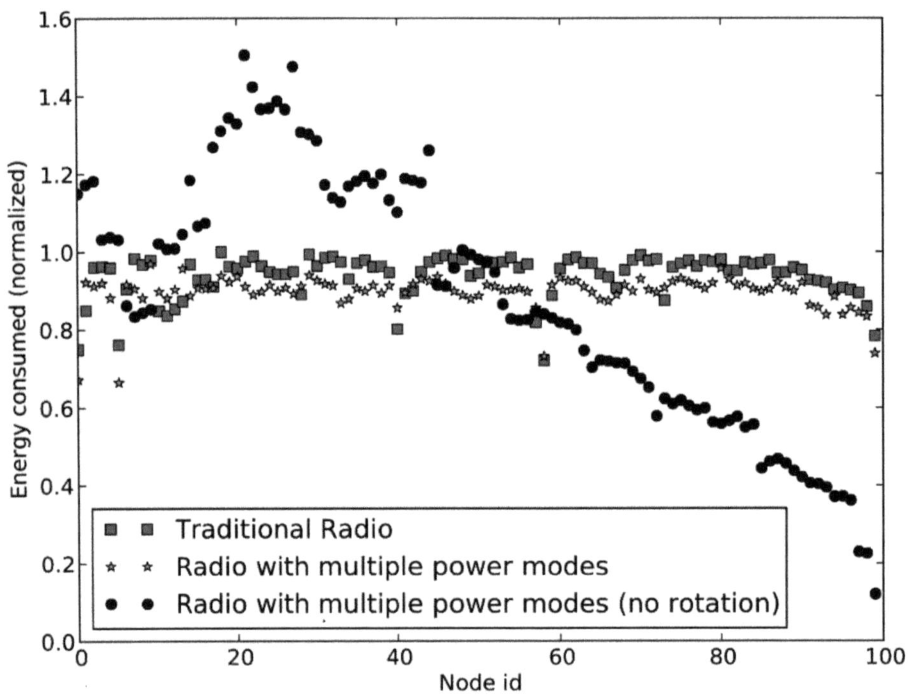

also validate our mathematical models against simulation results. The results of our model and simulations show a 25% reduction in latency and a 65% reduction in energy consumption for the same level of throughput for PD-MAC as compared to S-MAC.

3.1 Mathematical Analysis and Simulation Studies

We are interested in analytically modeling the performance of two wireless sensor networks: one that operates under PD-MAC and one that operates under S-MAC. We consider the same periodic data collection application as described in Section 2.3. Both designs operate under a fixed routing and scheduling strategy.

The differences between the two MAC protocols lie in the implementation of the network layer routes and schedules as well as the wake-up synchronization protocol. The synchronization handshake and communication of data and acknowledgment in an S-MAC based wireless channel occurs in a pairwise manner between a sender and a receiver node. If a receiver node has more than one upstream sender nodes, then the respective sender→receiver links are scheduled sequentially. Thus, the total communication time for data reception for a given receiver node is the sum of communication times (which includes the wake-up synchronization time) corresponding to each of its upstream neighbors. On the other hand, in PD-MAC, a receiver node synchronizes all of its upstream sender nodes with a common ping signal.

As explained in Section 2.4 above, PD-MAC uses the multiple radio power modes to conserve energy during the wake-up synchronization phase. While traditional S-MAC may use multiple transmission levels to adjust the range of communication, it does not vary the receiver sensitivity or the transmission levels during the wake-up synchronization phase with the goal of conserving total energy.

Consider the establishment of two links - each linking a downstream receiver node with its two upstream sender nodes - using PD-MAC and a modified version of S-MAC. Our modification to S-MAC, for comparison purposes, is illustrated through the use of Figure 8. In S-MAC, the two senders are separately synchronized and then allowed a given number of attempts to transmit their data. When a node wakes up to establish a communication link with its neighbor, it listens for a predefined duration before transmitting a synchronization request packet. The duration of this period is chosen long enough to accommodate the worst case drift (2Δ) between the clocks of the two nodes. That is, the first node that wakes up does not start transmitting until it is guaranteed that the other node on the link is awake to receive the synchronization request. This period is computed to be $2\Delta + W$, where W denotes the total duration of the synchronization request and reply packets. During this period a newly awakened node must listen on the channel to discover any existing neighboring sleep-listen schedules. The node that receives the synchronization request packet replies with a synchronization reply packet. Keeping the primary goal of comparing energy savings, we make a simplifying assumption that the reply packet is delivered error free. In addition, in regular S-MAC the synchronization reply, which is the rebroadcast of the synchronization packet by the node that receives a synchronization request, happens after a random back-off to prevent multiple nodes from transmitting at the same time. In our application setting for S-MAC, only two nodes (i.e. sender and receiver) may be active at a time. Hence, a back-off is not needed. Once synchronization is established, the two nodes exchange data/ack packets for up to N_d attempts. In our comparison studies the nodes keep awake until they expend all synchronization and data transfer attempts or until data transmission succeeds. This modification ensures that S-MAC based network is not penalized by having the nodes to stay awake for longer than it is necessary. In our application, a sender node is guaranteed to have data to transmit when it wakes up eliminating the need for a continuously operating sleep-listen schedule.

In PD-MAC, as shown in Figure 3, a common ping signal is sent by the receiver to synchronize the two senders. Both Figures 3 and 8 illustrate equivalent scenarios for ease of comparison: The maximum allowed synchronization attempts and data transmission attempts are each fixed at 2. Sender 1 synchronizes in the first attempt and also successfully transmits its data in the first attempt. Sender 2 misses the first synchronization attempt. It synchronizes in the second attempt and then transmits its data in the first attempt. Thereafter, all three nodes are scheduled to sleep.

Based on the duration of time spent in various states (sleep, active receive, active transmit, ping or drowsy) by each node during its operation, we derive analytical models for throughput, latency and energy consumption. The next three sections present an overview of our approach. Please refer to Sahota et al. (2011) for details.

Figure 8. Synchronization and data phase is implemented in pairwise manner in S-MAC

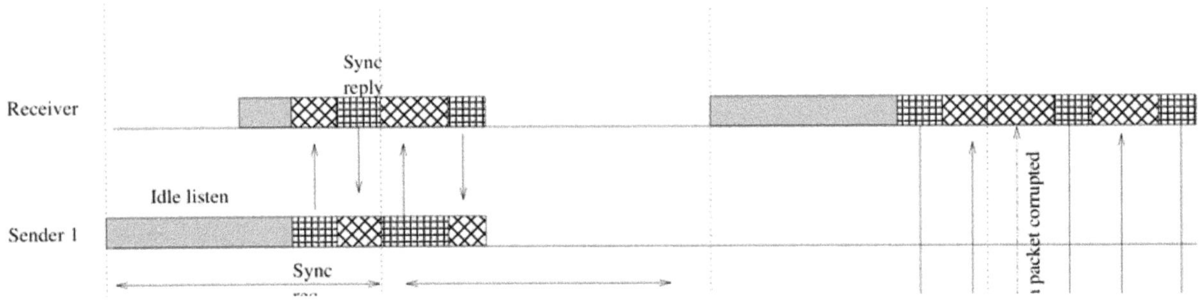

3.1.1 Throughput

Since the duration of the data collection round is fixed, we are interested in the number of data readings that are successfully received at the sink node from various sensor nodes in the network. Due to channel imperfections (noise, interference, fading, etc.) some data readings may be lost or corrupted. We model the number of data readings collected at the sink node at the end of one round. As it turns out, this value is dependent on the number of data readings that are successfully collected at each node in the network. For example, the number of data readings collected at a given node is dependent on the number of data readings successfully transmitted by all of its upstream sender nodes which, in turn, is dependent on the data readings that are collected at the upstream nodes and the channel conditions. A sink node's received data is, thus, dependent on the data collected by its upstream nodes, which are all of the nodes in the network.

Mathematically: $P_D^m(i)$, denoting the probability that node *m* has collected *i* data readings (including the locally generated reading) just after it has attempted to collect data from its upstream neighbors, is recursively modeled by Equation (1) (Box 1). When *m* is a leaf node in the routing tree, then $P_D^m(i) = 1$ for $i = 1$ and $P_D^m(i) = 0$ for $i \neq 1$.

The above expression computes the probability that a subset of upstream neighbors, $S_m \subseteq U_m$, successfully transmits $i - 1$ data units to node *m*, such that the nodes in S_m collectively contain $i - 1$ data units, for all such subsets S_m. The above expression is generic and applicable to any network with a periodic data collection application such as the one described in Section 2.3. For PD-MAC as well as modified S-MAC where nodes are awakened from sleep to transmit their data: $P_{suc}(i_k)$ denotes the probability that synchronization is established in at most N_s attempts and a data packet containing i_k data units is transmitted successfully in at most N_d attempts. $P_{suc}(i_k)$ can be calculated as shown in Box 2.where *q* is the probability of failure of a synchronization attempt, p_e is the bit error rate and $D(i_k)$ is the bit length of a data packet containing i_k data units. $P_{fail}(i_k)$ is simply the complement of $P_{suc}(i_k)$. Equation (1) holds for both PD-MAC and S-MAC, with the difference being the success probability of a synchronization attempt *(1 - q)*. For S-MAC, this is computed in accordance with the bit error probability (assuming no modulation), since the synchronization request packet contains encoded information about synchronization. For PD-MAC, this denotes the probability of the success of a ping, which is a much higher value since a ping has the sole purpose of awakening the sender nodes (which will receive the ping) and it contains no other useful information to be decoded.

See Figure 9 for the expected data count at node *0* versus the number of synchronization

Box 1.

$$P_D^m(i) = \sum_{S_m \subseteq U_m : \sum_{k \in S_m} i_k = i-1} \left\{ \left(\prod_{k \in S_m} P_D^k(i_k) P_{suc}(i_k) \right) \left(\prod_{k \in U_m \setminus S_m} P_D^k(i_k) P_{fail}(i_k) \right) \right\} \quad (1)$$

Box 2.

$$P_{suc}(i_k) = \sum_{s=1}^{N_s} (1-q) q^{s-1} \sum_{r=1}^{N_d} ((1-p_e)^{D(i_k)} (1-(1-p_e)^{D(i_k)})^{r-1}) \quad (2)$$

attempts allowed for the two protocols. The throughput performance of S-MAC approaches that of PD-MAC as we increase the number of allowed synchronization attempts, also confirmed by simulation results in the same figure. This is expected because the probability of a missed synchronization approaches *0* for both protocols as the number of synchronization attempts increases. Data transmission probability with a given number of attempts is the same for both protocols.

3.1.2 Latency

The period of communication within a fixed duration data collection round varies depending on the MAC protocol used. Since our network layer schedules data transmissions sequentially for both PD-MAC and S-MAC, the total period of communication can be estimated from the total communication time for every node in the network. We define the communication time for a node as the amount of time it spends attempting to collect data from all its upstream sender nodes, including any synchronization phase.

Latency in Modified S-MAC

Recall that the total communication time for a node is given by the sum of the communication times corresponding to each of its upstream neighbors. Let D_{km} denote the communication time for a receiver node m corresponding to an upstream neighbor *k*; that is, the duration of time it takes for node *m* to synchronize with, receive data from and transmit acknowledgement to its upstream neighbor *k*. We obtain an analytical model for D_{km}. Let $P_{sync}(n)$ denote the probability that synchronization succeeds in the nth attempt $(n \leq N_s)$ and $P_{sync-fail}$ denote the probability that synchronization fails in all N_s attempts. If synchronization fails N_s attempts, the data phase does not take place. However, if synchronization succeeds in at most N_s attempts, the two nodes attempt to exchange the data packet for up to a maximum of N_d attempts. Thus, the expected communication time for receiver node m and sender node *k* is given by:

$$E(D_{km}) = \sum_{s=1}^{N_s} P_{sync}(n) \times \left[T_{sync}(i) + T_{data}{}^{k} \right] + P_{sync-fail} \times T_{no-sync} \quad (3)$$

where $T_{sync}(i)$ denotes the expected duration of the synchronization phase given that synchronization succeeds in the ith attempt, $T_{data}{}^{k}$ denotes the expected data phase duration consisting of a maximum of N_d attempts, and $T_{no-sync}$ denotes the

Figure 9. Expected data count stored at sink

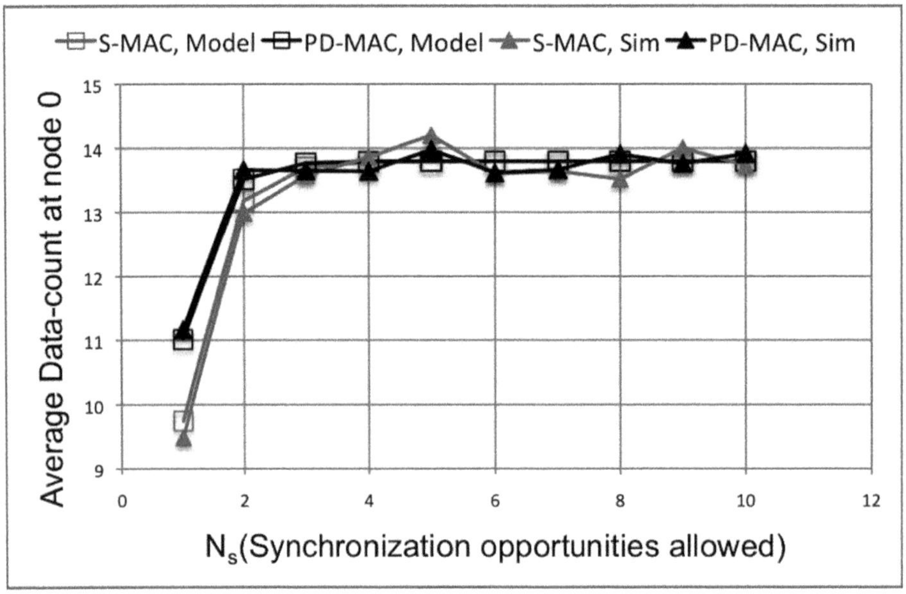

expected duration of the synchronization phase when synchronization fails all N_s attempts. In the latter case, no data phase takes place. For brevity of presentation, we refer the reader to Sahota et al. (2011) for detailed derivations of these quantities. The overall communication time for the downstream node *m* is a summation of the communication times corresponding to all its upstream sender nodes *k* and the overall communication time for the entire network is equal to the sum of the communication times for all the nodes in the network.

Latency in PD-MAC

In PD-MAC, all the upstream neighbors $k \in U_m$ of node *m* attempt to synchronize with node *m* by means of a common synchronization ping signal. The nodes that are successfully synchronized transmit their data in a time division manner. However, the total communication time of the entire network is still obtained by summing up the communication times of every node in the network. We follow a recursive approach to model the communication time of a receiver node *m* to synchronize and collect data from all its upstream sender nodes $k \in U_m$.

Let $T_m^{\,i}(S_m)$ denote the expected duration of time to synchronize the set of senders $S_m \subseteq U_m$ by node *m* and the associated data phases, given that *i - 1* synchronization attempts (and the associated data communication phases) have already occurred. The base case of the recursion is given by: $T_m^{\,i}(S_m) = 0$ for $i > N_s$ or $S_m = \phi$. $T_m^{\,1}(U_m)$ gives the communication time for node *m*. See Box 3.

where $P_{suc}^{S_j}$ is the probability that all nodes in the set S_j succeed in their data transmission and $T_{D+A}^{\,m}$ represents the duration of the entire data slot accounting for data transmissions from all senders and acknowledgement for the downstream node *m*. The first term in the above equation represents the case when at least one sender node is not synchronized in the current ping attempt. Of the senders that are synchronized, if all are successful in transmitting data in at most N_d attempts, an additional term $T_m^{\,i+1}(S_m - S_j)$ adds up to the communication time. However, if at least one node is unsuccessful in transmitting data in

Box 3.

$$T_m^{\,i}(S_m) = \sum_{j=0}^{|S_m|-1} \binom{|S_m|}{j} (1-q)^j q^{|S_m|-j} \left[\sum \begin{pmatrix} N_d . T_{D+A}{}^m + P_{suc}{}^{S_j} . T_m^{\,i+1}(S_m - S_j) + \\ (1 - P_{suc}{}^{S_j}).(N_s - i).N_d . T_{D+A}{}^m \end{pmatrix} \right] + (1-q)^{|S_m|} D_m^{\,1}(S_m) \quad (4)$$

N_d attempts, $(N_s - i)N_d T_{D+A}{}^m$ adds to the communication time. The final term in the above equation represents the case when all the upstream neighbors of S_m are synchronized in the ith ping. In that case, data transfer success may happen earlier (at $D_m^{\,1}(S_m)$) than when the next ping is scheduled to take place (at $N_d . T_{D+A}{}^m$). $D_m^{\,r}(S_m)$ denotes the time it takes to successfully deliver data from senders in the set $S_m \subseteq U_m$ that have successfully received the ping, given that *r - 1* data attempts have already been made. $D_m^{\,1}(S_m)$ is the portion of the communication time in which senders that have just received the ping try to send their data and are allowed a maximum of N_d data transmission attempts interspersed with collective ACK/NACK packets. See Figure 3.

The results of the model and simulation for both S-MAC and PD-MAC are shown in Figure 10. The figure compares the active communication time for the network for the two protocols versus the number of synchronization attempts allowed. PD-MAC achieves a 25% shorter round duration than S-MAC.

3.1.3 Energy Consumption

The energy consumption for S-MAC and PD-MAC can be computed as straightforward extensions of the models developed for the latency in Section 3.1.2. We divide the active time of a node into time spent in various modes (such as idle, drowsy, receive, transmit and ping). The current drawn in a given mode (Table 1), assuming that the battery voltage is a constant, gives an estimate of the power. Thus, to estimate the energy consumed in a given mode we model the time spent in the respective mode. The sum of the energy consumed in various radio modes gives the total energy consumed. We ask the reader to refer to Sahota et al. (2011) for detailed derivations.

The results of the energy consumption model are shown in Figure 11. Energy savings of approximately 65% are predicted for PD-MAC.

To compute the energy consumption, we used the values for the current drawn from the battery as per the datasheet of CC1110 for various modes of the radio (Texas Instruments, 2014) as shown in Table 1. A bit error rate of 0.01, ping transmission error probability of 0.1, 8-bits per data unit, 8-header bits per packet, ping pulse duration of 0.1s, and a baud-rate of 1200bps was used for our performance evaluation studies. For the proof of concept, we have implemented PD-MAC for the case of two nodes, i.e., one uplink node and one downlink node, see (Sahota et al., 2010) for more details.

4. FUTURE WORK

When a set of nodes in the routing tree fail, the network must have a way to gracefully recover and minimize the effect on the data collected from the fully functional nodes. We partly address this issue in our network stack by rotating the role of the sink among the four satellite nodes across data collection rounds. However, there is potential to

Figure 10. Expected round duration vs. number of synchronization opportunities

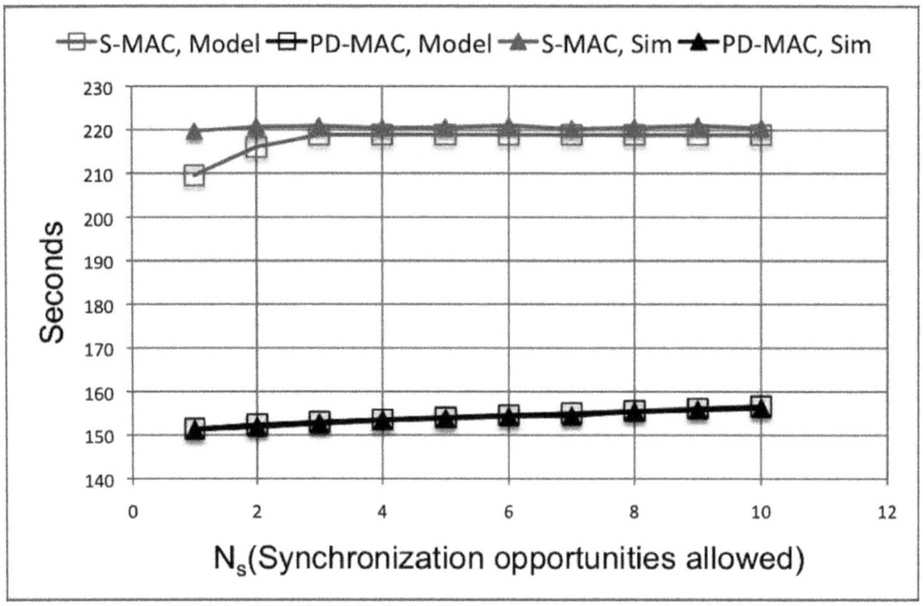

Figure 11. Expected energy consumption per node

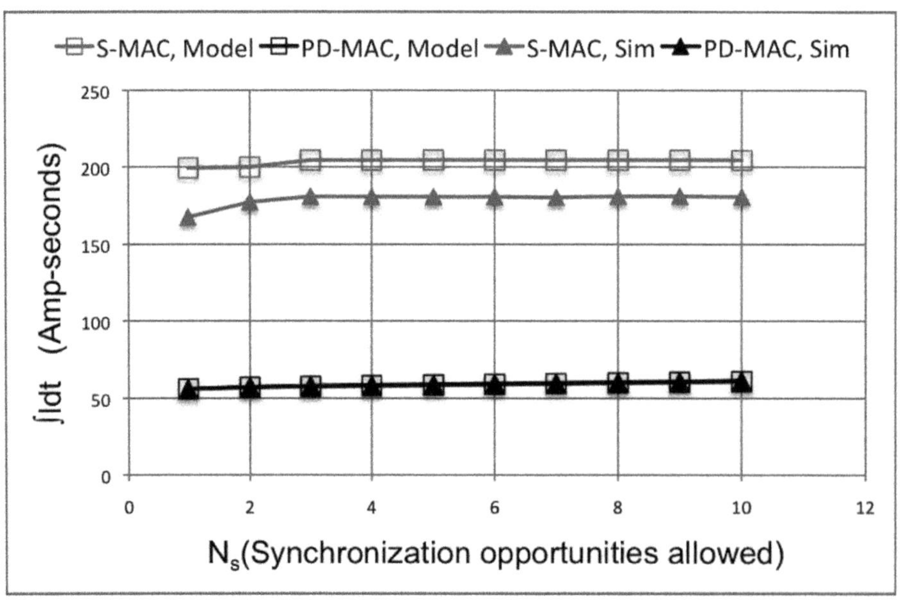

further improve the fault tolerance of the system by more adaptive routing techniques within data collection rounds. Both detection of faulty nodes and mitigation of the fault's effects are required.

Node localization to estimate the nodes' three dimensional location coordinates is also an essential requirement. Our routing strategy chooses the next hop based on the location of the neighboring nodes relative to the sink. Location information is also essential for the application layer to determine where the control inputs need to be applied.

Underground transmission loss is sensitive to the frequency of the signal used. The higher frequencies available in the CC1110 hardware have an unacceptably high attenuation rate. However, as the frequency decreases the antenna size increases. The design of a transceiver antenna of acceptable size, which operates at base frequency of 27 MHz, is the subject of future research.

5. CONCLUSION

The application of wireless sensor networks in precision agriculture is an active research field. This chapter presents the design of MAC and network layers for a custom periodic data collection application in precision agriculture. We presented our research on the design of an energy-efficient wireless sensor network for a periodic soil data collection application in precision agriculture. We introduced the design of cross-layered medium access control (PD-MAC) and network protocols for our sensor network with a novel wake-up synchronization mechanism with the goal of energy-efficiency. We utilize the event-driven simulation framework of TOSSIM to model the behavior of our PHY/MAC and network layers. Simulation results show that employing our design can help reduce the energy consumption and also balance the energy consumption profile across all the nodes in a sensor field, reducing the probability of a network partition.

We also presented analytical models to evaluation the performance to compare the performance of PD-MAC with a modified form of S-MAC under a given routing strategy. The approach is based on a probabilistic behavior modeling to compare metrics such as energy consumption, communication delay and throughput of a network employing PD-MAC with one employing S-MAC, both using the same routing strategy. While the two protocols can achieve the same packet success rates, the proposed new protocol is able to accomplish this in 25% less time and 65% less energy, as analyzed and simulated for a simple sensor field of 5×5 nodes. We successfully tested our MAC protocol in hardware employing the new wake-up synchronization strategy for one sender and one receiver case. We also validated our routing and scheduling strategies by deploying a network of nine nodes arranged in a grid fashion that transmit data to a central node over multiple hops.

ACKNOWLEDGMENT

This work was supported in part by the National Science Foundation under the grants NSF-ECCS-0801763, NSFECCS-0926029 and NSF-CCF-1331390.

REFERENCES

Adamchuk, V. I., Hummel, J. W., Morgan, M. T., & Upadhyaya, S. K. (2004). On-the-go soil sensors for precision agriculture. *Computers and Electronics in Agriculture*, *44*(1), 71–91. doi:10.1016/j.compag.2004.03.002

Beckwith, R., Teibel, D., & Bowen, P. (2004, October). Unwired wine: Sensor networks in vineyards. In Sensors, 2004. Proceedings of IEEE (pp. 561-564). IEEE.

Borms, J., Steenhaut, K., & Lemmens, B. (2010). Low-overhead dynamic multi-channel mac for wireless sensor networks. In *Wireless Sensor Networks* (pp. 81–96). Springer Berlin Heidelberg. doi:10.1007/978-3-642-11917-0_6

Buettner, M., Yee, G. V., Anderson, E., & Han, R. (2006, October). X-MAC: a short preamble MAC protocol for duty-cycled wireless sensor networks. In*Proceedings of the 4th international conference on Embedded networked sensor systems* (pp. 307-320). ACM. doi:10.1145/1182807.1182838

Burrell, J., Brooke, T., & Beckwith, R. (2004). Vineyard computing: Sensor networks in agricultural production. *Pervasive Computing, IEEE, 3*(1), 38–45. doi:10.1109/MPRV.2004.1269130

Cai, J., & Goodman, D. (1997). General packet radio service in GSM. *Communications Magazine, IEEE, 35*(10), 122–131. doi:10.1109/35.623996

El-Hoiydi, A., & Decotignie, J. D. (2004). WiseMAC: An ultra low power MAC protocol for multi-hop wireless sensor networks. In *Algorithmic Aspects of Wireless Sensor Networks* (pp. 18–31). Springer Berlin Heidelberg. doi:10.1007/978-3-540-27820-7_4

Gobriel, S., Mosse, D., & Cleric, R. (2009, June). TDMA-ASAP: Sensor network TDMA scheduling with adaptive slot-stealing and parallelism. In *29th IEEE International Conference on Distributed Computing Systems, 2009.* (pp. 458-465). IEEE.

Lal, R., Reicosky, D. C., & Hanson, J. D. (2007). Evolution of the plow over 10,000 years and the rationale for no-till farming. *Soil & Tillage Research, 93*(1), 1–12. doi:10.1016/j.still.2006.11.004

Levis, P., Lee, N., Welsh, M., & Culler, D. (2003, November). TOSSIM: Accurate and scalable simulation of entire TinyOS applications. In *Proceedings of the 1st international conference on Embedded networked sensor systems*(pp. 126-137). ACM. doi:10.1145/958491.958506

Levis, P., Madden, S., Polastre, J., Szewczyk, R., Whitehouse, K., Woo, A., . . . Culler, D. (2005). TinyOS: An operating system for sensor networks. In Ambient intelligence (pp. 115-148). Springer Berlin Heidelberg.

Lim, S., Ji, Y., Cho, J., & An, S. (2006). An ultra low power medium access control protocol with the divided preamble sampling. In *Ubiquitous Computing Systems* (pp. 210–224). Springer Berlin Heidelberg. doi:10.1007/11890348_17

Lim, S., Kim, S., Cho, J., & An, S. (2006, January). Medium access control with an energy-efficient algorithm for wireless sensor networks. In *Personal Wireless Communications* (pp. 334–343). Springer Berlin Heidelberg. doi:10.1007/11872153_29

Lin, E. Y., Rabaey, J. M., & Wolisz, A. (2004, June). Power-efficient rendez-vous schemes for dense wireless sensor networks. In *IEEE International Conference on Communications* (Vol. 7, pp. 3769-3776). IEEE. doi:10.1109/ICC.2004.1313259

Ma, J., Zhou, X., Li, S., & Li, Z. (2011, October). Connecting agriculture to the Internet of Things through sensor networks. In *Internet of Things (iThings/CPSCom), 2011 International Conference on and 4th International Conference on Cyber, Physical and Social Computing* (pp. 184-187). IEEE. doi:10.1109/iThings/CPSCom.2011.32

Polastre, J., Hill, J., & Culler, D. (2004, November). Versatile low power media access for wireless sensor networks. In *Proceedings of the 2nd international conference on Embedded networked sensor systems* (pp. 95-107). ACM. doi:10.1145/1031495.1031508

Polastre, J., Szewczyk, R., & Culler, D. (2005, April). Telos: enabling ultra-low power wireless research. In *IPSN 2005. Fourth International Symposium on Information Processing in Sensor Networks* (pp. 364-369). IEEE. doi:10.1109/IPSN.2005.1440950

Purvis, M., Sambells, J., & Turner, C. (2006). *Beginning Google maps applications with PHP and Ajax*. Michael Purvis, Jeffrey Sambells, and Cameron Turner.

Rajendran, V., Obraczka, K., & Garcia-Luna-Aceves, J. J. (2006). Energy-efficient, collision-free medium access control for wireless sensor networks. *Wireless Networks*, *12*(1), 63–78. doi:10.1007/s11276-006-6151-z

Rhee, I., Warrier, A., Aia, M., Min, J., & Sichitiu, M. L. (2008). Z-MAC: A hybrid MAC for wireless sensor networks. [TON]. *IEEE/ACM Transactions on Networking*, *16*(3), 511–524. doi:10.1109/TNET.2007.900704

Ruzzelli, A. G., Jurdak, R., & O'Hare, G. M. (2007). *On the RFID wake-up impulse for multi-hop sensor networks.*

Sahota, H., Kumar, R., & Kamal, A. (2011). A wireless sensor network for precision agriculture and its performance. *Wireless Communications and Mobile Computing*, *11*(12), 1628–1645. doi:10.1002/wcm.1229

Sahota, H., Kumar, R., Kamal, A., & Huang, J. (2010, June). An energy-efficient wireless sensor network for precision agriculture. In *IEEE Symposium on Computers and Communications (ISCC)* (pp. 347-350). IEEE. doi:10.1109/ISCC.2010.5546508

Schurgers, C., Tsiatsis, V., Ganeriwal, S., & Srivastava, M. (2002). Optimizing sensor networks in the energy-latency-density design space. *Mobile Computing. IEEE Transactions on*, *1*(1), 70–80.

Texas Instruments. (2014, September). *Low-Power SoC (System-on-Chip) with MCU, Memory, Sub-1 GHz RF Transceiver, and USB Controller*. http://www.ti.com/lit/gpn/cc1110f32

Van Dam, T., & Langendoen, K. (2003, November). An adaptive energy-efficient MAC protocol for wireless sensor networks. In *Proceedings of the 1st international conference on Embedded networked sensor systems* (pp. 171-180). ACM. doi:10.1145/958491.958512

Verma, S., Chug, N., & Gadre, D. V. (2010, March). Wireless sensor network for crop field monitoring. In *2010 International Conference on Recent Trends in Information, Telecommunication and Computing (ITC)* (pp. 207-211). IEEE. doi:10.1109/ITC.2010.21

Wong, K. J., & Arvind, D. K. (2006, May). SpeckMAC: low-power decentralised MAC protocols for low data rate transmissions in specknets. In *Proceedings of the 2nd international workshop on Multi-hop ad hoc networks: from theory to reality* (pp. 71-78). ACM. doi:10.1145/1132983.1132996

Xia, J., Tang, Z., Shi, X., Fan, L., & Li, H. (2011, December). An environment monitoring system for precise agriculture based on wireless sensor networks. In *Seventh International Conference on Mobile Ad-hoc and Sensor Networks (MSN)* (pp. 28-35). IEEE. doi:10.1109/MSN.2011.16

Ye, W., Heidemann, J., & Estrin, D. (2002). An energy-efficient MAC protocol for wireless sensor networks. In *INFOCOM 2002. Twenty-First Annual Joint Conference of the IEEE Computer and Communications Societies. Proceedings of the IEEE, 3*, 1567–1576.

Ye, W., Heidemann, J., & Estrin, D. (2004). Medium access control with coordinated adaptive sleeping for wireless sensor networks. *IEEE/ACM Transactions on Networking, 12*(3), 493-506.

KEY TERMS AND DEFINITIONS

Convergecast Network Traffic: The flow of data in a network where each node has one or more upstream neighbor but at most one downstream neighbor.

Drowsy Mode of Reception: The reduced sensitivity mode of a sensor's radio used during the wake-up synchronization phase.

MAC Layer Schedule: A schedule by which sender nodes in a local neighborhood of the routing tree are instructed to transmit packets to their one hop downstream node.

Network Layer Schedule: A schedule by which groups of nodes in the network are instructed to forward data to their downstream neighbor, after ensuring that all data from upstream nodes has been collected.

Ping Mode of Transmit: The higher powered mode of signal transmission used during the wake-up synchronization phase.

Precision Agriculture: The use of modern technology in agriclture with the goal to improve productivity and reduce the impact on the environment.

Wake-Up Synchronization: The handshake mechanism used to synchronize two sensor nodes that have just transitioned their transceivers to active mode from sleep.

Chapter 4
Active RFID System with Embedded Wireless Sensor Network for Reliable Data Communication

Raed Abdulla
Asia Pacific University of Technology & Innovation (APU), Malaysia

Widad Ismail
Universiti Sains Malaysia (USM), Malaysia

ABSTRACT

Reducing power consumption and per-hop latency are major challenges in Wireless Sensor Networks (WSNs), when network nodes or "tags" must be battery-powered. On the other hand, reducing transition overhead in RFID technology is crucial for efficient data collection and quick recognition of multiple tags. This chapter presents the method of analysis performed and explains the calculations used to evaluate the various performance behaviors based on the following mechanisms: (1) a duty cycle complying with the IEEE802.15.4 standard when assuming a fixed set of tags and free operating conditions. The system can be regarded as Reader Talk First (RTF), which is implemented during a tag collection process executed at the beginning of the system activation. (2) Wake-Up Mechanism, the tag transmits an identification packet to the reader and then waits for an Acknowledgment signal from the reader.

INTRODUCTION

RFID is an automatic identification system that stores and remotely retrieves data from devices called RFID tags or transponders by using an interrogator (or reader) and a computer network (Wang et al., 2008). The emerging technology of RFID promises to be a more comprehensive approach to data collection, and one of its potential applications is the improvement of supply-chain operations, offering a more automated and informative alternative to barcode. More recently, the evident benefits of an RFID system over barcode systems have formed a preference for RFID in the retail industry. These benefits give increased automation because of the ability of RFID tags

DOI: 10.4018/978-1-4666-8251-1.ch004

to be read without a visual line-of-sight, as well as their ability to accumulate more information than barcodes. Additionally, due to the suitability of RFID technology for locating and tracking objects, RFID has been recently applied to many location identification systems to detect the presence of tagged objects and/or people. Monitoring the system by way of an RFID reader is important for providing better and more efficient context-awareness services (Siadat & Selamat, 2008).

Furthermore, because reliable, sufficiently small and adequately low cost RFID tags can now be built, the reputation of RFIDs has lately been on the increase. Although the promise of RFID technology is thought to be great, its development has been slow because of anticipated implementation difficulties in areas such as signal quality and mobility in terms of spatiality when monitoring and tracking items in a large warehouse. In these cases the RFID mechanism is intrinsically unreliable and the system requires multiple read cycles for tag read efficiency. The necessity for modified RFID tag and reader devices to be made available is due to the high demand for these devices, with the goal of such a modification being the design of an RFID system capable of achieving a perfect network while being virtually invisible to the user.

However, research on efficient communication protocols for RFID-based systems has been rather limited in networked tag/reader environments, examples include systems for supply chain management, monitoring and controlling of large warehouses, and smart homes. Figure 1 (Lei & Zhi, 2006) shows architecture of networked smart nodes.

The last decade has seen new developments for overcoming the limitations of RFID, enabling the technology to meet the growing demands of the modern world. Recent advances in wireless communication, processing capability, and memory technology have fueled increasing research and industrial activity on the subject of wireless sensor networks (WSN). RFID does not provide information about the condition of the detected objects; in contrast, WSNs not only provide information about the condition of the objects and the

Figure 1. Architecture of networked smart nodes

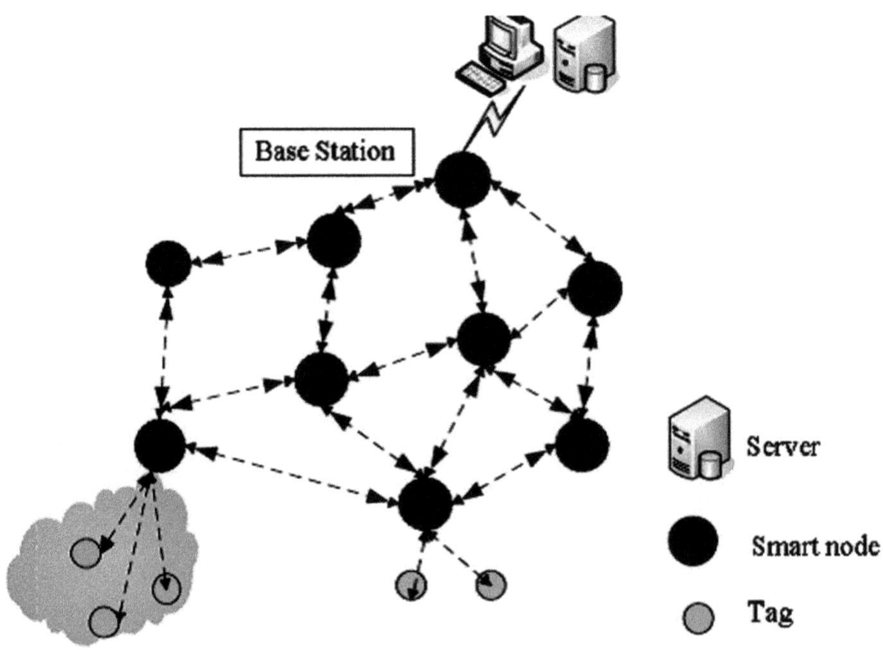

environment, but also enable multi-hop wireless communications (Callaway et al., 2002). Wireless sensors, also known as wireless sensor nodes or sensor nodes, have been making a significant impact on human daily life. Wireless sensors are the key that enables the technology for emerging cyber-physical systems, which would ultimately improve the quality of life (Shah &- Kumar, 2007). The RFID and WSNs are two complementary technologies; hence, an integrated implementation of these technologies expands their overall functionality in obtaining long-range and real-time information on the location and properties of objects and people. This extended coverage and effectively improved reliability can dramatically improve the read performance of an RFID system. Since WSN operates with low power consumption and low complexity, and since signal path loss is inversely related to range or distance.

In view of the above, there is a need for essential research on protocols and algorithms that can improve RFID-based WSN performance and thereby significantly benefit the emerging identification and monitoring applications.

This research addresses the problem currently plaguing the RFID systems that are used in large buildings, which are due to the amount of attenuation experienced by the signals transmitted from the tag to the reader and vice versa. When a tag sends transmission frames that include relevant information about the tag, such as addressing information and identifier strings (IDs), this long packet will increase the transmission overhead and thus increase the packet delivery latency, since time is wasted waiting for the whole packet to transmit. The aim of this research is to establish and develop a robust 2.4 GHz active RFID system using a standards-based WSN, preferably ZigBee, and apply it to a reliable asset identification system. The contribution is summarized as follows:

1. Duty-cycle scheme protocol is implemented based on IEEE 802.15.4 protocol. In this protocol, the result confirmed a high cost in term of effective data rate, occurring during idle listening and the necessity of retransmitting long identification packets.
2. The design of the above scheme mechanism is extended and enhanced to include a wake up scheme protocol, in order to minimize effective data rate by avoiding idle listening.

ZigBee and RFID are currently making a great impact on the telecommunications industry. A general overview of RFID and ZigBee will be given in this section.

ZigBee

In typical WSN applications, low transmission overhead and minimum power consumption are fundamental themes pervading the design of communication protocols developed, including routing and medium access control (MAC) layer protocols. One of the primary mechanisms for achieving low transmission overhead by the radio resource in WSNs is duty cycling (Wei et al., 2004, Demirkol et al., 2006), which reduces (but does not eliminate) idle listening. In this approach, the radio is turned on and off periodically according to a predefined scheme, specific to each MAC protocol. The energy consumed during the on and the off states, is the key parameter that characterizes the duty cycle which includes sleep time and wake time; the period of a duty cycle is equivalent to its sleep time plus wake time (Kumar et al., 2010, Jae-Han and Byung Tae, 2008). ZigBee (IEEE 802.15.4) is a wireless technology that builds upon the Institute of Electrical and Electronics Engineers (IEEE) standard 802.15.4, which defines the physical (PHY) and Medium Access Control (MAC) layers, and works on a low data rate standard (Kinney 2003; Sarijari et al. 2008). The ZigBee Alliance is an industry association that supports the use of the technology, enabling it for WSNs (ZigBee Alliance, 2006). Then, the ZigBee Alliance and the IEEE decided to join forces and ZigBee is the commercial name for this technology (IEEE, 2009). IEEE and ZigBee Alliance have been working continually to specify the entire protocol stack and establish layers for security, wireless networks, and application framework (Ondrej et al., 2006). Figure 2 (ZIgBee Alliance, 2006) shows ZigBee using the IEEE 802.15.4 physical and MAC layers.

IEEE 802.15.4 defines specifications for the lower two layers of the protocol, PHY and MAC; meanwhile ZigBee Alliance intends to provide the upper layers of the protocol stack, from the network layer to the application layer (IEEE, 2009). ZigBee standard supports low-speed, low-cost wireless networking with long device battery lifetimes and devices as well as low latency (Wheeler, 2007).

ZigBee technology is suitable for a wide range of monitoring activities, including building automation, health monitoring, automated meter reading equipment, grain storage, remote controls, heating and cooling control devices, fans, and structural integrity (Lin et al., 2006; Kim & Ayurzana, 2009; Kim et al. 2009; Zhoul et al., 2009).

Because the ZigBee standard is based on IEEE 802.15.4, implementing a ZigBee network has similar requirements to an 802.15.4 network. Figure 3 shows how ZigBee can be implemented in active RFID devices. The technology consists of three different physical device types that operate in a self-organizing network application: ZigBee Coordinators (ZC), ZigBee Routers (ZR), which are Full-Function Devices (FFDs), and ZigBee End Devices (ZED), which are Reduced-Function Devices (RFDs). These physical devices are used to create a ZigBee network (Wotton, 2006). The ZC acts as the network head and is typically mains powered, and there can only be one PAN (Personal Area Network) coordinator in a network. The ZR, which is an FFD, can perform the function of rout-

Figure 2. The ZigBee standard and IEEE 802.15.4

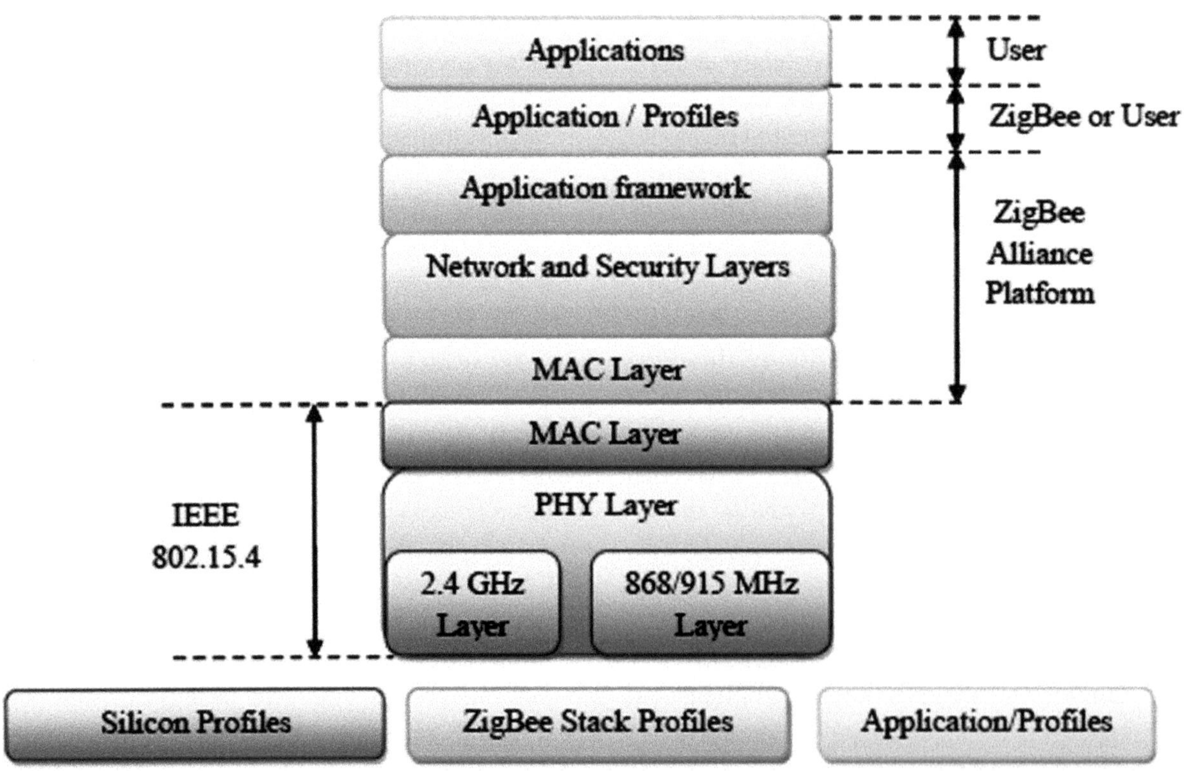

ing the data between the nodes. ZEDs are the leaf nodes, and can check the availability of the nodes to send data, have minimum functionality, and as reduced-function devices are always battery-powered (Baronti, Pillai et al. 2007). The ZC starts by sending a beacon frame to all its neighbors, and when a node accepts the framework of the beacon, it can be sent with a request to join the network (Gislason, 2008).

The ZigBee standard supports four multiple topology network configurations, namely, star, peer-to-peer, mesh and cluster tree topology. All topologies are established by only one coordinator (Howitt & Gutierrez 2003; Armholt, et al., 2007). Figure 4 illustrates all the ZigBee topologies. Advantages and disadvantages of each kind of topology depend on the individual application or situation.

RFID

Providing an optimized and a more effective context-awareness service are important in automatically tracking the location of objects in indoors. For several years, this has been the area of interest in various fields. For an outdoor localization system, Global Positioning System (GPS) was developed in the early 1970s for military applications and has increasingly used as a commercial location system today. However, the GPS signal is not sufficiently strong to work satisfactorily indoors, since, due to the multipath effect, signals cannot be received under dense canopies, which makes the process too expensive or excessively power-intensive (Sahinoglu et al., 2008). Thus, GPS still lacks positioning accuracy in indoor-based location tracking and cannot offer a solution for indoor location-awareness applications (Guillemette et al., 2008). Because of this

Figure 3. Implementing ZigBee in active RFID devices

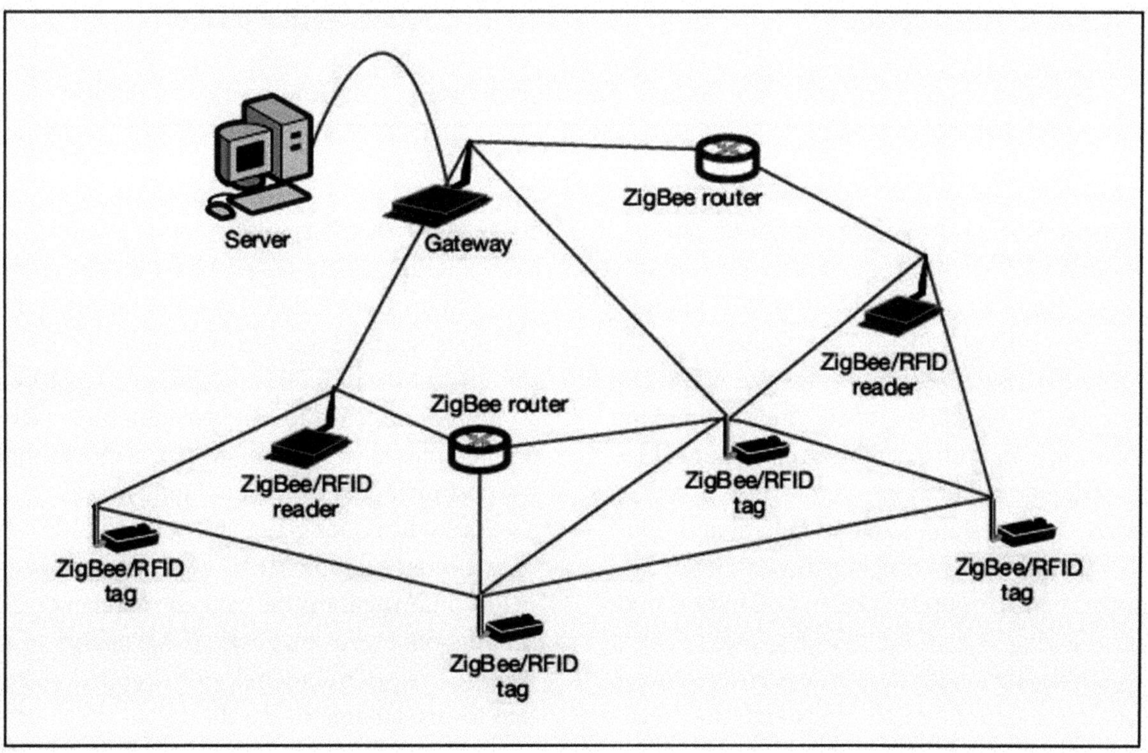

Figure 4. ZigBee topologies

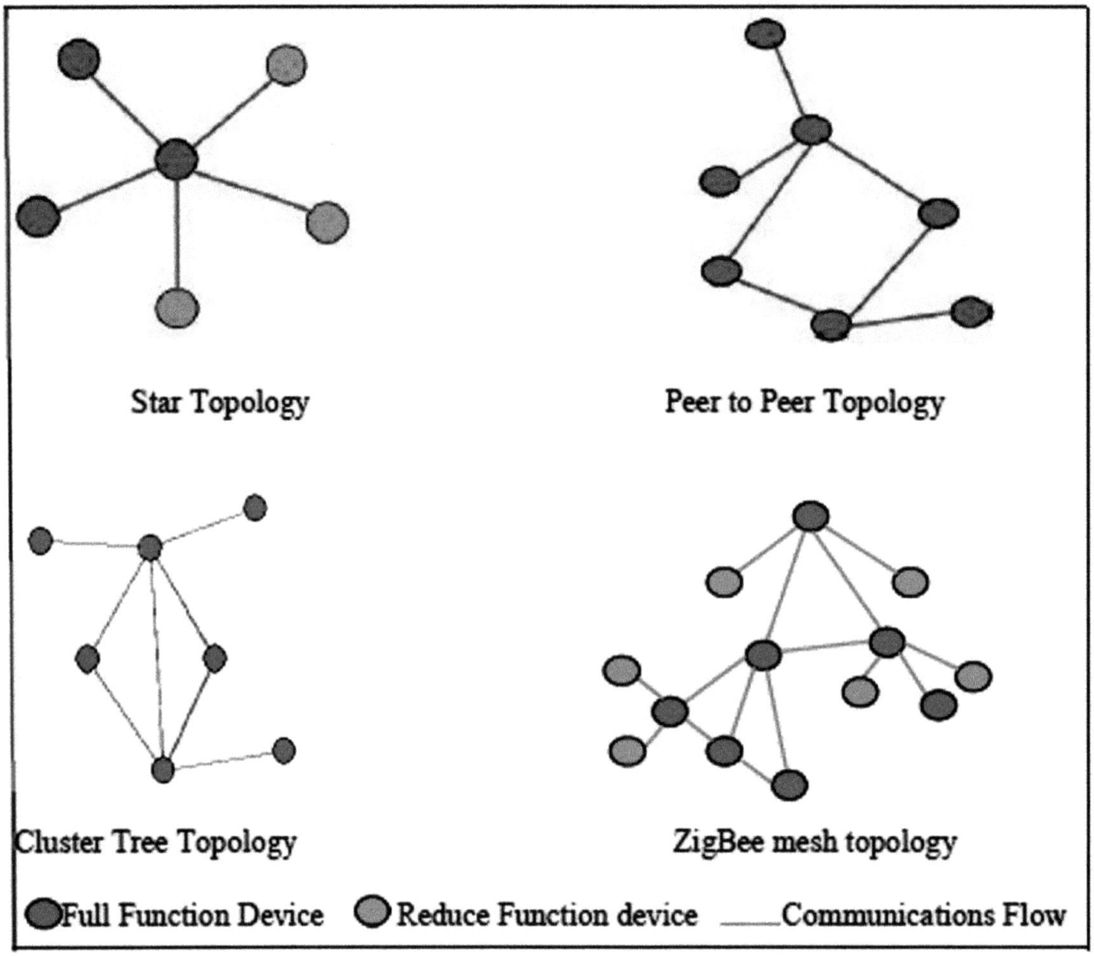

weakness, RFID is more suitable for indoor use. Although RFID has become widely-known only in recent years due to the widening of its applications, the concept of RFID has actually existed for decades. There are currently many commercial applications of RFID for monitoring and tracking systems, which in turn require improved accuracy and efficiency. Literature shows that RFID is already driving shifts in supply chain and retail capabilities by improving logistic efficiency and reducing costs (Fagui et al., 2009). In addition, both a car-tracking system for charging and a central car-parking tracking system have been developed, which can monitor the efficiency and functionality of RFID-enabled parking-lots (Pala and Inanc, 2007). RFID tags (e-tickets) have also been used to speed up identification in conferences and exhibitions, and the authors provide evidence that the application of RFID to Exhibition Logistics is workable (Ke et al., 2008). Some intelligent RFID applications have been suggested, such as collecting customer data in the retail industry, as well as access control (Dong-Liang, Ng et al., 2009). The operation of a simple RFID system begins with a reader interrogating the tagged merchandise by sending out electromagnetic (EM) energy to and from the tag (a microchip with a coiled antenna that forms a magnetic field).

Target of the Active RFID

An active tag has its own power source (battery on board), typically a battery, to run the chip and to transmit the data to the reader. An active tag allows very low-level signals to be received and can still generate high-level signal to be transmitted back to the reader. Data cards and RFID are being manufactured in increasingly smaller sizes, and it is now even possible to hide the data card in such a way that the consumer may be unaware of its presence. Because the active tag has a battery onboard they has finite lifetime and can transmit data without requiring a reader to power them. They have therefore much longer reading range then a passive tag. The sensor only has power when the tag is in a reader's interrogation zone. Sensor tags are mostly active. These sensors can sound off an alarm in the event of unauthorized inventory movement by emitting an impulse of high frequency signals on a regular basis; signals may include the tag ID, tag. The RFID reader can read data emitted from RFID tags. RFID readers and tags use a defined radio frequency and protocol to transmit and receive data status, and other tag information. The RFID-based sensor system developed clearly demonstrates the effectiveness of RFID technology (Ni et al., 2004). Active tags have longer reading ranges, high memory, and better noise protection. However, these tags are larger and heavier, more expensive, and have a shorter life (3 – 10 years) than passive tags. Read-only tags are used for simple identification purposes because they can only store a limited amount of information that cannot be altered. Such tags may be used to identify a package of nails or screws because they have many applications and are not designated to a particular item or activity.

RFID, similarly to WSN, is an extremely low-power and low-cost technology implemented in a number of possible methods, each one of them characterized by different communication ranges and transmission bandwidth. (Salvatore et al., 2012). Integration of RFID and WSN provides a new feature and improves their functionalities (Mirshahi et al., 2013).

(Salvatore et al. (2012) proposed fully integrated approach between RFID and WSN targeted to develop an advanced safety system in industrial plants with low-cost equipment by choose a real-time software component able to decrease the system latency to effectively help workers to access to specific safety conditions determined such as wearing safety helmet or wearing safety boots. Chen (2010) presented a wireless localization system for monitoring children position in theme park. In this work, an RFID reader is installed on each WSN node, hence integrating a hybrid localization system capable to guess the child position with a maximum error of 3 meters. (Xiong et al., 2011) enhanced the positioning accuracy reached by standard well-known received signal strength indicator (RSSI) based on WSN localization algorithms. In this work the integration is reached again by installing RFID readers on WSN units. Sensors of smart nodes are allocated to identify objects which may be a concern to the environment. Mirshahi et al. (2013) presented an intelligent supply chain management system by applying smart nodes which provides integration of RFID with WSN. In this work, WSN is used for observing and monitoring the environment of the products, while RFID for identifying the type of products in the number of rooms (warehouse). In order to show the system performance, the authors applied a case study and simulated in the Petri Net Toolbox in MATLAB environment. Rajesh (2013) proposed a low cost method to the development of a system that can monitor the real time parameters of a patient using an active RFID system along with the respective parameter monitoring wireless sensor. In this work, the detailed record of the patient which can be used by the doctors to observe the state later as well. The real time values are plotted on a graph using Visual Studio 10.Thus the combination of an RFID along with health monitoring sensors to uniquely distinguish each patient in a ward with its measuring data is achieved. A survey of WSN technology based reliable and efficient active RFID is presented by (Abdulla & Ismail, 2013). In this work, the accessible wireless standards are evaluated based on preferred criteria such as reliable end-to-end communication, low power consumption, latency, throughput and long range are crucial requirements. Then, the authors presented the advantages and performance issues of each wireless standard. Also, the authors compared and analyzed the available wireless protocols for the purpose of building an active RFID system, and they found that the ZigBee is preferred due to its mesh network capabilities and its flexibility for accessing more communication paths than Wi-Fi, Bluetooth or Ultra Wide Band.

LITERATURE REVIEW

How to achieve high throughput with low delay time is one of the major challenges faced by communication protocol designers. Many approaches were motivated by the desire to enhance reduce efficient energy by reducing transmission overhead and idle listening (the time that the node is awake and listening to the medium even though no packets are being transmitted to that node). Wei et al. (2004) proposed a medium access control protocol called S-MAC which applied few techniques to support self-configuration and to reduce contention per-hop latency and energy consumption. This enabled each node to adaptively switch the mode according to the traffic, in the operation of a multihop network. Adaptive duty cycling was performed by Van Dam & Langendoen (2003) who used shortening the wake period when the channel was idle for variable workloads and proved that this technique used one fifth of the energy used by S-MAC. Polastre et al. (2014) proposed the technique which employs extended preamble to achieve better packet delivery rates, throughput, latency, and energy consumption. In this technique, during the wake period, a node samples the medium and if a preamble is detected it remains awake to receive the data. All mention techniques use RTS/CTS, which causes high overhead due to the resultant small size of data payloads in sensor networks. Jurdak et al. (2008a) presented a comprehensive evaluation of MERLIN (MAC and efficient routing integrated with support for localization), and employed a controlled multi-path for reducing transmission overhead and packet redundancy by shifting the idle listening energy overhead to the transmitter. In fact, the duty cycling MAC protocols are suboptimal in terms of energy consumption at both the sender and receiver, and they introduce excess latency at each hop in data communication. Lopez et al. (2013) discuss the importance of the duty cycle window size and present an expression to determine it according to the particularities of the IEEE 802.15.4 specification. The proposed technique attempts to decrease electromagnetic pollution in order not to interrupt the normal operation of other radio devices in the environment. The experimental results determine that the algorithm implemented is a very good approach to sense and transmit information between nodes randomly distributed in an indoor, noisily area. Castagnetti et al. (2014) propose a global power management method for energy harvesting sensor nodes. This method is based on shared a transmission power control and duty-cycle optimization. The authors in this work taking into account the energy harvesting level in addition to the wireless channel conditions, and then they simultaneously manage both the transmission power and the transmission period.

Many researchers proposed radio triggered wake-up mechanism that allows maintaining the radio module completely switched off until it is required to be used for data. This approach requires additional hardware assistance to utilize a low power radio as a trigger to wake up the radio. Mishra et al., (2006) implemented a prototype system using low-power 802.15.4-based sensor motes and proposed a novel on-demand power on/off for the networking equipment at a remote site. The radio Triggered Wake-up with Addressing Capabilities (RTWAC) was proposed by Ansari et al. (2009) which consisted of an external low-cost hardware wakeup circuit attached to the microcontroller of a sensor node. They found that combining RTWAC with a duty cycle MAC protocol on TelosB resulted in remarkably small idle-mode power consumption and low latency in data communication. Le et al. (2007) focused on enhancement of 802.11by introducing 'low latency media access control' (LLMAC) for event-driven WSN; the proposed LLMAC scheme significantly improved the latency transmission compared to 802.11. Jurdak et al. (2008) proposed an on-demand mechanism called RFID Impulses that enabled nodes to sleep until they had to send or receive packets by using RFID technology as wake-up channel for sensor networks.

A sleep/wakeup protocol with a discrete-time multiple vacation queueing model is presented by Jin et al. (2012), the authors taking into account the switching procedure between the sleep mode and the active mode at sensor units. Furthermore, the authors presented a new effective analysis method to provide the formulas for performance measures in terms of the latency, handover ratio and the energy consumption of the sensor node. In addition, the authors presented numerical results to discuss the trade-off between the modulation constellation size and the system performance. Guo et al. (2012) proposed a novel sleep scheduling method to reduce the delay of alarm broadcasting from any sensor unit in WSNs. In this work, when a critical event happens, an alarm is quickly transmitted along one of the traffic paths to a center node, and then it is immediately broadcast by the center node along another path without collision. The proposed sleeping scheme could reduce overall power consumption and making the broadcasting delay is independent of the density of nodes. Ram et al. (2014) presented a mathematical energy model to evaluate the energy consumption, based on current traffic conditions. In addition, the authors developed the performance of different MAC protocols IEEE 802.15.4, and Data–Gathering MAC (DMAC) with RFID Impulse using the proposed mathematical energy model under variable data rates. The results confirmed that the RFID impulse outperforms the IEEE 802.15.4 and DMAC in terms of energy consumption.

This research presents the method of analysis performed and explains the calculations used to evaluate the various performance behaviors based on the following mechanisms:

1. A duty cycle complying with the IEEE 802.15.4 standard when assuming a fixed set of tags and free operating conditions. In this system, the reader manages most of the data structures and maintains the data for all the tags in its range; therefore, the system can be regarded as Reader Talk First (RTF), which is implemented during a tag collection process executed at the beginning of the system activation.
2. Wake-Up Mechanism. The tag transmits an identification packet to the reader and then waits for an acknowledgment (ACK) signal from the reader. This process, known as Tag Talk First (TTF), is implemented during the interrupt step; hence, the tags have no need to be synchronized or to listen to the RFID reader.

This research also discusses the effects of changing parameters in conventional IEEE802.15.4 protocol (RTF) and in the RFID system (TTF), and predicts the behavior of sensor network applications using comparisons between them, including the effect of TTF and RTF raw and effective data transmission rates. The choice and implementation of protocols, and the analysis used to compare them, are significant contributions of the present work. The aim of this research is to establish and develop a robust 2.4 GHz active RFID system using a standards-based WSN, preferably ZigBee, in a large building, and apply it to a reliable asset identification system.

REVIEW OF IEEE 802.15.4 PROTOCOL

IEEE 802.15.4 PHY Layer

IEEE 802.15.4 wireless protocol provides sufficient flexibility for different requirements of WSN applications by effectively tuning its parameters. In fact, low-power consumption, low-rate and low-cost wireless networking are the key features of the IEEE 802.15.4 protocol, which normally fits the requirements of WSNs (Wei, Heidemann et al., 2002). The IEEE 802.15.4 PHY layer provides two services: the PHY data service and the PHY management service. The PHY data service activates the transmission and reception of PHY Protocol Data Units (PPDU) through the physical radio channel. There are many features of the PHY layer, such as activation and deactivation of the radio transceiver, Receiver Energy Detection (ED), Link Quality Indication (LQI), channel selection, Clear Channel Assessment (CCA), and transmitting as well as receiving packets across the physical medium. Wireless links under 802.15.4 can operate in three license-free industrial/scientific/medical (ISM) frequency bands. These accommodate over-air data rates of 250 kbps in the 2.4 GHz band, 40 kbps in the 915 MHz band, and 20 kbps in the 868 MHz. A total of 27 channels are allocated in 802.15.4, including 16 channels in the 2.4 GHz band, 10 channels in the 915 MHz band, and 1 channel in the 868 MHz band (IEEE, 2009).

IEEE 802.15.4 PHY layer packet (PPDU) structure is illustrated in Figure 5 (IEEE, 2009), and consists of the following basic components:

1. **Synchronization header (SHR):** Which allows a receiving device to synchronize and lock into the bit stream.
2. **PHY Header (PHR):** Which includes the frame length information.
3. **PHY Service Data Unit (PSDU):** Which is a variable length payload that carries the MAC Sublayer frame.

IEEE 802.15.4 MAC Layer

The IEEE 802.15.4 MAC layer provides two services: the MAC data and the MAC management services. The MAC data service activates the transmission and reception of MAC protocol data units (MPDU) through the PHY data service. There are many features of MAC layer, including beacon management, Channel Access, Guaranteed Time Slot (GTS) management, Frame Validation, Carrier Senses Multiple Accesses with Collision Avoidance (CSMA-CA) mechanism, Acknowledged frame delivery, and association and disassociation.

IEEE 802.15.4 devices operate in two main modes: non-beacon and beacon. In beacon mode,

Figure 5. IEEE802.15.4 PHY frame format

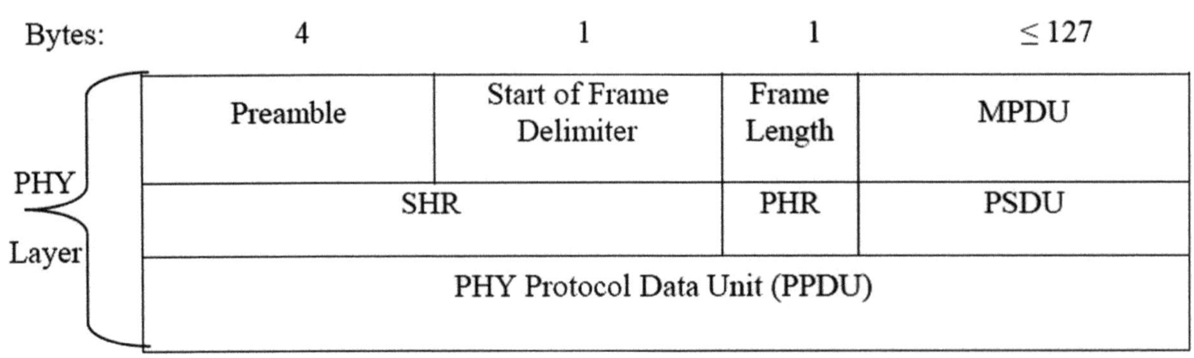

the network coordinator periodically sends out a beacon signal to devices within a network while controlling power consumption, with the network appearing like a cluster tree configuration. In contrast, the non-beacon mode is simpler, using mesh and peer-to-peer networks (Armholt et al., 2007), and operating like a two-way radio network (Callaway, et al., 2002; Buratti and Verdone, 2009). The IEEE 802.15.4 protocol uses (CSMA-CA) algorithm to manage and adjust the data transmission or MAC command frames; the algorithm is implemented via units of time named backoff periods (Lee and Wu, 2009), and uses two types of channel access mechanism, Slotted and Unslotted CSMA-CA, depending on the network configuration.

In Unslotted CSMA-CA, nonbeacon-enabled networks are employed, where devices use a binary exponential backoff mechanism to determine collisions, with the variable factor (BE) defining the number of slots during each backoff period. (BE) can have a value between (Min=3) and (Max=5) (Rohm & Goyal, 2009). Figure 6 shows the exponential backoff structure of the IEEE 802.15.4 system with BE initially set to 3. Thus, any device that wishes to transmit data frames must select a random time slot during the first ($2^{BE} - 1 = 7$) time slots, and then perform (CCA) during the first backoff period. If the channel is found to be idle, the device assumes the channel is free of carriers and transmits its data.

Otherwise, if the channel is found to be busy during the first backoff period, the device increments (BE) by 1, and selects a second random backoff period during the next ($2^{BE} - 1 = 15$) time slots. The CCA process is repeated, and if the second backoff period is also busy, the device changes the BE to 5 in order to select a third backoff period. If the third backoff period is free, then the device sends its data. Otherwise, the device waits for another random period before trying to access the channel again (IEEE, 2009).

In Slotted CSMA-CA, beacon-enabled networks are used, where the backoff slots are aligned with beacon transmission initiation. Thus, every time any device has data to send, it locates the boundary of the next backoff slot and then waits for a random number of backoff slots. If the channel is busy, it reschedules its packet for a later time, and then tries to access the channel again. If the channel is idle, the device can sends its data during the next available backoff slot boundary.

Data Transfer Models

The IEEE 802.15.4 protocol uses a data transfer transaction as a mechanism to transfer data between devices. Two types of data transfer transactions exist: direct and indirect data transmission.

For direct data transmission, the unslotted CSMA-CA mechanism is used to transfer data frames from a device directly to the coordinator. In a nonbeacon-enabled network, when a device has data to send to the coordinator, it simply sends its data frame to the coordinator using unslotted CSMA-CA. The coordinator then sends a MAC layer (ACK) back to the device, indicating the successful reception of the frame data, as shown in Figure 7 (a) (Jin-Shyan et al., 2007). On the other hand, in a beacon-enabled network, a device wishing to send data first listens for the network beacon, as shown in Figure 7 (b) (Jin-Shyan et al., 2007). When the beacon is found, the device transmits its data frame to the coordinator, using slotted CSMA-CA. The coordinator replies with a MAC layer ACK to the device that indicates the successful reception of the frame data.

For indirect data transmission, the unslotted CSMA-CA mechanism is used to transfer data frames from the coordinator to a device. In a nonbeacon-enabled network, when a coordinator has data to send to a device, it stores the data, waiting for the proper device to make contact and request the data. A device may make contact by

Figure 6. Exponential backoff in IEEE802.15.4

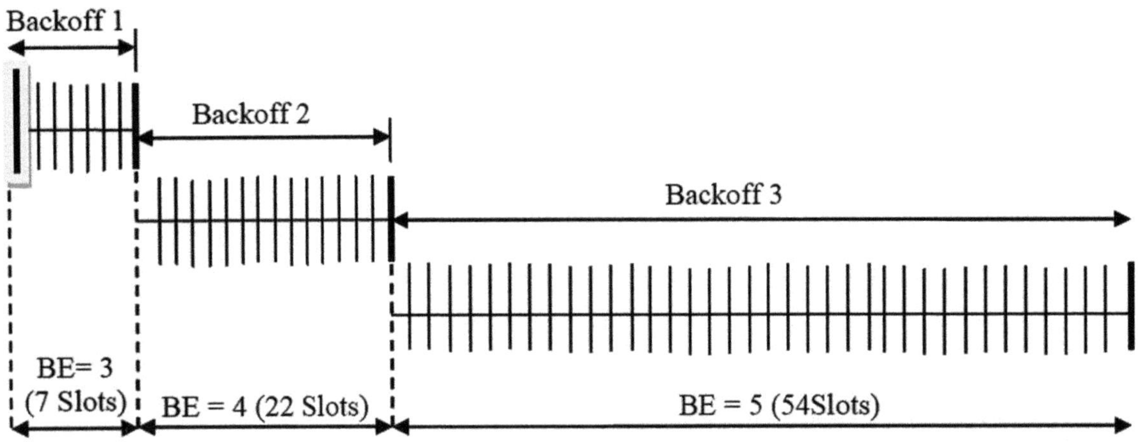

Figure 7. Direct data transmission in (a) non beacon-enabled and (b) beacon-enabled networks

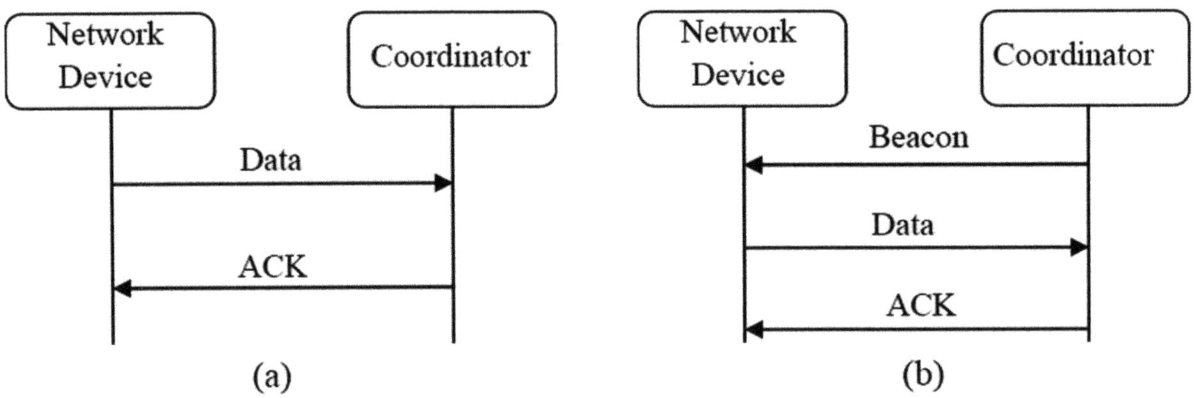

sending a MAC poll message to the coordinator, using unslotted CSMA-CA, as shown in Figure 8 (a) (Jin-Shyan et al., 2007). When the coordinator receives a poll request, it checks a packet queue to see if it has any buffered messages for the device; then it sends a MAC layer ACK back to the device indicating if it has data to send to the device or not. On the other hand, in a beacon-enabled network, if the coordinator has data to send to a device, it indicates in the network beacon that the data message is pending. The device periodically listens to the network beacon, and if a message is pending, sends a MAC command requesting the data, using slotted CSMA-CA, whereupon the coordinator sends a MAC layer ACK back to the device indicating the successful reception of the frame data.

The pending data frame is then sent using slotted CSMACA, and the device responds with a MAC layer ACK back to the coordinator, confirming the successful reception of the frame data. Upon reception of the ACK, the message is removed from the list of pending messages in the beacon. This sequence is summarized in Figure 8 (b) (Jin-Shyan et al., 2007).

Figure 8. Indirect data transmission in (a) non beacon-enabled and (b) beacon-enabled networks

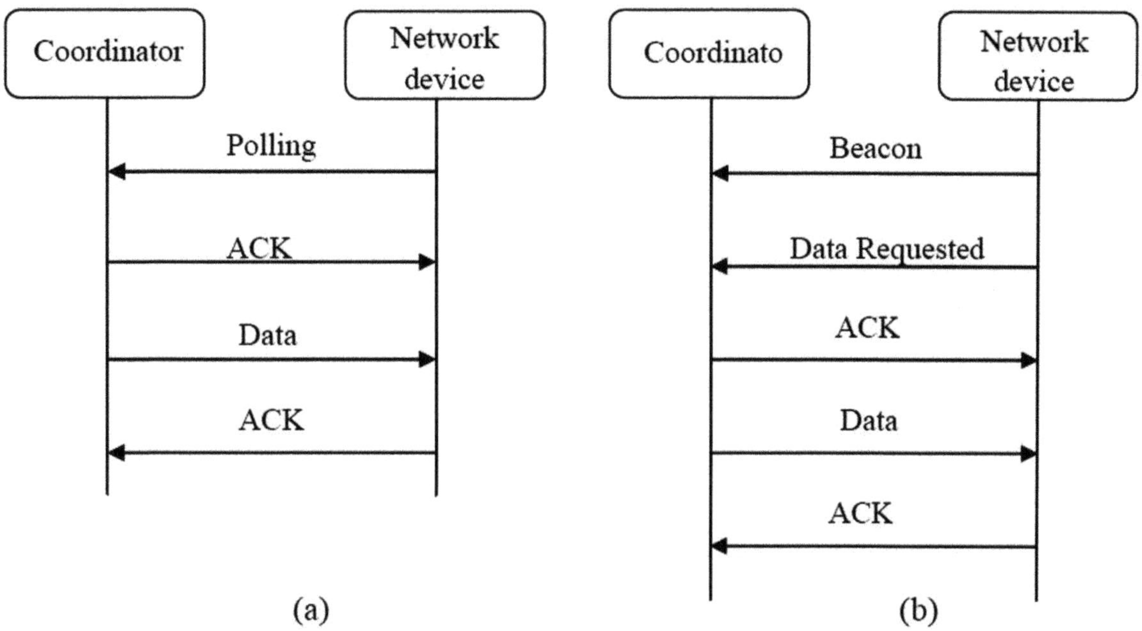

MAC Frame Formats

A device in 802.15.4 network can support both the 64-bit and the 16-bit PAN IDs address assigned during the association procedure; both PAN IDs are used to identify a network uniquely, and a single 802.15.4 network can accommodate up to 65535 devices (Callaway et al., 2002). The general MAC frame format is shown in Figure 9 (Jin-Shyan et al., 2007). Each MAC frame consists of the following main components:

1. **MAC Header (MHR):** Which is composed of three fields: Frame Control, Sequence Number and Addressing Field. The frame control field refers to the form of MAC frame being sent and specifies the format of the address field. Briefly, the frame control field specifies how the rest of the frame looks and what it contains. A data frame may contain multiple address types, including both source and destination information, with the size of the address field between 4 and 20 bytes.
2. **MAC Service Data Unit (MSDU) or Data Payload:** The data payload field, which contains information specific to the frame type, is of variable length and can be changed according to the size of the MAC frame.
3. **Frame Checks Sequence or MAC Footer (MFR):** Which contains the frame check sequence. The order of the fields of the data frame conforms to the order of the general MAC frame.

When data is transmitted from one device to another, a network-level ACK is transmitted back through the conventional route to the source node. This ACK packet indicates to the source device that the data packet was received by the destination device. If a network ACK is not received, the source device will re-transmit the data. Figure 10 (IEEE, 2009) shows the structure of the ACK frame, which is produced inside the MAC Sublayer. It consists of an MHR and an MFR, and has no MAC payload. Figure 11 (IEEE, 2009) gives a schematic view of the command frame.

Figure 9. General IEEE802.15.4 MAC frame format

Bytes:	2	1	4-20	0-102	2
MAC Sublayer	Frame Control	Data Sequence Number	Addressing	Data Payload	Frame check Sequence
	MAC Header (MHR)			MAC Service Data Unit (MSDU)	MAC Footer (MFR)
	MAC Protocol Data Unit (MPDU) ≤ 127 Byte				

Design of RFID Tag Based on Duty-Cycle Mechanism

Since the duty-cycle mechanism is the default operation of the IEEE802.15.4 protocol the RFID tag algorithm in this work is originally designed using this mechanism. The algorithm is as follows: before tags are registered to the reader system, during start-up and prior to integration with the environment system, all tag modules enable low-duty-cycle operation in a multi-hop network, such that sensor nodes are put into sleep mode and woken up periodically according to its schedule while waiting for an incoming broadcast frame from the RFID reader. If a tag needs to transmit, it must wait until the idle period is activated during the poll request time. If nothing is received, tags are maintained in idle mode for most of the time. Figure 12 illustrates the implementation of the RFID tag duty cycle scheme. In this system, the reader manages most of the data structures and maintains the data for all the tags in its range; therefore, the system can be regarded as Reader Talk First (RTF), which is implemented during a tag collection process executed at the beginning of the system activation.

The RFID reader needs to use an identical amount of time for accessing the RFID tag in its communication region. The delay or waiting time approach can be used when waiting for a response from the RFID tags. In principle, the delay provides more flexibility in accessing the RFID tags, at the cost of requiring more storage space bits in the tags, and in addition more space bits in the RFID commands (Vaidya & Das, 2008). The predefined RFID tag identified data report packet based on IEEE802.15.4 is broadcasted into the air by the antenna of the reader, to cover most of reading region. If the corresponding tag is present in the reading region, it will respond back to the reader by adding the data with its own Acknowledgment (ACK). The fundamental system functional blocks have been defined in the literature, and consist of three major parameters: Static, Quasi-static, and Dynamic parameters (Behera & Maity, 2008). The RFID tags must be fault-tolerant and respond to commands from a reader as soon as possible. The RFID reader's capability of distinguishing and receiving information from tags is the major

Figure 10. Schematic view of the ACK frame

Bytes:	2	1	2
MAC Sublayer	Frame Control	Data Sequence Number	Frame Check Sequence
	MHR		MFR

Figure 11. Schematic view of the command frame

Bytes:	2	1	4-20	1	variable	2
	Frame Control	Data Sequence Number	Addressing	Command Type	Command Payload	Frame Check Sequence
	MAC Header		MAC Payload			MAC Footer

challenge of correct spatial operation in a multipath environment. Essentially, with a plurality of RFID tags, the number of tags that struggle for the reader's attention increases, and this in turn can increase the time necessary to access all tags that are affixed to objects of interest. To meet the system requirements of this application and to address the realistic issue of reducing the data sent over the network, the software layer must control a transmission tag's data packet overhead. In this modeling, the total delay time of a packet moving through the RFID system based on IEEE802.15.4, shows that the following time delays, dependent on three major time parameters: Constant Time Delay, Semi - Constant Time Delay and Traveling Time Delay. The Constant Time Delay which includes processing time and transmission time. Processing time includes data (processing, transmitting and receiving) time in both the active tag and reader. While Transmission Time includes Backoff delay (T_{BOFF}) happens when CSMA-CA fails because collision occurs and CCA delay (T_{CCA}) happens when a CCA is performed. Semi- Constant Time Delay which includes (Interrupt function time and time between start counters and data processing). Traveling Time Delay Data includes the traveling time in the air from the reader to the active tag ($T_{Tag\text{-}Reader}$) and the reverse communication ($T_{Reader\text{-}Tag}$), in addition to the propagation delay and queuing delay. All the traveling time delays depend on the distance between the active tag and the reader. Constant time is always fixed and cannot be altered for the same type of Microcontroller (MCU). Propaga-

Figure 12. Implementation scheme of reader – tag communication in duty cycle mode

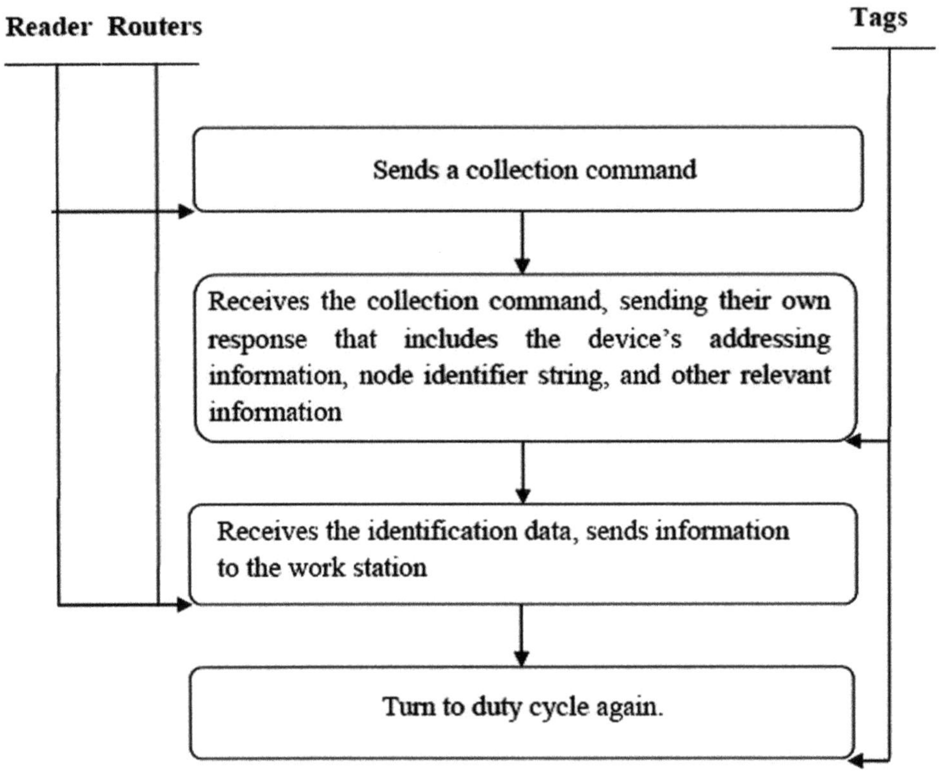

tion delay is below one microsecond for ranges under 300 meters (Maroti et al., 2004). Queuing delay becomes a dominant factor only in the case of heavy node traffic. Furthermore, in sensor networks, the propagation delay can be ignored because this time value is very low (Wei et al., 2002). Also, semi-constant time parameters are of secondary interest and are not taken into account in the RFID-based IEEE802.15.4 modeling time. Finally, all the constant parameters including the processing time delay and transmission delay are taken into account. Let the overall time (T) incurred by the reader to tag and the reverse communication process, when the tag is responding; thus,

$$T = T_{Proceeing} + T_{Boff} + T_{CCA} + 2\left(T_{Tag-Reader}\right) \qquad (1)$$

Control Tag Packet Overhead

The RFID tag architecture is composed of the MCU and RF transceiver. The target platform for this work simulations is the ZigBee (XBee Series2 module (Digi International, 2008) following the IEEE 802.15.4 duty cycle protocol, all the ZigBee end devices (ZEDs) (RFID tags) that receive the node-discover command broadcast from the reader must wait for a random time before sending their own data identification response. This response is sent in terms of a long message as a single packet that respectively includes: the source address (SA), serial number high (SH), serial number low (SL), network address (NA), node identifier string, and other relevant information, such as node type, node status, profile ID and manufacturer ID. Table 1 shows the tag discover response for an XBee Series2 module following the IEEE802.15.4 protocol. In this case, more data processing and transmitting time in the active tag is spent for processing and sending the packet containing all the tag's information than is necessary to identify only the tag identifier string. Thus, such a long packet in the duty cycle is suboptimal in terms of RFID tag energy consumption and introduces excess delay.

Since the tag typically has to wait the full period until all the tag data identification is finished, and at the reader because the reader must wait until the end of the extended message before deciding if the specific tag is the target or not. The "not taken part" prefix in the tag identification data may be because the information of the items to which the tags are attached has the same source address. Also, there is no need to take into account the common suffix part of the tag ID, since the tags have the same network address or manufacturer. When traffic load is low, the network contention is reduced by maximizing the time to send identification data reports. Hence, a small amount of queuing delay is caused by the periodic sleep for each of the IEEE 802.15.4 tags. Queuing delay becomes a dominant factor only in the case of heavy node traffic (Ali et al., 2007, Ali et al., 2008).

Design of RFID Tag Based on Wake up Mechanism

The RFID tag using the IEEE802.15.4 duty cycle mechanism, discussed in previous section, reduces idle listening but does not eliminate it. Moreover, active RFID mechanisms require that the tag to be synchronized to the RFID reader (or listen for the reader), and as a result the tag cannot go to sleep mode for long periods of time when inactive, which increases listening time and the end to end delay time of a packet. The RFID tag architecture is composed of the real-time clock alarm (RTC), MCU and RF transceiver. The architecture enables both the RF and MCU to go into sleep mode until there is a need to be actived. The real-time clock Alarm on the RTC generates an interrupt to wake up the MCU first from sleep mode, then the MCU wakes up the RF module to transmit data to the reader in a fixed interval, such as every second, minute or hour.

The RFID tag algorithm consists of two stages. At the first stage, before registration to the reader system, all tag modules are set to the low-power mode (hibernation mode), so tag nodes are put into deep sleep mode, where the tag sleeps all the time and only wakes up if triggered by an

Table 1. Node discover response of the XBee Series2 module

Index	Bytes Description	Number of Bytes
1	MY (Source Address)	2
2	SH (Serial Number High)	4
3	SL (Serial Number Low)	4
4	Node Identifier String	max 20 bytes
5	Parent Network Address	2
6	Node Type (0=Coordinator, 1=Router, 2=End Device)	1
7	Status (reserved)	1
8	Profile ID	2
9	Manufacturer ID	2
	Total when index 4 = 2 bytes	20

interrupt generated from the RTC alarm. At the second stage, the RTC alarm triggers the tag and periodically wakes it up from deep sleep mode so that it can send its data to the reader without listening for the reader command. The tag transmits an identification packet to the reader and then waits for an ACK signal from the reader. This process, known as Tag Talk First (TTF), is implemented during the interrupt step; hence, the tags have no need to be synchronized or to listen to the RFID reader.

In this mechanism, before a RFID tag is distributed to any area, a tag identifier string is implemented; to reduce the transmission overhead, only 2 random symbols of the RFID tag identifier string will be stored in the tag's memory (software layer) to distinguish between tags associated with their respective objects. In order to know the exact number of tags to be identified, the reader will discard the other irrelevant tag information when detecting the tag identifier string, by comparing it with the all the other tag identifier strings stored in its internal database. If a tag receives an ACK from the RFID reader portion, it has been successfully identified and does not need to send its identification again, even upon receiving another read cycle from the reader; hence the tag will return to the duty cycle or to the sleep mode. This method yields a significant drop in the number of bytes transmitted by the tag, and subsequently reduces tag and reader energy usage. In this modeling, the total delay time of a packet moving through the RFID system, shows that the following time delays: in addition to the data processing and transmitting time in active tag, the delay time in the active tag for deep sleep mode incurred when the tag is not active. Also, (data receiving, processing and transmitting) time in the reader. CCA delay (T_{CCA}) happens during the performance of a (CCA). Backoff delay (T_{BOFF}) happens when signal collision causes (CSMA-CA) to fail. The traveling time in the air from the reader to the active tag ($T_{Tag\text{-}Reader}$) and ($T_{Reader\text{-}Tag}$)

$$T = T_{SLEEP} + T_{Proceecing} + T_{BOFF} + T_{CCA} + 2\left(T_{Tag-Reader}\right) \tag{2}$$

METHODS OF ANALYSIS AND EVALUATION

This section presents the method of analysis performed and explains the calculations used to evaluate the various performance behaviors of the mechanisms that are developed; it also discusses the effects of changing parameters in conventional IEEE802.15.4 protocol (RTF) and in the RFID system (TTF) (designs and implementations in section 4 and 5 respectively), and predicts the behavior of sensor network applications using comparisons between them, including the effects of TTF and RTF raw and effective data transmission rates.

For all the calculations and the performance evaluation ahead, the focus will be on the non-beacon-enabled mode with unslotted IEEE802.15.4. In non-beacon enabled mode, no beacons are broadcast, so IEEE 802.15.4 is reduced to plain CSMA/CA. The unslotted IEEE 802.15.4 is the most suitable MAC choice; furthermore, durations are regularly expressed in numbers of symbols and not in seconds. The 2.4 GHz PHY layer duration of 1Byte = 2 symbols = 32 μs (IEEE, 2009).

Raw and Effective Data Rate for the TTF and RTF Systems

The IEEE 802.15.4 protocol supports a maximum over-the-air data rate of 250 kbps for the 2.4 GHz band. Any device that wishes to transmit data frames using IEEE 802.15.4 in non-beacon enabled networks and unslotted (CSMA-CA) must select a random time slot during the first ($2^{BE} - 1$). The associated protocols for both RTF and TTF systems are described in section 3 and 4 respectively.

In this section, to find the raw and effective data rate for both of the modified IEEE 802.15.4 and proposed RFID system, firstly suitable values should be found for the payload size and addressing field parameters. Secondly, the data frame transfer time (ms) should be determined through calculation; lastly, the effective data rate (Kbps) can be determined for the non-beacon enabled network. Initially, the back-off exponent (BE) is set to maximum (Refer to Figure 6), which means that the tag performs CCA three times before finding a free slot. Assuming the channel is found to be free, the worst-case channel access time can be calculated as: (Jennic, 2006)

$$CSMA\ /\ CA\left(dataframe\right) = T_{Boff} + T_{CCA} \tag{3}$$

where (T_{CCA} = 8 symbol times = 128 μs) (Ameer, P. M., et al). So, the CSMA/CA (data frame) ideally takes 17.408 ms. Since the XBee Series2 modules communicate using the IEEE 802.15.4 standard for the PHY layer, the maximum number of RF payload bytes that can be sent is 84 bytes (Digi International, 2008).

According to the algorithm used for the RTF system, the data pay load size (MAC Protocol Data Unit (MSDU) length sent from the RFID tag is 20 bytes, which is illustrated in Table 1. However, according to the algorithm used for the TTF system, the data payload size sent from the RFID tag is only 1 byte (2 symbols).

Also it is assumed that the module uses short addressing (16-bit source and destination). Since the data transfer rate of the radio is 250 Kbps, the time for sending one byte over the radio will be 8bit/ 250 Kbps = 32 μs. Figure 13 shows a schematic view of the XBee Series2 data frame.

From Figure 13, the (MAX MPDU) is defined as (Jennic, 2006 ; Lee, 2006)

$$\text{Frame Control+Sequence Number+}\\ \text{Address Field+Data Payload+FCS} \tag{4}$$

Therefore, (T_{TX1}) = (MAX MPDU) $_{TTF}$ will take 14 bytes including 1 byte as a data payload size, while (T_{TTX}) = (MAX MPDU) $_{RTF}$ will take 33 bytes including 20 bytes as a data payload size.

Data frame transfer time or transmission rate (ms) is defined as (Jennic, 2006)

$$\frac{\text{MAX PPDU}}{\text{Data Rate}} = \frac{\text{MPDU + SHR + PHR}}{\text{Data Rate}} \tag{5}$$

So, the (Data frame transfer time)$_{TTF}$ ideally takes 0.64ms, while, (Data frame transfer time)$_{RTF}$ ideally takes 1.248 ms.

The transmission of an ACK frame does not use (CSMA/CA) in a non-beacon enabled network (as shown in Figure 10), and is equal to 11 bytes only. So, ACK frame time over the radio ideally takes 11 bit / 250 Kbps = 0.352 ms. Furthermore, a (radio Turnaround Time) symbol commences after the reception of the data frame, which equals to 0.192 ms (Jennic, 2006), and this gives the tag enough time to switch between transmit and receive, or vice versa. Table 2 shows the parameters that have been employed to calculate the effective data rate during transmission between tag and the reader (TTF) (referring to the direct transmission discussed in section 3.3).

Raw data rate (Kbps) is defined as (Lee, 2006)

[illegible] (6)

Assume only one packet is forwarded through the network at a time. So, the (raw data rate) $_{TTF}$ ideally takes only 8.60 Kbps. On the other hand, for the RTF system (the indirect transmission discussed in section 3.3) the data rate is significantly reduced because of the network (AT ZED) polling rate. In this calculation, the poll request message time is 100 ms, which applies to when the parent (ZC, ZRs) receives a poll request from the ZED (RFID tag) to check if it has any buffered messages for the end device (Digi International,

Figure 13. Schematic view of the XBee Series2 data frame

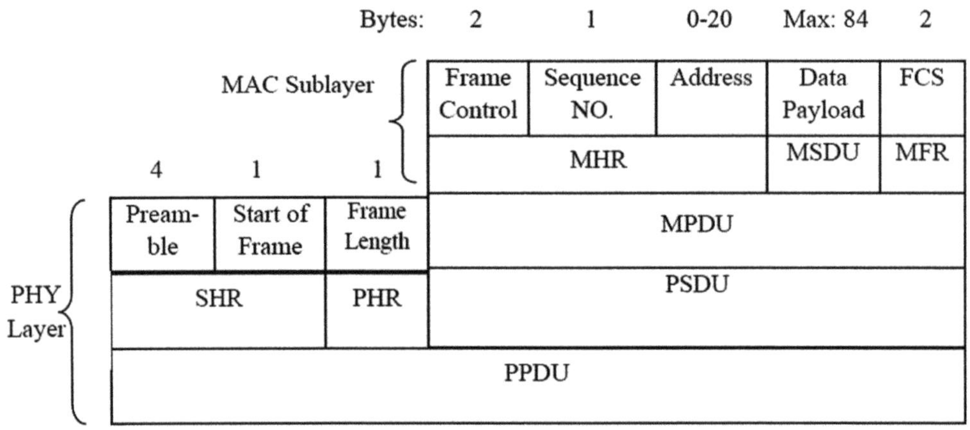

Table 2. TTF parameters for effective data rate

Parameters	Timing (ms)
CSMA/CA (data frame)	17.408
(Data frame transfer time) $_{TTF}$	0.640
Turnaround time	0.192
ACK transmission	0.352
Total	18.592

2008). This time is also called check time or check interval. Table 3 shows the parameters that have been used to calculate the raw and effective data rate for the transmission between the reader and the tag (RTF).

So, the (Raw data rate) $_{RTF}$ ideally takes only 2.617 Kbps. Based on a theoretical gross data rate of 250 kbps, the utilization rate for TTF is 3.44%, while the utilization rate for RTF is only 1.0468%. From the calculations, we can conclude that the data rate would be greatly affected by adding the pending messages to the performance of the RTF data transmission, because of the network device's polling rate (sending data requests periodically). Therefore, the TTF system shows higher throughput, and the RTF system has a lower effective data rate than the TTF system.

In practice, the raw data rate is somewhat lower because of the protocol built into the frame structure. The handling of the data rate protocol depends on the system architecture, the processor in use, and a variety of parameters that are engaged to ensure reliable data transmission; some of these include channel access time (T_{CCA}), maximum data payload, packet transmission time, and frame ACK time. Effective data rate (Kbps) is defined as (Jennic, 2006 ; Lee, 2006)

$$\frac{\text{Max Data Pay Load x The Number of Transmitted Packets}}{\text{Total Parameters Timing}} \qquad (7)$$

So, the (Effective data rate) $_{TTF}$ ideally takes 0.430 Kbps, while, (Effective data rate) $_{RTF}$ ideally takes only 1.342 Kbps respectively. The (Effective data rate) $_{TTF}$ is about 68% lower than for the (Effective data rate) $_{RTF}$.

Observations on the IEEE 802.15.4 Systems

Based on the study, even though a low duty cycle MAC protocol is energy efficient, it has many negative side-effects:

1. The RFID tag data rate is very low during idle mode because no data is transmitting or receiving. Therefore, it is not necessary

Table 3. RTF parameters for effective data rate

Parameters	Timing (ms)
CSMA/CA (data frame)	17.408
(Data frame transfer time) $_{RTF}$	1.248
Turnaround time	0.192
ACK transmission	0.352
Poll request time	100.000
Total	119.200

to keep tags listening all the time since this result in significant energy wastage.

2. The RFID tag responds to the reader command by transmitting a long message as a single packet, containing the device's address information, tag identifier string, and other relevant information. If only a few bits have been corrupted in the first transmission, such as in the prefix of the addressing information which comes before the tag identifier string, the long packet must be re-transmitted, and a big disadvantage results from the high cost in terms of energy and the end to end delay time of a packet.

CONCLUSION

This research explores the feasibility of a RFID-based WSN as an identification and monitoring system; the problem in question is how to create an RFID network using WSN principles and components, with the intention of optimizing the network to a defined data reporting rate. The RFID system performances in both duty-cycle and wake-up modes were compared and the theoretical approaches have been explained here to provide a conceptual framework for the research. The choice and implementation of protocols, and the analysis used to compare them, are significant contributions of the present work. It is observed that the ZigBee module standard in compliance with the IEEE802.15.4 is successfully employed and enabled inherent configuration, suitable for item tracking and human/object tracking. Raw and effective data rate are calculated according to XBee Series2 data frame. It is observed that the (Effective data rate)$_{TTF}$ is about 68% lower than for the (Effective data rate)$_{RTF}$.

ACKNOWLEDGMENT

The authors would like to thank the Malaysia Ministry of Higher Education (FRGS Fund - 6071306) for sponsoring the development of the in house built in system of RFID devices.

REFERENCES

Abdulla, R., & Ismail, W. (2013). Survey of WSN technology based reliable and efficient active RFID. Paper presented at the 2013 IEEE Malaysia International Conference on Communications (MICC) In IEEE, pp. 116-121. doi:10.1109/MICC.2013.6805810

Ali, A., Rashid, R. A., Arriffian, S. H. F., & Fisal, N. (2007). Optimal forwarding probability for real-time routing in wireless sensor network. Paper presented at the *2007 IEEE International Conference on Telecommunications and Malaysia International Conference on Communications, 2007. ICT-MICC* In IEEE, pp. 419-424. doi:10.1109/ICTMICC.2007.4448673

Ali, A., Rashid, R. A., & Fisal, N. (2008). *Cognitive Forwarding for Real-time Routing in Wireless Sensor Network.*

Ameer, P. M., Kumar, A., Manjunath, D., & Boyina, R. (2006, November). *Analysis of network architectures for Zigbee sensor clusters*. Paper presented at the *12th International Telecommunications Network Strategy and Planning Symposium, 2006. Networks, 2006*, 1–6.

Ansari, J., Pankin, D., & Mahonen, P. (2009). Radio-triggered wake-ups with addressing capabilities for extremely low power sensor network applications. *International Journal of Wireless Information Networks*, *16*(3), 118–130. doi:10.1007/s10776-009-0100-6

Armholt, M., Junnila, S., & Defee, I. (2007). *A Non-beaconing ZigBee Network Implementation and Performance Study*. Paper presented at the 4th *IEEE International Conference on Communications (ICC '07)*, Glasgow, Scotland, pp. 3232-3236. doi:10.1109/ICC.2007.536

Baronti, P., Pillai, P., Chook, V. W. C., Chessa, S., Gotta, A., & Hu, Y. F. (2007). Wireless sensor networks: A survey on the state of the art and the 802.15. 4 and ZigBee standards. *Computer Communications*, *30*(7), 1655–1695. doi:10.1016/j.comcom.2006.12.020

Behera, S., & Maity, C. (2008). *Active RFID tag in Real Time Location System*. Paper presented at the 5th International Multi-Conference on Systems, Signals and DevicesAmman (IEEE SSD 2008), pp. 1-7. doi:10.1109/SSD.2008.4632779

Buratti, C., & Verdone, R. (2009). Performance Analysis of IEEE 802.15.4 Non Beacon-Enabled Mode. *IEEE Transactions on Vehicular Technology*, *58*(7), 3480–3493. doi:10.1109/TVT.2009.2014956

Callaway, E., Gorday, P., Hester, L., Gutierrez, J. A., Naeve, M., Heile, B., & Bahl, V. (2002). Home networking with IEEE 802.15.4: A developing standard for low-rate wireless personal area networks. *IEEE Communications Magazine*, *40*(8), 70–77. doi:10.1109/MCOM.2002.1024418

Castagnetti, A., Pegatoquet, A., Le, T. N., & Auguin, M. (2014). *A Joint Duty-Cycle and Transmission Power Management for Energy Harvesting WSN* (pp. 928–936). IEEE Transactions on Industrial Informatics.

Chen, C. (2010). Design of a Child Localization System on RFID and Wireless Sensor Networks. *Journal of Sensors*, *2010*, 1–8. doi:10.1155/2010/450392

Cho, H., Choi, H., Lee, W., Jung, Y., & Baek, Y. (2006). LITeTag: Design and implementation of an RFID system for IT-based port logistics. *Journal of Communication*, *1*(4), 48–57.

Demirkol, I., Ersoy, C., & Alagoz, F. (2006). MAC protocols for wireless sensor networks: A survey. *IEEE Communications Magazine*, *44*(4), 115–121. doi:10.1109/MCOM.2006.1632658

Digi International. (2008). XBee/XBee-PRO - ZB OEM RF Modules datasheet. Retrrieved March 2 2014, from http://ftp1.digi.com/support/documentation/90000976_a.pdf

Dong-Liang, W., Ng, W. W. Y., Yeung, D. S. & Hai-Lan, D. (2009) In *International Conference on Machine Learning and Cybernetics.*, Vol. 4 Baoding, pp. 2330-2335.

Fagui, L., Yang, Z., & Wenhui, L. (2009) In. *Standard System Framework of RFID Application in Logistics*. Paper presented at the *Third International Symposium on Intelligent Information Technology Application(IITA 2009)*, Vol. 3 Nanchang, pp. 44-47.

Gislason, D. (2008). *Zigbee wireless networking*. Newnes.

Guillemette, M. G., Fontaine, I., & Caron, C. (2008). *Hybrid RFID-GPS Real-Time Location System for Human Resources: Development, Impacts and Perspectives*. Paper presented at the *Proceedings of the 41st Annual Hawaii International Conference on System Sciences*Waikoloa, HI, pp. 406-406. doi:10.1109/HICSS.2008.195

Guo, P., Jiang, T., Zhang, Q., & Zhang, K. (2012). Sleep scheduling for critical event monitoring in wireless sensor networks. *IEEE Transactions on Parallel and Distributed Systems*, *23*(2), 345–352. doi:10.1109/TPDS.2011.165

Howitt, I., & Gutierrez, J. A. (2003). IEEE 802.15.4 low rate: Wireless personal area network coexistence issues. In 2003 IEEE Wireless Communications and Networking (Vol. 3, pp. 1481-1486). New Orleans, LA, USA.

IEEE (2009). IEEE Standard for Information technology-Telecommunications and information exchange between systems-Local and metropolitan area networks-Specific requirements Part 15.4: Wireless Medium Access Control (MAC) and Physical Layer (PHY) Specifications for Low-Rate Wireless Personal Area Networks (WPANs) Amendment 2: Alternative Physical Layer Extension to support one or more of the Chinese 314-316 MHz, 430-434 MHz, and 779-787 MHz bands. *IEEE Std 802.15.4c-2009 (Amendment to IEEE Std 802.15.4-2006)*, c1-21.

Jae-Han, L., & Byung Tae, J. (2008) In. *Dynamic Duty Cycle Adaptation to Real-Time Data in IEEE 802.15.4 Based WSN*. Paper presented at the 5th IEEE Consumer Communications and Networking ConferenceLas Vegas (CCNC 2008), NV, pp. 353-357.

Jennic (2006). Application Note: JN-AN-1035, Calculating 802.15.4 Data Rates. Retrieved March 2 2014, from: http://www.jennic.com/files/support_files/JN-AN-1035%20Calculating%20802-15-4%20Data%20Rates-1v0.pdf

Jin, S., Yue, W., & Sun, Q. (2012). Performance analysis of the sleep/wakeup protocol in a wireless sensor network. *International Journal of Innovative Computing, Information, & Control*, *8*(5), 3833–3844.

Jin-Shyan, L., Yu-Wei, S., & Chung-Chou, S. (2007). A Comparative Study of Wireless Protocols: Bluetooth, UWB, ZigBee, and Wi-Fi. Paper presented at the *33rd Annual Conference of the IEEE Industrial Electronics Society (IECON 2007)*. Taipei, Taiwan.

Jurdak, R., Ruzzelli, A. G., & O'Hare, G. (2008a). *Adaptive Radio Modes in Sensor Networks: How Deep to Sleep?* Paper presented at the *Annual IEEE Communications Society Conference on Sensor, Mesh and Ad Hoc Communications and Networks*San Francisco *(SECON '08) 5th*, CA, pp. 386-394. doi:10.1109/SAHCN.2008.54

Jurdak, R., Ruzzelli, A. G., & O'Hare, G. M. P. (2008, August 3-7). Multi-Hop RFID Wake-Up Radio: Design, Evaluation and Energy Tradeoffs. In *Proceedings of 17th International Conference on Computer Communications and Networks, 2008 (ICCCN '08)* (pp. 1-8). doi:10.1109/ICCCN.2008.ECP.124

Ke, R., Peng, G., & Wang, X. (2008). *Further Research of RFID Applying on Exhibition Logistics*. Paper presented at the *International Conference on Cyberworlds*Hangzhou (pp. 631-634).

Kim, H., & Ayurzana, O. (2009). Improvement of data receive ratio in remote water meter system by upgrading sensor. *International Journal of Control. Automation and Systems*, *7*(1), 145–150. doi:10.1007/s12555-009-0118-7

Kim, H., Chung, J. M., & Kim, C. H. (2009). Secured communication protocol for internetworking ZigBee cluster networks. *Computer Communications*, *32*(13-14), 1531–1540. doi:10.1016/j.comcom.2009.05.014

Kinney, P. (2003). *Wireless Control that Simply Works* In *Communications design*. Paper presented at the *Communications design conference*, Vol. 2. Kinney Consulting LLC.

Kumar, P., Gu, X., Nes, M., Mushtaq, Q., & Schiller, J. (2010). *Optimizing Duty-Cycle for Delay and Energy Bound WSN Applications*. Paper presented at the *IEEE 24th International Conference on Advanced Information Networking and Applications Workshops Perth*, (WAINA) (pp. 692-697).

Le, H. C., Guyennet, H., Felea, V., & Zerhouni, N. (2007). *A low latency MAC scheme for event-driven wireless sensor networks* (pp. 291–301). Mobile Ad-Hoc and Sensor Networks.

Lee, B. H., & Wu, H. K. (2009). Study on a delayed backoff algorithm for IEEE 802.15.4 low-rate wireless personal area networks. *IET Communications*, *3*(7), 1089–1096. doi:10.1049/iet-com.2007.0517

Lei, Z., & Zhi, W. (2006). *Integration of RFID into Wireless Sensor Networks: Architectures, Opportunities and Challenging Problems*. Paper presented at the Fifth International Conference on Grid and Cooperative Computing WorkshopsHunan (GCCW '06), pp. 463-469.

Lin, Y. B., Chang, K. C., Chen, J. C., Wang, P. H., & Li, L. S. (2006). *ZigBee Sensor Network for Structural Health Monitoring*. Paper presented at the 4th International Conference on Earthquake Engineering.Taipei, Taiwan

Lopez, M., Sabater, J., Daemitabalvandani, M., Sabater, J., Gómez, J. M., Carmona, M., & Herms, A. (2013). Software management of power consumption in WSN based on duty cycle algorithms. Paper presented at the EUROCON, 2013 IEEE, and pp. 399-406. doi:10.1109/EUROCON.2013.6625014

Mirshahi, S., Uysal, S., & Akbari, A. (2013). Integration of RFID and WSN for supply chain intelligence system. Paper presented at the 2013 International Conference on Electronics, Computers and Artificial Intelligence (ECAI), pp.1, 6, 27-29. doi:10.1109/ECAI.2013.6636160

Mishra, N., Chebrolu, K., Raman, B., & Pathak, A. (2006). *Wake-on-WLAN*. Paper presented at the International World Wide Web Conference Committee Edinburgh (IW3C2), Scotland (pp. 761-769).

Ni, L. M., Liu, Y., Lau, Y. C., & Patil, A. P. (2004). LANDMARC: Indoor Location Sensing Using Active RFID. *Wireless Networks*, 701–710.

Ondrej, S., Zdenek, B., Petr, F., & Ondrej, H. (2006). *ZigBee Technology and Device Design*. Paper presented at the International Conference on Networking, International Conference on Systems and International Conference on Mobile Communications and Learning Technologies *(ICN/ICONS/MCL 2006)* (pp. 129-129).

Pala, Z., & Inanc, N. (2007). *Smart Parking Applications Using RFID Technology*. Paper presented at the *1st Annual RFID Eurasia* Istanbul (pp. 1-3).

Polastre, J., Hill, J., & Culler, D. (2004). *Versatile low power media access for wireless sensor networks*. Paper presented at the Conference on Embedded Networked Sensor Systems (SenSys; ACM) (pp. 3-5). Baltimore, Maryland, USA.

Rajesh, S. M. (2013). *Integration of Active RFID and WSN for real time low-cost data monitoring of patients in hospitals*. Paper presented at the 2013 International Conference on Control, Automation, Robotics and Embedded Systems (CARE) (pp. 16-18). doi:10.1109/CARE.2013.6733768

Ram, M., & Kumar, S. (2014). *Analytical energy consumption model for MAC protocols in wireless sensor networks*. Paper presented at the In 2014 International Conference on IEEE Signal Processing and Integrated Networks (SPIN) (pp. 444-447). doi:10.1109/SPIN.2014.6776994

Rashid, R. A., Sarijari, M. A. B., Rahim, M. R. A., & Skudai, J. (2008). *Wireless Home Security and Automation System Utilizing ZigBee based Multi-hop Communication*. Paper presented at the National Conference on Telecommunication Technology (NCTT). Putrajaya, Malaysia.

Rohm, D., Goyal, M., Xie, W., Polepalli, B., Hosseini, H., Divjak, A., & Bashir, Y. (2009). *Dynamic Reconfiguration in Beaconless IEEE 802.15.4 Networks Under Varying Traffic Loads*. Paper presented at the Global Telecommunications Conference, (GLOBECOM 2009). IEEE, Honolulu, Hawaii.

Sahinoglu, Z., Gezici, S., & Guvenc, I. (2008). *Ultra-wideband positioning systems: Theoretical limits, ranging algorithms, and protocols*. Cambridge university press. doi:10.1017/CBO9780511541056

Salvatore, C., Bocchino, S., Petracca, M., Pelliccia, R., Ghibaudi, M., & Pagano, P. (2012). WSN and RFID integrated solution for advanced safety systems in industrial plants. Paper presented at the *20th International Conference on Telecommunications and Computer Networks (SoftCOM)* (pp. 11-13).

Shah, K., & Kumar, M. (2007). *Distributed Independent Reinforcement Learning (DIRL) Approach to Resource Management in Wireless Sensor Networks*. Paper presented at the IEEE Internatonal Conference on Mobile Adhoc and Sensor SystemsPisa (MASS 2007), Italy (pp. 1-9). doi:10.1109/MOBHOC.2007.4428658

Siadat, S. H., & Selamat, A. (2008). *Location-based system for mobile devices using RFID*. Paper presented at the Second Asia International Conference on Modelling & SimulationKuala Lumpur, Malaysia (pp. 291-296). doi:10.1109/AMS.2008.44

Vaidya, N., & Das, S. R. (2008). Rfid-based networks: Exploiting diversity and redundancy. *Mobile Computing and Communications Review*, *12*(1), 2–14. doi:10.1145/1374512.1374514

Van Dam, T., & Langendoen, K. (2003). An adaptive energy efficient MAC protocol for wireless sensor networks. In *Proceedings of the 1st international conference on Embedded networked sensor systems*, Los Angeles, CA, USA doi:10.1145/958491.958512

Wang, Y., Zhao, X., Wu, Y., & Xu, P. (2008) In. *The research of RFID middleware's data management model*. Paper presented at the IEEE International Conference on Automation and Logistics Qingdao (ICAL 2008) (pp. 2565-2568).

Wei, Y., Heidemann, J., & Estrin, D. (2002) In *INFOCOM 2002. An energy-efficient MAC protocol for wireless sensor networks.* Paper presented at the *Twenty-First Annual Joint Conference of the IEEE Computer and Communications Societies*, (Vol. 3, pp. 1567-1576) New York, USA.

Wei, Y., Heidemann, J., & Estrin, D. (2004). Medium access control with coordinated adaptive sleeping for wireless sensor networks. *IEEE/ACM Transactions on Networking*, *12*(3), 493–506. doi:10.1109/TNET.2004.828953

Wheeler, A. (2007). Commercial Applications of Wireless Sensor Networks Using ZigBee. *IEEE Communications Magazine*, *45*(4), 70–77. doi:10.1109/MCOM.2007.343615

Wotton, P. (2006). Providing reliable sensing and control using ZigBee wireless networks. *RF DESIGN magazine, 29*(7), 18.

Xiong, Z., Sottile, F., Spirito, M. A., & Garello, R. (2011). Hybrid indoor positioning approaches based on WSN and RFID. Paper presented at the *2011 4th IFIP International Conference on* IEEE in *New Technologies, Mobility and Security (NTMS)* (pp. 1-5).

Zhoul, H., Chen, X., Liu, X. & Yang, J. (2009). Applications of Zigbee wireless technology to measurement system in grain storage. *Computer and Computing Technologies in Agriculture II,* 3, 2021-2029.

Zigbee Alliance. (2006). *Zigbee-2006 specification.* Retrieved March 2 2014, from http://www.zigbee.org/

KEY TERMS AND DEFINITIONS

IEEE802.15.4 Standard: Is a standard which specifies the physical layer and media access control for low-rate wireless personal area networks.

Line-of-Sight Propagation: Is a characteristic of electromagnetic radiation or acoustic wave propagation.

Medium Access Control Layer: In the seven-layer OSI model of computer networking, media access control (MAC) data communication protocol is a sublayer of the data link layer (layer 2).

Mesh Network: Is a network topology in which each node relays data for the network. All nodes cooperate in the distribution of data in the network.

Path Loss: Is the reduction in power density (attenuation) of an electromagnetic wave as it propagates through space.

Peer-To-Peer Networks: Is distributed application architecture that partitions tasks or work loads between peers.

Received Signal Strength Indicator: Is a measurement of the power present in a received radio signal.

Chapter 5
Adaptive IEEE 802.15.4 MAC Protocol for Wireless Sensor Networks

Yousef S. Kavian
Shahid Chamran University of Ahvaz, Iran

Hadi Rasouli
Shahid Chamran University of Ahvaz, Iran

ABSTRACT

The energy efficiency is a main challenging issue for employing wireless sensor networks (WSNs) in extreme environments where the media access progress consumes the main part of network energy. The IEEE 802.15.4 is adopted in low complexity, ultra-low power and low data rate wireless sensor applications where the energy consumption of nodes should be managed carefully in harsh and inaccessible environments. The beacon-enabled mode of the IEEE 802.15.4 provides a power management scheme. When the network traffic is variable, this mode does not work as well and the coordinator is not capable for estimating the network traffic and adjusting proper duty cycle dynamically. In this chapter an approach for estimating network traffic in star topology is proposed to overcome this issue where the coordinator could estimate the network traffic and dynamically adjusts duty cycle proportion to the variation of network traffic. The simulation results demonstrate the superiority of proposed approach for improving the energy consumption, throughput and delay in comparison with the IEEE 802.15.4 under different traffic conditions.

INTRODUCTION

As wireless sensor networks is deployed in harsh and inaccessible environments for collecting data and monitoring applications, connecting the sensor nodes in simple and efficient manner is an important and challenging issue (Jung, et al., 2009). The IEEE 802.15.4 standard for low-rate wireless personal area networks (LR-WPANs) has been introduced to achieve these requirements (IEEE Standard 802.15.4, Part 15.4, 2006). In many applications using this standard, devices will be battery powered (IEEE Standard 802.15.4, Part 15.4, 2006). Therefore to extend the network

DOI: 10.4018/978-1-4666-8251-1.ch005

lifetime, energy consumption for devices should be managed carefully. Beacon-enabled mode of IEEE 802.15.4 provides a power management mechanism based on duty-cycle. The MAC sub-layer of this mode provides various duty cycles by setting two system parameters, beacon order (BO) and superframe order (SO) (Chung, et al., July 2006; Sheu, et al., 2005). The nodes communicate with coordinator to send their packets in active period while during inactive period they enter in low power state to save energy. The standard IEEE 802.15.4 allocates further portion of active period in beacon-enabled mode to the contention access period (CAP) and if the contention free period length be zero, the CAP will be completed at the end of the active portion of the superframe. So the CAP has very important role in the standard performance.

Generally, contention based protocols like S-MAC (Ye, et al., 2004), B-MAC (Polastre, et al., 2004) and T-MAC (Dam, et al., 2003) have advantages such as scalability, flexibility and self-organization but in other hand they have disadvantages such as idle listening and collision which increase the devices consumption energy (Gilani, et al., 2011).

The throughput and energy consumption are major performance metrics for 802.15 (Park, et al., 2005). In case that the CAP length is longer than the time interval which is required for network traffic, the devices will spend more time in idle mode which increases the energy consumption. In the other side, if CAP length is shorter than the time interval required for network traffic, lots of packets will be lost because of collision and as a result the network throughput will be reduced.

The devices number or packet generation varies with the time in some networks. Also, in some networks, traffic is not predictable. In such conditions to optimize nodes energy consumption and achieve high throughput, the CAP length must be selected proportional to the network traffic. Therefore the coordinator should be able to estimate network traffic, so that the network traffic varies, it can dynamically adjust CAP length proportional to the traffic network variation.

In this study, in order to enhance coordinator capability of estimating the network traffic, the devices inform the coordinator about the number of packets inside its queue along with sending their packets. The coordinator also evaluates network traffic using devices queue state and channel idle time intervals at the end of active period of each superframe then selects CAP length for the next superframe based on network throughput.

RELATED WORKS

In order to evaluate the CAP, according to significance of CAP in this standard, recent literatures provide a lot of analytical models (Huang, et al., 2009; Ashrafuzzaman & Kwak, et al 2011; Ramachandran, et al., 2007; Tao, et al., 2006). The most studies conducted on CAP concentrate on slotted Carrier Sense Multiple Access with Collision Avoidance (CSMA/CA) algorithm, So that many accurate models have been provided which can evaluate the CSMA/CA algorithm (Pollin, et al., 2008; He, et al., 2008). Chung *et al.* (Chung, et al., July 2006) presented a new model for the slotted CSMA/CA, and evaluated its throughput limit.

Some other works concentrated on energy consumption and throughput. In order to minimize the energy consumption, a new model to determine the optimal BO and SO values has been provided regarding to the number of successful packet transitions and maximum delay constrain (Shu, et al., 2006). Suh *et al.* (Shu, et al., 2008) presented a new model in which the nodes that can't send their packet in active period can send packets in time intervals of inactive period called sentinel duration. Xiao *et al.* (Xiao, et al., 2010), proposed a new analytical model for IEEE 802.15.4 MAC with sleep mechanism to compute the energy consumption and throughput in single-hop networks in real-time applications. It has been shown that using of one CCA mechanism instead of two CCA mechanisms in CSMA/CA cause decreasing probity of packet discarding and consequently increases the network throughput (Patro & Kumar, et al., 2007). Neugebauer *et al.* (Neugebauer, et al., 2005) proposed an algorithm for BO adaption in IEEE 802.15.4 star-topology with beacon enabled mode that allows power saving with a trade-off according to delay.

THE IEEE802.15.4 MAC PROTOCOL: AN OVERVIEW

In this section, the IEEE802.15.4 standard is briefly reviewed. More investigation about the standard can be found in (IEEE Standard 802.15.4, Part 15.4, 2006) and (Zheng & Lee, et al., 2004). The IEEE 802.15.4 has been adopted in many low complexities, ultra-low power consumption, and low data rate wireless networks such as wireless sensor networks (IEEE Standard 802.15.4, Part 15.4, 2006) in recent years. This standard supports two star and peer-to-peer topologies. In the star topology a central node, named pan coordinator, plays the role of a pan controller and other devices can only communicate with that. In peer-to-peer topology nodes can communicate with each other as long as they are in radio range of each other.

The IEEE 802.15.4 standard defines the PHY and MAC sub-layer specifications for low-rate WPANs (LR-WPANs). The features of the MAC sub-layer are beacon management, channel access, guaranteed time slot (GTS) management, frame validation and acknowledged frame delivery.

The MAC sub-layer specifies two medium-access operation modes: the non-beacon-enabled mode and beacon-enabled mode. In non-beacon-enabled mode, each node contends with other nodes using the non-slotted CSMA/CA algorithm to access medium. In beacon-enabled mode, the network operates with a super frame structure, which its format is defined by the coordinator. The coordinator transmits special formatted frame called "beacon frame" periodically, to form superframe structure. The time interval between two beacon frames is called the beacon interval (BI). The superframe has two active and optional inactive periods. Because of saving the energy, the devices go to sleep mode during inactive period. The length of the active portion of the superframe, called superframe duration (SD), is divided into 16 equal slots including three parts: beacon period, contention access period (CAP) and contention-free period (CFP) respectively, shown in Figure 1. The coordinator transmits beacon at the beginning of active period in the first slot of each superframe. The CAP starts immediately following the beacon and continues until the beginning of the CFP on a superframe slot boundary. According to the slotted CSMA/CA algorithm, each node contends with others to medium-access in the CAP duration.

If the CFP length is equal to zero, the CAP will be completed at the end of the active period. The PAN coordinator allocates portions of active superframe for the applications which need low-latency and specific data bandwidth. These portions are named guaranteed time slots (GTS).

Figure 1. The structure of superframe in beacon enabled mode

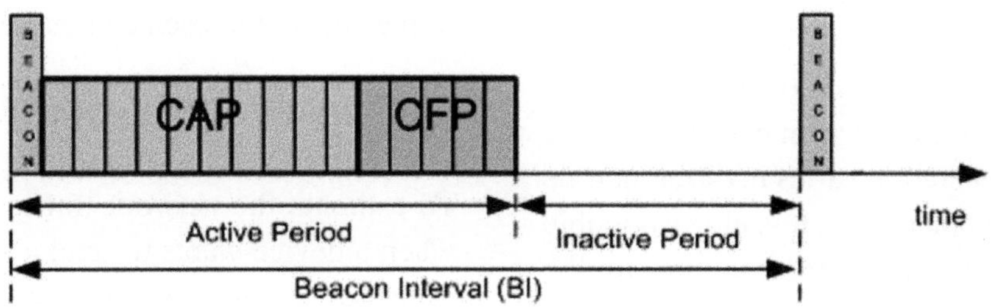

The CFP formed by GTSs, starts immediately following the CAP on a slot boundary, and complete before the end of the superframe active portion.

By determining the values of beacon order (BO) and superframe order (SO) in the beacon frame, the coordinator specifies the BI and SD duration as follows:

$$BI = aBaseSuperframeDuration \times 2^{BO} \ symbols \quad (1)$$

$$SD = aBaseSuperframeDuration \times 2^{SO} \ symbols \quad (2)$$

where:

$$aBaseSuperframeDuration = aBaseSlotDuration \times aNumSuperframeSlots (= 16) \quad (3)$$

The *aBaseSlotDuration* is number of symbols which forms a superframe slot when the superframe order is equal to zero and its default value is 60 symbols. The SO and BO parameters follow the $0 \le SO \le BO \le 15$ constraint.

DYNAMIC CONTENSION ACCESS PERIOD

The main idea of this chapter is to reduce the energy consumption in CAP using dynamic duty cycle in accordance with network traffic. Traffic may be changed as the time passes or even it may not be predetermined in some networks and applications. Thus, the network should adapt its CAP length with the volume of traffic. The changes in the volume of network traffic over the time may have different reasons. In some networks, the ratio of packet generation varies in different time intervals. On the other hand, it is either possible to add some new device in order to improve the network performance, it can respectively increase the network traffic, or over the time some of network nodes can be expired due to damages or exhausting the energy resources, this reduces the number of packets generated in the network.

After fixing BO and determining BI, selecting SO, plays a main role in the throughput and lifetime of the network at which traffic changes over time. Selecting a large SO increases throughput but decreases the network lifetime, on the contrary, selecting a small SO increases lifetime but decreases throughput due to the collision. In this situation, in order to decreasing the energy consumption and to have a high throughput, the CAP length must be changed proportional with the network traffic. Two scenarios may occur in the networks which their traffics change over the time:

Scenario1: if the network traffic increases in a time interval, the network will need to increase the duty cycle proportional with the network traffic.

Scenario 2: if the number of packets generated by devices decrease, selecting small and proportional SO not only sends the packets to their destination, but also by decreasing duty cycle, the lifetime of network increases. But if the selected SO is larger than proportional SO of the network traffic, it causes loss of energy resources of nodes due to their idleness.

Regarding to the mentioned scenarios, by selecting SO proportional with the network traffic, a trade-off between power consummation, throughput and delay can be reached.

In this standard, the coordinator is responsible for setting SO in each superframe. In proposed algorithm, to select an appropriate SO, the coordinator should have correct information about network traffic, throughput and idle time intervals of the channel in each contention access period.

Estimation of Network Traffic

To estimate the network traffic by coordinator, when a device wants to send a packet to the coordinator, the information about number of other packets in its queue along with the packet is sent to the coordinator. On the other hand, the coordinator defines an array which its size is equal to the number of network nodes. Each array cell is allocated to one node. The coordinator saves the queue state information of each node in the cell which is allocated to that. By receiving each packet, the coordinator updates the queue state information of the packet sender node. The coordinator uses this information to estimate the number of pending data packets at the end of each CAP and uses that to choose SO for the next superframe.

In this method, in order to have no changes in the standard frames nor define a new frame, each node sends the information of its queue state through the reserved three bits in standard data packet header. These three bits are illustrated in Figure 2.a and 2.b.

Using 3 bits, only the eight level of queue state can be displayed. Considering this constraint, the proposed algorithm should use these three bits appropriately. The sending packets in low network traffic are less than high network traffic. In other words, in time intervals that the network traffic is low, the number of packets in queue of nodes is few. Otherwise, the number of packets in queue of nodes is high when the network traffic is high. Thus in this algorithm, the coordinator needs more accurate information about the number of packets in queue of nodes when the network traffic is low compare to high traffic. Therefore in order to optimize using 3 bits, the nodes send queue state information to the coordinator according to Equation (3):

$$queueState = [\log_2{}^{(numOfPktsInQueue)}] + 1 \qquad (4)$$

Also if a node has no pending packet sets its queue state equal to zero. According to table 1, using 3 bits at most 64 pending packets can be displayed and if nodes maximum queue of size is more than 64, Equation (4) is rewritten as follows:

Figure 2. a. General MAC frame format, 2b. Format of the frame control field

Octets: 2	1	0/2	0/2/8	0/2	0/2/8	0/5/6/10/14	Variable	2
Frame Control	Sequence Number	Destination PAN Address	Destination Address	Source PAN Identifier	Source Address	Auxiliary Security Header	Frame Payload	FCS
		Addressing fields						
MHR							MAC Payload	MFR

A

Bits: 0-2	3	4	5	6	7-9	10-11	12-13	14-15
Frame Type	Security Enabled	Frame Pending	Ack. Request	PAN ID Compression	Reserved	Destination Addressing Mode	Frame Version	Source Addressing Mode

B

Table 1. The queue state values

numPktInQueue/a	QueueState
0	0
1	1
2	2
3-4	3
5-8	4
9-16	5
17-32	6
33-64	7

$$queueState = [\log_2^{(\frac{numOfPktsInQueue}{a})}] + 1 \quad (5)$$

Where:

$$a = [\frac{\max QueueSize}{64}] + 1 \quad (6)$$

The values allocated as queue state for different number of packets is shown in Table 1.

For example if the network nodes have a maximum queue of size 50 and a sending node has 10 pending packets in its queue, in this case, $\alpha=1$ and *queuestate* equals 5.

Calculating the Channel Idle Time

In the star topology, each device can only communicate with the coordinator. So the coordinator can calculate the sum of time intervals in which the channel is idle in CAP for each superframe based on Equation (7).

$$ChannelIdleTimeInCAP = \sum_{t=startCAP}^{t=endCAP} channelIdleTime \quad (7)$$

where $channelIdleTimeinCAP$ is the sum of channel idle time intervals in CAP of each superframe.

In this algorithm, the idle time intervals which their size is more than the minimum required time for sending a packet and receiving it's acknowledge packet, are considered. This minimum time is calculated as Equation (8).

$$\min TimeToSendPkt = \frac{PktSize}{channelRate} + AckWaitDuration + IFS \quad (8)$$

where IFS is a finite amount of time that the MAC sub-layer needs to process the data that are received by the PHY. These concepts are illustrated in Figure 3.

Dynamic Duty Cycle Algorithm

The coordinator selects SO for the next superframe according to $channelIdleTimeInCAP$, throughput of CAP and queue state of nodes at the end of CAP of each superframe. In this situation, three cases can be happened as below:

Case 1: In this case, the $channelIdleTimeInCAP$ length is more than half of the CAP length in current superframe. So the coordinator subtracts one unit from SO and selects it as the SO for the next superframe.

Figure 3. The minimum time for acknowledged transmission

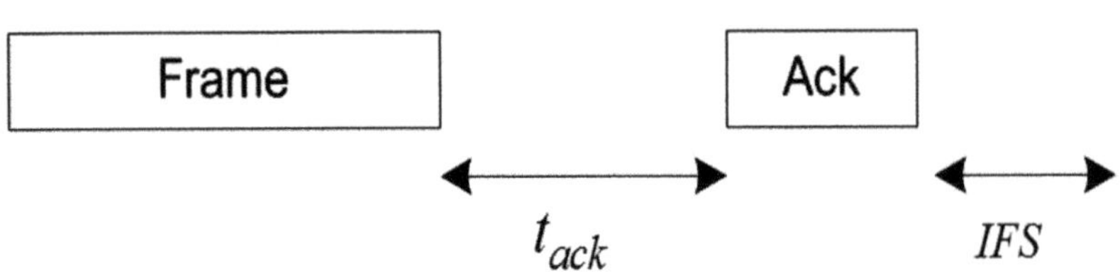

Case 2: If the sum of the channel idle time intervals is less than half of the CAP, the coordinator determines the SO regarding to the next superframe based on Equation (9).

$$numPkt = ChannelIdleTimeInCAP \times troughtInCAP \quad (9)$$

where:

$$troughputInCAP = \left(\frac{numSuccPktInCAP}{CAPlength} \right) \quad (10)$$

where $numPkt$ is the number of packets that can be sent in $ChannelIdleTimeInCAP$ based on network throughput in CAP and $numSuccPktInCAP$ is also the number of successfully sent packets to coordinator in each superframe.

Based on the nodes *queuestate* array if the remained pending packets be more than the $numPkt$ value, the active period of current superframe is not enough for network traffic. Therefore, the coordinator increases SO of the current superframe based on Equation (11) and selects it as the SO of the next superframe.

$$newSO = currentSO + \left\lceil \log_2^{\left\lceil (\frac{numPendingPkt}{numSuccPktInCAP})+1 \right\rceil} \right\rceil + 1 \quad (11)$$

Case 3: Based on the nodes *queuestate* array, if the remained pending packets are less than the $numPkt$ value, the coordinator selects the SO of the current superframe as the SO of the next superframe.

Inverse relation between network traffic and channel idle time leads to proper performance of the algorithm. Indeed, increasing of the network traffic reduces the channel idle time and vice versa. The pseudo code of the proposed algorithm is shown in Figure 4.

Beacon Frame Modification

The coordinator sets SO value in the Superframe Specification field of beacon frame through 4 bits in IEEE 892.15.4 standard. Since in proposed algorithm the value of SO varies with the variation of network traffic, if a portion of active period is being allocated to CFP, the SO value of CFP must be set independently of SO value of CAP. To do this, in the proposed algorithm in order to not changing the beacon frame, the coordinator sets the SO value of CFP independently through four reserved bits in GTS Specification field of GTS fields. The GTS Specification field and the GTS field format are illustrated in Figures 5a and 5b.

So Equation (2) modified to:

$$SD = aBaseSlotDuration \times \begin{pmatrix} \left(aNameCAPSlots \times 2^{SO_{CAP}}\right) + \\ \left(aNameCFPSlots \times 2^{SO_{CFP}}\right) \end{pmatrix} \quad (12)$$

where $aNnmCAPSlots$ and $aNumCFPSlots$ are the number of slots allocated to CAP and CFP respectively.

SIMULATION RESULTS

In this section results of the simulation of proposed algorithm are compared to IEEE 802.15.4 standard. The simulations have been performed based on the work in (Chen & Dressler, et al., 2007), and OMNeT++ (a public-source, component-based network simulator) has been used. The topology of the network is star and nodes communicate with coordinator in beacon enable mode. The coordinator is approximately placed in center of the network, and the nodes are distributed randomly in the coverage area so that all the nodes are in

Figure 4. Dynamic duty-cycle algorithm

$$\mathit{if}\left(\left(channelIdleTimeInCAP \geq \left(\frac{SD}{2}\right)\right) \&\& SO > 0\right)$$

$$SO_{next} = SO_{current} - 1;$$

$$\mathit{else\,if}\left(\left(ChannelIdleTimeInCAP < \left(\frac{SD}{2}\right)\right) \&\& \left(numPendingPkt > numPkt\right)\right)$$

$$SO = SO_{current} + \left[\log_2^{\left[\left(\frac{numPendingPck}{numSuccPktInCAP}\right)+1\right]}\right] + 1;$$

$$\mathit{else}$$

$$SO_{next} = SO_{current};$$

Figure 5. a. Format of the GTS Specification field, b. Format of the GTS field

Bits: 0-2	3-6	7
GTS Descriptor count	Reserved	GTS Permit

A

Octets: 1	0/1	Variable
GTS Specification	GTS Direction	GTS List

B

each other's radio range areas. In the beginning of superframe, the coordinator sends its beacon to the network nodes. In the active period, in order to access channel, nodes are contend with each other by using CSMA/CA algorithm. Then, nodes go to sleep mode at the end of active period to save energy. Receiving each package, coordinator sends an acknowledgement packet to its sender. Since the proposed algorithm is provided to adjust dynamic active period, the consumed energy of radio and CPU of nodes is just considered in the simulations.

Also, in order to have more realistic simulation results, the energy model of nodes is calibrated based on the measurements for MICAz motes. The MICAz is a 2.4 GHz mote module with 250 Kbps data rate which is designed based on IEEE 802.15.4 standard. This module has a CC2420 radio and a ATmega-128L microcontroller. The deployed power parameters for the radio and the microcontroller are listed in Tables 2 and 3 in the form of current draw.

The network has three traffic levels: low, moderate and high. The devices generate data randomly with an average of 0.25 packets per second in the low, 0.05 in the moderate, and 2 in the high level. The queue size of nodes is considered to be 50 and the packets have a MAC payload size of 50 bytes. The simulations are operated with different number of nodes, and each simulation run time is 500s along with simulation runs. During the simulations, the network traffic has been randomly changed between the three mentioned levels in 20BI interval times.

Simulations were performed in five separate conditions of (BO, SO) = (7, 5), (BO, SO) = (7, 4), (BO, SO) = (7, 3), (BO, SO) = (7, 2) and (BO, SO) = (7, dynamic SO). Comparisons with IEEE 802.15.4 are based on throughput, energy consumption, end-to-end delay and duty-cycle.

Performance Evaluation of Proposed Algorithm

Figure 6 shows the percentage of different SOs that the coordinator selects in the proposed algorithm based on the network traffic during simulation. In each column of the figure, the larger SOs belongs to the time interval in which the traffic is high. On the contrary, the smaller SOs belongs to the time interval in which the traffic is low. According to the Figure 6, by increasing the network nodes, the percentage of the larger SOs increases.

Figure 7 shows the average of duty cycle in the proposed algorithm. The results show that the average of duty cycle increases by increasing the network nodes. In other words, the active period length for sending packets increases by increasing the network traffic.

Figures 6 and 7 obviously show the dynamism of the proposed algorithm according to the variation of network traffic.

Table 2. Current consumption of CC2420

Mode	Current Consumptions
Receive Mode	18.8 mA
Transmit Mode	17.3 mA
Idle Mode	426 μA
Sleep Mode	20 μA

Table 3. Current consumption of MCU (Kramer & Geraldy, et al., 2006)

Mode	Current Consumptions
Active mode	8.65 mA
Power Save Mode	250 μA

Energy Consumption and Throughput

Figures 8 and 9 compare per packet energy consumption and throughput with the proposed algorithm and different conditions of IEEE 802.15.4, respectively. The stimulation results show the decreasing of energy consumption in the proposed method compared to IEEE 802.15.4 standard. According to Figure 8, the energy consumption decreases seriously in some cases. The average of consumed energy in the proposed algorithm decreases 44.6%, 25.7%, 19.8% and 30.4% compared to (BO, SO) = (7, 5), (BO, SO) = (7, 4), (BO, SO) = (7, 3), and (BO, SO) = (7,2) respectively.

Figure 9 shows the incensement of the proposed method throughput compared to the standard. The average of network throughput in the propose method increases 6% and 22% compared to (BO, SO) = (7, 5) and (BO, SO) = (7, 4). Also, the average of network throughput in the propose method increases 1.7 times and 3 compared to (BO, SO) = (7, 3) and (BO, SO) = (7, 2).

According to the results of the standard performance, although selecting a large SO such as (BO, SO) = (7, 5) and (BO, SO) = (7, 4) increases the network throughput in the high traffic time intervals, the per packet energy consumption increases in the low traffic time intervals due to the idleness of the nodes. Although by selecting

Figure 6. The SO ratio in proposed algorithm

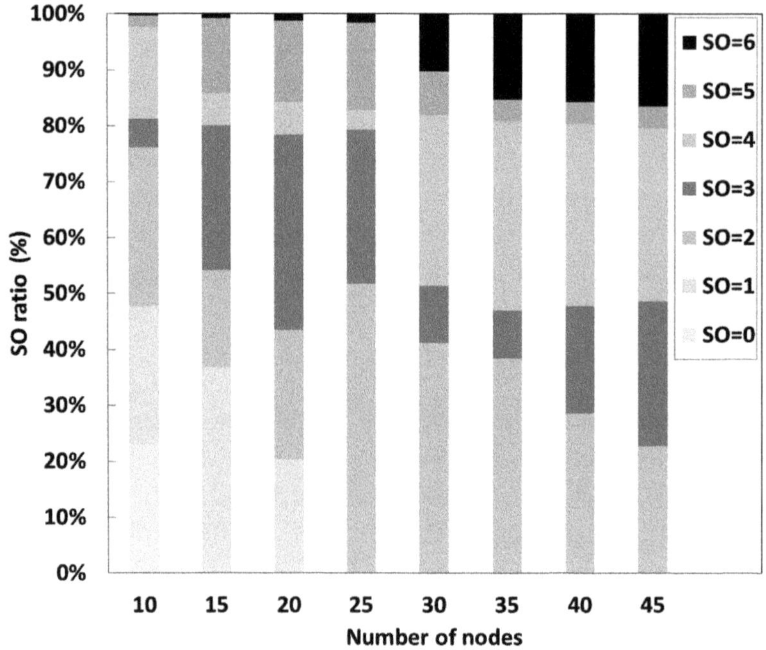

Figure 7. The average duty cycle of proposed algorithm

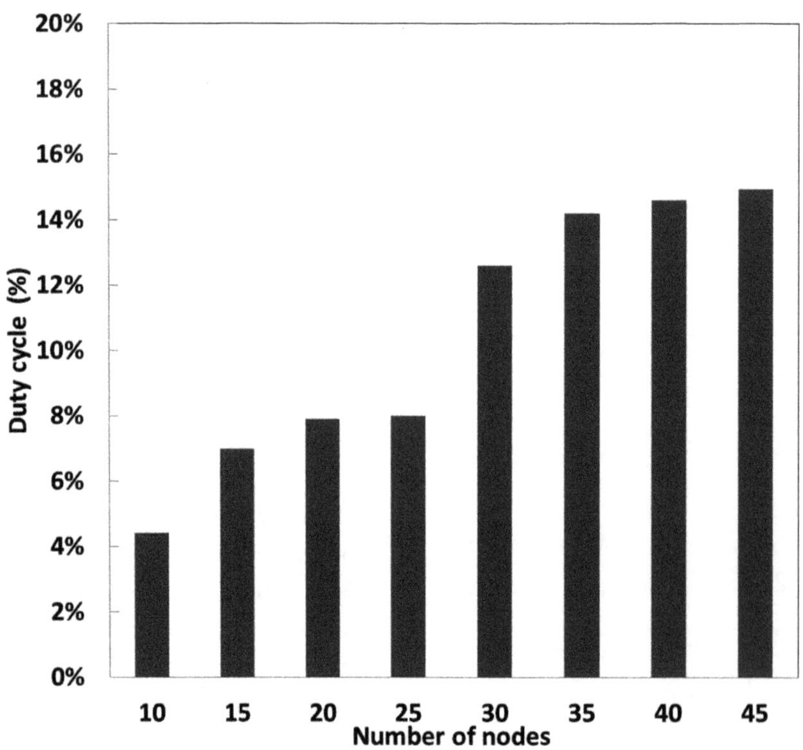

Figure 8. Comparison of average energy consumption

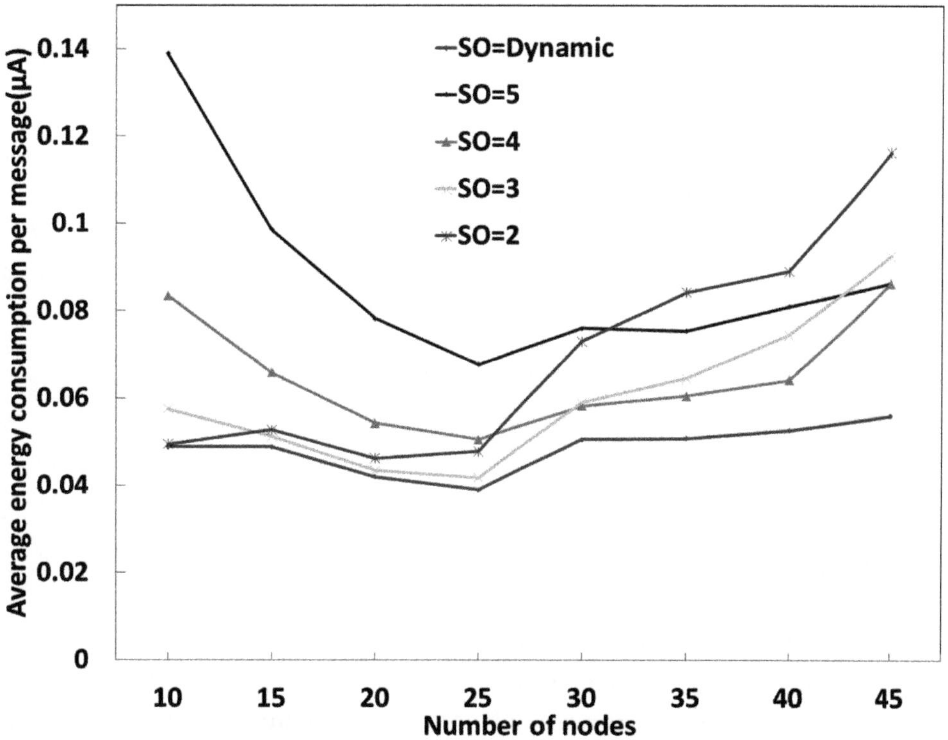

Figure 9. Comparison of network throughput

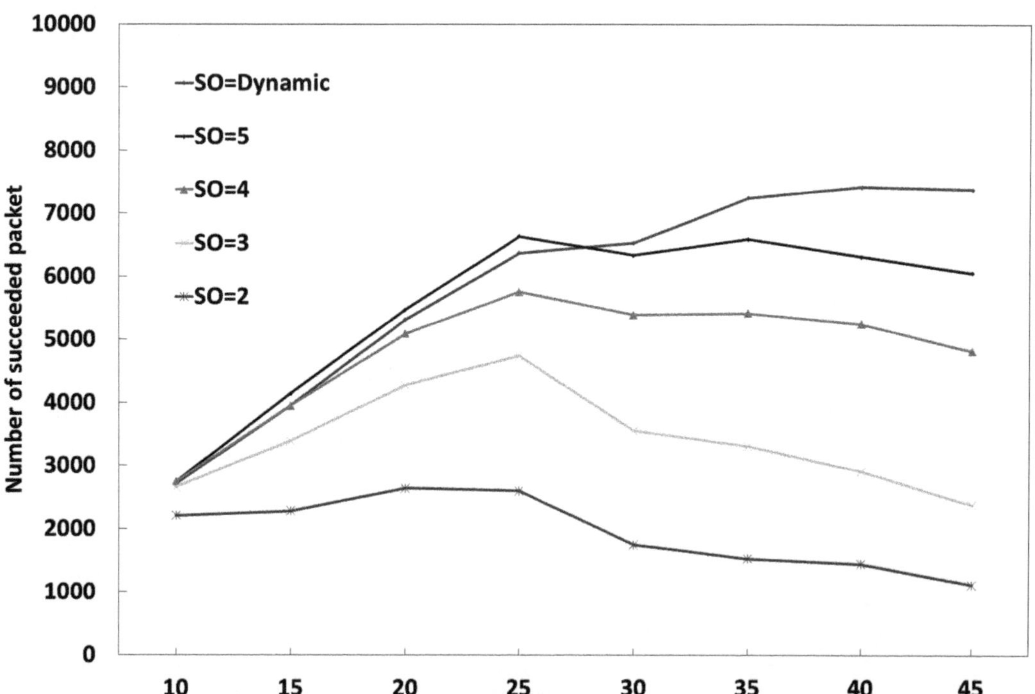

a small SO such as (BO, SO)=(7, 2) and (BO, SO)=(7, 3), the per packet energy consumption decreases in the low traffic time intervals, but the great number of packets will fail due to collision and discarding in the time intervals in which the traffic is high. When the network traffic is low, the proposed method decreases the per packet energy consumption by decreasing the active period. On the other hand, when the network traffic is high, the network throughput increases by increasing the active period. Therefore, the proposed method achieves a high throughput compared to the IEEE 802.15.4 standard by consuming lesser energy in the networks in which the traffic changes in different time intervals.

End to End Delay

In star topology, the end-to-end delay is measured as the time interval in which a packet is sent from one node, and received by the coordinator at the other side. Figures 10 and 11 show the end to end delay and duty cycle of the proposed method compared to the mentioned standard conditions respectively. The proposed algorithm adjusts the active period proportional to the network traffic. Thus, the packets in the queue of notes wait a short time to be sent. The end to end delay in the proposed method decreases seriously comparing to the standard conditions that have duty cycles close to those of the proposed algorithm. According to the results, the average end to end delay in the proposed method in comparison with (BO, SO) = (7, 2), (BO, SO) = (7, 3) and (BO, SO) = (7, 4) is 50, 20.5, and 6.5 times less. In average, the delay of the proposed algorithm in comparison with the (BO, SO) = (7, 5) increases 20%.

The end to end delay of the proposed algorithm, as shown in Figure 10, in states that the number of network nodes is high (35, 40, 45) is lesser than (BO, SO) = (7, 5). Because in these states when the network traffic is high, the coordinator select high SO such as SO=5, SO=6 in the proposed algorithm. But the end to end delay in the proposed algorithm is more than the (BO, SO) = (7, 5) in states that the number of network nodes is low (10, 15, 20). In this states the network traffic generally is low and the network doesn't need high CAP length. Therefore although big active period such as (BO, SO) = (7, 5) decreases the end to end delay, but as it seriously increases the energy consumption of the nodes which is not reasonable.

CONCLUSION

In this chapter, a new algorithm was provided to allocate dynamic active period proportional to the network traffic in the IEEE 802.15.4 standard. This algorithm is applicable in star topology and beacon enabled mode of standard for the networks in which the traffic changes in different time intervals. In order to optimize the energy consumption of nodes with high throughput and low end to end delay, this algorithm determines the active period dynamically and proportional with the network traffic. In proposed algorithm, the coordinator selects the active period proportional to the network traffic by using the information of the queue state of nodes. The simulation results show the improvement in the energy consumption of nodes along with high throughput and low delay in the proposed algorithm compared to the IEEE 802.15.4 standard.

Figure 10. Comparison of network end-to-end delay

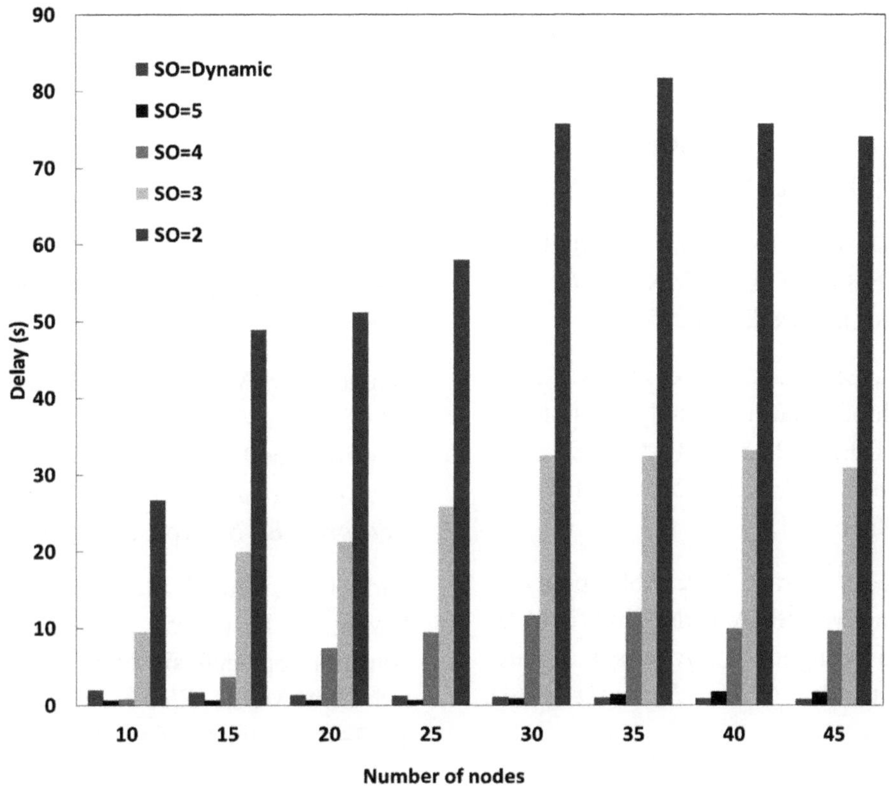

Figure 11. Comparison of network duty cycle

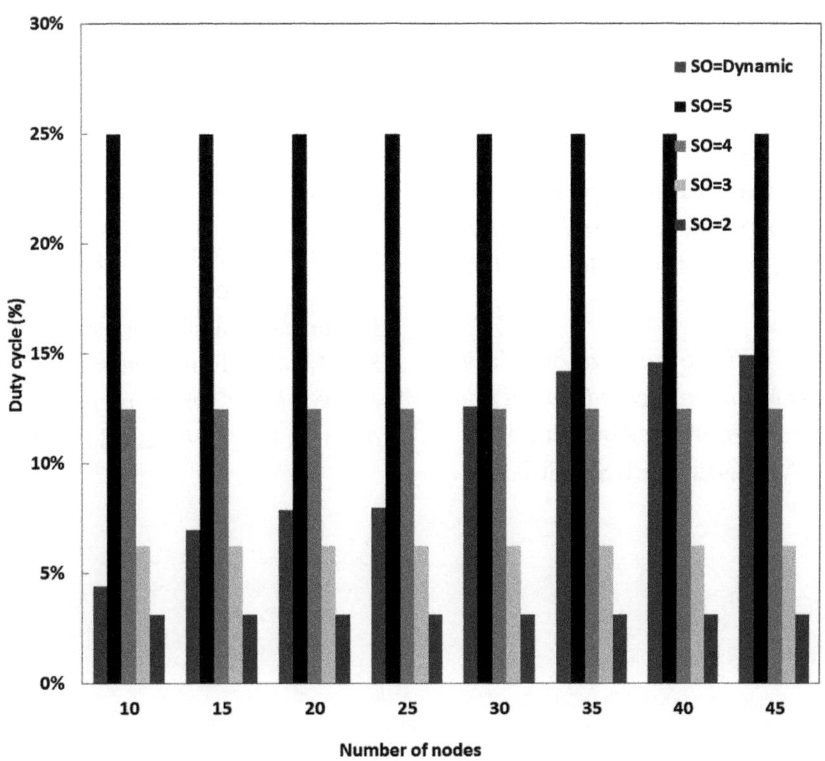

REFERENCES

Ashrafuzzaman, K., & Kwak, K. S. (2011). On the performance analysis of the contention access period of IEEE 802.15. 4 MAC. *IEEE Communications Letters*, *15*(9), 986–988. doi:10.1109/LCOMM.2011.071211.111271

Chen, F., & Dressler, F. (2007). A Simulation Model of IEEE 802.15. 4 in OMNeT+. 6. Fachgespräch Sensor netzwerke. pp. 35.

Chung, M. Y., Lee, T. J., & Lee, H. R. (2006, July). MAC throughput limit analysis of slotted CSMA/CA in IEEE 802.15.4 WPAN. *IEEE Communications Letters*, *10*(7), 561–563. doi:10.1109/LCOMM.2006.1673013

Gilani, M. H. S., Sarrafi, I., & Abbaspour, M. (2011). An adaptive CSMA/TDMA hybrid MAC for energy and throughput improvement of wireless sensor networks. *Ad Hoc Networks*, *11*(4), 1290–1304.

He, J., Tang, Z., Chen, H. H., & Wang, S. (2008). An accurate Markov model for slotted CSMA/CA algorithm in IEEE 802.15.4 networks. *IEEE Communications Letters*, *12*(6), 420–422. doi:10.1109/LCOMM.2008.071460

Huang, Y. K., Pang, A. C., & Hung, H. N. (2009). A comprehensive analysis of low-power operation for beacon-enabled IEEE 802.15. 4 wireless networks. *IEEE Transactions on Wireless Communications*, *8*(11), 5601–5611. doi:10.1109/TWC.2009.081485

IEEE-SA. (2006). *IEEE Standard 802.15.4, Part 15.4: Wireless Medium Access Control and Physical Layer Specification for Low Rate Wireless Personal Area Networks*. IEEE Std. 802.15.4.

Jung, C. Y., Hwang, H. Y., Sung, D. K., & Hwang, G. U. (2009). Enhanced Markov chain model and throughput analysis of the slotted CSMA/CA for IEEE 802.15. 4 under unsaturated traffic conditions. *IEEE Transactions on Vehicular Technology*, *58*(1), 473–478. doi:10.1109/TVT.2008.923669

Kramer, M., & Geraldy, A. (2006). Energy measurements for micaz node (Tech. Rep. KrGe06). University of Kaiserslautern, Kaiserslautern, Germany.

Neugebauer, M., Plonnigs, J., & Kabitzsch, K. (2005 January). A new beacon order adaptation algorithm for IEEE 802.15.4 networks. In *Proceedings of European Workshop on Wireless Sensor Networks (EWSN)* (pp. 302-311).

Park, T. R., Kim, T. H., Choi, J. Y., Choi, S., & Kwon, W. H. (2005). Throughput and energy consumption analysis of IEEE 802.15. 4 slotted CSMA/CA. *Electronics Letters*, *41*(18), 1017–1019. doi:10.1049/el:20051662

Patro. Kumar, R. (2007). Analysis and improvement of contention access protocol in IEEE 802.15. 4 star network. *Mobile Adhoc and Sensor Systems, 2007. MASS 2007. IEEE Internatonal Conference on. IEEE.*

Polastre, J. Hill, J. and Culler, D (2004). *Versatile low power media access for wireless sensor networks* (pp. 95-107).

Pollin, S., Ergen, M., Ergen, S., Bougard, B., Der Perre, L., Moerman, I., & Catthoor, F. (2008). Performance analysis of slotted carrier sense IEEE 802.15.4 medium access layer. *IEEE Transactions on Wireless Communications*, *7*(9), 3359–3371. doi:10.1109/TWC.2008.060057

Ramachandran, I., Das, A. K., & Roy, S. (2007). Analysis of the contention access period of IEEE 802.15. 4 MAC. *ACM Transactions on Sensor Networks*, *3*(1). doi:10.1145/1210669.1210673

Sheu, S. T., Shih, Y. Y., & Chen, L. W. (2005). An adaptive interleaving access scheme (IAS) for IEEE 802.15.4 WPANs. In *Proceedings of IEEE Vehicular Technology Conference (VTC)*, 3, 1523-1527.

Shu, F., Sakurai, T., Vu, H. L., & Zukerman, M. (2006). Optimizing the IEEE 802.15. 4 MAC. In TENCON 2006. 2006 IEEE Region 10 Conference (pp. 1-4).

Suh, C., Mir, Z. H., & Ko, Y. B. (2008). Design and implementation of enhanced IEEE 802.15. 4 for supporting multimedia service in Wireless Sensor Networks. *Computer Networks*, *52*(13), 2568–2581. doi:10.1016/j.comnet.2008.03.011

Tao, Z., Panwar, S., Gu, D., & Zhang, J. (2006). Performance analysis and a propose improvement for the IEEE 802.15.4 contention access period. *Wireless Communications and Networking Conference*. 4, 1811-1818.

van Dam, T., & Langendoen, K. (2003). An adaptive energy-efficient mac protocol for wireless sensor networks. *Proceedings of the 1st international conference on Embedded networked sensor systems*, 171-180. doi:10.1145/958491.958512

Xiao, Z., He, C., & Jiang, L. (2010). Slot-based model for IEEE 802.15. 4 MAC with sleep mechanism. *IEEE Communications Letters*, *14*(2), 154–156. doi:10.1109/LCOMM.2010.02.091981

Ye, W., Heidemann, J., & Estrin, D. (2004). Medium access control with coordinated adaptive sleeping for wireless sensor networks. *IEEE/ACM Transactions on Networking*, *12*(3), 493–506. doi:10.1109/TNET.2004.828953

Zheng, J., & Lee, M. J. (2004, June). Will IEEE 802.15.4 make ubiquitous networking a reality? A discussion on a potential low power, low bit rate standard. *IEEE Communications Magazine*, *42*(6), 140–146. doi:10.1109/MCOM.2004.1304251

KEY TERMS AND DEFINITIONS

Beacon-Enabled: A mode that network operates with a super frame structure, which its format is defined by the coordinator.

Dynamic Duty Cycle: Adjustable duty cycle proportion to the sensor network traffic.

Energy Efficient MAC: The MAC protocol that considers energy efficiency issues in wireless sensor networks.

IEEE 802.15.4: The IEEE 802.15.4 is a standard that specifies the *physical layer* and *media access control* for low-rate wireless *personal area networks* and is suitable for employing in wireless sensor networks.

MAC Sub-Layer: The sub-layer of wireless sensor network that is responsible for sharing the media between sensor nodes in right manner.

Star Topology: A topology in that all sensor nodes communicate to coordinator directly.

Wireless Sensor Networks: Some sensor nodes which sense the environment and communicate wirelessly to transmit the sensed data.

Chapter 6
Prevalence of Anomalies in Real World Sensor Network Deployments:
The Need for Detection Mechanisms

Giovani Rimon Abuaitah
Wright State University, USA

Bin Wang
Wright State University, USA

ABSTRACT

Sensor network deployments have enabled automated data collection at a finer granularity compared to human-centric sparse deployments of traditional telemetric data loggers. However, this fine granularity cannot be achieved in the presence of failing nodes, frequent network failures, and/or potentially malicious network behaviors. Moreover, given a perfect and secure network with long-lived sensor nodes, the collected sensor data should be fault-free to guarantee quality of reasoning. Unlike traditional devices, wireless sensor motes are resource-constrained battery-powered devices equipped with few on-board sensors which are susceptible to failures, and batteries cannot always be reliably recharged. Over the years, sensor network deployments have materialized a large number of monitoring and event-detection applications. Researchers and decision makers have observed several anomalous patterns ranging from functional failures to pervasive sensing faults in most of these deployments. In this chapter, we discuss a few of these deployments to emphasize the need for an anomaly detection mechanism.

INTRODUCTION

Over the last decade, real-life deployments of sensor networks have grown in numbers for various applications. Examples include monitoring weather conditions (Tolle et al., 2005; Szewczyk et al., 2004a), habitat monitoring and wildlife tracking (Szewczyk et al., 2004a; Sorber et al., 2007; Anthony et al., 2012), monitoring of health conditions, flood detection (Ingelrest et al., 2010), fire detection (Liu et al., 2011), volcano monitoring (Werner-Allen et al., 2006), structural health

DOI: 10.4018/978-1-4666-8251-1.ch006

monitoring (Kim et al., 2006), monitoring the micro-climate inside greenhouses (Ahonen et al., 2008), monitoring voltage usages at the micro-grid (Dawson-Haggerty et al., 2012), monitoring water quality (Ramanathan et al., 2006), measuring the quality of perishable food and medicine (Bijwaard et al., 2011), and tracing and verification of goods at different phases of the supply chain (Bijwaard et al., 2011). Most of these deployments have helped decision makers better understand their targeted environments by analyzing the data collected from a large number of unattended sensors. Other deployments have enabled event-driven applications by reporting significant events such as fires in large forests (Liu et al., 2011). A typical deployment of a wireless sensor network (WSN) consists of several tens or hundreds of low-power low-cost tiny devices equipped with sensor boards capable of measuring several phenomena. These devices, usually called sensor motes, are also equipped with wireless communication modules that enable them to communicate with one another. Table 1 shows the characteristics of some sensor platforms (i.e. motes) popularly used in sensor network deployments. Two types of communication paradigms are possible: (a) a simplified single-hop communication where every sensor mote communicates directly with a base station. A base station is a more powerful node which may further process and then relay messages towards a connected PC, and (b) a more complex multi-hop communication paradigm that involves forwarding messages from sensing nodes to the base station with the help of other intermediate sensor nodes in the network (see Figure 1). The base station node is sometimes called the "sink" while other nodes in the network are referred to as "source nodes."

Sensor data yield and quality were two major concerns of many of these sensor network deployments (Szewczyk et al., 2004a; Tolle et al., 2005; Barrenetxea et al., 2008a; Liu et al., 2011). Data yield is the ratio between the number of packets successfully delivered to the base station and the expected number of packets generated by all nodes in the network. Besides a shortened lifetime, several anomalies such as network failures may account for a very low yield (Szewczyk et al., 2004a), resulting in a very high number of missing data points that cannot be simply interpolated. Data quality, on the other hand, describes the amount of deviation of the observed data points from the expected phenomena. Lower deviation results in higher accuracy. To achieve higher fidelity in the data, appropriate packaging is necessary to protect the sensors from harsh environmental conditions (Szewczyk et al., 2004a; Barrenetxea et al.; 2008a, Liu et al., 2011). Without packaging, batteries may rapidly discharge or sensor circuitries may fail due to their exposure to chemicals such as carbon dioxide, introducing sensor data faults. Even with appropriate packaging, sensors might still experience natural unknown faults or calibration drifts or might generate uncorrelated readings unexpectedly (Barrenetxea et al., 2008a).

The rest of the chapter studies a number of sensor network deployments and identifies the various anomalies prevalent in these deployments.

GREAT DUCK ISLAND (GDI)

Overview

The GDI project (Szewczyk et al., 2004a) serves as one of the earliest proof-of-concept sensor network deployments. The main purpose of this deployment was to help ecologists study the distribution and abundance of Leach's Storm Petrels (a type of seabird species) at the Great Duck Island, Maine. The system deploys a multi-tier architecture (Figure 2) where one or more sensor patches are located at the lowest tier and the higher tier forms a transit network of gateways connecting the sensor patches to the remote base station(s). Each sensor patch is comprised of burrow motes equipped with temperature, humidity, and non-contact infrared temperature sensors to monitor

Table 1. Popular commercial sensor platforms

Name	Manufacturer	Processor	Program/Flash Memory	Data Memory	External Flash	Radio Frequency	Battery Type	OS/VM	Prog. Languages
TelosB	MEMSIC	8MHz TI MSP430	48Kbytes	10Kbytes	1Mbyte	2.4GHz	2x AA	TinyOS	nesC
MICA2		ATmega128L	128Kbytes	4K bytes	512Kbytes	868/916 MHz			
MICAz						2.4GHz			
IRIS		8MHz 8bit ATmega1281		8Kbytes				TinyOS, MoteRunner VM	nesC, Java/C#
XM1000	advanticsys	8MHz TI MSP430f2618	116Kbytes		1Mbyte			TinyOS, Contiki	nesC, C
Tmote Sky	Sentilla	8MHz TI MSP430	48Kbytes	10Kbytes					
SunSPOT	ORACLE	400MHz 32bit Atmel AT91SAM9G20	8Mbytes	1Mbyte	-		Li-Ion Rechargeable	Squawk VM	Java ME
Waspmote	Libelium	8MHz ATmega1281	128Kbytes	8Kbytes + 4Kbytes	2Gbytes SD Card	868/900MHz, 2.4GHz	Auxiliary Rechargea-ble	None	C++
Imote2	MEMSIC (Intel-licensed)	13-416MHz PXA271 Xscale	32Mbytes	256Kbytes + 32Mbytes	-	2.4GHz	3x AAA / Li-Lon	TinyOS	nesC
iSense Core Module 3	Coalesenses	4-32MHz 32bit RISC Controller	512Kbytes Shared	128Kbytes Shared	-		Pluggable Resources	iSense Firmware	C++
TinyNode	Shockfish	TI MSP430	48Kbytes	10Kbytes	512Kbytes	433 MHz, 868MHz, 915 MHz	Pluggable Resources	TinyOS	nesC

Figure 1. A typical multi-hop sensor network

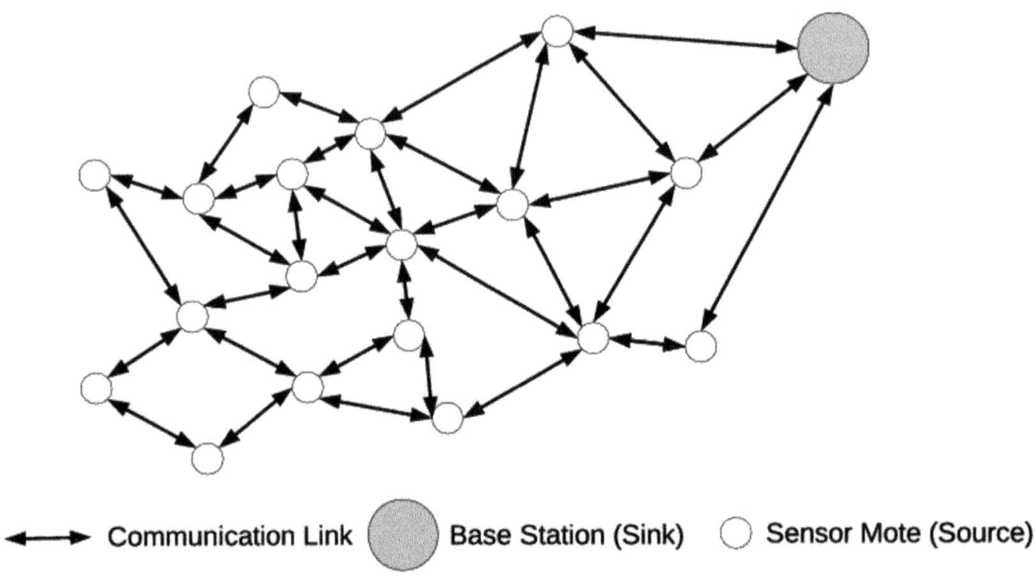

the occupancy of nesting burrows, and weather motes to measure the surface micro-climate using temperature, humidity, and pressure sensors. There were two types of deployments for sensor patches (with incremental installments): a single hop network of 49 Mica2Dot (MEMSIC, 2014) motes running TinyOS (TinyOS, 2014) each of which samples readings every 5 minutes, and a multi-hop network of 98 motes with data sampled every 20 minutes. In addition, a verification network of in-burrow cameras was also deployed to obtain ground truth for occupancy verification by collecting 15-second movies every 15 minutes. The deployment operated over 115 days during the summer and autumn of 2003 and covered an ellipsoidal area of 221 by 71 squared meters. To protect the electronics from severe environmental conditions such as flooding or rain, the internal structure of motes was entirely covered with O-rings and conformal coatings. On-board sensors, however, needed to be exposed to the outside world to preserve their sensing capabilities.

Observations

Data Yield

Data yield is the ratio between the number of data points successfully delivered to the base station and the expected number of data points generated by all nodes in the network. During the multi-hop deployment, several failures (anomalies) were spotted that resulted in a low data yield. For instance, five burrow motes stopped reporting only after 5 days of the deployment. This was due to rapid depletion of their batteries. Another example was the failure of gateways or the transit network accounting for a sensor data yield of merely 28%. In addition, the GDI deployment indicated that adequate packaging and efficient mote placement is essential: only 78 motes were recovered out of 150 devices as some were moved by animals and most of the recovered motes had their antennas removed or shortened by animals, eventually causing the node to fail to report its readings.

Measurement Faults

The GDI multi-hop deployment and an earlier deployment of a single hop network in the year 2002 (Szewczyk et al., 2004a), in which sensor motes also measured ambient light levels and sensor readings were sampled every 70 seconds, revealed that sensors may frequently malfunction and provide erroneous or abnormal measurements due to various reasons. For example, during the

Figure 2. Great Duck Island (GDI) deployment

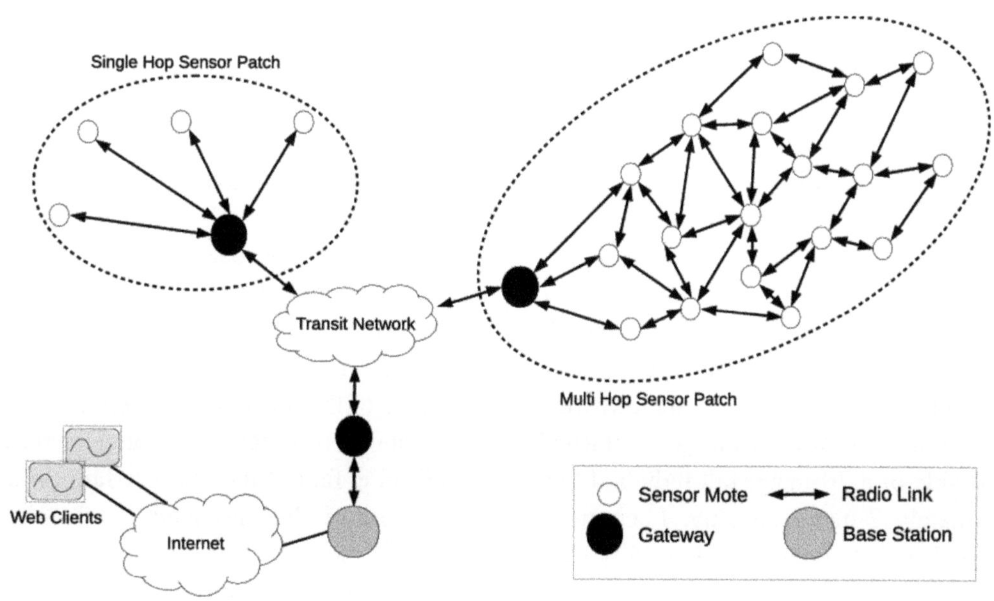

2002 deployment, about half of the temperature and humidity sensors reported faulty readings. This was mostly due to water contacts resulting in short circuits. A faulty temperature reading manifested itself in a persistent reading of 0° C, whereas an anomalous humidity reading was either higher than 150% (outlier) or very small. The temperature sensor most often never recovered while humidity sensors which reported high spikes recovered after they dried up. Humidity sensors with very small values were also correlated with permanent node outages: 55% of motes exhibiting this behavior failed within 2 days (Szewczyk et al., 2004b). In addition, wet sensors in general caused the battery level to drop drastically. Light sensors, on the other hand, reported reliable readings most of the time with the few exceptions when there was a spike usually correlated with temperature/humidity faults. Both deployments—single and multi-hop—showed that the operating voltage threshold is 2.7V when using 2-AA alkaline batteries while using a lithium battery maintained a constant voltage, and sensor calibration, although time consuming, is essential before deployment. Some of these facts have motivated researchers to design new mote architecture (Polastre et al., 2005) which is more solid in face of sensor failures (e.g., cuts the power line to the failed humidity sensor when detected) and lasts longer (e.g., the operating voltage threshold is 1.8V).

REDWOOD FORESTS

Overview

A multi-hop network of Mica2Dot TinyOS-powered sensor motes (Tolle et al., 2005) have replaced traditional data loggers in the redwood forests of Sonoma, California. Motes were attached to the west side of a 70-meter tall redwood tree at approximately 2-meter spacing. Each mote was equipped with temperature, humidity, and two PAR sensors (photo-diodes) to measure the micro-climate surrounding the tree. The network operated over 44 days during the Spring of 2004 and collected readings at a sampling rate of 5 minutes. The collected data helped biologists draw conclusions regarding the moving gradients inside the structure of the tree due to variations of the different phenomena over the tree volume. Packaging included a white skirt cover to protect the internal electronics from water and wind but still exposed the sensors to the environment to preserve their sensing capabilities. To be able to access the data from the outside world, a Stargate (Stargate, 2014) gateway was placed at the bottom of the tree which stored sensor readings received over the multi-hop network into a TASK (Buonadonna et al., 2005) database server.

Observations

Data Yield

The entire network collected only 49% of the total sampled data points. Network and node failures were the underlying reasons behind such a low yield. The gateway node suffered outages at the beginning of the deployment while five nodes stopped reporting during the second week. Researchers of the Redwood Forests deployment emphasized the need for network monitoring tools that can detect and report an abnormal behavior once it occurs.

Measurement Faults

By analyzing the data, researchers have found several anomalous readings. For example, some sensor motes running on a low battery produced erroneous readings where others reported readings outside of the normal range (e.g., humidity readings above 100%RH). The latter is an example of a clipping fault. Since battery failures accounted for most of the anomalies, a simple anomaly detection and filtration algorithm was applied. Sensor readings, reported when the mote's voltage was outside the range 2.4V-3.0V, were deleted. In addition, humans were involved to remove motes which reported unexpected readings. This deployment has shown that physical installation (e.g., orientation) of sensor motes is very critical and calibration is inadequate to improve the fidelity of the data and other techniques are needed. A time consuming calibration of the humidity and temperature sensors resulted in only slight improvements.

INTEL BERKELEY RESEARCH LAB (IBRL)

Overview

Researchers at the Intel Berkeley Research Laboratory (IBRL) deployed an indoor network of 54 Mica2 and Mica2Dot motes in March 2004 (Buonadonna et al., 2005). Every 30 seconds, motes collected temperature, relative humidity, light, voltage, and topology information. The intention behind this deployment, which lasted for about 30 days (720 hours), was to verify the capability of the Tiny Application Sensor Kit (TASK) services in keeping the network alive over extended periods of time, and to provide the sensor network community with a publicly accessible sensor dataset allowing developers and other researchers to experiment with. The dataset acts as a benchmark to evaluate and compare anomaly detection algorithms against each other. The deployed motes formed the multi-hop network shown in Figure 3. The base station is the node with ID 0 and the farthest nodes at the upper left corner (e.g., node 51) are about 10 hops away from the base station. Sensor nodes ran TinyDB which is part of the TASK sensor kit. TASK also provided visual tools for inspecting the quality of data and the status of the network.

Observations

Data Yield

The entire IBRL sensor dataset (Intel Lab Data, 2012) contained 2,313,153 entries pertaining to only about half the expected number of entries. Each entry corresponds to a packet or sample received from one of the deployed sensors. Each packet contains the sampling date, sampling time, sequence number (epoch), node id, temperature reading, humidity reading, light reading, and the voltage level.

IBRL researchers found that 4 nodes failed after the first few days for unknown reasons. Even before failing, a high number of packet loss was observed. For example, for a little more than 8 hours starting 11:24 pm on March 1, packets from node 15 were not delivered to the base station most likely due to network contention or interference. The plot of node 15 data in Figure 4 shows missing sensor values over the lifetime of the node[1]. The rest of the motes were able to successfully transmit 30% to 82% of their packets over the lifetime of the deployment. In addition to the previous findings, we observe that the entire network has gone offline for two intermittent days. Although this may seem as a correlation of packet losses across sensors, this is not the case over the entire deployment (Buonadonna et al., 2005). The two intermittent days of complete loss may be due to a very high network contention or a base station breakage. The loss of sensor readings, however, seems to be highly correlated (a lost packet would result in total absence of all sensor readings). It is also worth mentioning that in the IBRL deployment, three nodes were potentially deployed later after the very first deployment of the original 54 nodes. They may also have been deployed initially but never reported any packets until too late in the deployment time for unknown reasons. It seems like the three nodes rebooted themselves (or were physically reset) as the sequence numbers did not always follow a sequential order.

Figure 3. Intel Berkeley Research Lab (IBRL) deployment (Buonadonna et al., 2005)

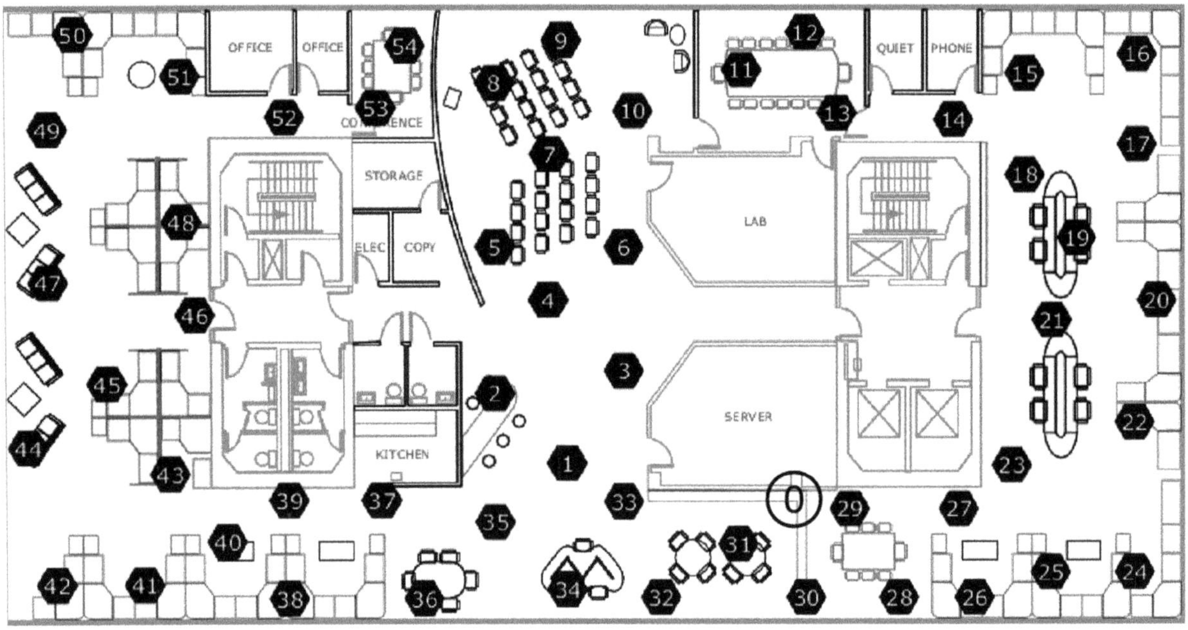

In addition to total loss of packets, some nodes have reported packets with missing sensor values. Several packets (i.e., samples) from node 5 were delivered to the base station intermittently until March 3. However, the sensor readings were completely missing due to unknown reasons. From node 57, only 3 packets were delivered during the entire monitoring period, one of which had only a single sensor reading (humidity reading).

Measurement Faults

Rajasegarar et al (Rajasegarar et al., 2007) have analyzed the IBRL set for data anomalies. They found that node 14 reported few anomalous readings (i.e., anomalous temperature and humidity) within a four-hour window starting midnight March 1. Within the same time window, node 37 reported anomalous readings at all times (i.e., anomalous node). It is believed that the reason behind the total anomalies of node 37 is its physical placement close to the kitchen (always higher temperature and humidity) and the unusual interference at the corner where node 14 is located accounts for its slightly erroneous behavior. In addition to anomalies resulted from nodes 14 and 37, other forms of fault were observed in (Ni et al., 2009). For example, nodes 24 and 32 as well as other spatially correlated sensor nodes experienced clipping faults in the light readings over the course of the month. Clipping occurred in the middle of the day when the light sensor could not read values exceeding the limits of the analog to digital converter (ADC). As in the redwood forests project, due to low battery levels, some sensors have also shown abnormal behaviors that manifested themselves as spikes followed by a "stuck-at" fault as referred to in (Ni et al., 2009). In the latter work, researchers have found that 20% of the total temperature readings were faulty. Figure 5 plots the temperature data measured by 15 nodes in the IBRL deployment. The figure reveals a number of faulty readings that manifest themselves as spikes, noise, outliers, and stuck-at faults. We will discuss the characteristics of such faults in the coming chapter. Since temperature and

Figure 4. Sensor data for node 15. Missing readings for over 5 minutes are left blank.

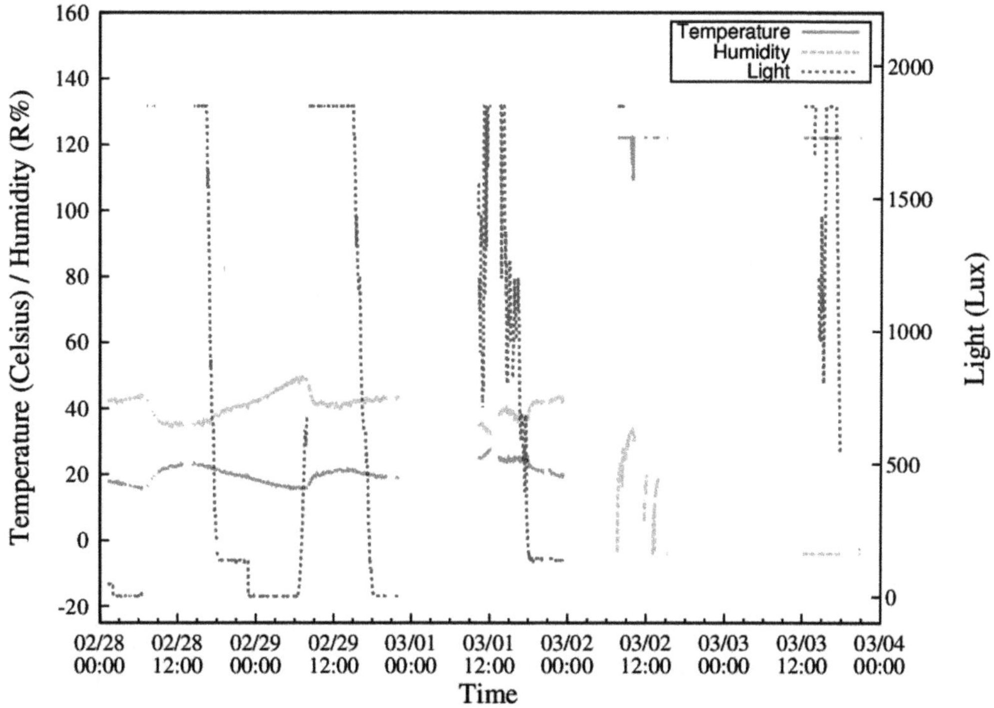

humidity readings are very highly correlated, one can imagine a similar plot for the humidity data. Notice that the majority of the temperature readings are spatially correlated (a node's temperature reading is close to its neighbors' readings at a time). Temperature readings are also temporally correlated within a short time window.

Figure 6 measures the frequency of faulty readings in the IBRL dataset. A number of sensor nodes (node 18, 29, 30, 32, 46, and 50) can only leverage 47% to 57% of its overall collected temperature and humidity readings. The rest of the readings are faulty and may be harmful to the decision making process. Temperature and humidity sensors exhibit spike, stuck-at, and noise faults whereas light sensors suffer from clipping. Figure 7 shows the actual number of clipping faults in the IBRL dataset. About half of the sensor nodes exhibit this behavior.

SENSORSCOPE

Overview

The SensorScope system (Ingelrest et al., 2010), developed at EPFL (Swiss Federal Institute of Technology Lausanne), underwent 7 outdoor deployments over a span of three years (2006-2008) in different locations in Switzerland at a scale ranging from 6 to 97 weather stations. Each weather station is equipped with a well-packaged Shockfish TinyNode (TinyNode, 2014) and nine environmental sensors which measure a handful of environmental observations every 2 minutes (e.g., air humidity and temperature, soil moisture, rain precipitation, solar radiation, watermark, and wind direction/speed). Sensor motes receive power from an NiMH battery that is rechargeable via a solar panel located at the top of the station. The two most significant deployments took place at: (i) a rock glacier located at 2500-meter high on

Figure 5. Temperature data of 15 nodes from the IBRL deployment

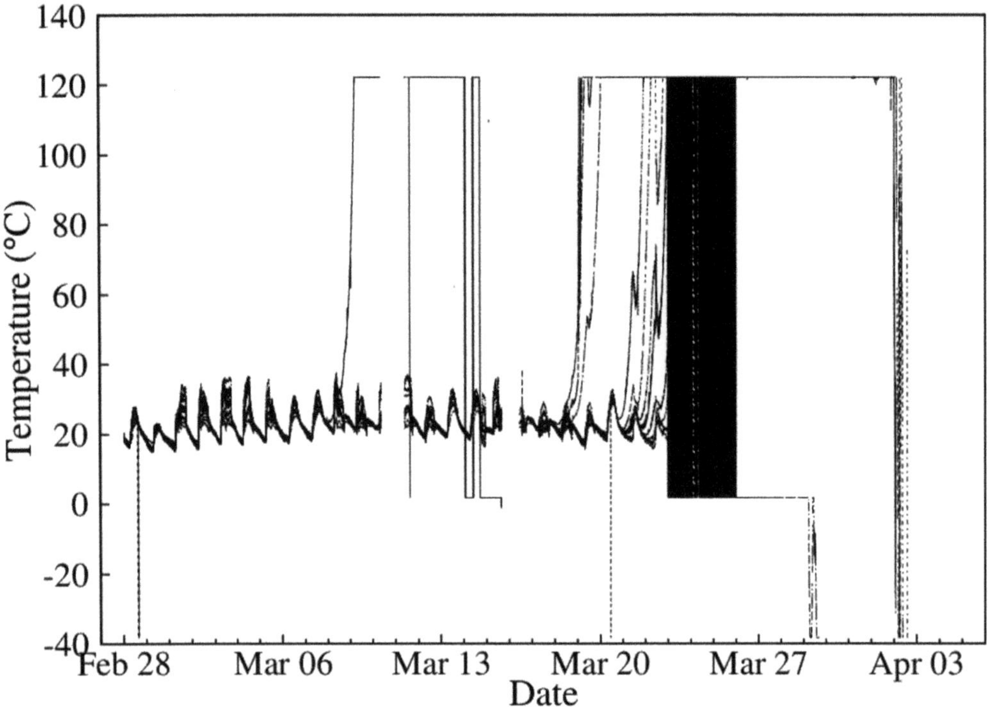

top of a mountain in Génépi, Switzerland, and; (ii) the 2400 meter high Grand Saint Bernard pass located between Switzerland and Italy. In both cases, weather data were collected over a multi-hop network of weather stations and transmitted by a GPRS-enabled gateway to a database server. Data are publicly accessible over the Internet and can be viewed using popular Web interfaces (i.e., Google Maps, Microsoft's SensorMap). The Génépi deployment had 16 stations scattered over 500x500 meters area reporting weather data during August-October 2007 (60 days). Detecting floods and rock falls in the area was the main reason behind this deployment as the mountain is a source of mud streams and rock releases due to the melting of underlying permafrosts. The second deployment consisted of 17 stations deployed in September 2007 for 45 days on a 900 meter long line over the Bernard pass to prevent avalanches (Barrenetxea et al., 2008b).

Observations

Data Yield

There were a few missing data points during both the Génépi and Grand St. Bernard deployments. As an example, during the former deployment, station 7 did not report any readings over more than a week during the middle stage of the deployment. Mostly, hardware failures due to short circuits accounted for this behavior. As reported in (Barrenetxea et al., 2008a), the total number of data points collected from the two deployments was 5800000 and 4300000 respectively. This results in a data yield of approximately 93% and 87% respectively. In order to be able to monitor the network, SensorScope collected status packets in addition to data packets. Status packets carried energy, network, and topology information that can help network administrators identify potential network and node failures.

Figure 6. Percentage of sensor data faults in the IBRL dataset. Faults mainly include spike faults, stuck-at faults, and noise faults exhibited by the temperature and humidity sensors as well as clipping faults in the light readings. The number of outliers is negligible.

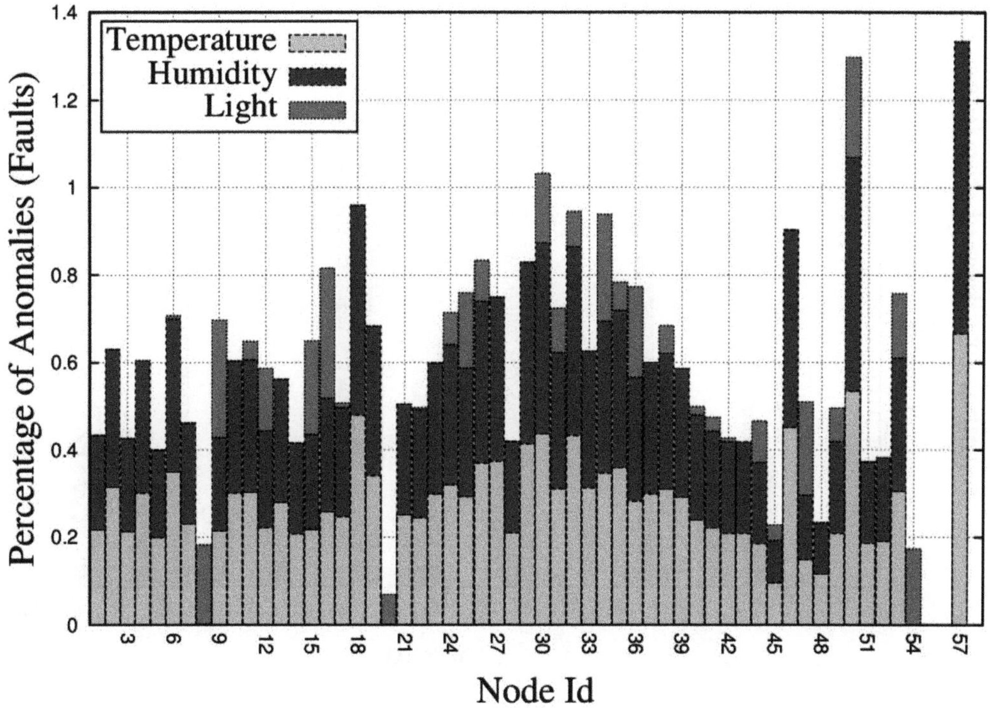

Measurement Faults

SensorScope researchers have pointed out several challenges associated with sensor network deployments. Some of the major challenges were: (i) packaging, (ii) calibration, (iii) detecting software bugs during deployment, and (iv) tracing unexpected failures. Sensor packaging is important but difficult. In both deployments, erroneous sensor readings were reported during cold and humid periods due to inadequate packaging. The latter allowed corrosions to form on the connector to the sensor board causing short circuits. This resulted in useless readings from all humidity sensors at Génépi mountain (Barrenetxea et al., 2008a) on October 6 as seen in Figure 8. It also caused the rain sensor of station 14 at the Bernard pass (Ingelrest et al., 2010) to report faulty precipitation values. The humidity sensor, a Sensirion SHT75 sensor, was embedded within a stacked-plates structure and a small adapting circuit. Whereas, the wires were protected using epoxy resin. Calibration is required before, during, and post deployment. After packaging, sensors were pre-calibrated by comparing their values to a reference station over several days. During pre-calibration, some temperature sensors reported an offset of over than 2° C, which is much higher than the expected offset of 0.3° C, and needed to be discarded. Once deployed, some sensors such as the wind direction sensor should also be calibrated to avoid significant errors in their measurements. Finally, it is important to re-calibrate the sensors after deployment to flag faulty readings.

Despite the fact that the majority of software bugs are easily detected and fixed before deployment and during indoor testing, some code bugs go undetected especially if configuration parameters

Figure 7. Number of clipping faults in the IBRL dataset. The histogram shows the number of anomalous readings caused by a clipping fault for each node.

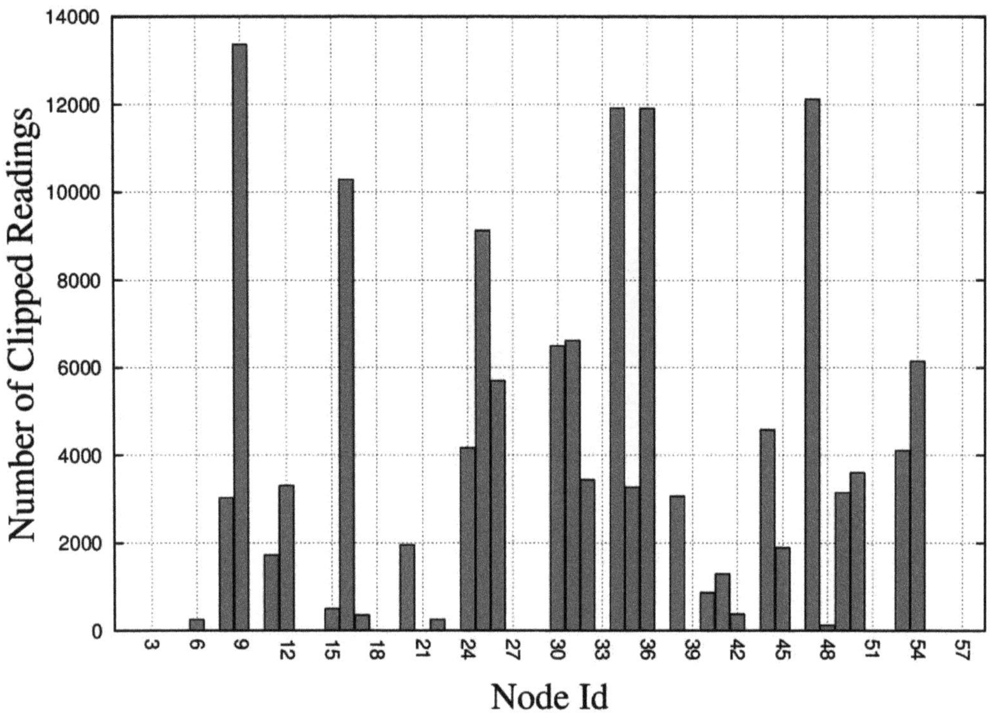

are modified right before or during deployment. As an example, by observing the wind speed data of station 6 during the Génépi deployment, developers discovered a subtle bug in the wind sensor's driver: an 8-bit counter was not adequate to store the number of revolutions the anemometer completes during one sampling period. The bug was not detected during pre-deployment testing since the sampling period was set much shorter (30 seconds). This bug affected all wind sensors alike, rendering the wind speed readings unusable over the period during which counter overflows were encountered. Figure 9 shows wind speed readings collected from two stations during the Génépi deployment. Clearly, all data before October 19 are useless.

Finally, uncorrelated sensor readings may be reported unexpectedly. In (Barrenetxea et al., 2008a), authors joked about one occasion where one of the weather stations continuously generated uncorrelated sensor readings (anomalous node), to find out that it was entirely covered with plastic wraps by construction workers to protect it from a near-by construction.

We have seen that, although sensors were carefully calibrated, anomalies still occurred due to other issues mentioned above (i.e., inadequate packaging, software bugs, unexpected behaviors). In (Sharma et al., 2007), for example, authors have identified SHORT faults (outliers) in three of the 31 weather stations deployed outdoor on the EPFL campus and it is not clear what caused such faults. Flagging and removing anomalies is usually done manually with human intervention. Manual anomaly filtering is time consuming, therefore, SensorScope researchers have called for better approaches that rely on powerful off-line analysis and rapid on-line detection algorithms to filter out outliers and hence to ameliorate sensor data quality.

Figure 8. Humidity readings spanning 3 days in October. Corrosions forming on the connector to the sensor board on October 6 due to high humidity accounts for the erroneous readings.

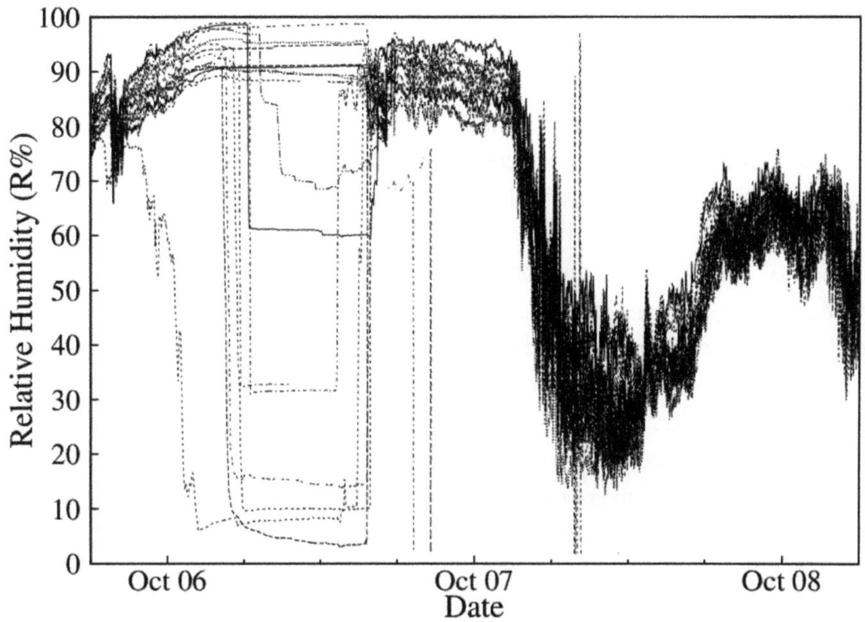

Figure 9. Wind speed readings collected from two stations during the Génépi deployment. Data prior to October 19 are unusable due to a software bug in the wind sensor's driver.

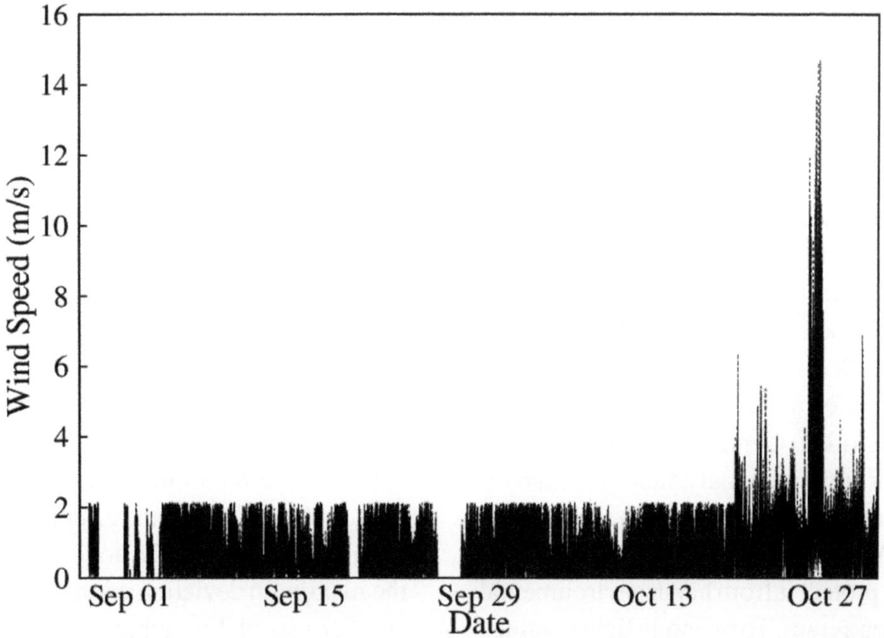

GREENORBS

Overview

The GreenOrbs project (Mo et al., 2009; Liu et al., 2011) aimed at monitoring the microclimate and ecological trends in the forest. The collected sensor data support several applications including fire risk evaluation, biodiversity and forestry research, monitoring of carbon dioxide absorbed by trees during the photosynthesis process, and estimation of canopy closure (Mo et al., 2009). GreenOrbs deployments were carried out at several locations in the forest at a scale ranging from 50 nodes to 349 nodes during a recent deployment (Shi et al., 2011). The long-term goal is to achieve a year-round kilo-scale sensor network deployment with 1000+ motes. The first deployment operated over 30 days starting July 2008 and consisted of 50 TinyOS-powered TelosB (Polastre et al., 2005) motes scattered in a 20,000 meter squared forest at the campus of Zhejiang Agriculture & Forestry University, Hangzhou, China. The deployed motes formed a multi-hop network of 6-hop diameter. The second deployment took place in March 2009 with 120 motes forming a 10-hop network. Since then, the network has been continuously expanded. The system employs the TinyOS conventional services for network-wide synchronization to enable duty cycling (FTSP (Maróti et al., 2004)), reconfiguration (DRIP (Tolle and Culler, 2005)), and sensor data collection (CTP (Gnawali et al., 2009)). Motes collect temperature, humidity, light, voltage, and carbon dioxide (CO2) levels. The network also collects routing, neighborhood, and node statistics information. The latter helps diagnosing the network to identify root causes such as link failures or any functional abnormal behavior in the network (Shi et al., 2011). Each sensor mote runs on 2 AA batteries and is enclosed inside a plastic box to protect it from harsh environmental conditions such as rain. To maintain light readings close to the true illuminance values, the top of the box is designed to be transparent, whereas holes on the sides of the box maintain temperature and humidity values close to those outside of the box. Finally, the box is mounted on a bracket to protect it from wild animals.

Observations

Data Yield

At a scale of 330 nodes, GreenOrbs researchers found that the network yield dropped from 60% to 10% when changing the sampling frequency from 20 minutes to 66 seconds (Liu et al., 2011). Further investigations revealed that the main reason behind packet losses was packet drops due to poor wireless channels or severe collisions. This accounted for 61.08% of all packet losses. Due to this severe case, about 10% of the nodes were forced to drop a packet 20 to 50 times after 30 failed retransmission attempts. The rest of the packet drops were due to forwarding queue overflows at the receiver side (i.e. severe data congestion). Ingress drops could also take place due to a software bug that leads to a locked memory state of the forwarding queue in CTP. All the ingress drops occurred on less than 5% of the nodes. In addition, a number of nodes never successfully reported data to the sink (Liu et al., 2011) due to software or hardware problems (Miao et al., 2011). Routing loops, on the other hand, were considered silent failures that are hard to detect but can drastically hinder the network performance.

Measurement Faults

Similar to previous deployments discussed above, calibration turned out to be very important but hard and time consuming. For instance, during a deployment test of 21 nodes (Mo et al., 2009), the maximum deviation of one light reading from another was 6KLux when all motes were placed under the same illuminance level and 8.41KLux when placed under different illuminance levels. Light readings were used to estimate the canopy closure (i.e., the percentage of the ground area vertically shaded by overhead foliage). This shows how important it is to pre-calibrate the sensors before deployment to eliminate instrumental errors and deviations in the light readings between correlated sensors, and hence arrive at an accurate estimation. Despite packaging and calibration efforts, GreenOrbs researchers still identified frequent pervasive sensing faults (Liu et al., 2011).

VIGILNET

Overview

As one last real-world example, we consider VigilNet, a sensor network deployment for target tracking and surveillance (He et al., 2006). It is an example of an event-driven application. The major challenge of VigilNet is to accurately estimate the speed and location of an intruder (e.g., a vehicle) while keeping the entire system stealthy and energy efficient. Stealthiness is very important when motes are deployed in hostile environments and therefore, are vulnerable to attacks. The size of a sensor node should be unobtrusively small and the network traffic should be minimized. Positioning sensor motes in hidden areas protects them from node compromise while minimizing network traffic reduces the chance of interception of RF signals and packet sniffing. In VigilNet, 70 Mica2 motes running TinyOS were deployed over a 280-feet long grassy road, 35 motes on each side of the road. The road may typically resemble a hostile area to be monitored. A sensor board is mounted on top of the mote and is capable of measuring magnetic fields generated by the movement of target objects, acoustic signals, motion, and distance of the object. The dual-axis magnetometer can sense slowly moving vehicles at a distance of approximately 8-10 feet. In addition, few long-range gateways are deployed in the sensor field which are responsible for relaying traffic to a distant base station. The base station is attached to a portable device such as a laptop which, in turn, triggers two cameras in the sensor field when a movement is detected.

Observations

In order to precisely estimate the position of the moving vehicle, the number of faulty readings generated by the magnetometers, due to potential change of power state or hardware failures, should be minimized. VigilNet employs aggregation where a leader node fuses correlated magnetic readings from its group members and discards readings from deviating members. The degree of aggregation (DOA), i.e., the number of group members, is a key parameter which controls the percentage of false positives (i.e., reporting an event while there is no vehicle in the vicinity of the sensor field) and false negatives (i.e., missing reports of moving vehicles). Without employing aggregation, the false positive rate is very high (60%). False positives are eliminated if DOA is set to 3, i.e., a leader node fuses readings from three members before reporting an event to the base station. In addition to on-line fault detection and elimination via in-network aggregation, VigilNet also applies off-line fault detection at the base station by analyzing the spatiotemporal correlation among consecutive events. Consequently, the base station sends power-off signals to shut down misbehaving nodes.

Similar to the deployments discussed previously, VigilNet researchers have also advocated for software calibration to eliminate sensor drifts as much as possible before deployment as well as repeated calibrations over time to adapt to changes in the environment. Unlike the deployments discussed above, however, sensor nodes in

VigilNet are more vulnerable to security attacks due to the nature of the application. For some experiments, which required a long duration of time, VigilNet researchers could not afford to deploy the system unattended due to security issues (He et al., 2006). Nodes in VigilNet can also be easily compromised. Despite precise calibration and fault detection via on-line fusion and off-line analysis, node compromise can drastically affect the fidelity and trustworthiness of the vehicle detection estimate. Once a node is compromised, injecting falsified readings in a smart way may harm the aggregation process or even worse these readings become hard to detect off-line. Hence, anomaly detection algorithms should also be able to identify such behavior and potentially eliminate readings from compromised nodes. Furthermore, these algorithms should detect malicious network behaviors or security attacks, such as the Denial of Service (DoS) attack, which may hinder the network performance (e.g., data yield, latency, etc.) or shorten its lifetime.

A SMARTER SUPPLY CHAIN

As can be seen, anomaly detection in existing sensor network deployments is a crucial challenge which needs to be addressed. Researchers have also envisioned future deployments at a very large scale. As an example, Franklin et al (Franklin et al., 2005) have envisioned a supply chain management (SCM) system, an example of a high fan-in (HiFi) system, where sensor data are initially collected and cleaned at the receptor-level which constitutes a number of field deployable units (FDUs). A FDU has a network of sensors and a set of RFID readers. FDUs may be located at different places in the supply chain, including store shelves, store checkouts, and manufacturing lines. Data are then aggregated and smoothed by Stargate-like devices (Stargate, 2014) at the second level (i.e., dock doors level). At the third level (i.e., warehouse level), sensor data are arbitrated and sensor streams are correlated across all devices by full-fledged centralized servers. The interior of the system constitutes the regional centers (fourth level) where data are validated, and the headquarters (fifth-level) at which data mining algorithms are run over the processed sensor data streams on the fly to foster the business decision making process and to provide better sales recommendations. A HiFi proof-of-concept prototype was proposed by the authors in (Franklin et al., 2005) which constitutes three-levels of the main system's hierarchy and uses TinyDB (Madden et al., 2005) for querying as well as TelegraphCQ (Chandrasekaran et al., 2003), an adaptive data stream processor. The authors have also demonstrated the significance of pushing data cleaning and data reduction (i.e., aggregation) functionalities to the edges of the HiFi system. This calls for anomaly detection algorithms at the integrated WSN-RFID (Roy et al., 2010) end. The CSAVA (clean, smooth, arbitrate, validate, and analyze) process discussed in the paper incorporates several data processing tasks at the various levels of the HiFi system: cleaning, fault detection, anomaly detection, calibration, stream correlation, and event monitoring. Cleaning and smoothing are achieved by simple filtering queries; sensor readings with an RSSI (Received Signal Strength Indicator) higher than a certain threshold will be discarded, and if the number of readings over a time window is less than a user-specified threshold, these readings are dropped. This calls for more advanced anomaly detection techniques at the receptor-level which may further improve accuracy against erroneous and noisy sensor data, detect abnormal sensor activities, and avoid fail-dirty cases where sensor nodes fail to report correct readings due to a possibly drained battery.

FUTURE RESEARCH DIRECTIONS

Rapid advancements in sensor technologies and the emergence of Internet of Things (IoT) are constantly allowing new types of sensor network deployments at various scales. As a future work, we plan to study recent deployments that employ state-of-the-art sensor devices and identify anomalous behaviors in such deployments. Regardless of how advanced sensor devices will become in the coming years, we believe that such devices will still be failure-prone. Moreover, despite the continuous improvements in networking protocols, networks are still susceptible to various failures that could dramatically decrease the percentage of sensor data yield. Finally, even if we could achieve a perfect network and use fault-free sensor devices, malicious behaviors are inevitable in certain environments.

CONCLUSION

This chapter studied six popular sensor network deployments established during the last decade: (i) Great Duck Island (GDI); (ii) Redwood Forests; (iii) Intel Berkeley Research Lab (IBRL); (iv) SensorScope; (v) GreenOrbs, and (vi) VigilNet deployments. Table 2 compares these deployments in terms of the type and number of motes being deployed, the sampling frequency used by every mote to emit a new reading, the duration and size of the deployment, and the packaging being used to cover the motes' circuitry. The chapter also identified a number of different anomalies prevalent in each of these deployments. Such anomalies hindered both the total data yield of the deployed network as well as the quality of the sensor data. Table 3 summarizes the different incidents of anomalies that took place in each deployment, their impact on data yield or sensor data quality, and the category of each incident being a network/node failure, a measurement fault, an actual event, or a malicious behavior. The table also summarizes the percentage of total data yield of each deployed network. Given these observations and the way sensor network deployments are being envisioned in the near future, it is critical for a sensor network to self-identify and then eliminate the different types of anomalies at earlier stages to improve data yield and/or enhance the quality of its collected data.

Table 2. Summary of the existing deployments discussed in this chapter

Name	Num of Motes	Sampling Freq	Mote Type	Duration	Area	Packaging
Great Duck Island	49	5 Minutes	Mica2Dot	115 Days	221x71m^2	O-rings/conformal coating
	98	20 Minutes				
Redwood Forests	35	5 Minutes	Mica2Dot	44 Days	70-meter	White skirt cover
IBRL	54	30 Seconds	Mica2/Mica2Dot	30 Days	Intel Lab	No packaging
SensorScope	16	2 Minutes	TinyNode	60 Days	500x500m^2	Stacked-plates structure/ epoxyresin
	17			45 Days	900-meter	
GreenOrbs	50-349	10 Minutes	TelosB	30 Days-1 year	20,000 m^2	Plastic box
VigilNet	70	n/a	Mica2	n/a	280-feet	No packaging

Table 3. Summary of anomalies prevalent in the deployments discussed in this chapter

Nae	Data Yield	Observed Anomalies		
		Anomaly	**Impact**	**Category**
Great Duck Island	28%	gateway outage	main cause of the very low data yield	network failure
		rapid depletion of node's battery	5 motes failed after 5 days	node failure
		permanent node outages due to very low humidity	55% of nodes failed within 2 days	
		broken antennas	only 78 out of 150 motes were recovered	
		outliers, spikes, calibration errors	about *half* of the temperature and humidity readings were faulty	measurement faults
Redwood Forests	49%	gateway outage	outages at the beginning of the deployment accounted for most of the missing packets	network failure
		rapid depletion of node's battery	5 nodes failed during the second week	node failure
		low battery, out-of-range/clipping faults, calibration errors	resulted in several erroneous readings	measurement faults
IBRL	≈50%	unknown network failure	4 nodes failed after the first few days	network failure
		network contention or interference	high number of packet loss, nodes transmitted only 30% to 82% of their packets	
		base station breakage	entire network went offline for two intermittent days	
		node reboots	three nodes reported packets too late in the deployment lifetime	node failure
		missing data points	two nodes reported packets with missing sensor data points	unknown
		low battery, spikes, stuck-at faults, clipping faults, noise, outliers	20% of the total temperature readings were faulty, some nodes leveraged only 47% to 57% of their overall collected temperature and humidity readings, about half of the sensor nodes suffered clipping in light readings	measurement faults
SensorScope (Génépi)	≈93%	hardware failure due to short circuit	station 7 did not report readings for over a week	node failure
		stuck-at faults, noise, outliers	all humidity sensors reported faulty readings on October 6	measurement faults
		calibration errors	temperature sensors reported a calibration offset of more than 2°C	
		software bug	a bug in wind sensor's driver rendered wind data from all stations useless	
		unexpected event	one station covered by plastic wraps continuously generating erroneous readings	
SensorScope (St. Bernard)	≈87%	noise, outliers	faulty rain precipitation values reported by station 14	measurement faults
GreenOrbs	60% - 10%	packet drops due to link failures	accounted for 61.08% of all packet losses	network failure
		ingress drops	all the ingress drops (38.92% of all packet drops) occur on less than 5% nodes	
		routing loops	accounted for a good number of silent failures	
		software bugs, hardware problems	a number of nodes never reported data to the sink	node failure
		calibration errors	deviation of 6KLux and 8.41KLux in light readings	measurement faults
VigilNet	n/a	a moving vehicle	if not estimated correctly, the deployment does not achieve its purpose	event
		low battery, hardware failure	false positive rate is very high (i.e., 60%)	measurement faults
		calibration errors	sensor drifts pre-deployment and over the deployment time	
		security attacks	could not afford to deploy the system unattended	malicious behavior

REFERENCES

Ahonen, T., Virrankoski, R., & Elmusrati, M. (2008). Greenhouse monitoring with wireless sensor network. In *Proceedings of the IEEE/ASME International Conference on Mechtronic and Embedded Systems and Applications* (pp. 403–408). doi:10.1109/MESA.2008.4735744

Anthony, D., Bennett, W. P., Vuran, M. C., Dwyer, M. B., Elbaum, S., Lacy, A., & Wehtje, W. (2012). Sensing through the continent: towards monitoring migratory birds using cellular sensor networks. In *Proceedings of the 11th international conference on Information Processing in Sensor Networks* (pp. 329–340).

Barrenetxea, G., Ingelrest, F., Schaefer, G., & Vetterli, M. (2008a). The hitchhiker's guide to successful wireless sensor network deployments. In *Proceedings of the 6th ACM conference on Embedded network sensor systems* (pp. 43–56). doi:10.1145/1460412.1460418

Barrenetxea, G., Ingelrest, F., Schaefer, G., & Vetterli, M. (2008b). Wireless sensor networks for environmental monitoring: The sensorscope experience. In *IEEE International Zurich Seminar on Communications* (pp. 98–101).

Bijwaard, D. J. A., van Kleunen, W. A. P., Havinga, P. J. M., Kleiboer, L., & Bijl, M. J. J. (2011). Industry: using dynamic wsns in smart logistics for fruits and pharmacy. In *Proceedings of the 9th ACM Conference on Embedded Networked Sensor Systems* (pp. 218–231). doi:10.1145/2070942.2070965

Buonadonna, P., Gay, D., Hellerstein, J., Hong, W., & Madden, S. (2005). Task: sensor network in a box. In *Proceedings of the Second European Workshop on Wireless Sensor Networks*, (pp. 133–144). doi:10.1109/EWSN.2005.1462005

Chandrasekaran, S., Cooper, O., Deshpande, A., Franklin, M. J., Hellerstein, J. M., Hong, W., & Shah, M. A. (2003). Telegraphic: Continuous dataflow processing for an uncertain world. In *Proceedings of the CIDR Conference*.

Dawson-Haggerty, S., Lanzisera, S., Taneja, J., Brown, R., & Culler, D. (2012). @scale: Insights from a large, long-lived appliance energy wsn. In *Proceedings of the 11th international conference on Information Processing in Sensor Networks*, (pp. 37–48). doi:10.1145/2185677.2185683

Franklin, M. J., Jeffery, S. R., Krishnamurthy, S., & Reiss, F. (2005). Design considerations for high fan-in systems: The HiFi approach. In *Proceedings of the CIDR Conference* (pp. 290–304).

Gnawali, O., Fonseca, R., Jamieson, K., Moss, D., & Levis, P. (2009). Collection tree protocol. In *Proceedings of the 7th ACM Conference on Embedded Networked Sensor Systems* (pp. 1–14). doi:10.1145/1644038.1644040

He, T., Krishnamurthy, S., Luo, L., Yan, T., Gu, L., Stoleru, R., & Krogh, B. (2006). VigilNet: An integrated sensor network system for energy-efficient surveillance. *ACM Transactions on Sensor Networks*, *2*(1), 1–38. doi:10.1145/1138127.1138128

Ingelrest, F., Barrenetxea, G., Schaefer, G., Vetterli, M., Couach, O., & Parlange, M. (2010). SensorScope: Application-specific sensor network for environmental monitoring. *ACM Transactions on Sensor Networks*, *6*(2), 17:1–17:32.

Intel Lab Data. (2012). *Intel lab data*. Retrieved December 2, 2012, from http://db.csail.mit.edu/labdata/labdata.html

Kim, S., Pakzad, S., Culler, D., Demmel, J., Fenves, G., Glaser, S., & Turon, M. (2006). Wireless sensor networks for structural health monitoring. In *Proceedings of the 4th international conference on Embedded networked sensor systems* (pp. 427–428). doi:10.1145/1182807.1182889

Liu, Y., He, Y., Li, M., Wang, J., Liu, K., Mo, L., & Li, X.-Y. (2011). Does wireless sensor network scale? A measurement study on greenorbs. In *Proceedings of IEEE INFOCOM* (pp. 873 –881).

Madden, S. R., Franklin, M. J., Hellerstein, J. M., & Hong, W. (2005). TinyDB: An acquisitional query processing system for sensor networks. *ACM Transactions on Database Systems*, *30*(1), 122–173. doi:10.1145/1061318.1061322

Maróti, M., Kusy, B., Simon, G., & Lédeczi, A. (2004). The flooding time synchronization protocol. In *Proceedings of the 2nd international conference on Embedded networked sensor systems*, (pp. 39–49). doi:10.1145/1031495.1031501

MEMSIC. (2014). *Memsic wireless sensor networks*. Retrieved October 10, 2014, http://www.memsic.com/wireless-sensor-networks/

Miao, X., Liu, K., He, Y., Liu, Y., & Papadias, D. (2011). Agnostic diagnosis: Discovering silent failures in wireless sensor networks. In *Proceedings of IEEE INFOCOM* (pp. 1548–1556). doi:10.1109/INFCOM.2011.5934945

Mo, L., He, Y., Liu, Y., Zhao, J., Tang, S.-J., Li, X.-Y., & Dai, G. (2009). Canopy closure estimates with greenorbs: sustainable sensing in the forest. In *Proceedings of the 7th ACM Conference on Embedded Networked Sensor Systems* (pp. 99–112). doi:10.1145/1644038.1644049

Ni, K., Ramanathan, N., Chehade, M. N. H., Balzano, L., Nair, S., Zahedi, S., ... Srivastava, M. (2009). Sensor network data fault types. *ACM Transactions of Sensor Networks, 5*(3), 25:1–25:29.

Polastre, J., Szewczyk, R., & Culler, D. (2005). Telos: enabling ultra-low power wireless research. In *Proceedings of the 4th international symposium on Information processing in sensor networks.*

Rajasegarar, S., Bezdek, J. C., Leckie, C., & Palaniswami, M. (2007). Analysis of anomalies in IBRL data from a wireless sensor network deployment. In *Proceedings of the 2007 International Conference on Sensor Technologies and Applications (SENSORCOMM)*, (pp. 158–163). Washington, DC, USA. doi:10.1109/SENSORCOMM.2007.4394914

Ramanathan, N., Balzano, L., Burt, M., Estrin, D., Harmon, T., Harvey, C.Srivastava, M. (2006). *Rapid deployment with confidence: Calibration and fault detection in environmental sensor networks* (Tech. Rep.). Center for Embedded Networked Sensing, UCLA and Department of Civil and Environmental Engineering, MIT.

Roy, S., Jandhyala, V., Smith, J., Wetherall, D., Otis, B., Chakraborty, R., & Sample, A. (2010). RFID: From supply chains to sensor nets. *Proceedings of the IEEE, 98*(9), 1583–1592. doi:10.1109/JPROC.2010.2053690

Sharma, A., Golubchik, L., & Govindan, R. (2007). On the prevalence of sensor faults in real-world deployments. In *Proceedings of the 4th Annual IEEE Communications Society Conference on Sensor, Mesh and Ad Hoc Communications and Networks*, (pp. 213–222). doi:10.1109/SAHCN.2007.4292833

Shi, L., Liao, Q., He, Y., Li, R., Striegel, A., & Su, Z. (2011). SAVE: Sensor anomaly visualization engine. In *Proceedings of the IEEE Conference on Visual Analytics Science and Technology*, (pp. 201–210). doi:10.1109/VAST.2011.6102458

Sorber, J., Kostadinov, A., Garber, M., Brennan, M., Corner, M. D., & Berger, E. D. (2007). Eon: a language and runtime system for perpetual systems. In *Proceedings of the 5th international conference on Embedded networked sensor systems* (pp. 161–174). New York, NY, USA. doi:10.1145/1322263.1322279

Stargate (2014). *Stargate: Distinguishing capabilities*. Retrieved October 10, 2014, from http://platformx.sourceforge.net/home.html

Szewczyk, R., Mainwaring, A., Polastre, J., Anderson, J., & Culler, D. (2004a). An analysis of a large scale habitat monitoring application. In *Proceedings of the 2nd international conference on Embedded networked sensor systems*, (pp. 214–226). doi:10.1145/1031495.1031521

Szewczyk, R., Polastre, J., Mainwaring, A. M., & Culler, D. E. (2004b). Lessons from a sensor network expedition. In *Proceedings of the First European Workshop on Wireless Sensor Networks* (pp. 307–322).

Tiny, O. S. (2014). *Tinyos home page*. Retrieved October 10, 2014, from http://www.tinyos.net/

TinyNode. (2014). *Tinynode: Real life wireless sensor networks*. Retrieved October 10, 2014, from http://www.tinynode.com/

Tolle, G., & Culler, D. (2005). Design of an application-cooperative management system for wireless sensor networks. In *Proceedings of the Second European Workshop on Wireless Sensor Networks*, (pp. 121–132). doi:10.1109/EWSN.2005.1462004

Tolle, G., Polastre, J., Szewczyk, R., Culler, D., Turner, N., Tu, K., & Hong, W. (2005). A macroscope in the redwoods. In *Proceedings of the 3rd international conference on Embedded networked sensor systems (SenSys)* (pp. 51–63). doi:10.1145/1098918.1098925

Werner-Allen, G., Lorincz, K., Johnson, J., Lees, J., & Welsh, M. (2006). Fidelity and yield in a volcano monitoring sensor network. In *Proceedings of the 7th symposium on Operating systems design and implementation (OSDI)* (pp. 381–396).

ADDITIONAL READING

Barrenetxea, G., Ingelrest, F., Schaefer, G., & Vetterli, M. (2008). The hitchhiker's guide to successful wireless sensor network deployments. In *Proceedings of the 6th ACM conference on Embedded network sensor systems* (pp. 43–56). doi:10.1145/1460412.1460418

Buonadonna, P., Gay, D., Hellerstein, J., Hong, W., & Madden, S. (2005). Task: sensor network in a box. In *Proceedings of the Second European Workshop on Wireless Sensor Networks*, (pp. 133–144). doi:10.1109/EWSN.2005.1462005

Franklin, M. J., Jeffery, S. R., Krishnamurthy, S., & Reiss, F. (2005). Design considerations for high fan-in systems: The HiFi approach. In *Proceedings of the CIDR Conference* (pp. 290–304).

He, T., Krishnamurthy, S., Luo, L., Yan, T., Gu, L., Stoleru, R., & Krogh, B. (2006). Vigilnet: An integrated sensor network system for energy-efficient surveillance. *ACM Transactions on Sensor Networks*, *2*(1), 1–38. doi:10.1145/1138127.1138128

Ingelrest, F., Barrenetxea, G., Schaefer, G., Vetterli, M., Couach, O., & Parlange, M. (2010). Sensorscope: Application-specific sensor network for environmental monitoring. *ACM Transactions on Sensor Networks, 6*(2), 17:1–17:32.

Liu, Y., He, Y., Li, M., Wang, J., Liu, K., Mo, L., & Li, X.-Y. et al. (2011). Does wireless sensor network scale? a measurement study on greenorbs. In *Proceedings of IEEE INFOCOM* (pp. 873 –881).

Miao, X., Liu, K., He, Y., Liu, Y., & Papadias, D. (2011). Agnostic diagnosis: Discovering silent failures in wireless sensor networks. In *Proceedings of IEEE INFOCOM* (pp. 1548–1556). doi:10.1109/INFCOM.2011.5934945

Mo, L., He, Y., Liu, Y., Zhao, J., Tang, S.-J., Li, X.-Y., & Dai, G. (2009). Canopy closure estimates with greenorbs: sustainable sensing in the forest. In *Proceedings of the 7th ACM Conference on Embedded Networked Sensor Systems* (pp. 99–112). doi:10.1145/1644038.1644049

Ni, K., Ramanathan, N., Chehade, M. N. H., Balzano, L., Nair, S., Zahedi, S., Srivastava, M. (2009). Sensor network data fault types. *ACM Transactions on Sensor Networks, 5*(3), 25:1–25:29.

Rajasegarar, S., Bezdek, J. C., Leckie, C., & Palaniswami, M. (2007). Analysis of anomalies in ibrl data from a wireless sensor network deployment. In *Proceedings of the 2007 International Conference on Sensor Technologies and Applications (SENSORCOMM)* (pp. 158–163). Washington, DC, USA. doi:10.1109/SENSORCOMM.2007.4394914

Sharma, A., Golubchik, L., & Govindan, R. (2007). On the prevalence of sensor faults in real-world deployments. In *Proceedings of the 4th Annual IEEE Communications Society Conference on Sensor, Mesh and Ad Hoc Communications and Networks* (pp. 213–222). doi:10.1109/SAHCN.2007.4292833

Szewczyk, R., Mainwaring, A., Polastre, J., Anderson, J., & Culler, D. (2004a). An analysis of a large scale habitat monitoring application. In *Proceedings of the 2nd international conference on Embedded networked sensor systems* (pp. 214–226). doi:10.1145/1031495.1031521

Szewczyk, R., Polastre, J., Mainwaring, A. M., & Culler, D. E. (2004b). Lessons from a sensor network expedition. In *Proceedings of the First European Workshop on Wireless Sensor Networks* (pp. 307–322).

Tolle, G., Polastre, J., Szewczyk, R., Culler, D., Turner, N., Tu, K., & Hong, W. (2005). A macroscope in the redwoods. In *Proceedings of the 3rd international conference on Embedded networked sensor systems* (pp. 51–63). doi:10.1145/1098918.1098925

KEY TERMS AND DEFINITIONS

Anomaly: A behavior that is inconsistent with the expected normal behavior. In sensor networks, network and node failures, measurement faults, and malicious network or data behaviors are all considered anomalies. Sensor literature also refers to anomalous sensor readings as outliers, abnormal, faulty, or malicious data.

Clipping Fault: A series of sensor readings that maintain steadiness over a period of time during which those readings exceed the maximum value allowed by the Analog to Digital converter (ADC).

Sensor Data Quality: The amount of deviation of the observed sensor data points from the expected phenomena.

Sensor Data Yield: The ratio between the number of packets successfully delivered to the base station and the expected number of packets generated by all sensor nodes in the network.

Sensor Network Deployment: An installment of a network of sensors at a certain area in a single or multi hop fashion.

Spike Fault: A reading highly deviated from the expected sensor reading which spans over more than one reading consecutively and may or may not return to the normal behavior afterwards.

Stuck-At Fault: A series of sensor readings that maintain steadiness for the rest of the node's lifetime or may return back to normal after a significant amount of time.

ENDNOTES

1. Missing periods are measured using a time gap of 5 minutes.

Chapter 7
Reliability Evaluation Methods for Resilient Wireless Sensor Networks

M. Saeed Ansari
Iran University of Science and Technology, Iran

Ali Mahani
Shahid Bahonar University of Kerman, Iran

ABSTRACT

Wireless sensor networks (WSNs) are deployed for various applications such as military applications, environmental monitoring, security and surveillance, health care applications and so on. In most of these applications, reliable data transport is of great importance and also a facet of quality of service (QoS). This chapter discusses reliable data transport approaches and protocols. The presented protocols should be energy efficient besides they guarantee reliability in data transmission. The authors review protocols mostly based on packet and event reliability taking advantage of retransmission and redundancy mechanisms. In addition, network component failures and fault tolerant techniques should be also considered in reliable WSN designs. Since different reliability evaluation modeling approaches is presented. Finally the authors highlight the challenges to WSNs reliability enhancement and some future research directions are addressed in particular.

INTRODUCTION

Wireless sensor network (WSN) is a well-studied research area and it has drawn the attention of many researchers in the last two decades. WSNs are probably the best candidate for the new generation of control and monitoring systems (Akyildiz et al., 2002, Yick et al., 2008). Recent advances in wireless communication technology and micro-electro-mechanical systems (MEMS) have enabled taking advantage of tiny, low-power and low-cost sensors in wireless sensor networks. This reduction in size and cost of sensors has made them vulnerable to environmental noises which increases information inaccuracy. The other significant issue which is actually one of the major challenges in WSNs is to reliably transport of successfully sensed data over multiple hops toward sink(s)

DOI: 10.4018/978-1-4666-8251-1.ch007

(Willig et al., 2005). As is said in (Mahmood et al., 2011, Pereora et al., 2007), reliability can be classified into two different categories i.e.

- Packet reliability
- Event reliability
- Hop-by-Hop reliability
- End-to-End reliability

Packet reliability deals with reliably transport of collected data to the sink(s) and event reliability is a factor to determine how observable an occurred event in the environment is. In fact the difference between packet reliability and event reliability is that when we talk about event reliability there is no need that all the messages reach to the sink(s), just one notification of each event is enough. Since there are strong limitations in terms of processing, memory, power and etc. energy efficient methods are presented in order to increase network's lifetime. These approaches mostly put the sensors in the sleep mode to optimize their power consumption which leads to multi-hop data transmission. As mentioned before, successful delivery of the sensed data despite collisions, congestions and channel errors is another problem in WSNs. Both Hop-by-Hop reliability and End-to-End reliability deal with this problem. In Hop-by-Hop, all the nodes in the path are responsible for a reliable transmission whereas, in End-to-End only the source node and destination node (mostly the sink) are responsible. To increase packet or event reliability retransmission and redundancy mechanisms are offered. As is clear through End-to-End reliability and Hop-by-Hop reliability definitions, in End-to-End reliability just the source node retransmit the lost data whereas, in Hop-by-Hop reliability every single node in the path is able to retransmit the lost information by holding the data in its buffers (Mahmood et al., 2011). Redundancy mechanisms can be used instead of retransmission techniques. What a redundancy technique does is to add some overhead data to the original one which allows the receiver to recover the lost data. So to guarantee the reliability the lost packets should be retransmitted or some redundancy technique should be taken which both are an overhead on the energy consumption of the network. So to enhance the reliability, network resources should be considered besides reliable data transmission.

Consider network resources, one of the key steps offered in the design of a WSN is reliability modeling. It can be used to estimate networks elements performances in order to advance the design of the network. Usually to find out the reliability of a WSN, the designers do a three steps procedure. They first develop the network measure reliability mathematical model. Then they figure out the model's parameters and finally based on the model and measured parameters they compute network reliability (Venkatesan et al., 2013).

BACKGROUND

Based on the nature of the WSNs, each node in the network tends to send their collected information to the sink(s). Clearly there would be traffic of transmitted packets around the sink(s) and as there is strong limitation in the buffering space in sensor nodes, this lead to packet loss. Of course this is not the only reason of packet losses in the network. Other reasons such as collisions between packets, node failures (due to software/hardware errors, energy depletion and etc.) and others may cause packet losses. As mentioned before despite all these faults, the applications expect packet reliability in WSNs.

In general to increase network reliability, WSNs constraints should be taken into account. WSNs constraints are power, transmission range, sensing radius, processing and others. As mentioned before the three important duties in WSNs is to sense the environment, processing data and also data transmission. Each of these three phases, are supposed to meet different requirements.

For example in sensing phase, coverage plays an important role; so is connectivity in communication phase. As communication is done through unreliable wireless links, several approaches have been proposed.

One of the offered techniques to guarantee packet delivery is retransmission. The retransmission mechanism is achieved by the use of acknowledgment which means the transmitter node waits for the acknowledgment of the next hop node in the path between itself and the sink. The most common acknowledgment mechanisms are:

- Explicit Acknowledgment
- Implicit Acknowledgment
- Negative Acknowledgment

In the explicit acknowledgment (EACK) mechanism, the receiver node sends back a message to the transmitter node after successfully achieving of the packet. It is the same for the negative acknowledgment (NACK). In fact in NACK mechanism the receiver can notify the sender node that a missing has occurred by looking for holes in the sequence number space (Willig et al., 2005). Although these tow mechanisms seem to be able to recover packet losses, they put a high overhead on the network traffic and due to their high energy consumption may not be a good candidate for most of the WSN applications which care about energy efficiency. The other existing mechanism which is actually more energy efficient is implicit acknowledgment (IACK). In this mechanism if the sent packet overheard being forwarded twice, it means that the packet has successfully received and this is a simultaneous acknowledgment mechanism which reduces message overhead (Woo et al., 2004). But what is common between these three mechanisms is the need to caching the packets for probable retransmission requests. Based on sensor nodes memory limitation caching and buffering a huge number of packets is not possible. This means that the presented retransmission-based reliability protocols are not scalable and not suitable for large networks consist of the huge number of sensor nodes. Another mechanism to achieve reliable transmission is redundancy-based reliability. The main idea of this technique is to transmit multiple copies of each packet (Rizzo, 1997). The packet is divided into m fragments by the sender node; these m fragments get encoded by adding k redundant fragments and the packet which is now made up of $m + k$ fragments is sent to the sink. If the total number of received fragments by the sink is equal or greater than m, the receiver can reconstruct the original data. Finally the last offered approach is multipath routing. The main idea of this approach is using multipath between sender and receiver nodes. The data packet is divided into k parts and each part is sending through one of the k founded. It worth noting that the selected paths are actually the first shortest k paths. These paths may be partially or fully disjoint. Obviously this technique will cause some overhead and consumes more energy than single path routing.

Here a brief description of existing data transfer protocols based on retransmission and redundancy mechanisms is presented i.e.

Retransmission-Based Reliability Protocols: This is an accepted fact that End-to-End reliability model is not a practical model for new wireless sensor networks. This model is supposed to guarantee a successful path between source and destination (sink) so uses all retransmission and other existing techniques which make strong limitations on bandwidth besides energy inefficiency (Pereora, 2007. So Hop-by-Hop reliability model which has proven to be more energy efficient is used instead. It is why traditional transport protocols such as TCP (transport control protocol) which was used for reliable data delivery by providing End-to-End reliability (Postel, 1981) is not a suitable mechanism for modern WSNs with constrained resources anymore. Off course there are still applications need guaranteed delivery so researchers have tried to advance TCP. In addition transport protocols can also be divided into two categories from another aspect i.e.Upstream and Downstream.

In the upstream, collected information by sensor nodes is transmitted to the sink(s) whereas in the downstream the sink(s) generated data is transmitted to the sensor nodes (Zheng et al., 2009). As an example of downstream data delivery consider sensor nodes software updating process. Since upstream data delivery is more common, it has drawn the attention of many researchers where downstream data delivery has not been focused on the researches as much as upstream one (Levis et al., 2004, Wan et al., 2005 and Park et al., 2008).

Here we are going to present taxonomy of reliable data transport protocols for wireless sensor networks but not in much detail. The data delivery direction, acknowledgment and other mentioned concepts is presented for each of the presented protocols (Mahmood et al., 2011, Pereora et al., 2007).

Reliable Multi-Segment Transport (RMST): Authors in (Stann et al., 2003) introduced a reliable data transport protocol. Their presented protocol uses diffusion routing. They called their protocol Reliable Multi-Segment Transport (RMST). RMST is able to work in two modes i.e. caching and non-caching mode. While working in caching mode the packets are stored for probable retransmission requests. Their proposed protocol suffers from extra overhead. In fact they retransmit packets every lost packet to guarantee reliability where there is no need for that. If $n-1$ of n similar packets are lost, no data collection inaccuracy has occurred. This protocol is suitable for upstream data delivery direction as it is not responsible for any detected holes in packets sequence in a downstream direction.

Event-To-Sink Reliable Transport (ESRT): The concept of event reliability was first presented in 2005 when Akan and Akyildiz introduced Event-to-Sink-Reliable-Transport (ESRT). This protocol is an end-to-end upstream transport protocol which guarantees transmission of every occurred event in the network to the sink while it is not responsible for the delivery of each packet. In other word, not all the packets are guaranteed to be delivered to the sink(s) but events are certainly reported. This protocol is actually the first proposed protocol which cares congestion while providing reliability. Based on the number of received packets within a time interval the sink calculates how reliable an event successful detection is. In fact that a CN bit is added to the data packet header which is set when an intermediate node in the path detects congestion (congestion is detected based upon nodes buffer). This packet sends to the sink; sink as the network manager decides about the selected path based on the received packets. The sensor nodes reporting frequency is adjust by the sink in terms of current and desired event reliabilities (Akan et al., 2005). Energy inefficiency is the main disadvantage of ESRT. Of course this should be considered in the application using ESRT that this protocol doesn't guarantee desired event reliability in case of congestion occurrence.

Sensor Transmission Control Protocol (STCP): In the same time Iyler, Gandham and Venkatesan proposed their protocol named Sensor Transmission Control Protocol (STCP). STCP is a transport protocol which provides both congestion control and reliability just like ESRT. It is an End-to-End upstream transport retransmission based protocol. They add a bit to the packet header to determine congestion. When the packet is transmitting through the network, it passes multiple nodes, in each node, if the length of the queue is more than a threshold, the bit is set. The sink tries to find a path for the notified sensor node to guarantee data delivery and to reduce congestion (Iyer et al., 2005).

Reliable Bursty Convergecast (RBC): Authors in (Zhang et al., 2005) proposed Reliable Bursty Convergecast (RBC) protocol. This protocol uses an implicit acknowledgment Hop-by-Hop reliable data transform mechanism. It stores the packets for probable retransmission requests in each hop of the path between source and destination. The mentioned protocol employs a windowless block-acknowledgment mechanism in order to detect packet losses more efficient. In this method information of all successfully received packets is added to the data packet header to let the sender retransmit (if required) packets based on this information. In the case of not delivering expected information to the sender in a period of time (this period is set based upon the next hop node queue), the set timer will be expired and Hop-by-Hop retransmission starts. In this method each node has a virtual queue. This virtual queue is designed to let the smaller number of retransmitted packets a higher priority; this is called intra-node scheduling. RBC also allows a node with a fuller buffer to have a higher priority; hence it can seize the channel with more chances. This ability of RBC is called inter-node scheduling. Intra-node and inter-node scheduling help RBC to avoid congestion.

Asymmetric and Reliable Transport (ART): Asymmetric and Reliable Transport (ART) is presented by Tezcan and Wang (Tezcan et al., 2007). It uses clustering technique to achieve upstream End-to-End reliability and downstream query reliability. Hence it provides bidirectional reliability. They select some of the sensor nodes within the network based on the residual battery power of them and name them essential nodes (e-node). The e-nodes have the best energy level among all the distributed nodes. The selected e-nodes would be cluster heads. They have considered an event notification bit in the packet header which can be enabled by e-nodes. When event notification bit is enabled, the first event message is sent to the sink and a timer is set simultaneously. The timer works till an acknowledgment receives. The timers have a predefined expiration time that the ACK signal should be received in that time, else the sensor node, consider the sent packet as a lost packet and retransmit it. As mentioned the event notification bit exists in e-nodes, so packet loss at non-essential nodes is not recoverable. As mentioned ART provide downstream query reliability. In fact the sink sends query to the e-node. The e-nodes are supposed to check the sequence order and in case of fragment loss detection send back a negative acknowledgment to the sink. Following above procedure leads to congestion control.

GARUDA: Park et al. (Park et al., 2008) proposed GARUDA. This protocol which is also designed with Hop-by-Hop reliability provides both downstream multicast and broadcast reliability. Since their proposed protocol uses Wait-for-First-Packet (WFP) pulse, it can send single packet data however generation WFP pulses put hardware overhead on the sensor node design and increases power consumption. This protocol selects some of sensor nodes as core nodes and allows them retransmit lost packets. These core nodes selections create a high latency and this make GARUDA a non-practical solution for large networks with huge number of sensor nodes. Hence the reliability is achieved by these retransmission capabilities of core nodes (in this protocol no congestion control system is predicted). It also use a type of bit map, called A-map, to recognize a core node packets. The other advantage of using A-map is to prevent NACK implosion.

We just mentioned some of the most common retransmission based reliability protocols. More algorithms are presented in (Wanet al., 2005, Zhou et al., 2008, Shaikh et al., 2010, Le et al., 2009 andMahmood et al., 2011).Table 1 summarizes the above description.

Redundancy-Based Reliability Protocols: As mentioned before retransmission despite the overhead it put on the network is the most common technique to achieve reliability. In wired networks a bit level corruption is considered as a whole packet loss whereas in wireless networks a small bit errors in a packet (caused by noisy communication channel (Rizzo, 1997)) is acceptable and there are recovery mechanisms. The lost data can be recovered using Forward Error Correction (FEC) techniques. Using error correction techniques reduces transmission overhead caused

Table 1. Reliable retransmission based transfer protocols for WSN

Transport Protocol	WSN Direction	Acknowledgment	Loss Recovery	Type of Reliability
RMST	Upstream	NACK	Hop-by-Hop	Packet Driven
ESRT	Upstream	Implicit	End-to-End	Event Driven
STCP	Upstream	ACK\NACK	End-to-End	Packet\Event Driven
RBC	Upstream	Implicit	End-to-End	Packet Driven
ART	Upstream (events) & Downstream (queries)	ACK (event) & NACK (query)	End-to-End	Packet Driven
GARUDA	Downstream	NACK	Hop-by-Hop	Packet Driven

by mentioned retransmission techniques. In fact adding some correcting bits is used instead of exchanging acknowledgment signals. Efficiency and importance of FEC techniques is demonstrated in (Akyildiz, 2002). In addition retransmission techniques are not energy efficient at all while FEC methods are more suitable for modern WSN due to design constraints.

Erasure coding is a method in which the transmitter divides the packet m fragments with k redundant fragments added to them. These $m+k$ fragments are sent instead of the original m fragments data (original packet). The sink, as a final receiver reconstructs data based on these received fragments. Imagine m' fragments have successfully delivered to the sink, just in case that $m' \geq m$ reconstruction is able. Number of redundant fragments depends on the type of loss recovery. For an End-to-End erasure coding the total number of redundant fragments is calculated in term of the probability of successful delivery of data through the whole path should be considered. In a Hop-by-Hop erasure coding the number of redundant fragments to be added to the node in the ith hop is calculated in term of the probability of successful delivery of data to the ith hop. Clearly number of redundant fragments is much more in the End-to-End erasure coding than in Hop-by-Hop erasure coding.Here is a brief discussion on the existing approaches using redundant information.

Authors in (Wen et al., 2007) proposed a theoretical analysis in order to evaluate redundancy and retransmission performances. By the use of this analysis they try to provide an energy efficient data transmission protocol. They provide energy efficiency by the successful packet delivery energy consumption and reliability is evaluated based upon the packet arrival probability. In the proposed protocol uses Gilbert model to address loss behavior (Bolot et al., 2005). Since Gilbert model is designed for one hop channels it is not suitable for multi-hop networks; so two-state Gilbert model is used. They compare their proposed model which is based on erasure code to retransmission based protocol. The result of this comparison is the more energy efficiency of erasure code than retransmission based protocol. Of course an unconditional loss probability (ULP) is introduced which means loss probability of packet i in the same condition which packet j has lost. The efficiency of erasure based protocol is upon a determined threshold for ULP. In general, due to analysis in a low loss area erasure code based protocols perform more reliable than retransmission based protocols besides their energy efficiency. A simple technique to advance this protocol in a noisy environment is to increase redundancy which clearly leads in energy inefficiency. In fact there is a tradeoff between energy consumption and reliability level.

One of the other proposed theoretical analyses which use erasure codes to provide reliability is introduced in (Ali et al., 2007). They use Reed-Solomon codes. In addition Genetic Algorithms (Gas) is used to estimate the best number of fragments. By the use of this technique they advance energy efficiency and also increase reliability, hence they also solve the tradeoff between energy efficiency and reliability in terms of redundant packet retransmission. Of course GAs mathematical calculations in energy consumption have been ignored. No downstream reliability mechanism is offered and in upstream reliability since coding and decoding is done in source and destination respectively, reliability level will decrease in case of hop counts.

A two mode protocol is introduced in (Marchi rt al., 2007). It includes full and differentiated reliability modes. In the full mode a full End-to-End packet delivery to the sink is guaranteed, whereas differentiated mode is used in applications where a fully End-to-End packet delivery is not expected. Full reliability is achieved by the use of ACK\ NACK mechanisms for loss packet recovery. They use Acknowledgment Window (AW) to determine the number of packets which the source node is allowed to send. After transmitting these packets an Explicit Acknowledgment Request (EAR) is sent to the destination in order to confirm packets delivery. The main problems of this protocol are high message overhead and extra expected memory to store and cache packets for probable retransmission requests. However a mechanism is proposed in which just the core part of data is sent but there is nothing mentioned about the details of core part of data selection. Hence coding and decoding is just done in the terminals (source and destinations) it seems not to be a good candidate protocol in large networks.

An implemented erasure code is proposed in (Kim et al., 2004). Of course to avoid delay and network traffic only the case is considered in which only one source and destination is designated. The systematic codes which are derived from Reed-Solomon codes are used in the proposed method in order to achieve three good points. First, sensor nodes processing power can be reduced which lead to low cost design of WSNs; second, node's memory usage reduces and finally a more reliable data transfer method is provided. These three advantages are obtained based on this fact that in systematic codes reconstruction process computation is decreased. In fact in systematic codes the message can be reconstructed without any recovery process just if a part of encoded message is the exact original message.

Srouji et al. (Srouji et al., 2011) proposed an erasure code based protocol. Erasure coding is considered in each hop to guarantee Hop-by-Hop reliability, whereas in all the other mentioned protocols just source and destination nodes did erasure coding. They named the proposed scheme RDTS which is more energy efficient and reliable besides its low traffic overhead in comparison with other hop-by-Hop and also End-to-End protocols. Unlike (Ali et al., 2007) RDTS calculate number of fragments based on the probability of successful data delivery and this probability is supposed to be a range of 0.7 to 0.95. RDTS also uses partial coding in order to reduce computational overhead. By the use of this technique erasure coding computation is reduced in the nodes within the path. To go deeper in detail, in partial coding mechanism, just in case of original fragments losses or when not adequate fragments are received, coding or decoding is performed.

It is approved that a practical WSN should be consists of huge number of sensor nodes deployed in an environment with a high density. It is to increase coverage and network lifetime. In general to increase sensor networks design efficiency a sleep mode is considered besides normal mode. At any time slot some of the sensor nodes are in this mode. The sensor nodes in sleep mode are inactive and don't participate in data transmission. The number of nodes in normal mode should be selected such to satisfy coverage and connectivity. Connectivity is a parameter to determine the ability of reporting to the sink node. The designers are supposed to find the minimum number of nodes to achieve connectivity and coverage which are not independent issues (Winston et al., 2011, Fan et al., 2010).

Besides connectivity and coverage, fault tolerant is also an important issue in WSNs. As sensor nodes are not accessible after deployment, detecting and recovering from faults is significant. Enhancing WSNs reliability, using fault tolerant techniques is discussed in (Kshirsagar et al., 2011, Mishra et al., 2012 and Paradis et al., 2007).

As mentioned before to design a reliable WSN three steps should be passes i.e.

- Reliability mathematical model consideration
- Find out model parameters
- Reliability estimation in terms of models parameter

The main issues in design of a sensor network is to collect information, process the gathered data and transmitting processed data to the sink(s) in such a way to satisfy coverage and connectivity issues. In general reliability can be measured in terms of different criteria such as mean time to repair (MTTR) and mean time to failure (MTTF) for hardware reliability measurements, coverage and connectivity reliability expression and etc. A set of factors which affect on WSNs reliability is listed below (Venkatesan, 2013):

- **Hardware Constraints:** Nodes energy depletion can be a good example of this type of factors.
- **Software Quality:** Sensor software manages hardware and is one of the most effective factors on reliability.
- **Environmental Conditions:** The parts of a sensor which are directly in contact with the environment such as sensing unit are highly sensitive to environmental conditions.
- **Deployment Strategy:** Sensors deployment affects on network coverage, connectivity, detection and communication. Communication itself affects on network lifetime so deployment is an effective factor on network lifetime.
- **Wireless Link Characteristics:** High distance between sender and receiver nodes decreases link quality. Other problems such as hidden terminal also affect on reliability (Patheja et al., 2012).

The mostly used techniques in order to model reliability are depicted in figure 1.

Deterministic methods are more accurate and need fewer calculations but they are not a good candidate for real time applications at all. There are probabilistic methods against deterministic ones. Simulation based probabilistic methods give an estimation of reliability whereas analytical based ones give an exact value of reliability. A system may fail because of dependency between system components. This is the case which combinational models consider. Reliability Block Diagram (RBD) is one of the simplest existing models which is widely used in WSNs data transport reliability models (Sha et al., 2007). Each physical component is represented by a block and represented blocks are arranged in a way to provide series, parallel and k-out-of-n configurations. A graphical model for reliability is also offered named fault tree. This technique is a top down failure analysis to analyze an undesired state of the system. In other words a fault tree is a combination of events which may lead to system undesired state.

RELIABILITY CHALLENGES AND RELATED ISSUES

A wireless sensor network is supposed to be reliable, scalable, power efficient and responsive in order to be a practical and real-world benefit platform. Power efficiency means operating at the

Figure 1. Reliability modeling technique

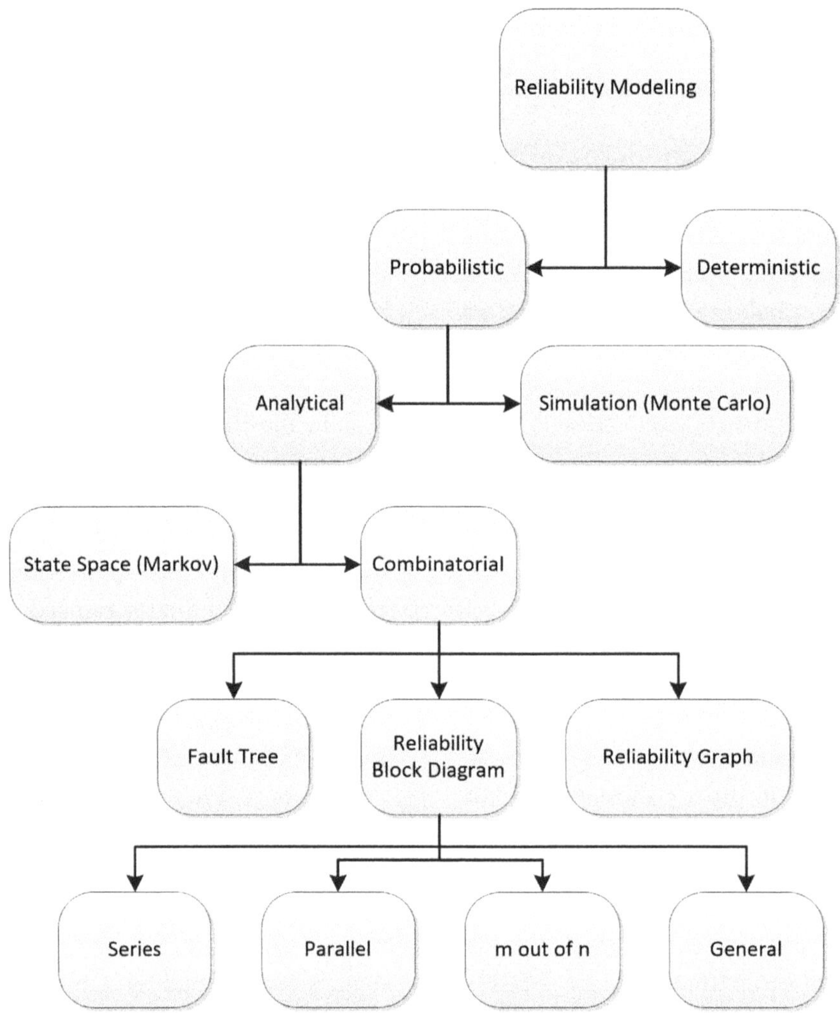

lowest allowable power level. Scalability means the ability of the network to adjust itself when number of nodes increases. In fact extra overhead should not be put on the network when number of nodes in the networks alters. Responsiveness means network tolerate the changes in topology and adapt itself. Finally reliability which is the main focus of this chapter means the ability of the network to transmit the gathered data reliably. Another aspect of reliability is to reliably sense and monitor the environment.In general reliability of a WSN (considering both aspects of reliability) is almost the most important factor which should be satisfied in the design of WSN (Islam et al., 2012).

To give a comprehensive view of WSNs reliability imagines a sample WSN. This network is consist ofthousands of sensor nodes which construct network basement and platform. To make a WSN reliable (considering both aspects of reliability) there are two significant factorswhich should be examined. The first factor this basement itself and the second one is how to use this platform. In fact in the first factor, the number of distributed nodes and the distribution method are critical issues. Many papers have been published which discuss the distribution of sensor nodes in a WSN effects on reliability and also other challenges such as connectivity, etc. Distribution of nodes is an application based issue which means it is highly dependent to the application for which a WSN is designed but in general an efficient distribution method is a method which is able to prolong network lifetime. As an example, authors in (Mahani et al., 2012) have proposed a deployment methodology to prolong their network lifetime.They have suggested Zipf distribution for sensor nodes. They have proven according to WSNs converge-cast nature, using a non-uniform deployment methodology help the network to save energy and to prolong network lifetime. Other researches have been done in the same area. For instance an optimization is done on the place of the sink node in a circular network in (He et al., 2010). Non-uniform distribution strategies have been applied to WSNs in (Naik et al., 2006, Chen et al., 2008) which worth to refer.

As mentioned before sensor nodes have a constrained energy resource and it is important to do considerations in order to save their energy. Hence it is clear that the number of nodes is highly dependent to the distribution method and a good distribution methodology guarantee the number of active nodes. As the time passes the number of active nodes reduces gradually, this event lead to connectivity and reliability reduction. On the other hand, there are some protocols which prolong network lifetime. Almost all of these protocols which are actually designed for MAC layer, use sleep mode. In fact these protocols prolong network lifetime through putting sensor nodes in sleep mode. The nodes in sleep mode are inactive and don't participate in any of operations. As the total number of nodes is also reduced in this technique the problem of coverage, reliability etc. would come into existence (Yang rt al., 2011). According to what discussed, it is of great importance to find out the best method of using sleep mode in order to solve the trades between energy efficiency and coverage, reliability etc. (Liang et al., 2011). A proper solution for this problem is presented in the next section "SOLUTIONS".

The second factors deals with the method of optimized usage of this basement. Assume the nodes are deployed in the sensing area, these nodes are supposed to connect each other to transfer their sensed data to a sink (network manager). Finding a path from sensing node to the target is of great importance and plays a significant role in the network reliability.

Routing protocols are used to provide reliability in wireless sensor networks. What makes a routing protocol practical is its energy efficiency (Pantazis et al., 2013). Besides energy efficiency routing protocols should be robust. It means that the network works properly despite probable links and nodes failures. One of the most common techniques is using multipath routing. The idea of this technique is to use multipath between the transmitter node and the receiver one. The offered paths may be completely or partially disjoint. Obviously this technique will cause some overhead and consumes more energy than single path routing (Zheng, 2009). Of course in static networks (in which sensor nodes and the sink don't move and are fixed) this overhead is not that much and also can be reduced by learning the location of nodes and sinks in the initialization phase. In general to satisfy reliability the network designers should consider more sensor nodes in the distribution phase than what is needed and also trades-off between reliability and yielded energy consumption.

It is worth noting that assuming same characteristics for sensor nodes is not practical in real world and in most of the real applications heterogeneous nodes are assumed in order to enhance reliability (Yarvis et al., 2005).

Routing has drawn the attention of many researchers. Flooding is one of the first offered routing protocols in which the sensed data is broadcasted to all the other nodes in the network. As is clear this method is not energy efficient at all besides adding extensive overhead. The researchers tried to advance routing protocols and the gossip was offered. This protocol chooses one of the neighbor nodes randomly and transmits the sensed data to it. The same procedure is used until the packet is delivered to the destination. This method cause significant delay.

A brief taxonomy of existing routing protocols is discussed here (Zheng, 2009).More energy efficient routing protocols is surveyed in (Lonare et al., 2012) which are worth referring.

Layered and In-Network Processing-Based Protocols: These protocols consider the networks as a set of clusters. This is done to advance energy consumption. There are several nodes in each cluster and one of them is the cluster head (CH). Each node in the cluster sends its data to the CH and the CH sends it to the destination (sink). In fact each cluster itself seems to be a network which the cluster head acts as its sink node. Since in these protocols just the luster sends the data to the sink, it would be more energy efficient. On the other hand as the total traffic is reduced, the collision probability reduces which leads to a more reliable communication. Low-Energy Adaptive Clustering Hierarchy (LEACH) is a well-known example of these kinds of protocols.

Data Centric Protocols: The protocols which involve this category are different from address-centric protocols. The intermediate nodes in these protocols aggregate the receiving data from their neighbor nodes and send the aggregated data to sink. Since fewer transmissions are required in these techniques, energy is saved and reliability is improved. Sensor Protocols for Information via Negotiation (SPIN) is one of the protocols which are classified in this category.

Multipath Based Protocols: As mentioned before multipath routing protocols are against single path routing protocols since these protocols select the first k paths to the destination. The main idea of this technique is to divide the load into k parts evenly and send each part to the destination through one of the founded k shortest paths. Shortest paths are select to increase reliability since long path are face to more bit error rate. Examples of these protocols are discussed in (Ganesan et al., 2001 and Lou et al., 2005).

QoS Based Protocols: In all the mentioned categories energy consumption was the key factor for the routing process whereas other quality of service (QoS) parameters in terms of reliability and delay should be considered. In this category some trades between energy efficiency and delay, energy efficiency and robustness and traffic overhead and reliability (such in multipath protocols) is discussed.

Heterogeneity-Based Protocols: As discussed before, homogeneous networks are not as practical as heterogeneous ones as they are not like real models. Recently heterogeneous networks are the researchers interesting topic which in them nodes are considered to have different characteristics. The main idea of this technique (using heterogeneous networks instead of homogeneous networks) is to use two different types of nodes. One type of sensor nodes is those which just sense the environment. The sensors of this type are cheap and are the majority of the distributed sensors. The other type of sensor nodes is the expensive nodes as they are able to process data and transmit them to the destination.

Since not all the nodes transmit their gathered information to the sink and just a few numbers of nodes are able of doing this, they are somehow like clustering protocols from this point of view.

SOLUTION

As mentioned before although putting sensor nodes in sleep mode increase network lifetime, it causes some negative effects on other network parameters such as reliability, connectivity, coverage, delay, etc. Hence a trade comes into existence between these factors. A mathematical model will help designers to deal with this trade and to find the best point of network operation. Before going through equations there are some useful definition which are necessary to be introduced. In a common classification, effective factors on WSNs reliability are categorized as follows:

- **Node Reliability:** Include software and hardware failures which may change the proper operation of sensor nodes.
- **Link Reliability:** Include environmental conditions affecting link quality such as signal to noise ratio. In addition collision probability which may cause packet loss is put in this category. As technologoly minimum feature size is shrinking, lower power supply voltages are getting bold in sensor nodes design which lead to a lower link reliability in comparison to traditional WSNs. Hence using efficient retransmission techniques and routing protocols is recommended (Øvsthus et al., 2014).

Authors in (Ansari et al., 2012) have proposed a mathematical model to find the optimal point of network operation. To simplify the conditions they have ignored node reliability. The link reliability itself is affected by two main factors i.e. bit error rate (BER) and collision probability. In fact the bit error rate affects link reliability through changing packets bits. As is related to the environmental conditions, BER can be modeled using signal to noise ratio as such in the formula below:

$$\gamma(d) = 10 \times \log(\frac{4d}{l}) \quad (1)$$

1 stands for wavelength, for common applications the frequency is set to 2.4 GHz and by the use of the famous $c = f \times »$. As is designated in (1) signal to noise ratio (SNR) is high in large distances (d). Based on SNR, bits in the packet may change. The flip probability of one bit is given by:

$$P_{bit\ error} = 0.5e^{-\frac{\gamma}{2}} \quad (2)$$

Since a successful communication in achieved in terms of no bit changes in the transmitted packet, (Ansari et al., 2012) gives the probability of a fault free communication as in (3):

$$P_{no\ bit\ change} = (1 - P_{bit\ error})^{8F} \quad (3)$$

The network payload, preamble and cyclic redundancy code (CRC) is considered in F. Collision probability is the other factor which can reduce link reliability. When the number of nodes in normal mode increases in a network, as they all want to send their gathered data to the sink(s), there would be a high traffic around the sink(s). This high traffic leads to packets collision. In order to calculate collision probability the average contention window is first calculated as in (Ansari et al., 2012).

$$\overline{W} = \frac{1 - c - c \times (2 \times c)^m}{1 - 2 \times c} \times \frac{W_0}{2} \quad (4)$$

Number of allowed retransmission is determined by m and W_0 is the initial size of contention window. Since not all the nodes in normal mode have packets to send, nodes transmission probability (A) seems to be important in collision probability calculations.

$$\tau = \frac{b_0}{\overline{W}} \quad (5)$$

where b_0 the probability of a randomly selected node is has at least one packet to send. Then collision probability would be as follows:

$$c = 1 - (1 - \tau)^{n-1} \quad (6)$$

Number of neighbor nodes and sender node itself is gathered in n. So the probability of a packet transmission among the node itself and all its neighbors would be:

$$P_{t,1} = 1-(1-\tau)^n \quad (7)$$

And the probability of a successful transmission in term of collision is:

$$P_{s,1} = \frac{n \times \tau \times (1-\tau)^{n-1}}{1-(1-\tau)^n} \quad (8)$$

A successful transmission and a collision occupied time, T_s and T_c respectively, are given as follows:

$$T_s = t_{PHY} + t_{MAC} + t_{DIFS} + t_{SIFS} + t_{RTS} + t_{CTS} + t_{ACK} + t_{SIFS} + \sigma + E[p] \quad (9)$$

$$T_c = t_{DIFS} + t_{SIFS} + t_{RTS} + t_{CTS} \quad (10)$$

Average packets length and time slots are presented by $E[p]$ and A respectively. Based on T_s and T_c the mean time duration between two following backoff counter is determined by:

$$t_{slot} = \sigma \times (1-P_{t,1}) + (P_{t,1} \times P_{s,1} \times T_s) + (P_{t,1} \times T_c \times (1-P_{s,1})) \quad (11)$$

According to mentioned formulas expected time to complete a transmission is given in (12).

$$T_{RES} = T_{slot} \times \overline{W} \quad (12)$$

So a packet transmission delay B and updated value of b_0 are given in (13) and (14) respectively.

$$B = T_{RES} + T_s \quad (13)$$

$$b_{0,new} = B \times \lambda \quad (14)$$

In addition » stands for unsaturated total traffic. By the use of these formulas a recursive computation is performed to obtain an approximation of collision probability. After modeling the most common faults, link reliability can be obtained as a combination of collision probability and bit error rate. The probability of a link to be a fault free one would be:

$$P_{link} = P_{no\ bit\ change} \times (1-c) \quad (15)$$

Figure 2 depicts the effects of link probability on two terminal network reliability. As it is shown in this figure and as is expected the higher link reliability leads to a more reliable two terminal reliability and on the other side increasing sleep probability make the network two terminal reliability worse.

As mentioned before the node failures have been neglected in the proposed method in (Ansari et al., 2012) reliability can be estimated in term of link reliability. To estimate two terminal reliability (End-to-End reliability), the network is considered as a graph with a set of vertexes which are connected to each other by some edges. Nodes are the graph vertexes and those which are in the radio range of each node are connected to that vertex. Monte Carlo simulation is used for reliability estimation which is a good candidate for large statistical problems (Yeh et al., 2010). Monte Carlo Simulation (MCS) needs a set of paths between source and destination nodes. Since finding all existing paths is not possible in large networks, an algorithm is expected to find a set of paths. The finding path algorithm is demonstrated in (Ansari et al., 2012, Yeh et al., 2010). The found path set is then used to achieve reli-

Figure 2. Two terminal network reliability versus link reliability and sleep probability

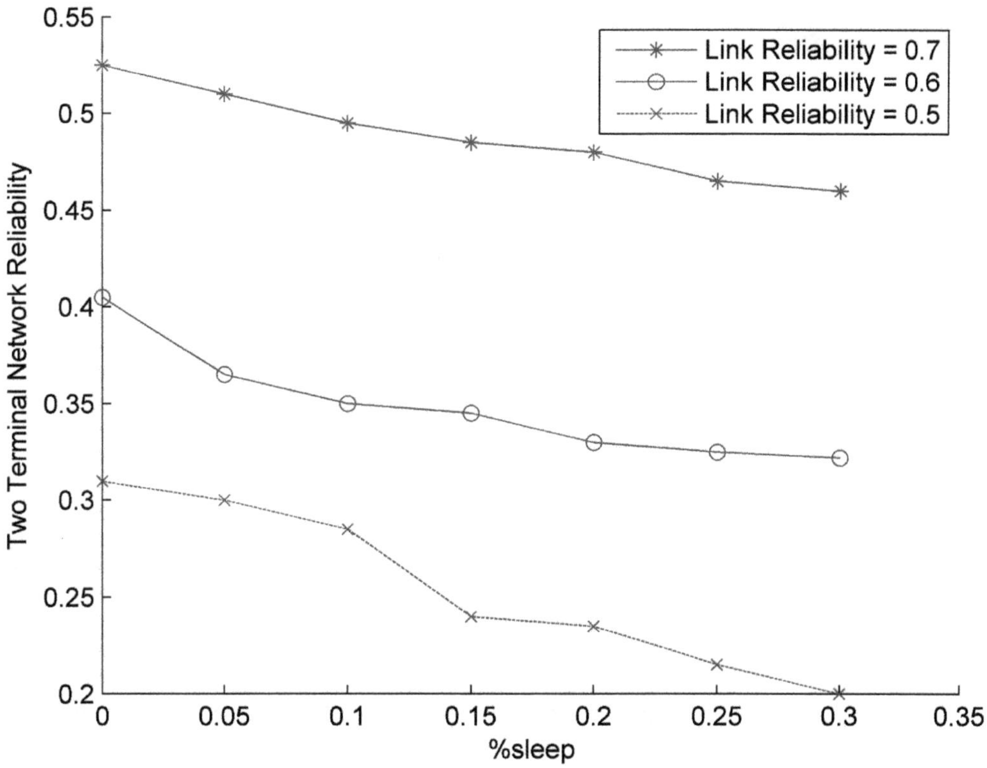

ability. The reliability evaluation process is as depicted in figure 3. In the first step the reliability matrix (R) is created using (15). $R(i,j)$ presents the link reliability between node i and node j. Second the adjacency matrix (A) which indicates connectivity between nodes is created. In fact A is a two state matrix that $A(i,j)$ is one if node j is in the radio range of node i else it would be zero. As the simulation is based on statistics, a random matrix (U) is generated to model reliability. If $U(i,j) \le R(i,j)$ then $U(i,j)=1$ else $U(i,j)=0$. This random matrix is used to update adjacency matrix according to the following formula:

$$B = A \wedge U \tag{16}$$

After creating the new adjacency matrix, all the founded paths are checked again to determine if they are acceptable path or not. Since the adjacency matrix has changed, the connections between nodes have altered and it is possible that a previous path from the source to the sink may be not valid any more.

Based on (Mahani et al., 2011) the End-to-End delay is calculated. Three different probabilities are introduced in order to make delay calculation easier.

$$P_{is} = (1-\tau)^{n'} \tag{17}$$

$$P_{ss} = \binom{n'}{1} \times \tau \times (1-\tau)^{n'} \tag{18}$$

Figure 3. Reliability estimation procedure

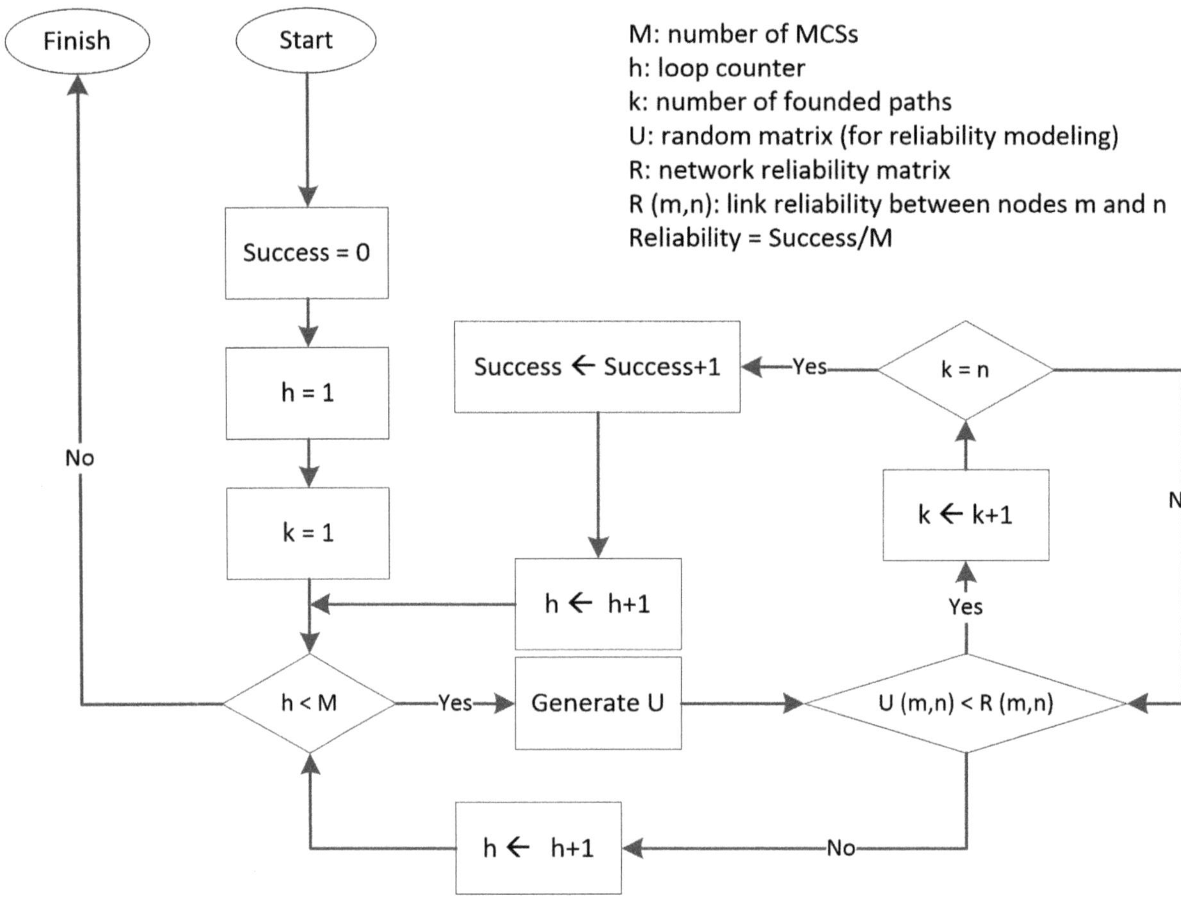

$$P_{cs} = 1 - P_{ss} - P_{is} \quad (19)$$

In the three above formulas n' is the number of active neighbors of transmitter sensor node. P_{is} is the probability that the communication channel be in an idle state. P_{cs} is the probability of sensing channel busy in case of collision occurrence and finally P_{ss} is the probability of sensing the channel busy in case of a successful transmission happening.

A ± factor is introduced as the average length of the slot time and is computed in the following manner:

$$\alpha = (P_{is} \times \sigma) + (P_{cs} \times T_c) + (P_{ss} \times T_s) \quad (20)$$

Finally by the use of ±, number of hops between source and destination, average contention window, packet length and bandwidth the total End-to-End delay is achieved as in (21).

$$D = H \times ((\alpha \times \overline{W}) + \frac{packet\ length}{bandwidth}) \quad (21)$$

Figure 4 depicts the total end-to-end delay as a function of sleep probability and for various values of collision probability (A).

Authors in (Ansari et al., 2012) have tried to present a comprehensive model for their wireless sensor network so they used a method to model coverage intensity. As mentioned before one of the factors which make a wireless sensor network

Figure 4. Sleep probability effect on end-to-end delay

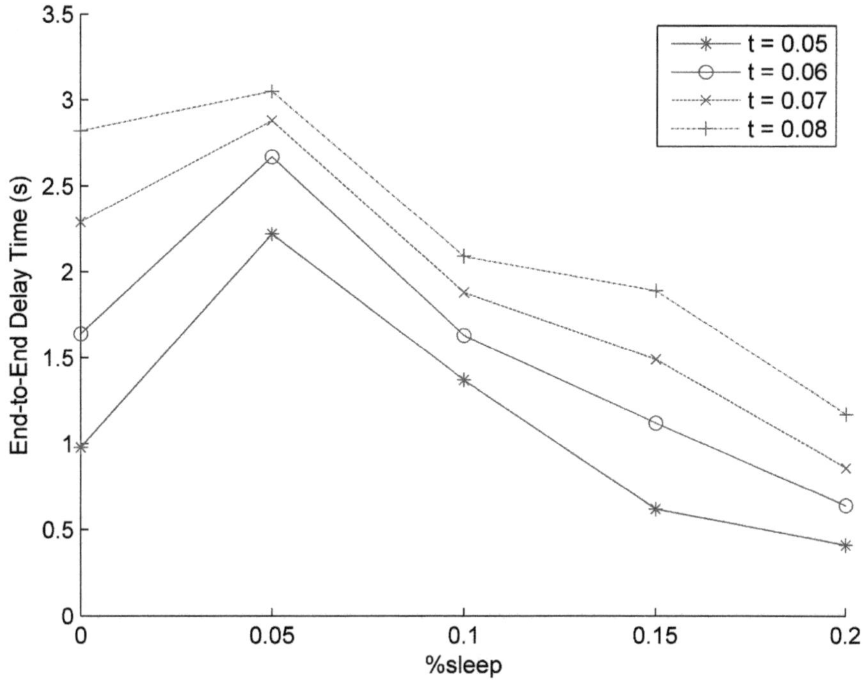

practical is coverage intensity. The application which is using WSN expects to control the whole area of the network. In fact coverage intensity is the probability that a random point in the network be covered at least by one sensor. In order to calculate coverage intensity a probability density function $f(x,y)$ is considered which is determined based upon deployment strategy. Since the sensor nodes are distributed randomly and uniformly in a circular area with a radius of R, coverage intensity would be obtained as follows (Peng et al., 2011):

$$P(a,b) = \iint_{(x-a)^2+(y-b)^2<r^2} f(x,y)dxdy = \frac{\pi r^2}{\pi R^2} = \frac{r^2}{R^2} \qquad (22)$$

In (22), r stands for sensing radius of sensor nodes. It is mostly considered equal for all the nodes in the network. According to what mentioned coverage intensity C is given in (23):

$$C = 1 - (1 - P(a,b))^{n'} \qquad (23)$$

Figure 5 gives a preview of coverage intensity as a function of sleep probability. As it is clear in this figure and also in (23), the ratio of r/R is a significant factor in coverage intensity.

In general authors in (Ansari et al., 2012) have proven through simulations that the End-to-End reliability will reduce as the sleep probability on sensor nodes increases. In fact incrementing sleep probability reduces number of awake nodes, this will lead to longer path between source and destination (Mahani et al., 2012) and according to what discussed in the link reliability definition, and longer paths are less reliable. On the other hand incrementing sleep probability reduces n' which lead to a lower coverage intensity. Besides all these disadvantages, network lifetime increase by the sleep probability incremention and this can

Figure 5. Effect of r/R ratio on network coverage intensity vs sleep probability

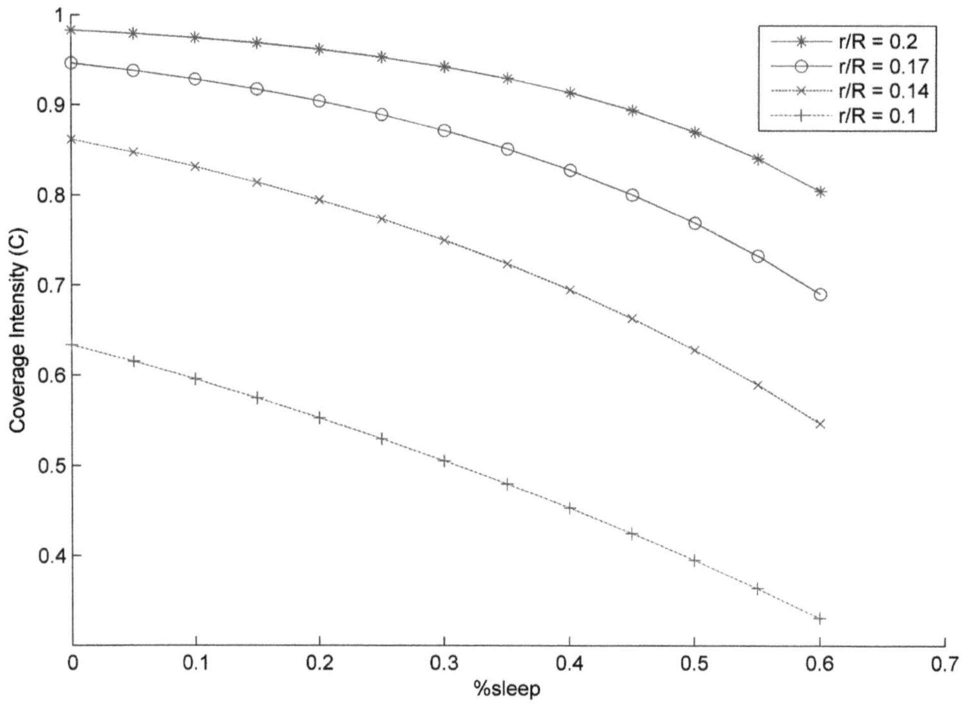

be known as an advantage. To find the best operating point of the network a figure of merit (FoM) is introduced to solve these trades (See Box 1).

The point in which the maximum of FoM happens is known as the best operating point of the network which gives the best sleep probability for the network in order to satisfy reliability, keep coverage intensity in an acceptable level and gives the minimum delay beside energy efficiency and prolonging network lifetime.

Figure 6 summarizes the above comments. As it is obvious for various values of link reliability and collision probability the best point for network operation is a fixed point so it can be a trusted point for network design.

Authors in (Cai et al., 2006) proposed a model to increase network lifetime and also in order to present dependency between reliability and sensor density. In the proposed model the sensing radius and transmitting radius differ from each other to give a more accurate model. They also introduced the following concepts:

- **K-Coverage:** It means for every single point in a query, there are at least k sensor nodes which cover it.
- **K-Coverage Ratio:** It is the ratio of K-Coverage region to the whole query region (Z).

Box 1.

$$FoM = \frac{(coverage\ intensity) \times (End-to-End\ Reliability)}{End-to-End\ Delay} \quad (24)$$

Figure 6. Figure of merit (FOM) vs. sleep probability

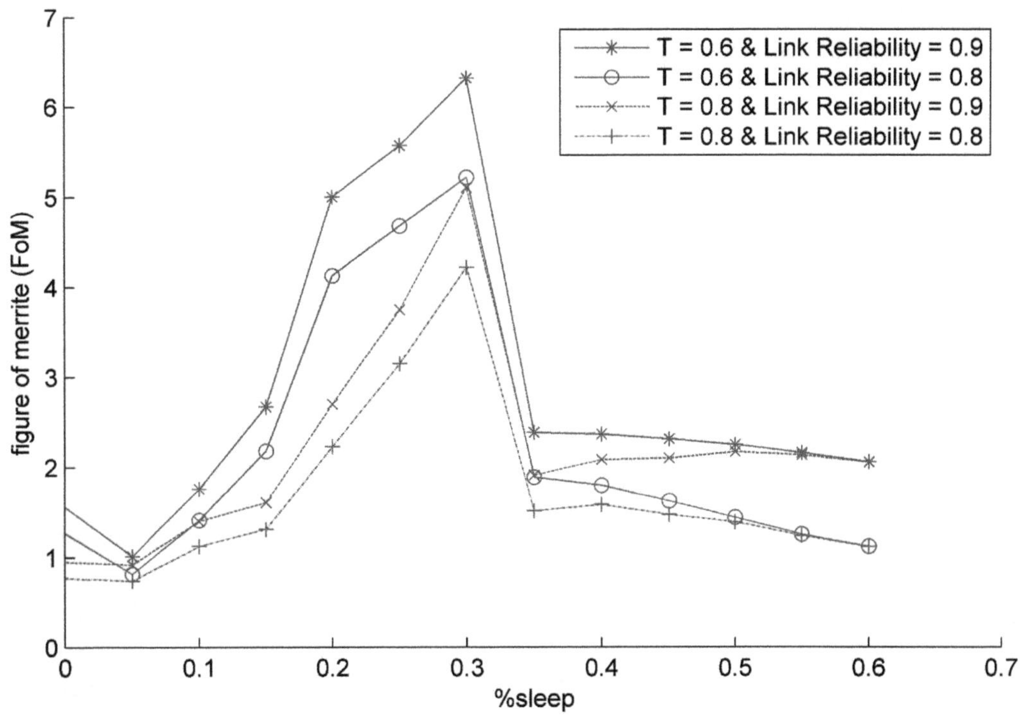

- **K-Connectivity:** The set of sensor nodes in a WSN is said to be K-Connectivity if between each two sensor nodes, k-node disjoint paths exists. As is proved in (Wang et al., 2003), K-Connectivity and K-Coverage are dependent to each other and if transmission radius is at least twice sensing radius, K-Connectivity is guaranteed in term of K-Coverage.
- **Normalized Lifetime:** It is the ratio of randomly removed sensor nodes such that K-Coverage gets its minimal value (network constraint) to the total number of nodes in the network.
- **Sensor Node's Degree:** For each sensor node, the number of nodes which are in the transmission radius of it are called the node's neighbor and also designated as sensor node's degree.
- **Normalized Invulnerability:** If the nodes with the higher degrees start to be eliminated till the K-Coverage gets its minimum acceptable value, the ratio of removed sensor nodes to the total number of nodes distributed in the area is called normalized invulnerability.
- **K-Reliability:** As is clear in (25), K-Reliability is the weighted average of normalized lifetime (R_L) and normalized invulnerability (R_I) which are restricted by K-Coverage ratio.

$$K - Reliability = w \times R_L \times (1 - w) \times R_I \quad (25)$$

K-Coverage ratio is itself dependent to the node's density. As the number of node increases in a fixed area, in the other word when the node's density increases, K-Coverage ratio behaves such as a descending exponential. So by increasing node's density, normalized lifetime and normalized invulnerability increase which lead to a higher

K-Reliability ratio. This is how node's density is related to the reliability.

Shin et al. (Mahani et al., 2012) discussed packet delivery reliability in dense wireless sensor networks. They have taken advantage of forward error correction (FEC) technique in order to reduce bit error rate due to noise wireless links. Their offered encoding mechanism is "single error correction double error detection" (SECDED). It is worse noting that in this mechanism every single byte is encoded into a 24-bit word. The successful delivery probability of one source byte is calculated using (26).

$$P_{success} = \sum_{i=0}^{e_c} \binom{K}{i} \times (1-e)^{K-i} \times e^{i} \quad (26)$$

In the above formula e is the bit error rate; K stands for the number of bits in the encoded data which is 24 in the proposed mechanism and finally e_c is the maximum number of bits which can be corrected. As mentioned before as they use SECDED, e_c. would be considered 1.

The reliability issue is a multi-faceted problem. In one hand event detection reliability should be guaranteed and on the other hand event detection should be able to be reliably transmitted to the networks manager(s) which are located sink(s) in the networks. Of course not all of the applications need a hundred percent reliability ensurement so an acceptable level of packet loss and also event masking would be tolerable depending on the application.

To address event detection reliability, almost all of the proposed protocols offer network coverage increment. Also K-Coverage concept was introduced to meet this problem. High level of coverage is achieved by increasing the number of active nodes. Reducing sleep probability lead to a more number of active nodes whereas energy consumption would increases. So there is a trade to solve. As random sleeping protocols put the nodes in idle mode in an accidental method and don't care the position of nodes, it is possible that a region in network get out of monitoring range of sensors so smart sleeping protocols which use location of sensor nodes to put them in idle mode seems to be a good solution.

To ensure communication reliability three different mechanisms discussed. As the packets are lost due to noisy links, collision probability and buffering limitations predicted for nodes, the only solutions seems to be retransmitting corrupted data, correcting altered data and sending data on a path with a lower noise. In retransmission methods, the lost packets are supposed to be sent until they get to the destination. This is done by the use of acknowledgment mechanisms. In the other method which is correcting altered packets of data (actually known as redundancy based data transmission protocols)based on the used technique, a number of occurred faults in the data packets is detectable and the protocol helps us to correct detected faults. Hence retransmission is not needed for each fault and this reduces network overhead. The last approach to achieve reliability is choosing the suitable rout for transmissions. Less congested routs, more reliable links and energy efficiency can be considered as the routing protocols effective factors.

FUTURE RESEARCH DIRECTIONS

As discussed in the chapter although accurate event detection is a significant issue in wireless sensor network it has been ignored in the major of done researches. Accurate event detection is important because by using it, not all same event will be reported to the sink(s) hence network traffic will be significantly reduced. The other issue which has been not considered much is network congestion which is an effective factor on reliability. In fact as the congestion level increases, coverage of the network behaves descending. So by improving congestion control mechanisms reliability can be enhanced. As mentioned in the chapter, there are protocols which consider congestion control and reliability but more researches seemto be needed. Retransmission process should be stopped when the found lost packets are in the same zone as other packets are supposed to be sent soon.

The other thing which is worth noting is real time implementation. Almost all of the offered protocols for data packet delivery reliability are based on simulation whereas real time implementations are required to evaluate proposed protocols.

For a future research direction cross layer design developments can be a good topic. In fact the media access control (MAC) can be helpful in improving channel contention difficulties while congestion problems can be solved using techniques in network layer such as routing.

Sensor nodes just gather the information and transmit it to the sink to process it. In other word sensor nodes themselves don't do any computation on the data, this causes an excessive amount of data transmission and as communication consumes more energy than processing, the whole WSN energy consumption increases.

Except retransmission based protocols, information redundancy techniques were discussed in this chapter. As said these technique are not as popular as retransmission based ones and only few of the protocols have used this technique and among those which have used this, almost all of them have evaluate it in real applications and there is lack of real implementation of this technique.

It seems an accurate combination of redundancy based data delivery reliability techniques and retransmission based ones is the main direction of future researches. Also data transport reliability in wireless multimedia sensor networks (WMSN) is unexplored.

Also another trend in wireless sensor networks, improvement of other QoS parameters can be studied. Also WSNs limitations and node failures effects on reliability would be an interesting topic. Concept of security and its dependency to reliability can be another interest.A classification of trends on wireless sensor networks is as follows:

- Most of the proposed protocols use single path routing while in case of using multi path routing; some unexplored issues such as fairness should be addressed.
- Almost all of the proposed protocols use sensors which are able to gather data of one type. STCP is among those few protocols which support node priority with different importance. In fact when different phenomena are sensed, different sensory data characteristics are generated. These various characteristics lead to different properties in terms of reliability, packet loss, delay, etc. which should be addressed in new data transport protocols designs.
- In the proposed protocols, except in STCP, congestion control and loss recovery have not been considered simultaneously. Reliability can be enhanced by using loss recovery technique while congestion control can improve throughput.In additionan effective transport protocol should satisfy performance metrics such as energy efficiency.
- As different tasks which are done in each layer of the network are not independent, cross layer interactions can be considered. As an example consider a WSN with unreliable links. As discussed prior bit error rate is an effective factor on reliability performance and transport layer protocol and also routing can affect on Hop-by-Hop retransmissions in the same way. Hence cross layer improvement seems to be required in transport protocols.

CONCLUSION

Wireless sensor network which is actually a new wave in sensor applications has been gaining interest as a platform that reduces human interventions due to its ad hoc nature. In WSNs huge number of sensor nodes are deployed independent of wiring limitations; hence a new world of sensing applications possibilities has been open up. As the network is able to learn from problems, it can adapt itself quickly and heals itself. The unique characteristics of WSNs make them a good candidate for monitoring applications but in order to advance WSNs performance all the characteristics should be considered simultaneously and also should work together in harmony. Of course all the applications are not designed for a same purpose but the common issue in all the applications is maintaining data delivery reliability which is an important challenge in WSN until now. By the advances in wireless sensor networks the applications which are using them are significantly increasing. Each application requires its own quality of services (QoS). But what makes a WSN practical is reliability. One of the aspects of reliability is reliable data transport which itself is a multi-faceted issue. Media Access Control (MAC) and routing protocols have been studied greatly in recent years but reliability is an issue which has not been well studied.

In this chapter taxonomy of some of the most common reliable data transport protocol was surveyed. Many of the discussed protocols take advantage of different technique on different layers. MAC layer is the layer which has taken the attention of many researchers and huge number of protocols has been proposed due to MAC layer properties. Other layers from MAC up to application layer have been also considered.

The addressed protocols use different techniques to ensure reliability issue. Different protocols achieve different levels of reliability. Some of them guarantee a full End-to-End packet or event reliability whereas the others which are used in the applications that don't need a hundred percent packet delivery are designed such to tolerate some degree of packet losses. The main techniques and existing methods to ensure reliability which were packet retransmissions and information redundancy discussed through the chapter in details. In general data transport protocols are supposed to have flexible reliability besides their energy efficiency. They also are supposed to guarantee application's quality of service (QoS). Among the different addressed mechanisms the retransmission based techniques were used much more than redundancy based ones which take advantage of FEC technique to ensure packet delivery reliability. In fact by the use of buffer management techniques and dynamic retransmission timeout they guarantee packet delivery reliability. Also it was proven that for wireless sensor networks, Hop-by-Hop retransmission based protocols are more practical than End-to-End retransmission based protocols. In fact Hop-by-Hop mechanisms were proved to be more efficient in environment with high level of noise in them. In addition for large network which the packets within them are supposed to pass high number of hops to get to the destination they do better than End-to-End mechanisms. Besides the mentioned advantages for Hop-by-Hop mechanisms this should be also considered that they are able to be advanced in the case that they are implemented for event reliability ensurement.

The importance of reliability in wireless sensor network applications is discussed in this chapter in details. Also different aspects of WSNs such as deployment, design and functionality were briefly addressed. Some the offered reliability modeling studied during the chapter and their performances measured. As said most of the researches which have done improvement on communication reliability are based on transport protocols and routing algorithms. Fault tolerant techniques in order to improve communication reliability mentioned and concepts of coverage, connectivity, energy efficiency and other parameters of quality of service studied.

One of the most important outcomes of this chapter is the importance of energy efficiency. In fact all the communication reliability techniques focus on energy efficiency besides satisfying reliability and network lifetime is considered as a high priority issue.

Several reliability modeling techniques presented and by the use of them coverage intensity, End-to-End delay and reliability modeled. According to the simulation results, the best point for network operation also founded and the effect of sleep probability as a power consumption reduction technique on coverage, delay and reliability studied.

REFERENCES

Akan, O. B., & Akyildiz, I. F. (2005). ESRT: Event-to-sink reliable transport in wireless sensor networks. *IEEE/ACM Transactions on Networking*, *13*(5), 1003–1016. doi:10.1109/TNET.2005.857076

Akyildiz, I. F., Su, W., Sankarasubramaniam, Y., & Cayirci, E. (2002). Wireless sensor networks: A survey. *Computer Networks*, *38*(4), 393–422. doi:10.1016/S1389-1286(01)00302-4

Ali, S., Fakoorian, A., & Taheri, H. (2007). Optimum Reed-Solomon erasure coding in fault tolerant sensor networks, in *Proceedings of the 4th International Symposium on Wireless Communication Systems (ISWCS)* (pp. 6-10). Trondheim, Norway.

Ansari, M. S., Mahani, A., & Kavian, Y. S. (2012). *Energy-efficient network design via modeling: optimal designing point for energy, reliability, coverage and end-to-end delay* (pp. 1–8). IET Network; doi:10.1049/iet-net.2012.0024

Bolot, J. C. (2005). End-to-end packet delay and loss behavior in the internet, in *Proceedings of the ACM SIGCOMM Conference on Communications Architectures, Protocols and Applications (SIGCOMM)* (pp.13-17). San Francisco, California, USA.

Cai, W., Jin, X., Zhang, Y., Chen, K., & Tang, J. (2006). *Research on Reliability Model of Large-Scale Wireless Sensor Networks, international conference on wireless communication, networking and mobile computing* (pp. 1–4).

Fan, G. J., & Jin, S. Y. (2010). Coverage Problem in Wireless Sensor Network: A Survey. *Journal of Networks*, *5*(9), 1033–1040. doi:10.4304/jnw.5.9.1033-1040

Ganesan, D., Govindan, R., Shenker, S., & Estrin, D. (2001). Highly-resilient, energy efficient multipath routing in wireless sensor networks, *Proceeding ACM MobiHoc'01*, Long Beach, CA, pp. 251-254. doi:10.1145/501416.501452

Ganesan, D., Govindan, R., Shenker, S., & Estrin, D. (2001). Highly-resilient, energy efficient multipath routing in wireless sensor networks. *Mobile Computing and Communications Review*, *5*(4), 10–24. doi:10.1145/509506.509514

He, Y., & Xu, T. The research of non-uniform node distribution inwireless sensor networks. *IEEE Int. Symp. Information Engineeringand Electronic Commerce*, July 2010

Islam, K., Shen, W., & Wang, X. (2012).Wireless sensor network reliability and security in factory automation. IEEE Transactions on Systems, Man AND Cybernetics, *42*(6).

Iyer, Y., Gandham, S., & Venkatesan, S. (2005). STCP: a generic transport layer protocol for wireless sensor networks, in *Proceedings of 14th International Conference on Computer Communications and Networks (ICCCN)* (pp. 449-454). San Diego, California, USA. doi:10.1109/ICCCN.2005.1523908

Kim, S., Fonseca, R., & Culler, D. (2004). Reliable transfer on wireless sensor networks. In *Proceedings of the 1st Annual IEEE Communications Society Conference on Sensor and Ad Hoc Communications and Networks (IEEE SECON)* (pp. 449-459). Santa Clara, California, USA.

Kshirsagar, R. V., & Jirapure, B. (2011). A Survey on Fault Detection and Fault Tolerance in Wireless Sensor Networks. *International Journal of Computers and Applications*, *3*(1), 130–138.

Le, T., Hu, W., Corke, P., & Jha, S. (2009). ERTP: Energy-efficient and reliable transport protocol for data streaming in wireless sensor networks. *Computer Communications*, *32*(7-10), 1154–1171. doi:10.1016/j.comcom.2008.12.045

Levis, P., Patel, N., Culler, D., & Shenker, S. (2004). Trickle: A self-regulating algorithm for code propagation and maintenance in wireless sensor networks, in *Proceedings of the First USENIX/ACM Symposium on Networked Systems Design and Implementation (NSDI)* (pp. 29-31). San Francisco, California.

Liang, Z., Ma, S., & Zhang, Q. (2011).Analysis of wireless sensor sleep mechanism with group arrival queuing model. *The Eighth IEEE Consumer Communications and Networking Conf.*, (pp. 570–574.) doi:10.1109/CCNC.2011.5766545

Lonare, S., & Wahane, G. (2012). A Survey on Energy Efficient Routing Protocols in Wireless Sensor Network. *IEEE Communications Survey and Tutorial, VOL*, *15*(2), 551–591.

Lou, W. (2005). "Am efficient N-to-1 multipath routing protocol in wireless sensor networks", Proceeding IEEE MASS'05 (pp. 1-8). Washington, DC.

Mahani, A., Ansari, M. S., & Kavian, Y. S. (2012). Reliability or performance: a tradeoff in wireless sensor networks, 8th IEEE, *IET International Symposium on Communication Systems, Networks and Digital Signal Processing*.

Mahani, A., Kargar, A., Kavian, Y. S., & Rashvand, H. F. (2012). Non-uniform distribution of multi-hop sensor networks: Performance improvement and energy hole mitigation. *IET Wireless Sensor Systems*, *2*(4), 302–308. doi:10.1049/iet-wss.2012.0032

Mahani, A., Tahmasebi, F., Nezamabadi-pour, H. (2011). *Two tiers wireless mesh networks: optimal configuration, ICCKE11* (pp. 235–239).

Mahmood, M. A., & Seah, W. (2011). Event reliability in wireless sensor networks. In International Conference on Intelligent Sensors, Sensor networks and Information Processing (pp. 377-382).

Mahmood, M. A., & Seah, W. K. G. (2011). Event Reliability in Wireless Sensor Networks. In *Proceedings of the seventh International Conference on Intelligent Sensors, Sensor Networks and Information Processing (ISSNIP)* (pp. 1-6). Adelaide, Australia.

Marchi, B., Grilo, A., & Nunes, M. (2007). DTSN: Distributed transport for sensor networks, in *Proceedings of the 12th IEEE Symposium on Computers and Communications (ISCC)*, Aveiro, Portugal, pp. 165-172.

Mishra, S., Jena, L., & Pradhan, A. (2012). Fault Tolerance in Wireless Sensor Networks. *International Journal of Advanced Research in Computer Science and Software Engineering*, *2*(10), 146–153.

Naik, L. N., & Agnew, G. (2006). Data capacity improvement of wireless sensornetworks using non-uniform sensor distribution,Int. J. Distrib. *SensorNetw.*, *2*(2), 121–145.

Øvsthus, K., & Kristensen, L. M. (2014). *An industerial perspective on wireless sensor networks – a survey of requirements, protocols and challenges* (pp. 1391–1412). IEEE Communication Surveys & Tutorials.

Pantazis, N. A., Nikolidakis, S. A., & Vergados, D. D. (2013). Energy efficient routing protocols in wireless sensor networks: A survey. *IEEE Communications Surveys and Tutorials, 15*(2), 551–591. doi:10.1109/SURV.2012.062612.00084

Paradis, L., & Han, Q. (2007). A Survey of Fault Management in Wireless Sensor Networks. *Journal of Network and Systems Management, 15*(2), 171–190. doi:10.1007/s10922-007-9062-0

Park, S. J., Sivakumar, R., Akyildiz, I., & Vedantham, R. (2008). GARUDA: Achieving effective reliability for downstream communication in wireless sensor networks. *IEEE Transactions on Mobile Computing, 07*(02), 214–230. doi:10.1109/TMC.2007.70707

Patheja, P. S., Waoo, A., & Shrivastava, P. (2012). Fault Tolerant, Energy Saving Method for Reliable Information Propagation in Sensor Network. *International Journal of Smart Sensors and Ad Hoc Networks, 1*(4), 34–38.

Peng, M., Chen, H., Xiao, Y., Ozdemir, S., Vasilakos, A. V., & Wu, J. (2011). Impact of sensor node distributions on coverage in sensor networks. *Elsevier Int. J. Parallel and Distrib. Comput., 71*(12), 1578–1591. doi:10.1016/j.jpdc.2011.04.011

Pereora, P. R., Grilo, A., Rocha, F., Nunes, M. S., Casaca, A., Chaudet, C., ... Johansson, M. (2007). *End-to-End Reliability in Wireless Sensor Networks: Survey and Research Challenges*. Workshop on IP QoS and Traffic Control, Lisbon, Portugal.

Postel. J. (1981). *Transmission Control Protocol* (IETF RFC 793).

Rizzo, L. (1997). Effective erasure codes for reliable computer communication protocols. *ACM Computer Communication Review, 27*(02), 24–36. doi:10.1145/263876.263881

Shaikh, F., Khelil, A., Ali, A., & Suri, V. (2010). Trccit: Tunable reliability with congestion control for information transport in wireless sensor networks, in *5th Annual International ICST Wireless Internet Conference (WICON)*, Singapore, pp. 1-9. doi:10.4108/ICST.WICON2010.8623

Shaikh, F. K., Khelil, A., & Suri, N. (2007). On Modeling the Reliability of Data Transport in Wireless Sensor Networks, *Proceedings of 15th EUROMICRO International Conference on Parallel, Distributed and Network-Based Processing*, pp. 395-402. doi:10.1109/PDP.2007.55

Srouji, M. S., Wang, Z., & Henkel, J. (2011). RDTS: A Reliable Erasure-Coding Based Data Transfer Scheme for Wireless Sensor Networks, in *Proceedings of the 17th International Conference on Parallel and Distributed Systems (ICPADS)*, Tainan, Taiwan, pp. 481-488. doi:10.1109/ICPADS.2011.104

Stann, F., & Heidemann, J. (2003). RMST: reliable data transport in sensor networks, in *Proceedings of the First IEEE International Workshop on Sensor Network Protocols and Applications (SNPA)*, Anchorage, Alaska, USA, pp. 102- 112. doi:10.1109/SNPA.2003.1203361

Tezcan, N., & Wang, W. (2007). ART: An asymmetric and reliable transport mechanism for wireless sensor networks. *International Journal of Sensor Networks, 02*(3-4), 188–200. doi:10.1504/IJSNET.2007.013199

Venkatesan. L., Shanmugavel. S., Subramaniam. C. (2013).A survey on modeling and enhancing reliability of wireless sensor network, pp. 41-51, doi.org/.<ALIGNMENT.qj></ALIGNMENT>10.4236/wsn.2013.53006

Wan, C. Y., Campbell, A., & Krishnamurthy, L. (2005). Pump-slowly, fetchquickly (PSFQ), a reliable transport protocol for sensor networks. *IEEE Journal on Selected Areas in Communications*, *23*(04), 862–872. doi:10.1109/JSAC.2005.843554

Wang, X., Xing, G., & Zhang, Y. et al.. (2003). Integrated Coverage and Connectivity Configuration in Wireless Sensor Networks, *1st ACM Conference on Embedded Networked Sensor Systems*. doi:10.1145/958491.958496

Wen, H., Lin, C., Ren, F., Yue, Y., & Huang, X. (2007). Retransmission or Redundancy: transmission reliability in wireless sensor networks. In *IEEE 4th International Conference on Mobile Adhoc and Sensor Systems (MASS)* (pp. 1-7). Pisa, Italy.

Willig, A., & Karl, H. (2005). Data transport reliability in wireless sensor networks: A survey of issues and solutions. *Praxis der Informationsverarbeitung und Kommunikation*, *28*(2), 86–92.

Winston, J. J., & Paramasivan, B. (2011). A Survey on Connectivity Maintenance and Preserving Coverage for Wireless Sensor Networks. *International Journal of Research and Review in Wireless Sensor Networks*, *1*(2), 11–18.

Woo, A., & Culler, D. C. (2004). A transmission control scheme for media access in sensor *annual international conference on mobile computing and networking* networks, in proceeding of 2001. ACM (mobicom'01), Rome, Italy, pp.221-235.

Wu, X., Chen, G. H., & Das, S. K. (2008). Avoiding energy holes in wirelesssensor networks with non-uniform node distribution. *IEEE Transactions on Parallel and Distributed Systems*, *19*(5), 1–11.

Yang, O., & Heinzelman, W. (2011). Sleeping multipath routing: a trade-off between reliability and lifetime in wireless sensor networks. In *Global Communication Conf., GlobeCom* (pp. 1–5).

Yarvis. M., Kushalnagar. N., Singh. H., Rangarajan. A., Liu. Y., & Singh. S. (2005). Exploiting heterogeneity in sensor networks. In *Proceeding IEEE INFOCOM'05*, (Vol. 2, pp. 878-890).

Yeh, W. C., Lin, Y. C., & Chung, Y. Y. (2010). Performance analysis of cellular automata Monte Carlo simulation for estimating network reliability. *International Journal of Expert Systems with Application*, *37*(5), 3537–3544. doi:10.1016/j.eswa.2009.09.070

Yick, J., Mukherjee, B., & Ghosal, D. (2008). Wireless sensor network survey. *Computer Networks*, *52*(12), 2292–2330. doi:10.1016/j.comnet.2008.04.002

Zhang, H., Arora, A., Choi, Y. R., & Gouda, M. G. (2005). RBC: Reliable bursty convergecast in wireless sensor networks, in *Proceedings of the 6th ACM international symposium on Mobile ad hoc networking and computing (MobiHoc)*, Urbana-Champaign, Illinois, USA, pp. 266-276.

Zheng, J., & Jamalipour, A. (2009). *Wireless sensor networks A networking perspective*. Hoboken, New Jersey: John Wiley.

Zhou, H., Guan, X., & Wu, C. (2008). RTMC: Reliable transport with memory consideration in wireless sensor networks, in *IEEE International Conference on Communications (ICC)*, Beijing, China, pp. 2819-2824.

KEY TERMS AND DEFINITIONS

Cluster-Based Network: A hierarchical network in which nodes are assigned to different clusters with periodically selected cluster heads in order to minimize energy consumption and achieve scalability and better network performance.

Energy Efficient Routing: Is a set of routing protocols which consider nodes remained energy while choosing the next step nodes in a packet transmission.

Network Coverage: Is the probability that all network's area be tracked and monitor by the deployed sensor nodes.

Network Lifetime: Is the amount of time that a network disconnect due to one or more sensor nodes failure so the network is not fully operative anymore.

Node Deployment: The way in which sensor nodes are distributed in the sensing area. It has significant effects on network coverage.

Path Reliability: Is the probability of a secure and fault free packet transmission between source and destination nodes via multi hops.

Resilient Network: A network which can tolerate sensors nodes failures and works properly in the presence of faulty nodes.

Chapter 8
A Taxonomy of Sensor Network Anomalies and Their Detection Approaches

Giovani Rimon Abuaitah
Wright State University, USA

Bin Wang
Wright State University, USA

ABSTRACT

Over the course of operation, sensor network deployments usually exhibit numerous types of anomalies that can result in missing, erroneous, malicious, or rather meaningful data points indicating an important event. Sensor data points can go missing due to network performance degradation as a result of a failing node, a network failure, or a malicious network behavior. Erroneous points are faulty measurements potentially generated by malfunctioning sensors. Malicious data points are usually injected into the network, replayed, or altered by an adversary. Finally, events manifest themselves as abnormal data points reported by sensor nodes expectedly. The significance of such anomalies calls for detection mechanisms that can identify an abnormal behavior in a sensor network deployment. In this chapter, the authors systematically analyze the different types of anomalies prevalent in a sensor network, construct a taxonomy of such anomalies, and discuss potential detection mechanisms necessary to reveal the root causes behind each anomaly.

INTRODUCTION

In the past few years, sensing data generated by various sensor devices have helped researchers and data scientists draw better conclusions about targeted environments. However, sometimes a sensor network deployed in an environment may fail to report a large number of its sensor readings, defeating the purpose of the deployment in the first place. Sensor networks may also generate inconsistent[1] data, affecting the quality of decision making as a consequence. This necessitates running an anomaly detection algorithm to identify inconsistencies and possibly filter them to

DOI: 10.4018/978-1-4666-8251-1.ch008

improve quality. In sensor network deployments which aim at detecting events of interest (Liu et al., 2011b; Ingelrest et al., 2010), sensor nodes may choose to report an observation only if it deviates from the normal behavior. Therefore, an event can also be perceived by an anomaly detection mechanism running at every node in the network. Sensor networks are also susceptible to malicious threats if deployed in a hostile environment. A malicious threat can vary from a simple passive eavesdropping attack to a total compromise of a sensor node. Sensor nodes, once compromised, can inject falsified readings into the network. Any malicious data behavior as a result from a security threat can degrade the trustworthiness, integrity, and/or accuracy of the sensor data. In addition, compromised nodes can launch smart attacks such as Denial of Service (DoS) attacks which may eventually hinder the performance and lifetime of the deployed sensor network.

This chapter provides a comprehensive taxonomy of the different types of anomalies prevalent in sensor network deployments and discusses their root causes. The chapter also surveys potential anomaly detection mechanisms that may help identify a single anomaly. Figure 1 illustrates three major types of anomalies that can exist in a sensor network deployment; (i) a natural fault; (ii) a malicious behavior; and (iii) an event. We discuss each category independently, pinpointing the possible root causes and the potential tools or mechanisms necessary to detect (and possibly recover from) all or a subset of these anomalies.

The chapter is organized as follows. Natural faults, malicious behaviors, and events are discussed in the next three sections. Having emphasized the demand for anomaly detection mechanisms in real-world applications and deployments of sensor networks, we then summarize the challenges of designing an anomaly detection algorithm for sensor networks. Finally, we discus related work and conclude at the end of the chapter.

NATURAL FAULTS

Natural faults are of high occurrence and are considered major anomalies that need to be addressed in almost every sensor network deployment. They can roughly be divided into: (a) network failures; (b) node failures; and (c) sensor data faults (i.e., measurement faults). Network and node failures highly impact the network performance and lifetime and are usually diagnosed either at the base station by specialized tools such as Sympathy (Ramanathan et al., 2005) and AD (Agnostic Diagnosis) (Miao et al., 2011) or in a decentralized manner by on-line self-diagnosis tools like the recently proposed TinyD2 (Liu et al., 2011a). These tools usually attempt to identify the root cause of the failure and resolve the issue immediately either via an automated action or by human intervention (e.g., changing node location or replacing a dead battery). A comparative survey of diagnosis and debugging tools for wireless sensor networks was compiled in (Rodrigues et al., 2012). Visualization tools (Jurdak et al., 2011a; Shi et al., 2011) can also be helpful in detecting network and node failures. Sensor data faults, on the other hand, degrade the quality of the overall collected dataset, resulting in imprecise conclusions or maybe meaningless inferences.

NETWORK FAILURES

Failures inside the network may occur due to several factors including interferences, routing inconsistencies, duplicate transmissions, lack of response from a subset of nodes, or even due to unexpected reasons.

Single Link Failures

In ad-hoc networks, a failing or broken wireless link can be roughly defined as a link with a very

Figure 1. Types of anomalies in sensor network deployments and their corresponding detection mechanisms

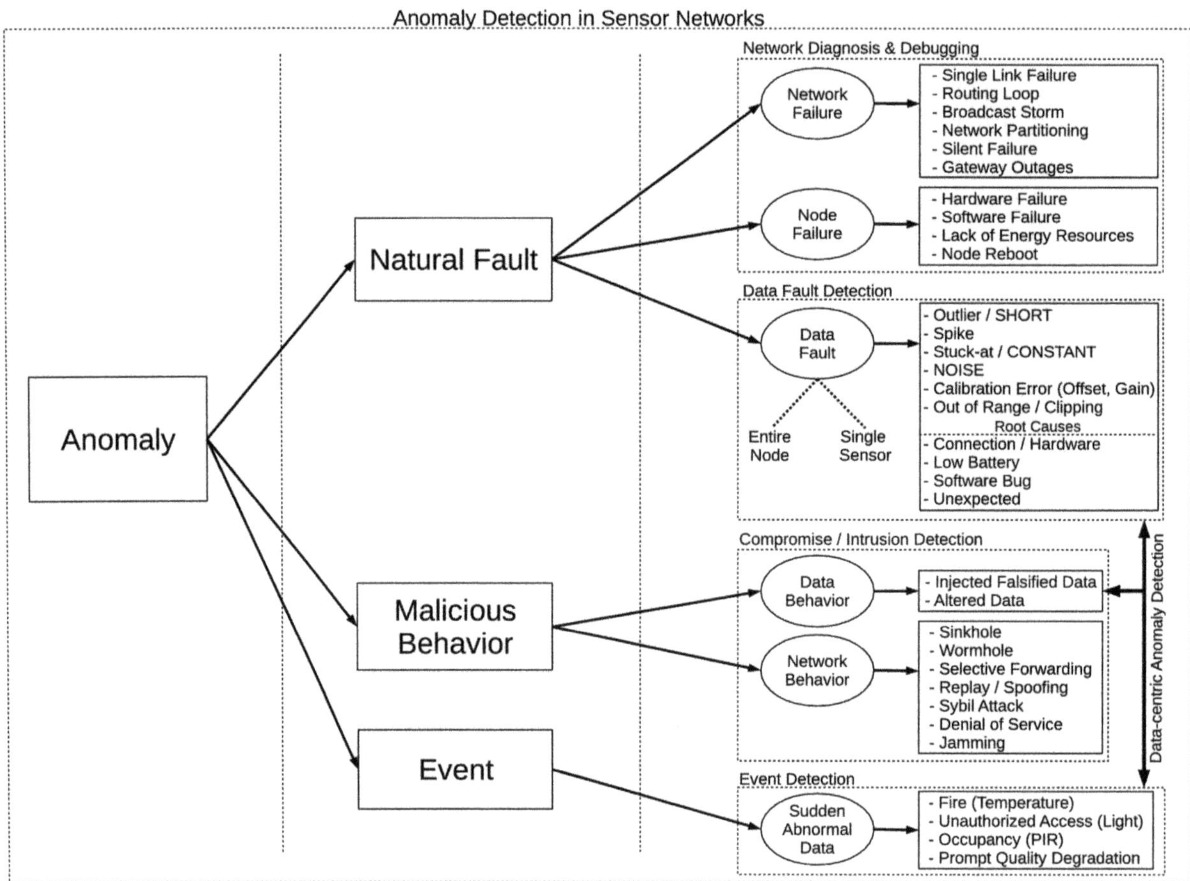

poor quality that it becomes impossible to reliably exchange packets between the two nodes at both ends (Lindhorst et al., 2010).

Root Causes. In a typical deployment, this transient failure is a result of harsh environmental conditions (e.g., wind), node mobility (Lenders et al., 2006), and/or presence of obstacles (Kim and Shin, 2011) (e.g., a herd of deers) in the vicinity of the two communicating end nodes. A long-term failure is also possible due to severe packet collisions, congestions at the receiving node, or heavy channel interferences caused by coexisting devices using the same channel. This dramatically increases the packet loss and/or corruption rate over that link.

Detection. Failing links are usually detected by collection or multi-hop protocols (Woo et al., 2003; The MultihopLQI Protocol (2012); Gnawali et al., 2009) using metrics like WMEWMA (Window Mean with Exponentially Weighted Moving Average) (Woo et al., 2003), LQI (Link Quality Indicator) (CC2420, 2012), ETX (expected transmissions) (De Couto et al., 2003), or a four bit link estimation (Fonseca et al., 2007). Once detected, the network quickly recovers by allowing a node to use a different link (if there exists one) for data transmissions. Link failures can also be identified by network diagnosis tools upon collecting enough statistics at the base station. In Sympathy, for instance, every node periodically sends connectivity metrics (i.e., node's routing table and neighbor list), flow metrics (i.e., PACKETS TRANSMITTED counter) as well as channel error metrics (i.e., BAD PACKETS

RECEIVED and GOOD PACKETS RECEIVED counters) to the base station (Ramanathan et al., 2005). The base station also keeps track of few metrics such as a SINK PACKETS RECEIVED counter which indicates the number of received packets from each node. Such statistics help discover disconnected paths to the sink, poor links, and congestions. AD follows a different approach for detecting link failures (Miao et al., 2011). It exploits the correlation of two metrics (i.e., RetransmitCounter and RadioOnTimeCounter) collected in the same way as in Sympathy. If the correlation score between the two metrics is high, this indicates that most of the radio-on time is spent for retransmitting packets and therefore the link between the node and its parent must be of low quality. Topology visualization tools such as Surge, MViz, and Octopus (Jurdak et al., 2011a) can further show a visual indication of a failure or quality degradation of a link upon tree topology changes.

It is worth mentioning that single link failures usually do not prevent the node from transmitting its sensor readings to the base station. Once the node updates its routing table, neighboring nodes with potentially healthy links to that node will still relay its readings to the sink.

Transient Routing Loops

Root Causes. In distance vector routing, a standard routing mechanism in sensor networks, failure or quality degradation of a single link calls for state updates at each node that needs to avoid the failed link. During state convergence, inconsistencies in routing tables may introduce transient loops (Pei et al., 2004).

Detection. Loops can be inherently detected by the underlying multi-hop routing protocol. For instance, the data path validation mechanism in CTP (Gnawali et al., 2009) allows a node to detect possible loop formations by comparing the path-to-sink cost (ETX) within the received data packet with its own path cost. If the former is smaller, the topology information of the transmitter must be stale, indicating a possible loop. Resolution of transient loops may also take quite a while. As for the current network diagnosis tools, it is difficult to detect routing loop anomalies at the back-end by relying solely on a single metric. AD can detect routing loops by exploiting the correlation between two metrics: (i) LoopCounter which measures the number of times a similar packet has passed a node; and (ii) SuccAckCounter, the number of successful transmissions by a node. If the two are highly correlated, the same packet must have passed through the current node multiple times. Jurdak et al (Jurdak et al., 2011b) discuss a few approaches to loop detection at a centralized base station (e.g., injecting diagnostic packets into the network, use of dynamic source routing, and maintenance of full topology). However, these approaches do not scale well with the network size.

Broadcast Storms

Root Causes. The broadcast storm problem was first defined in (Ni et al., 1999) as that of serious duplicates, collisions, and contentions caused by simple flooding and redundant rebroadcasts. Redundant broadcasts are very common in sensor networks and there is a need for mechanisms to reduce the number of duplicates once detected. As an example, due to link quality degradation, involved nodes in CTP periodically broadcast frequent probing beacons to quickly adjust their routing tables using the adaptive beaconing mechanism. This eventually congests the network with multiple duplicates since no duplicate suppression mechanism is adopted. On the other hand, a node in Trickle (Levis et al., 2004) (a dissemination mechanism used by other code and parameter dissemination protocols to spread updates in the network) suppresses its own transmission once it hears k similar summaries from its neighbors. This reduces the number of duplicates to avoid a broadcast storm.

Detection. Currently, there is no direct way to tell at the back-end if a broadcast storm has actually occurred. Nevertheless, keeping track of the number of duplicates received by each node may indicate that a storm is more likely to happen.

Network Partitioning

This is yet another network failure which is directly associated with link or node failures. Network partitioning is considered a significant failure since in some research work the network lifetime is based on the time till a network partition (Singh et al., 1998; Chang and Tassiulas, 2004).

Root Causes. Specifically, a failure of a number of critical nodes or a long-term failure in the link between such nodes and the other parts of the network results in a network partition or cut. Critical nodes are those few nodes that are geographically positioned close to the sinks rather than the network edges. This may be a rare event (i.e., failure) but once occurred, a large subset of sensor nodes can be disconnected. Another form of network partitions are temporal partitions (Herman et al., 2007) which are a result of displaced duty cycles at different subnetworks. For instance, one subnetwork is sleeping while the other is actively transmitting readings. The latter is a rare failure if a network-wide time synchronization protocol such as FTSP (Maróti et al., 2004) is in place.

Detection / Recovery. Detecting cuts at the back-end is an easy task by keeping track of the number of received packets at each node; lack of packet receptions from a group of sensor nodes over a long period of time indicates a potential network cut. Moreover, as proposed in (Shrivastava et al., 2005), only a small number of sensor nodes called sentinels can be used to detect ϵ-cuts where the parameter ϵ is the fraction of disconnected nodes and is specified by the end-user. The proposed algorithm can detect network partitions by solely relying on the dead/live state of these few sentinel nodes. To resolve network partitioning, either a human action is required (e.g., repositioning few sensor nodes or replacing batteries of dead nodes) or some other recovery mechanism such as the one proposed by Dini, et al (Dini et al., 2008) in which a mobile node (i.e., a robot) is sent to the location where the partitioning has occurred.

Silent Failures

A silent failure (Miao et al., 2011) is a type of network failures which can go undetected if we solely rely on a single metric. A single metric (e.g., number of packets transmitted) may appear to be stable despite the existence of an anomalous network behavior (e.g., routing loops). Examples of silent failures are ingress drops which occur when a portion of the incoming packets are dropped by the node, and frequent changes of parents in the routing tree. Routing loops can also be considered as silent failures since diagnosis tools may easily overlook such a failure as discussed earlier.

Detection. A detection of a silent failure requires diagnostic tools that exploit spatial and temporal correlations between several metrics periodically collected from each node in the network. For example, AD can detect ingress drops by tracking the correlation scores of two metrics: ReceiveCounter and TransmitCounter. A node may receive packets from its children in the routing tree but denies forwarding some of them. Similarly, frequent parent changes may be detected by AD using the correlation scores between the ParentChange-Counter and the RadioOnCounter metrics. A high correlation score indicates that a node changes its parent almost at every instance when the radio is turned on.

Other Causes of Network Failures

Although there are mechanisms to reveal the root causes discussed above, sometimes a network failure is hard to detect. Some sensor network deployments (Szewczyk et al., 2004; Ingelrest et al., 2010) form a network structure that is more complex than a single multi-hop network. Data packets can go missing due to gateway outages, lost GPRS signals, and failures in other parts of a larger network (e.g., the transit network in the GDI deployment) (Szewczyk et al., 2004).

Table 1 summaries the different types of network failures discussed in this section. The table shows possible root causes and existing mechanisms for detecting every type of failure.

NODE FAILURES

Node failures are another common faults in sensor networks which may severely hinder performance if not rapidly detected and agilely recovered from. A sensor node may fail due to hardware or software crashes, energy depletion, and local rebooting. We briefly outline each of these anomalies (as well as possible root causes) and present potential techniques to detect some of them.

Hardware Failures

Hardware modules of a single node including communication (radio and antenna), computation (CPU), and storage (memory) modules may fail or enter a lock state over the course of deployment due to poor manufacturing or harsh environments (i.e., storms, water floods, etc.). A bad radio chip, a broken antenna (Szewczyk et al., 2004; Khan et al., 2010), or a locked CPU or memory prevents a node, whose sensing modules may still be healthy, from transmitting its sensor data towards the sink and hence rendering such data useless. In other words, the sensor data never leave the node and are never used at the base station for finer-grained analysis.

Software Failures

In spite of the utilization of conventional simulation and hardware emulation tools (e.g., TOS-

Table 1. Network failures, their root causes, and possible detection approaches

Type	Root Causes	Existing Detection Mechanisms
single link failure (transient, permanent)	-wind, obstacle, mobility (transient) -severe collisions, congestions, channel interferences (permanent)	-inherently by collection protocols (e.g., CTP (Gnawali et al., 2009)) -network diagnostic tools (e.g., Sympathy (Ramanathan et al., 2005), AD (Miao et al., 2011)) -topology visualization tools (e.g., Surge, MViz, Octopus (Jurdak et al., 2011a))
transient routing loop	-routing table inconsistencies	-inherently by routing protocols (e.g., data path validation in CTP) -network diagnostic tools (e.g., AD)
broadcast storm	-flooding, redundant broadcasts -adaptive beaconing in CTP	none
network partitioning	-failure of critical nodes -permanent link failure between a critical node and rest of network -displaced duty cycles at different sub-networks (temporal partitions)	-using sentinels (Shrivastava et al., 2005)
silent failures (e.g., ingress packet drops, parent changes, routing loops)	-a buffer overflow (ingress packet drops) -old parent-child link degradation (parent changes) -routing table inconsistencies (routing loops)	-network diagnostic tools (e.g., AD)

SIM (Levis et al., 2003), AVR JTAG ICE (AVR JTAGICE mkII, 2014)) for debugging a sensor network before deployment, this combination may still overlook some significant software bugs once the nodes are pushed to the field and exposed to real-world conditions and dynamic environments. A major bug that leads to a node failure is the one that causes task queue overflows. In addition to task queue overflows, sensor nodes in a real-world deployment may also be prone to operating system (OS) crashes.

Root Causes. An example of a task queue overflow is observed in the CC1000 radio stack of TinyOS-1.x where two posts of fairly long execution path of 400 ms are constantly reposted causing other calls to fail. Consequently, the node enters a deadlock state and can no longer transmit readings to the sink.

Detection. Detecting software bugs require a source-level debugging tool such as Clairvoyant (Yang et al., 2007). Clairvoyant does not require any extra hardware and does not modify the source code on the sensor nodes. It is loaded into the boot sector of each node and is used only when needed. Each node can be debugged separately by using a GDB-like terminal running at the host machine to send typical debug commands (e.g., break, step, watch, backtrace). It is also possible to initiate network-wide debug commands addressing all nodes in the network. The latter commands are spread into the sensor network via simple flooding. Clairvoyant was able to detect the CC1000 bug by sending a led logging command to display the value of the head of the task queue on the three LEDs. The head value changes every time a task is successfully posted. It was observed that the LEDs never changed their values and hence concluding a task queue overflow. In addition to source-level debuggers, Agnostic Diagnosis (Miao et al., 2011) is also capable of detecting the latter bug by correlating two metrics: TaskPostCounter which measures the number of tasks posted by the sensor node and TaskExecuteCounter which is the number of executed tasks. If both metrics are weakly correlated, it is an indication of a potential task queue overflow which causes the node to eventually fail.

Lack of Energy Resources

A power outage may be caused by battery disconnection or lack of energy supplied by the harvesting system in place (e.g., solar panels). If no energy harvesting techniques are employed, the battery eventually dies due to energy depletion. Short circuits due to water contacts (e.g., floods, heavy rain) may also lead to eventual power depletion. Furthermore, although a rare event, a battery may suffer internal failure or it may be disconnected (e.g., loose connection) from the main node circuitry. Full depletion of the energy resource is a very common failure or anomaly in sensor networks as a significant number of deployments use non-rechargeable batteries to power up the sensor nodes (Tolle et al., 2005; Szewczyk et al., 2004; Buonadonna et al., 2005; Mo et al., 2009). Once the battery is discharged (i.e., battery level is below the operating voltage), a sensor node may not always be able to report its readings successfully (Szewczyk et al., 2004).

Node Reboots

For application purposes, sensor nodes may be configured as desired by the application designer to reboot periodically or there might be cases where a node enters infinite loops or suffers from stack overflows leading to node reboots.

Detection. Detecting node reboots by sink-based diagnosis tools is fairly simple. For instance, Sympathy investigates an UPTIME metric of each node. If this counter has rolled over, it indicates a node reboot since a node resets this metric to zero once a node has booted. Furthermore, source-level debuggers (Yang et al., 2007) may identify a subtle bug which leads to a node reboot. An example of this is a bug in the interrupt routine handler of the Microphone module in TinyOS-

1.x. Apparently, because of a constant fire up of the tone detector interrupt, handler routines are invoked several times eventually leading to a stack overflow. Clairvoyant addresses this bug by sending break, cond, and backtrace commands to the failing node to discover that the stack constitutes of several copies of the same handler routine that lead to continuous node reboots.

Node Failures Detection and Recovery

Detection of a node failure is usually performed by checking the failing node's neighbors (Staddon et al., 2002). Sympathy does so by first checking if the sink has received any messages from the failed node over one epoch. If not, the absence of the failed node in the NEIGHBOR LIST of each of its neighbors is an indicator of a total node crash. CTP follows the same approach by checking any stale entries corresponding to the failing node in the routing table of its neighbors. The off-line mechanism proposed by Khan et al (Khan et al., 2010) can further distinguish among the four root causes of a node failure mentioned above. The proposed approach requires an extra piece of hardware ($60), a power meter, that is attached to every node and acts as a tele-diagnostic power tracer. Each power meter periodically measures the current draw of the line between the battery module and the sensor mote as well as the voltage level. The base station uses the power traces obtained from each meter to train a simple classifier which accepts, as inputs, a number of static statistical features (e.g., mean and variance of power values over a sliding window) as well as a more complex data mining based classifier (Markov Model) to further improve the detection accuracy over the static classifier. Once the training process is complete, classification of every node is performed based on its current power trace reported within a window of the same size. Power profiles of a dead node should exhibit similar failure states and the closest state will be assigned as the failure type for that node. Finally, recovery from node failures is inherent in fault-tolerant sensor network protocols. CTP, for instance, can quickly reroute traffic around a failed node in less than a second (Gnawali et al., 2009) by applying the adaptive beaconing mechanism.

Table 2 summaries the different types of node failures discussed in this section. The table shows possible root causes and existing mechanisms for detecting every specific type of failure.

SENSOR DATA FAULTS

As mentioned in (Guo et al., 2009) and (Liu et al., 2011b), network and node failures are functional in that such a failure can drastically impact the network functionality and hence hinder its performance. Whereas, sensor data faults or measurement faults are considered non-functional faults since they only impact the fidelity of the reported data. Measurement faults or anomalies are those readings that deviate from the expected value and can be triggered by a single sensor (e.g., light or temperature sensor) or the entire node can experience a hardware failure or internal malfunction (e.g., loose connection with the sensor board) that drives the node to report false readings for all attached sensors. It is also possible that more than one sensor is attached to the same ADC channel. Thus, a single channel fault will impact all sensors connected to it (Ramanathan et al., 2006; Szewczyk et al., 2004). Sharma et al (2010) classify sensor data faults into: CONSTANT, SHORT, and NOISE faults. This naming convention was first suggested by Ramanathan et al (2006). Ni et al (2009), on the other hand, have compiled an excellent taxonomy of sensor data faults prevalent in sensor networks. For the sake of consistency with earlier fault classifications, we decide to adapt the same terminologies used in both of these works.

Network diagnosis tools are designed to detect functional faults rather than measurement faults,

Table 2. Node failures, their root causes, and possible detection approaches

Type	Root Causes	Existing Detection Mechanisms
hardware failure (bad radio, broken antenna, locked CPU/memory)	-poor manufacturing -harsh environmental conditions (e.g., storms)	-tele-diagnostic power tracer (Khan et al., 2010)
software failure (e.g., task queue overflow, OS crashes)	-constant reposting of multiple tasks with long execution paths	-tele-diagnostic power tracer -source-level debuggers (e.g., Clairvoyant (Yang et al., 2007)) -network diagnostic tools (e.g., AD (Miao et al., 2011))
lack of energy resource	-battery disconnection -battery depletion -water contacts -internal battery failure	-tele-diagnostic power tracer
node reboot	-infinite loop -stack overflow	-tele-diagnostic power tracer -source-level debuggers (e.g., Clairvoyant) -network diagnostic tools (e.g., Sympathy (Ramanathan et al., 2005))

focusing on data quantity rather than data quality (Ramanathan et al., 2005). Therefore, detection of the latter should take a different path. There have been several detection mechanisms proposed in the literature to identify sensor data faults. They range from rule-based mechanisms (Ramanathan et al., 2006; Sharma et al., 2007) to mechanisms that are based on statistics to techniques rooted in machine learning and data mining. In fact, some of these techniques may be able to detect events or malicious data behaviors and are discussed in both (Zhang et al., 2010) and (Xie et al., 2011). Due to the detection generality of these mechanisms, we discuss a single method (Ni et al., 2009) for identifying each individual fault. Notice that most of these faults can occur in both a temporal and spatial context. This means that a single sensor may experience an anomalous or faulty behavior within its own data over time, or that fault can be spatially uncorrelated with other sensors attached to neighboring nodes.

Outlier (SHORT)

Characteristics. An outlier, or a SHORT fault (Sharma et al., 2010; Ramanathan et al., 2006), is a single isolated sensor reading that is significantly far from the expected reading. This fault has been commonly seen in several sensor network deployments (Szewczyk et al., 2004; Ramanathan et al., 2006; Ingelrest et al., 2010; Kaiser et al., 2003) and its cause is usually unknown. In the IBRL deployment, for example, a few nodes exhibited one or a couple of outliers prior to the time where their battery voltage began to drop under the normal operating voltage of the temperature and humidity sensors. Outliers would appear in all sensing modalities (temperature, humidity, and light) concurrently. Figure 2(a) shows all instances of correlated outliers within the readings of node 20. It is believed that the correlation between temperature and humidity outliers is due to the fact that temperature and humidity are measured by the same Sensirion sensor (share the same ADC channel). In most cases, sensor data outliers also correlate well with an outlier in the voltage readings. The voltage outlier may be outside the operating voltage range of the sensor. The operating voltage of the temperature and humidity sensor in the IBRL deployment is about 2.35-2.36 Volts. Below this value, the sensor readings become anomalous. Voltage outliers can also lie within the normal operating voltage range as is the case with node 20. Figure 2(b) assures that the voltage values for node 20 are in fact outliers as they highly deviate from the voltage values in previous epochs. Other outliers may appear in the IBRL dataset but only within a period of a spike fault as we will discuss shortly.

Detection. Typically, the detection of this fault is the process of examining the rate of change between two successive readings, if the change is dramatically higher, then an outlier or SHORT fault has just occurred. Since outliers are useless/ insensitive data which is highly uncorrelated with the underlying phenomenon, it can simply be discarded by the sensor network application.

Spike Faults

Characteristics. Similar to an outlier, a spike is also a reading highly deviated from the expected reading. However, a spike spans over more than one reading consecutively and may or may not return to the normal behavior afterwards. Spike faults are very common in sensor network deployments. Figure 3(a) shows two instances of a spike fault exhibited by node 9 in the SensorScope St. Bernard deployment during the first 6 hours of its lifetime. Another example of a spike fault is seen in a few soil sensors that measured the ammonium concentration in a 48-sensors 12-day Bangladesh deployment (Ramanathan et al., 2006; Ni et al., 2009). This deployment took place in 2006 and aimed at finding the amount of arsenic in groundwater. In both examples, the data returned to its normal behavior after the spike fault. In IBRL dataset, spike faults are more frequent and have higher variance and gradient. Moreover, in the IBRL data, most nodes begin exhibiting a spike fault when approaching the end of the deployment (between March 19 and March 25) as their battery voltage level goes under 2.36 Volts (see Figure 3(b)). Node 18 is the only node that loses its battery power much earlier than the other nodes and begins exhibiting spike faults on March 8 (see Figure 7(a)). After a number of epochs (≈ 2910 epochs or 1 day), a spike fault transforms to a stuck-at fault where the sensor readings get stuck at a constant value as we will discuss shortly.

Low battery, failure of the sensor module, or a short-circuited connection are potential causes of a spike fault. Apart from being a fault, depending on the application, spikes can also be a reflection of the real phenomenon. For instance, light sensors generate large spikes when a room becomes suddenly dark (W¨alchli and Braun, 2009; Li and Parker, 2008).

Detection. A trivial way to detect this fault involves modeling the rate of change over a short period of time (i.e., within a pre-defined window size). Distinguishing between an anomalous event

Figure 2. Outlier readings of node 20 in the IBRL deployment

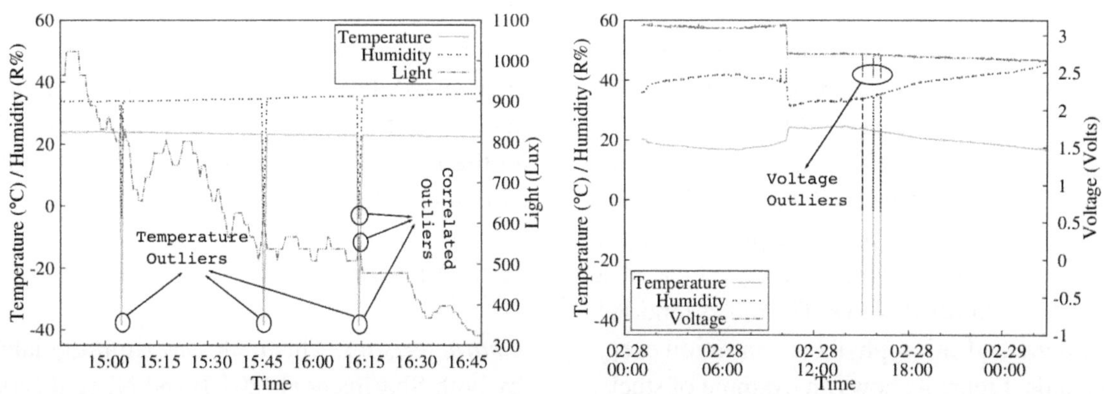

(a) Outliers in all sensing modalities (temperature, humidity, and light) seem to be highly correlated

(b) Voltage outliers within the operating voltage range of the temperature and humidity sensor

Figure 3. Spike faults in different deployments

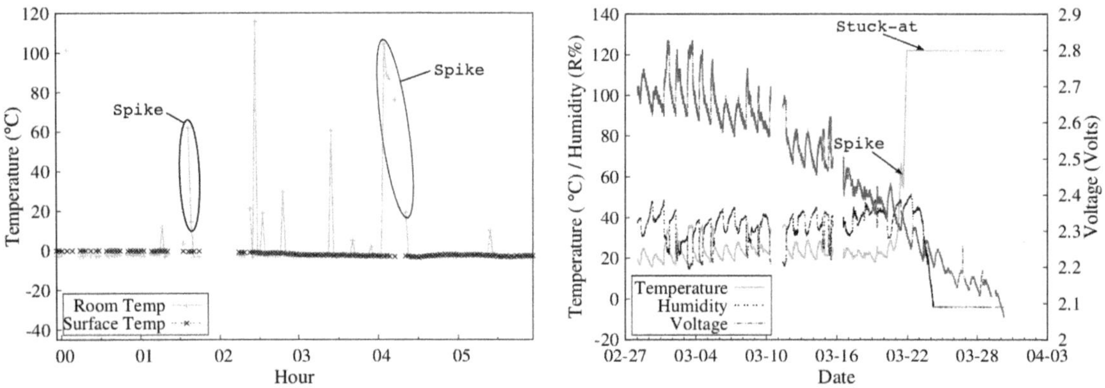

(a) The first 6 hours of readings from node 9 in SensorScope St. Bernard deployment. The node tends to recover after a spike

(b) Node 21 in the IBRL deployment maintains a constant behavior following the spike fault til the end of its lifetime

and a spike fault is a difficult task as we will discuss later in the 'Events' section. Therefore, it is not recommended to discard spike faults unless further examination is carried out to find whether it is a result of a potential failure or a true behavior of the measured phenomenon.

Stuck-At (CONSTANT) Faults

Characteristics. As the name implies, a "stuck-at" fault is a series of readings that maintain steadiness for the rest of the node's lifetime or may return back to normal after a significant amount of time. False readings may be in or outside the range of the measured phenomenon. Indeed, the latter is what Sharma et al (Sharma et al., 2010) refer to as a CONSTANT fault. The fault has been spotted in a few deployments (e.g., IBRL (Buonadonna et al., 2005), NAMOS (Namos, 2012), and SensorScope (Ingelrest et al., 2010)). The rapid 24-hour deployment of the Networked Aquatic Microbial Observing System (NAMOS) took place in August 2006 and consisted of 9 motes (buoys) located at Lake Fulmor, James Reserve. The motes reported temperature and chlorophyll concentration every 10 seconds. Figure 4 shows an example of stuck-at faults exhibited by the temperature sensor of node 6 in the SensorScope G´en´epi deployment. Notice that the faulty readings are outside the range of the expected temperature. On the other hand, Figure 5 shows that some of the humidity sensors exhibited a constant fault where the faulty readings were still considered in the range of the true phenomenon. In the IBRL deployment, almost all nodes exhibited a constant fault over their course of operation. Instances of a stuck-at fault are shown in Figure 3(b) following a spike fault.

Detection. Typically, variance is used as the main detection mechanism since a near-zero variance within a time window of a predefined size represents a constant behavior. As the case with spike faults, "stuck at" anomalies should not be discarded until we are certain that it is a case of failure (usually a sensor ADC malfunction) rather than an exhibition of the true phenomena (as we will see in the case of an expected clipping fault in the light data when we discuss out-of-range/clipping faults).

Noise Faults

Characteristics. This fault was similarly labeled by both Sharma et al (2010) and Ni et al (2009). In the former, it refers to high variance in the data

Figure 4. Constant faults exhibited by node 6 in the SensorScope Génépi deployment

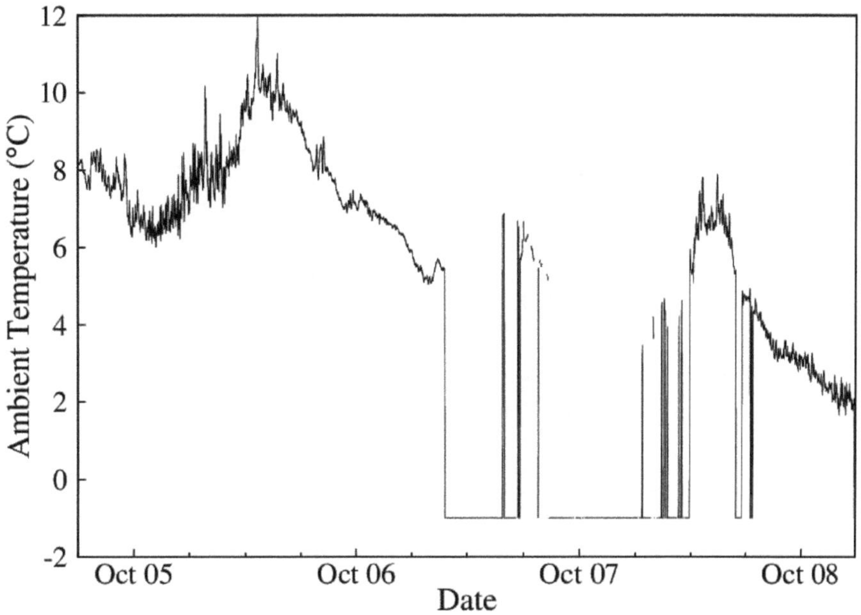

Figure 5. Humidity readings spanning 3 days in October. Corrosions forming on the connector to the sensor board on October 6 due to high humidity accounts for the erroneous readings

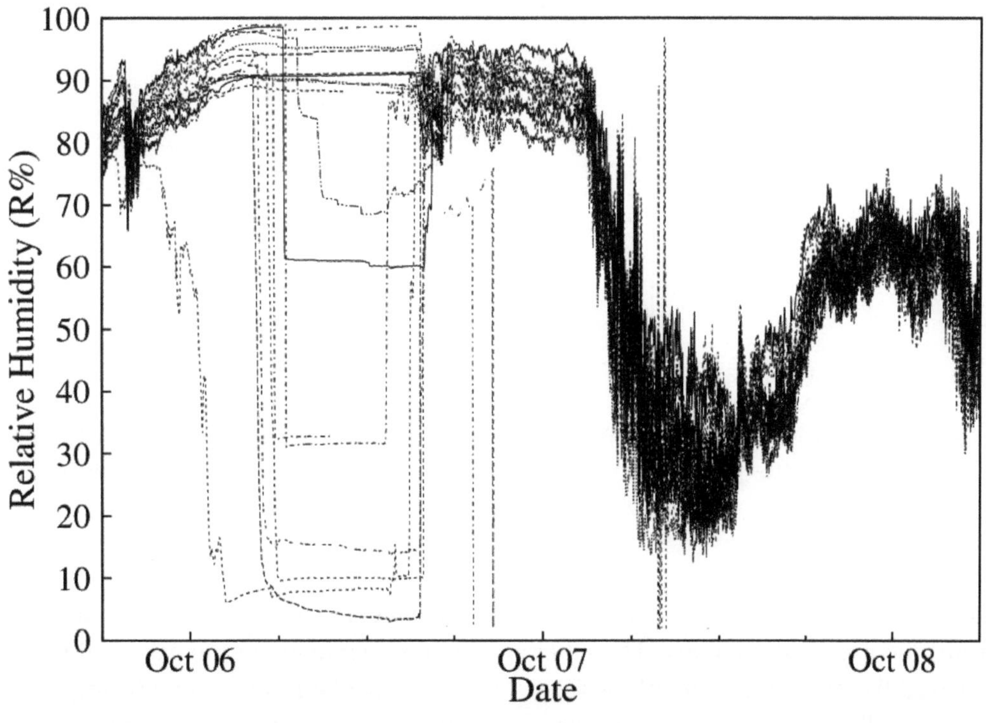

Figure 6. Rain precipitation measured by two nodes (node 4 and 14) in the SensorScope St. Bernard deployment. Node 14 exhibits a noise fault

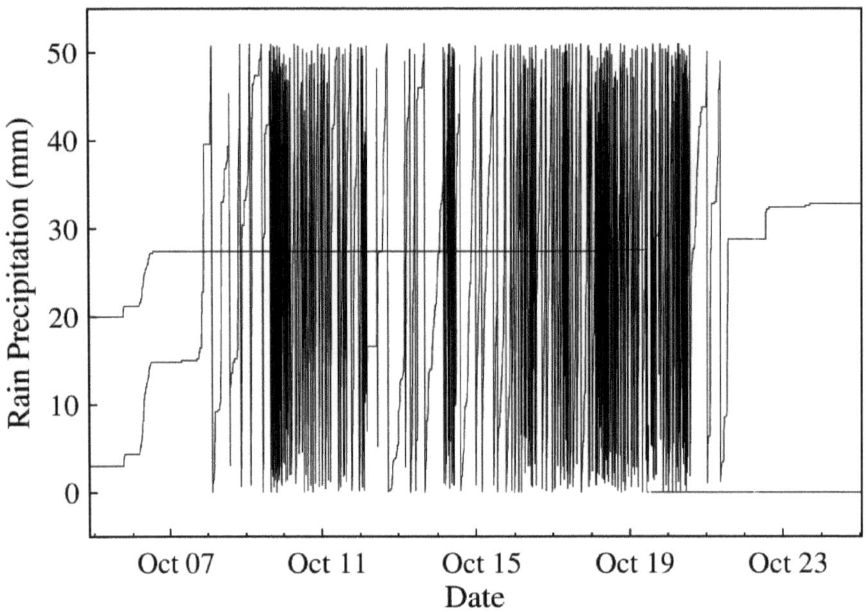

Figure 7. Noise faults in the IBRL deployment

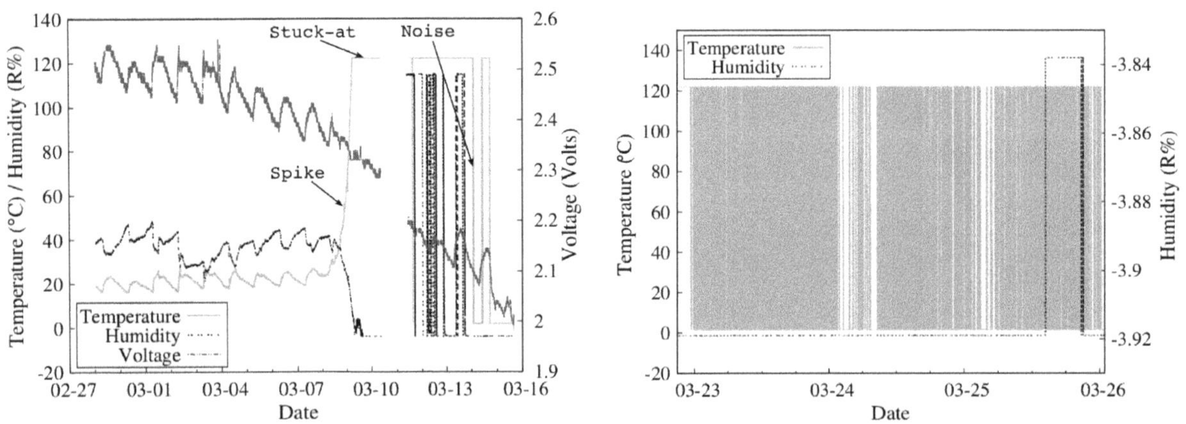

(a) Node 18 exhibits a noise fault at the very end of the deployment when the battery level becomes very low

(b) Noise fault in temperature and humidity readings of node 6. Both faults are not correlated as is the case with temperature and humidity outliers

and the data are usually highly uncorrelated with (distant from) the expected readings. The latter also emphasized the existence of highly variant data which can still resemble the true physical phenomenon. Henceforth, it is important to distinguish between an expected noise which is within the range of the expected behavior and a definitive fault due to low battery or a potential hardware failure (short circuit). One example of a noise fault was spotted at the SensorScope St.

Bernard deployment within the rain meter readings of node 14. As can be seen in Figure 6, during the noise fault, readings have high variance. In the IBRL deployment, we consider noise to be the fluctuation of highly deviated constant values that appear at the end of the node's lifetime when its battery is drained. This behavior is mixed with a stuck-at behavior as seen in Figure 7(a). Only few number of nodes (nodes 6, 18, and 23), in fact, produced highly noisy data. Figure 7(b) shows a closer look at the largest noise fault exhibited by node 6. The temperature reading rapidly fluctuates between two values 1.1731°C and 122.153°C and may get stuck on any of the two values for a few consecutive epochs. Unlike outliers, the noise faults in temperature readings are not correlated with noises in the humidity readings.

Detection. A primary identifier of a noise fault is the variance. For example, if the variance of reported sensor data is entirely different from the expected variance as given by the sensor data sheet, then it is a definitive fault. The data identified this way is invaluable and cannot be discarded once marked as a natural resemblance of the true phenomenon.

Calibration Errors

Characteristics. Calibration is becoming a highly prioritized requirement for every sensor network deployment (Szewczyk et al., 2004; Ingelrest et al., 2010; Mo et al., 2009; He et al., 2006; Ramanathan et al., 2006). Calibration is usually performed before, during, and after deployment to achieve the highest accuracy of the collected sensor data. Depending on the type of sensors, sometimes pre-deployment calibration is adequate. Nevertheless, some sensors need to be repeatedly calibrated over the course of deployment due to the change in their calibration equation and exposure to the environment. Usually higher cost sensors require less calibration and hence experience less calibration faults. One example of sensors that need to be continually calibrated is the ISE (Ion-selective Electrodes) sensor used in the Bangladesh deployment (Ramanathan et al., 2006). Calibration is the mapping from the actual sensor readings to the expected readings resembling the true phenomenon (Ramanathan et al., 2006). The mapping function contains two calibration parameters: (i) the bias or gain which relates to the value needed to drive the expected rate of change of the sensor readings; and (ii) the offset which is a constant distance from the true phenomena. The acceptable values of these two parameters are usually given in the sensor data sheet. Both the bias and the offset parameters can change over time given the dynamics of the environment and the manufacturing cost of the sensor. If the actual readings become distant from the true physical phenomena, then we say that sensor data has encountered an offset fault. On the other hand, if the readings bear variances higher than the expected rate of change, a gain fault can be observed within the data. Finally, if both parameters change, this change manifests itself as a drift fault. The three calibration faults were elaborately discussed in (Ni et al., 2009).

Detection. Apparently, it is difficult to detect calibration errors in most cases due to the difficulty of distinguishing between a potential calibration fault and a true measurement of the phenomenon. Apart from this, spatial correlation may be leveraged to identify miscalibrated sensors. Nonetheless, sensor readings from mis-calibrated sensors should not be discarded as an appropriate formula can be applied off-line to correct the faults.

Out of Range / Clipping Faults

Characteristics. Sometimes, sensor readings may lie outside the sensitivity range of the transducer or exceed the maximum value allowed by the Analog to Digital converter (ADC). The latter is a manifestation of a clipping fault. An example of an out-of-range fault was reported in the Bangladesh deployment (Ramanathan et al., 2006) while measuring the concentration of chloride. There were cases when the concentration measurements were high enough that could not be captured by the sensitivity range of the transducer. In (Ramanathan et al., 2006), the sensitivity range is referred to as the total detection range of the calibration curve and any sensors that report readings outside this range are considered BROKEN. As for clipping, although not necessarily a fault, it still implies an expected anomalous behavior. In the IBRL deployment, a clipping fault or anomaly is spotted when very high light intensities are clipped at the peak. True light values may even reach 100,000 Lux which corresponds to full sunlight. However, the light sensor in the IBRL deployment could only measure up to 1847.36 Lux (see Figure 8). Despite the fact that there are light readings very close to the lower ADC bound (zero), we still cannot conclude that these are indeed clipping faults (Ni et al., 2009). Ideally, clipping faults are similar to "stuck at" faults. The difference between the two is that, with clipping, the constant behavior is observed only at the peaks of the possible ADC range.

Detection. Clipping faults are indeed similar to "stuck at" faults and hence all detection mechanisms used for "stuck at" faults can be applied to identify clipping faults. The only difference between the two faults is that, with clipping, the constant behavior is observed only at the peaks of the data range. Leveraging spatial correlations, we can further identify a clipping fault which should not be discarded as it resembles an expected anomalous behavior.

Connection / Hardware Failures

This and the next fault in the list are in fact primary root causes of other sensor data faults (such as spike, "stuck at", and noise faults). We list them here for the sake of completeness and consistency

Figure 8. Clipping faults exhibited by node 20 in the IBRL deployment. The light readings clipped at the highest peak are outside the range of possible ADC values.

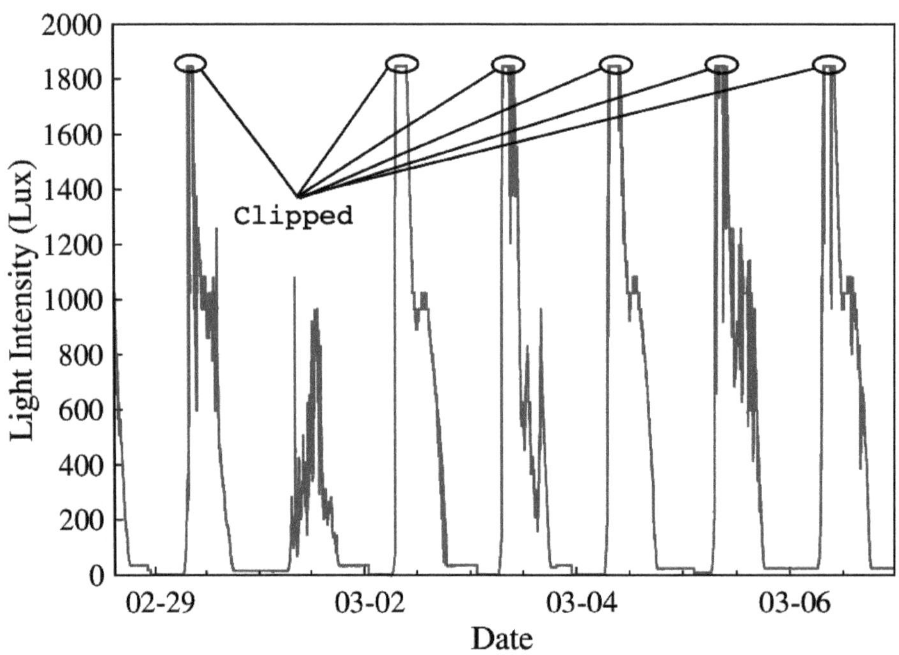

with the earlier work of Ni et al (2009). The primary cause of this fault is short circuits created due to direct contacts with the environment (e.g., water contact, mud, etc. (Szewczyk et al., 2004; Ingelrest et al., 2010; Ramanathan et al., 2006)). This fault manifests itself as very high or very low readings and could also resemble an out-of-range fault. Other causes of hardware failures include sensor aging and loose wire connections. It is important to note that it is not always feasible to identify the root cause of an unknown hardware failure, but once a hardware fault is spotted, generated sensor readings during the period of fault should be discarded.

Low Battery Levels

Just like hardware failures, low battery is a major cause of other sensor data failures (e.g., spike fault followed by a "stuck at" fault (Buonadonna et al., 2005), noise fault (Ramanathan et al., 2006) in several sensor network deployments (Szewczyk et al., 2004; Tolle et al., 2005; Buonadonna et al., 2005; Ramanathan et al., 2006)). In addition, hardware failures can be highly correlated with low battery faults. A short circuit can cause an unusual drop in battery level (Szewczyk et al., 2004). Maintaining power profiles using for example tele-diagnostic power tracers (Khan et al., 2010) may help identifying low battery failures. More often, data generated during a low battery period are harmful to the decision making process and should be discarded.

Sensor data faults are very common in sensor network deployments. Uncovering the root cause of such faults is a task that is sometimes very difficult. Software bugs such as the driver bug of the wind sensor in the SensorScope deployment (Ingelrest et al., 2010) and unexpected conditions such as unintentional covering of the entire sensor mote (Barrenetxea et al., 2008) are hard to track. In the case of an existing software bug in the sensor driver, spatial correlation among the sensor nodes does not help at all if the same software runs on all sensors. On the other hand, sensor readings of a covered sensor mote will always be spatially uncorrelated with its neighboring motes but the root cause of this uncorrelated behavior cannot be revealed unless physical investigation is carried out. Table 3 summaries the different types of measurement faults discussed in this section. The table shows example deployments where such faults occurred, possible root causes, and potential data features that can be used in order to detect measurement faults.

MALICIOUS BEHAVIORS

The second type of anomalies prevalent in sensor networks is malicious behaviors (network and data). This anomaly is more frequent in sensor networks deployed in hostile environments (He et al., 2006). Ideally, there are two types of adversaries that can introduce a malicious behavior in a sensor network: (i) insiders; and (ii) outsiders (Karlof and Wagner, 2003; Wang et al., 2006). The resource capability of outsiders can be unlimited; an outsider with a laptop-class device is capable of jamming the entire network using a long-range transmitter to emit signals that interfere with the same channel frequencies used by the sensor network (Wang et al., 2006). A laptop-class outsider can also eavesdrop on or capture every single message flowing into the targeted sensor network.

Moreover, he/she can replay messages or inject new data into the network. Nevertheless, outsiders cannot access the code of a sensor node and cannot modify (encrypted) data generated by each node. In contrast, insiders have a special access to the sensor network by compromising one or more nodes eventually laying hands on the security keys, code, and data stored in these nodes. Compromising a sensor node usually requires very little or no effort (Hartung et al., 2005). Once a node is compromised, it becomes easy to launch attacks that may affect either the overall network behavior or the data generated

Table 3. Measurement faults, their root causes, and detection features

Type	Example Deployments	Root Causes	Main Detection Feature
outlier (SHORT)	IBRL (Buonadonna et al., 2005), SensorScope (Ingelrest et al., 2010), Bangladesh (Ramanathan et al., 2006), NIMS (Kaiser et al., 2003), GDI (Szewczyk et al., 2004)	unknown	rate of change
spike	IBRL,SensorScope, Bangladesh	-low battery -sensor failure -short circuit	temporal rate of change
stuck-at (CONSTANT)	IBRL,SensorScope, NAMOS (Namos, 2012)	-sensor ADC malfunction	variance
NOISE	IBRL, SensorScope	-low battery -hardware failure (short circuit) -poor calibration	variance
calibration error	Bangladesh	-change in calibration equation -exposure to environment	spatial correlation
clipping /out of range	IBRL, Bangladesh	-reading outside the range of the ADC	variance accompanied by spatial correlation

by that node. It is further possible to use another more powerful laptop-class device to attack the network (i.e., generate a malicious behavior) after stealing all the information and identity of the compromised node.

Existing link layer encryption and authentication techniques (Perrig et al., 2002; Karlof et al., 2004) can serve as preventive measures against outsider attacks. Most of the possible security attacks that are launched by an outsider can be prevented by denying access of an adversary to join the network using authentication with globally shared keys. However, these mechanisms are rendered useless in the face of an attack launched by a laptop-class insider or a compromised node. Even with encryption and authentication techniques in place, several anomalous behaviors due to an insider attack may still exist in a sensor network deployment. Intrusion detection systems (ID-Ses) (da Silva et al., 2005; Van Phuong et al., 2006) take one step ahead of preventive measures in an attempt to detect such anomalies. An IDS is a system that monitors/audits host or network data for suspicious activities. Based on the audit data, an IDS for sensor networks should be able to detect both the malicious network behaviors that influence the overall performance and the forged data that hinder the fidelity of sensor data streams. There are two types of IDSes: (i) misused or signature-based; and (ii) anomaly-based. In the former, a database of attack signatures is established a priori from the audit data. Any attack that matches a signature in the database can be identified. On the other hand, anomaly detection based IDSes typically mark those behaviors that deviate from the normal behaviors as anomalies (attacks).

In the following discussion, we present the major attacks against sensor networks that either affect the data or introduce a malicious network behavior. These attacks can be generated by either an outsider or an insider. However as mentioned above, a simple authentication and encryption technique is adequate to stop most outsider attacks (e.g., passive eavesdropping, sinkhole, Sybil, selective forwarding). Therefore, we mainly focus on compromise behaviors or active attacks launched by a compromised node or laptop-class insider.

MALICIOUS DATA BEHAVIORS

Attack Characteristics. When an adversary captures the sensor node, he/she can reprogram it to inject falsified readings into the network. If a significant number of nodes is compromised, data forged by these nodes will highly impact the decision making process at the base station. An adversary can also alter readings sent by neighboring nodes before forwarding the data messages towards the sink. This, in turn, compromises the integrity of the sensor data. If a sensor network employs in-network processing or aggregation (Madden et al., 2002; Madden et al., 2005), false data can further mislead the aggregated results computed by the aggregation points including the base station.

Compromise behaviors injected in sensor networks can be classified as either (i) naive or (ii) smart behaviors. Naive compromised nodes always send falsified readings, whereas smart compromised nodes inject falsified readings in an intermittent manner along with legitimate data. In both scenarios, a compromised node may send highly deviated readings or readings which are closer to the normal behavior of other sensor nodes, varying the amount of harm to the fidelity of the data. Figure 9 shows both naive and smart compromise behaviors with different degrees of deviations exhibited by three nodes (nodes 12, 17, and 20) during two hours of our lab deployment.

During the process of data falsification, an adversary may also decide to change the behavior of the data to be identical to any of the natural sensor data faults discussed in the previous section. Sometimes fault models themselves can be close in characteristics to a naive or smart compromise behavior. An example of this behavior is the accelerometer readings in (Sharma et al., 2007). In this case, it is very hard to distinguish between a fault and a malicious attack. In addition, an adversary can manipulate its data or its neighbors' data in a manner that is much harder to detect.

Detection. Just like with measurement faults, there are several proposed mechanisms such as reputation-based (e.g., Ganeriwal et al., 2008) and machine learning-based (e.g., Abuaitah and Wang, 2014) mechanisms that detect malicious data behaviors caused by a compromised node. In general, detection is carried out by using statistical techniques and leveraging temporal correlations as well as spatial correlations among neighboring nodes. Rather than addressing complete compromise of a sensor network, all these techniques hope that less than half of the neighboring nodes are compromised and the detection accuracy degrades with the rates of false positives and false negatives increasing no faster than a rate proportional to the ratio of compromised nodes to the total number of nodes in the network (Karlof and Wagner, 2003).

MALICIOUS NETWORK BEHAVIORS

Unlike malicious data behaviors, malicious network behaviors do not directly influence the fidelity of the sensor data, rather, they impact the overall network performance. The most common security attacks launched by a compromised node or a laptop-class insider which introduce malicious behaviors into the network are (Wang et al., 2006; Xing et al., 2010): selective forwarding attacks; sinkholes; wormholes; Sybil attacks; spoofing and replay attacks; and Denial of Service (DoS). A significant number of proposed IDSes (da Silva et al., 2005; Loo et al., 2006; Yu and Tsai, 2008) and compromise detection mechanisms can identify malicious network behaviors (such as rule-based, machine-learningbased, etc.). IDSes offer much higher detection generality than techniques targeting a specific spectrum of attacks. A typical IDS may run on the base station, at every node in the network, or at selected nodes (called monitor nodes). For a survey of proposed IDSes in sensor networks, the reader is referred to (Farooqi and Khan, 2009). In this section, we describe each attack separately and discuss potential ways to detect it.

Figure 9. Compromise data behaviors during lab deployment. Node 8 emits normal temperature readings and is shown as a reference to compare against compromise nodes (nodes 12, 17, and 20).

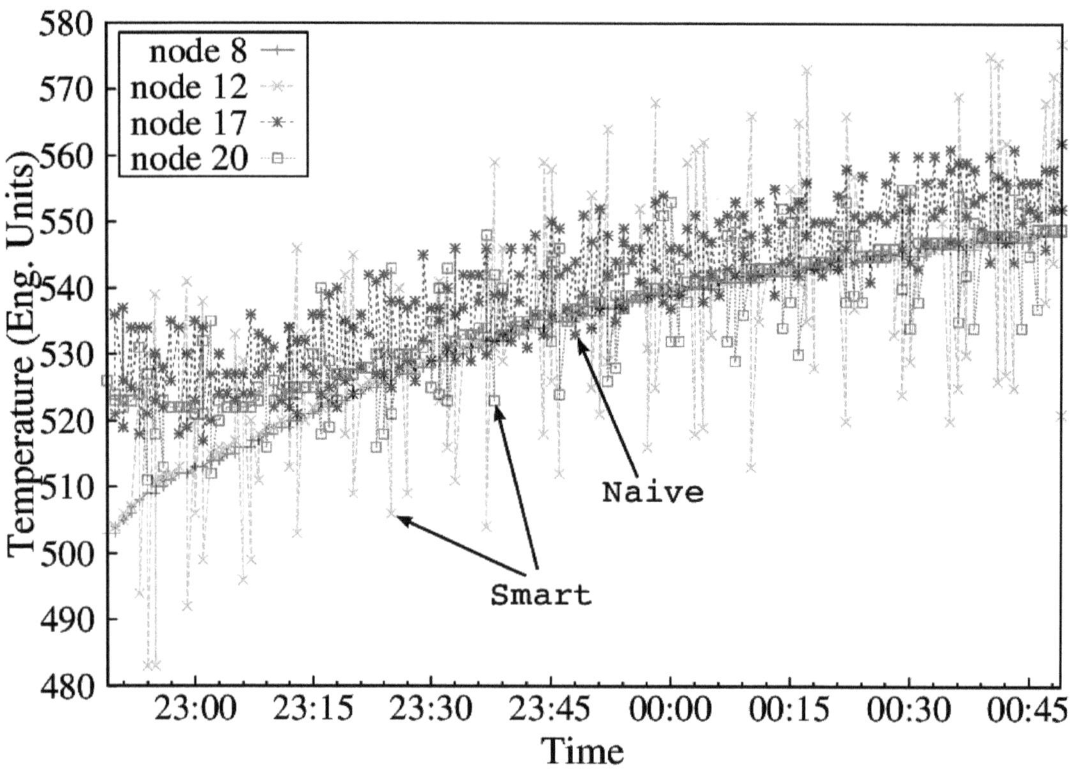

Selective Forwarding

Attack Characteristics. Once a sensor node is compromised, it becomes trivial for the adversary to selectively forward flowing traffic. This means that a compromised node may drop a portion of the packets (control and data) passing through it. Adversaries may either choose to drop all packets of the same type or deny packets from a number of source nodes. If every packet is being dropped, the compromised node is said to form a blackhole.

Detection. A selective forwarding attack (or a blackhole) can be detected by a more generic IDS or by a detection mechanism specifically designed for this type of attack (i.e., assuming no other attacks will be launched by the adversary). An example of the latter is the multi-hop acknowledgment based scheme proposed by Yu and Xiao (2006). In this scheme, a number of selected intermediate nodes called ACK nodes over the path from the source to the base station are responsible for notifying the source of any suspicious drop. Upon detecting an event, the source node sends a report packet containing a predefined ACK counter (which is decremented by one at every hop) towards the base station. The report packet propagates downstream by every hop and only ACK nodes reply with an acknowledgment. If an intermediate node does not hear a specific number of ACKs after some time, it suspects its downstream neighbor and sends an alarm packet upstream notifying the source node of this suspicious activity. After collecting a number of alarm packets, a source node filters out false alarms and identifies the malicious nodes where the dropping might have taken place. A downstream detection mechanism was also proposed where the base station can reveal packet losses based on the range of lost sequence numbers communicated by intermediate nodes downstream. Once malicious nodes are identified, both the source and the base station can notify the routing layer to exclude such nodes from the forwarding path.

IDSes follow a different approach. The IDS proposed in (da Silva et al., 2005), for instance, runs on monitor nodes. A monitor node listens to neighboring traffic and applies a set of rules to identify misbehaviors. If the number of failures is higher than a certain threshold (i.e., expected amount of occasional failures), a flag is raised. The "retransmission rule" is used to detect selective forwarding attacks (or blackholes). This rule states that if the monitor node, in its promiscuous mode, did not receive the same packet intended for its next hop, a failure counter is incremented.

Sinkhole

Attack Characteristics. This is one of the most severe routing layer attacks in sensor networks (Karlof and Wagner, 2003; Krontiris et al., 2008). Figure 10(a) illustrates a typical form of a sinkhole attack. The compromised node is equipped with a long range transceiver that can directly reach the base station. The adversary launches the attack by advertising this link to the neighbors. Inherent to most of the underlying routing protocols such as CTP, the neighboring nodes choose the compromised node as their next hop and re-advertise their updated routing information (e.g., path quality). In turn, nodes distant from the compromised node may elect the advertised path and update their routing information accordingly. This eventually enables the compromised node to suck the entire network traffic in a certain region towards itself. The adversary can also use a wormhole as discussed later to launch a sinkhole attack (see Figure 10(b)). Once a sinkhole is created, it becomes trivial to launch other attacks such as the selective forwarding attack discussed earlier.

Detection. Ngai et al (2006) proposed an intrusion detection algorithm that defends against a sinkhole attack. In their work, the base station is assumed to have the location information of all nodes in the network. The algorithm first identifies a number of suspected nodes. Suspected nodes are those that report data which are inconsistent with the other nodes in the network. The algorithm then forms a circle covering all suspected nodes. All nodes inside the circle are called the affected nodes. Once the list of affected nodes is formed by the base station, the base station forms a request message containing a timestamp, the list of all affected nodes, and the base station's private key and floods it into the network. The timestamp and the key are used to avoid replay attacks. Once an affected node receives the request, it checks to see if its id is in the list. If so, it sends a reply message, over the reverse path, to the base station containing its own id, the id of the next hop, and the path cost (e.g., link quality of the path from the node to the base station). Such information indicates the network flow (routing patterns) and is maintained in the node's routing table. Upon receiving all replies, the base station constructs a tree using the next hop information and discovers the sink hole represented by the root of the tree since all information flow towards the root node. Moreover, an algorithm is proposed to identify the sinkhole in the presence of a few compromised nodes that attempt to trick the base station into believing that a good node is the sinkhole hiding the identity of the real sinkhole. This algorithm exploits encryption and authentication and uses redundant paths to send the reply messages to the base station to avoid selective forwarding attacks initiated by the malicious nodes. It also attempts to identify the malicious nodes by exploiting inconsistencies within the collected network flow information at the base station.

Wormhole

Attack Characteristics. The wormhole attack was first introduced by Hu et al (2002) for adhoc networks. In this attack, a tunnel is typically

Figure 10. A demonstration of a sinkhole attack in sensor networks

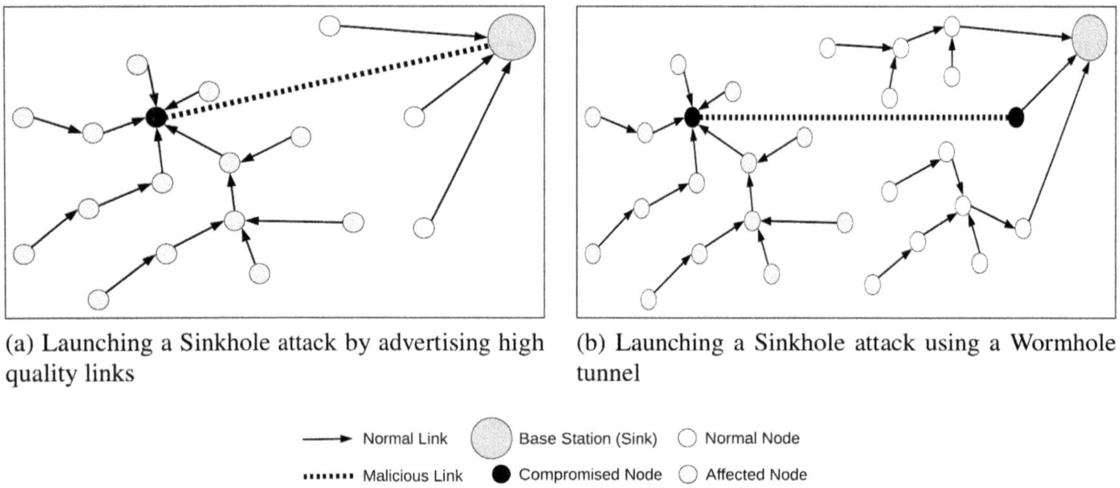

established between two distant malicious nodes (e.g., compromised nodes in Figure 10(b)). Each malicious node uses a higher-than-usual transmission power to relay the network traffic flowing into it without modifying the contents of the forwarded packets. This forces nodes at two opposite parts of the network to believe they are neighbors. In a similar attack called the "Hello flood" attack, a form of a Denial of Service, the adversary attempts to convince every node in the entire network that it is a neighbor of the compromised node by broadcasting beacon messages using an extremely high transmission power. As a result, all messages generated by nodes that are at least two hops away from the malicious node will be lost. Similar to a sinkhole, a wormhole can be used to launch other attacks in the network such as a sinkhole (Figure 10(b)), a selective forwarding attack, or a Sybil attack discussed later.

Detection. An RSSI-based mechanism was proposed in (Pires et al., 2004) which detects wormholes (and Hello floods) in the presence of no other attacks in the network. This scheme assumes that every node is aware of its geographical location and the sensor network is symmetric, static, and homogeneous. Upon reception of a message, each node computes the expected signal strength of the received signal using both the predefined transmission power and the distance between itself and the node that transmitted the message. If the difference between this value and the actual RSSI value reported by the radio chip is greater than a certain threshold, the received message is considered suspicious and the node transmitting the message is labeled as "suspicious." The receiving node then broadcasts its findings to its neighbors who, in turn, reply with their opinions about the same suspicious node. The receiving node uses these opinions (votes) to determine if the suspicious node is launching a wormhole (or a Hello flood) attack.

The rule-based IDS in (da Silva et al., 2005) uses the "radio transmission range rule" to detect a wormhole (or a Hello flood). This rule states that "all messages received by the monitor node in a promiscuous mode must be originated from one of its neighbors."

Spoofing / Replay

Attack Characteristics. An adversary can change the compromised node's routing information, spoof its own identity, or fabricate snooped link-layer acknowledgments. As for acknowledgment spoofing, most link-layer protocols used in today's sensor network deployments use implicit and/or explicit acknowledgments. By fabricating an acknowledgment upon listening to neighborhood traffic, an adversary may be able to persuade the sending node to think that the destination node is alive while in fact it is being disabled by the adversary or has already failed. The adversary can also convince the sending node that a weak link is strong leading to selective forwarding since the link cannot deliver the packet. On the other hand, modifying/replaying routing information or impersonating another node's identity may result in network failures such as routing loops and network partitioning. With identity spoofing in sensor networks, since a node can hear its neighborhood traffic, it can simply use one of its neighbors identities as its own to deteriorate the network functionality. For instance, an adversary (i.e., a compromised node) can spoof routing updates in CTP (Gnawali et al., 2009) in order to create a routing loop between two nodes A and B. The adversary first picks up node A's identity by listening on neighboring traffic. It then sends a fake routing update message to node B containing node A's address as the source address. Node B then marks node A as its parent and rebroadcasts the update. In turn, node A snoops (overhears) node B's update and marks node B as its parent (Karlof and Wagner, 2003). This causes traffic to circulate between both nodes endlessly until the loop is detected. The same methods discussed earlier which were used to identify network failures can be used to detect such a failure as well. However, there will be no clear distinction between a natural network failure or a failure caused by spoofing.

Detection. Chen et al (2007) have proposed a scheme that exploits the spatial correlation property of the received signal strength (RSS) to detect and localize identity spoofing and thus to avoid routing loops. In this scheme, the RSS readings of every identity are clustered in a signal space using the K-means clustering algorithm. The RSS readings from the same physical location should belong to one cluster. If a malicious node claimed the identity of another node, its RSS readings would belong to a different cluster since it is distant from the other normal node. Based on the distance between the centroids of the two clusters, the scheme can tell that the network is under a spoofing attack. The accuracy of this detection scheme is improved if the compromised node is located further away from the node with the stolen identity.

Sybil Attack

Attack Characteristics. Sybil attacks were first discussed in the context of peer-to-peer networks (Douceur, 2002). It is an example of an identity-based attack where the compromised node claims multiple identities. A taxonomy of the different forms of a Sybil attack on sensor networks was established in (Newsome et al., 2004). The authors distinguished between Sybil attacks using direct vs. indirect communication, stolen vs. fabricated identities, and simultaneous vs. non-simultaneous use of identities.

Detection. The authors in (Newsome et al., 2004) proposed five techniques to verify the identity of a Sybil node (i.e., an imaginary node with a stolen or fabricated identity). These are: (i) radio resource verification; (ii) cryptographic key validation; (iii) identity registration; (iv) position verification; and (v) code attestation. Each one of these techniques may miss one or two types of the classified attacks except code attestation which is always capable of detecting any form of a Sybil attack.

Denial of Service

Attack Characteristics. A Denial of Service (DoS) attack eliminates a network's capability to perform its expected functionality. All previous attacks discussed so far are mainly routing layer attacks. DoS attacks, however, can affect any of the five layers (physical, link, routing, transport, and application layer). Wood and Stankovic (2002) constituted a comprehensive survey of DoS attacks. In their survey, they have shown that most of these attacks can be countered using cryptography, authentication, or fairly simple preventive measures except for jamming. Jamming is a physical layer attack whereby a laptop-class insider can use a long-range transceiver to interfere with the radio frequencies used by the legitimate sensor nodes (Wang et al., 2006). This may shut the entire network down in a short period of time. This can also happen if a small number of compromised nodes with similar radio capabilities as legitimate nodes that are randomly distributed across the network launch a collaborative jamming attack (Wang et al., 2006). In addition, the Hello flood discussed earlier is another example of a DoS attack. It uses higher power to transmit "Hello" messages to the entire network. Finally, notice that network and node failures may also cause a DoS attack since such failures may render parts of the network non-functional.

Detection. The work in (Mpitziopoulos et al., 2009) surveys a number of detection techniques proposed for jamming. Typically, these techniques use signal strength, location information, or an estimate of the transmitter's interference strength as the detection measure. As for Hello floods, techniques used to detect wormholes can similarly be used to reveal Hello floods as well. Finally, the rule-based IDS in (da Silva et al., 2005) uses the "repetition rule" to detect a generic DoS attack. The repetition rule states that "the same message can be retransmitted by the same neighbor only a limited number of time."

In summary, there are instances where proactive measures may be adequate to prevent malicious behaviors. Nevertheless, a total compromise of a sensor node or a small number of nodes can break these measures. Rapidly detecting a malicious behavior when the adversary launches the attack is another line of defense. Once the malicious behavior is identified, further actions can be taken to eliminate the compromised node. Table 4 summaries the different types of malicious network behaviors. The table pinpoints the main characteristics of each attack and lists potential mechanisms to detect the attack.

EVENTS

The last type of anomalies in sensor network deployments is events. In event-driven application deployments such as VigilNet, the application administrators are interested in these abnormal instances that deviate from the normal setting. They would like the deployed network to notify them of any inconsistent behavior within the monitored area. Figure 11 illustrates few examples of potential events which occurred in the IBRL deployment. Usually, events manifest themselves as sudden abnormal jumps in the data. However, the amount of jump and the behavior of the data after this differs from one situation to another. Due to the high uncertainty involved in predicting the behavior of an event, it is important to study the types of events desired to be detected by the deployed sensornet. Often, the characteristics of the event are known a priori or by experimentations (Wittenburg et al., 2010).

Detection. Three approaches are common for event detection: (i) fully centralized; (ii) loosely distributed; and (iii) fully distributed. In the centralized approach, all data points are first gathered at the base station and then a detection algorithm is applied over the collected data. This incurs high detection delay and high energy cost. In the second approach, i.e., loosely decentralized, events are tracked at each sensor node individually. When an event is detected, it is reported to the base station for further verification with other nodes' suggestions. Finally, the fully decentralized approach requires the sensor nodes to exchange information in order to detect a potential event in their vicinity without direct help from the base station. Wittenburg et al (2010) proposed a fully decentralized scheme for event detection and qualitatively compared it

Table 4. Malicious network behaviors, attack characteristics, and possible detection approaches

Type	Attack Characteristics	Potential Existing Detection Mechanisms
selective forwarding	-an adversary (i.e., compromised node) may drop some or all packets (control and data) passing through it. -if all packets are being dropped, the node becomes a blackhole	-rule-based IDS (da Silva et al., 2005) -multi-hop ack based scheme (Yu and Xiao, 2006)
sinkhole	-adversary possesses a long range transceiver that can reach the base station -adversary sucks all traffic in certain region towards itself by advertising its high quality link	-intrusion detection algorithm proposed by Ngai et al (2006)
wormhole	-adversary establishes a tunnel between two distant compromised nodes using a higher-than-usual transmission power -nodes at opposite sides of the network think they are neighbors resulting in packet losses -if all nodes become neighbors of the compromised node, this leads to a Hello flood	-rule-based IDS -RSSI-based mechanism (Pires et al., 2004)
spoofing /replay	-change compromised node's routing information -spoof node's own identity -fabricate snooped link-layer acknowledgments	-data path validation in CTP -network diagnostic tools (e.g. AD) -exploiting spatial correlation property of RSS (Chen et al., 2007)
sybil attack	-compromised node claims multiple identities	-radio resource verification, cryptographic key validation, identity registration, position verification, and code attestation (Newsome et al., 2004)
denial of service	-eliminate network's capability to perform its expected functionality -shut down the entire network in a short period by jamming with a long-range transceiver -hello floods	-rule-based IDS -RSSI-based mechanism (Pires et al., 2004) -techniques relying on signal strength, location estimation, or transmitter's interference strength (Mpitziopoulos et al., 2009)

Figure 11. Potential events prevalent in the IBRL deployment

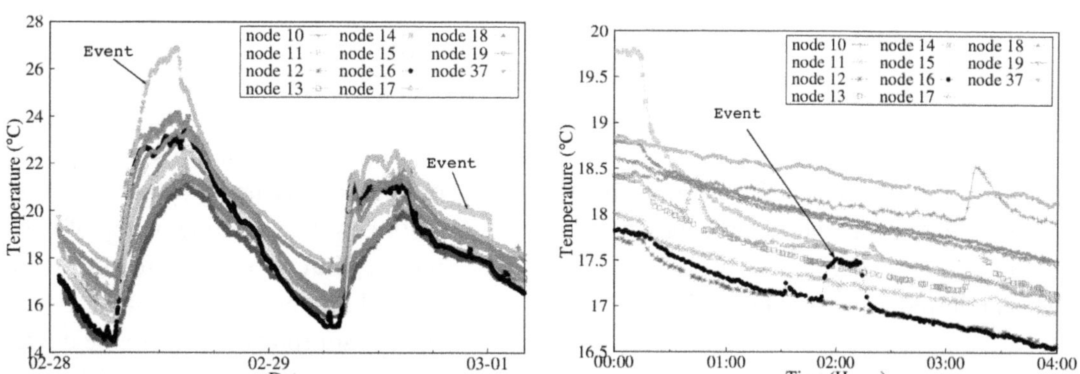

(a) Node 37 shows higher temperature values due to its physical placement close to the kitchen and node 14 is believed to exhibit deviated behavior at the end due to unusual interference [54]

(b) The first four hours starting midnight March 1. During short periods, nodes 10 and 16 were potentially exposed to regions of higher heat than the other regions where the rest of the nodes are deployed.

with other schemes. Their scheme is proven to be superior to others as it consumes less processing and memory resources, does not require any cluster-like infrastructure, leverages pre-training to improve accuracy, and allows sensor nodes to collaborate.

A key factor in detecting events is to be able to disambiguate between a faulty measurement and an event that has happened in a specific region (Gupchup et al., 2008). Fault-event disambiguation is not a trivial task. However, if spatial correlation is exploited, geographically independent faults can be distinguished from events which are based on correlated readings observed by a number of sensors in a region or event boundary. The work of Ding et al. (2005) proposed a spatial data mining based mechanism to detect event boundaries in presence of faulty measurements.

DATA-CENTRIC ANOMALIES

Having classified anomalies into three major categories (natural faults, malicious behaviors, events), we now identify these anomalies that only influence the collected data samples (data-centric). Measurement faults, malicious data behaviors, and events are all examples of a data-centric anomaly. Data faults and malicious data behaviors hinder the quality, integrity, and/or trustworthiness of the sensor data whereas events manifest as abnormal data. Some detection mechanisms (e.g., Abuaitah and Wang, 2014) focus on data-centric anomalies in general and do not necessarily distinguish a measurement fault from an event from a malicious data behavior. For a survey of such mechanisms, the reader is referred to (Zhang et al., 2010).

CHARACTERISTICS OF ANOMALY DETECTION ALGORITHMS

After establishing a comprehensive taxonomy of the different types of anomalies in sensor network deployments, we now discuss the challenges in designing an anomaly detection mechanism or algorithm. These challenges are common to all types of anomalies discussed in this chapter. An anomaly detection algorithm should be:

- **Resource-Aware:** The most prevalent resource constraint in the majority of sensor network deployments is "energy." Communication usually accounts for the highest amount of energy consumption. Processing power can also be high if advanced and more complex operations are performed at the sensor node (e.g., surveillance (He et al., 2006)). Memory, on the other hand, may not be a constraining resource in recent days (Tsiftes and Dunkels, 2011). However, writing to flash memory is still very expensive in terms of power consumption. Given these constraints, an anomaly detection algorithm should try to eliminate any additional burdens on the battery capacity and, if possible, utilize the remaining space in the data memory before attempting to write to much larger flash memory.
- **Simple:** The cost of computation may not be the major concern in sensor network deployments. However, algorithms that run on low-power resource-constrained embedded devices such as sensor motes should not incur a high computational complexity. Complex algorithms can incur delay and may unexpectedly crash the operating system running on top of the sensor motes. Moreover, it is difficult to debug or track the operation of a large network of tiny sensor motes to ensure the correctness of the detection algorithm before deployment. Hence, simplicity of algorithms may ease the debugging process.
- **Accurate:** As any other detection algorithm, an anomaly detection algorithm designed, for sensor networks should achieve

a high detection accuracy. Anomaly detection schemes that yield high rates of false positives and false negatives are not desirable.

- **Rapid:** Detection latency is an important measure of the effectiveness of an anomaly detection algorithm. A low detection latency is desirable in all cases. In an event-driven application, for example, events should be identified in a timely manner. Natural faults and malicious behaviors should also be revealed before the entire network goes down or the sensor data is severely contaminated with inconsistent data points.
- **Scalable:** Most recent deployments constituted a descent number of sensor nodes. A recent deployment of GreenOrbs (Shi et al., 2011) consisted of 349 nodes and is still in expansion. Therefore, detection algorithms that run perfectly in a lab environment with few sensor nodes may not necessarily adapt to networks at large scales.
- **Adaptable:** WSN deployments usually involve incremental installation of new nodes in the field over time. The detection algorithm should tolerate any addition of new nodes or removal of dead nodes.
- **Robust:** Although we have put down a comprehensive taxonomy of anomalies, sometimes it is difficult to anticipate the true behavior of an anomaly (e.g., a new attack on sensor networks). Therefore, it is highly desirable that a detection algorithm identifies unknown anomalies such as novel events, unexplored future faults, or newly emerged malicious activities.
- **Secure:** This challenge is perhaps more relevant to sensor networks deployed in hostile environments. An algorithm designed for revealing a malicious behavior should itself be resilient to attacks.
- **Practical:** In this chapter, we discussed a number of detection and mitigation techniques against anomalies in sensor networks. However, most existing techniques have not been tested in real-world deployments. This calls for conducting real-world experiments to prove the effectiveness of such techniques under realistic environments. Moreover, in recent sensor network deployments for monitoring applications, several other services are employed such as sensor data collection, parameter dissemination, etc. Rather than interrupting the normal operation of the sensor network, an anomaly detection algorithm or service should run alongside existing protocols.

RELATED WORK

An earlier attempt was made by Jurdak et al (2011b) to classify anomalies in sensor networks. However, the authors only addressed natural faults and did not elaborate on measurement faults. A measurement fault, as one type of a natural fault, is a very significant type of anomaly and was extensively described in our work for further classification following the same terminologies used in (Ni et al., 2009) and (Sharma et al., 2007). Malicious behaviors and events can also be considered as anomalies in sensor network deployments and hence are an essential part in our taxonomy.

Ni et al. (2009) compiled a study of sensor data faults prevalent in sensor network deployments and presented helpful features that can be used in the detection process. Both Xing et al. (2010) and Wang et al. (2006) surveyed the different types of malicious network behaviors and discussed a number of countermeasures. Our work borrowed the terminologies presented in these works to maintain consistency. However, we deal with anomalies in general rather than single measurement faults or malicious network behaviors. This work is more general in scope by providing an insight to the reader about "anything" that can go wrong in a sensor network whether deployed in a friendly or a hostile environment by constructing a comprehensive taxonomy of such anomalies.

A number of surveys (Rodrigues et al., 2012; Zhang et al., 2010; Xie et al., 2011; Farooqi and Khan, 2009) exist in the literature that address the detection of a specific scope of anomalies discussed in this chapter. In (Rodrigues et al., 2012), the authors compared among the various diagnosis and debugging tools which handle functional faults (network and node failures). The work of Zhang et al. (2010) surveyed the different mechanisms for detecting data-centric anomalies. Intrusion detection systems for identifying malicious network behaviors were discussed in (Farooqi and Khan, 2009). Finally, Xie et al. (2011) studied a number of detection mechanisms for data-centric anomalies as well as malicious network behaviors. Our work in this chapter differs from these surveys in two aspects. First, the main focus of this chapter is the analysis rather than the detection of different types of anomalies that are prevalent in recent sensor network deployments. Second, the discussion in these surveys does not necessarily show how each detection mechanism deal with every individual anomaly. In contrast, our work gives the reader an insight on how to do so by presenting some existing solutions.

CONCLUSION

In this chapter, we established a taxonomy of different types of anomalies prevalent in sensor network deployments. We discussed each anomalous behavior and gave insights on how to potentially identify such a behavior using existing detection mechanisms. Finally, having advocated for anomaly detection mechanisms, we presented the challenges of designing a new mechanism suitable for real-world sensor network deployments. This comprehensive study can also help researchers narrow their focus to a specific type of anomaly or choose to implement a more generalized anomaly detection mechanism.

REFERENCES

Abuaitah, G. R., & Wang, B. (2014). Online Data-Centric Anomaly Detection Framework For Sensor Network Deployments. In *The 2014 IEEE International Conference on Computing, Networking and Communications (ICNC)*, (pp. 599-604). doi:10.1109/ICCNC.2014.6785404

Atmel. (2014). *AVR JTAGICE mkII.* Retrieved October 10, 2014, from http://www.atmel.com/tools/AVRJTAGICEMKII.aspx

Barrenetxea, G., Ingelrest, F., & Schaefer, G., & Vetterli (2008). The hitchhiker's guide to successful wireless sensor network deployments. In *Proceedings of the 6th ACM conference on Embedded network sensor systems (SenSys)* (pp. 43–56). doi:10.1145/1460412.1460418

Buonadonna, P., Gay, D., Hellerstein, J., Hong, W., & Madden, S. (2005). Task: sensor network in a box. In *Proceedings of the Second European Workshop on Wireless Sensor Networks (EWSN)* (pp. 133–144). doi:10.1109/EWSN.2005.1462005

CC2420. (2012). *Cc2420 datasheet.* Retrieved June 16, 2012, from http://inst.eecs.berkeley.edu/cs150/Documents/CC2420.pdf

Chang, J.-H., & Tassiulas, L. (2004). Maximum lifetime routing in wireless sensor networks. *IEEE/ACM Transactions on Networking*, *12*(4), 609–619. doi:10.1109/TNET.2004.833122

Chen, Y., Trappe, W., & Martin, R. (2007). Detecting and localizing wireless spoofing attacks. In *Proceedings of the 4th Annual IEEE Communications Society Conference on Sensor, Mesh and Ad Hoc Communications and Networks (SECON)* (pp. 193–202).

da Silva, A. P. R., Martins, M. H. T., Rocha, B. P. S., Loureiro, A. A. F., Ruiz, L. B., & Wong, H. C. (2005). Decentralized intrusion detection in wireless sensor networks. In *Proceedings of the 1st ACM international workshop on Quality of service & security in wireless and mobile networks (Q2SWinet)* (pp. 16–23).

De Couto, D. S. J., Aguayo, D., Bicket, J., & Morris, R. (2003). A high-throughput path metric for multi-hop wireless routing. In *Proceedings of the 9th annual international conference on Mobile computing and networking (MobiCom)* (pp. 134–146).

Ding, M., Chen, D., Xing, K., & Cheng, X. (2005). Localized fault-tolerant event boundary detection in sensor networks. In *Proceedings of the 24th IEEE International Conference on Computer Communications (INFOCOM)*, (volume 2, pp. 902–913).

Dini, G., Pelagatti, M., & Savino, I. M. (2008). An algorithm for reconnecting wireless sensor network partitions. In *Proceedings of the 5th European conference on Wireless sensor networks (EWSN)* (pp. 253–267). doi:10.1007/978-3-540-77690-1_16

Douceur, J. R. (2002). The sybil attack. In *Revised Papers from the First International Workshop on Peer-to-Peer Systems (IPTPS)* (pp. 251–260).

Farooqi, A. H., & Khan, F. A. (2009). Intrusion detection systems for wireless sensor networks: A survey. In Communication and Networking, (vol. 56, pp. 234–241). doi:10.1007/978-3-642-10844-0_29

Fonseca, R., Gnawali, O., Jamieson, K., & Levis, P. (2007). Four-bit wireless link estimation. In Proceedings of HotNets VI.

Ganeriwal, S., Balzano, L. K., & Srivastava, M. B. (2008). Reputation-based framework for high integrity sensor networks. *ACM Transactions on Sensor Networks, 4*(3), 15:1–15:37.

Gnawali, O., Fonseca, R., Jamieson, K., Moss, D., & Levis, P. (2009). Collection tree protocol. In *Proceedings of the 7th ACM Conference on Embedded Networked Sensor Systems (SenSys)*, (pp. 1–14). doi:10.1145/1644038.1644040

Guo, S., Zhong, Z., & He, T. (2009). Find: faulty node detection for wireless sensor networks. In *Proceedings of the 7th ACM Conference on Embedded Networked Sensor Systems (SenSys)*, (pp. 253–266). doi:10.1145/1644038.1644064

Gupchup, J., Sharma, A., Terzis, A., Burns, R., & Szalay, A. (2008). The perils of detecting measurement faults in environmental monitoring networks (Invited Paper). In *Proceedings of DCoSS*.

Hartung, C., Balasalle, J., Han, R., Hartung, C., Balasalle, J., & Han, R. (2005). *Node compromise in sensor networks: The need for secure systems* (Tech. Rep. CU-CS-990-05), University of Colorado, Boulder, CO.

He, T., Krishnamurthy, S., Luo, L., Yan, T., Gu, L., Stoleru, R., & Krogh, B. (2006). Vigilnet: An integrated sensor network system for energy-efficient surveillance. *ACM Transactions on Sensor Networks, 2*(1), 1–38. doi:10.1145/1138127.1138128

Herman, T., Pemmaraju, S., Pilard, L., & Mjelde, M. (2007). Temporal partition in sensor networks. *In Proceedings of the 9th International Conference on Stabilization, Safety, and Security of Distributed Systems (SSS)* (pp. 325–339). doi:10.1007/978-3-540-76627-8_25

Hu, Y., Perrig, A., & Johnson, D. B. (2002). *Wormhole detection in wireless ad hoc networks. Technical report*. Department of Computer Science, Rice University.

Ingelrest, F., Barrenetxea, G., Schaefer, G., Vetterli, M., Couach, O., & Parlange, M. (2010). Sensorscope: Application-specific sensor network for environmental monitoring. *ACM Transactions on Sensor Networks, 6*(2), 17:1–17:32.

Jurdak, R., Ruzzelli, A. G., Barbirato, A., & Boivineau, S. (2011a). Octopus: Monitoring, visualization, and control of sensor networks. *Wireless Communication & Mobile Computing*, *11*(8), 1073–1091. doi:10.1002/wcm.826

Jurdak, R., Wang, X., Obst, O., & Valencia, P. (2011b). Wireless sensor network anomalies: Diagnosis and detection strategies. In Intelligence-Based. *Systems Engineering*, *10*, 309–325.

Kaiser, W. J., Pottie, G. J., Srivastava, M., Sukhatme, G. S., Villasenor, J., & Estrin, D. (2003). Networked infomechanical systems (nims) for ambient intelligence (Technical Report 31). University of California Los Angeles, Networked Embedded Systems Lab, Los Angeles, CA.

Karlof, C., Sastry, N., & Wagner, D. (2004). Tinysec: A link layer security architecture for wireless sensor networks. In *Proceedings of the 2nd international conference on Embedded networked sensor systems (SenSys)*, (pp. 162–175). doi:10.1145/1031495.1031515

Karlof, C., & Wagner, D. (2003). Secure routing in wireless sensor networks: Attacks and countermeasures. In *Proceedings of the First IEEE International Workshop on Sensor Network Protocols and Applications*, (pp. 113 – 127). doi:10.1109/SNPA.2003.1203362

Khan, M. M. H., Le, H. K., LeMay, M., Moinzadeh, P., Wang, L., & Yang, Y., … Jin, X. (2010). Diagnostic powertracing for sensor node failure analysis. In *Proceedings of the 9th ACM/IEEE International Conference on Information Processing in Sensor Networks (IPSN)*, (pp. 117–128). doi:10.1145/1791212.1791227

Kim, K.-H., & Shin, K. G. (2011). Self-reconfigurable wireless mesh networks. *IEEE/ACM Transactions on Networking*, *19*(2), 393–404. doi:10.1109/TNET.2010.2096431

Krontiris, I., Giannetsos, T., & Dimitriou, T. (2008). Launching a sinkhole attack in wireless sensor networks; the intruder side. In *Proceedings of the 4th IEEE International Conference on Wireless and Mobile Computing, Networking and Communications (WIMOB)*, (pp. 526–531). doi:10.1109/WiMob.2008.83

Lenders, V., Wagner, J., & May, M. (2006). Analyzing the impact of mobility in ad hoc networks. In *Proceedings of the 2nd international workshop on Multi-hop ad hoc networks: from theory to reality* (REALMAN) (pp. 39–46). doi:10.1145/1132983.1132991

Levis, P., Lee, N., Welsh, M., & Culler, D. (2003). TOSSIM: accurate and scalable simulation of entire tinyos applications. In *Proceedings of the 1st international conference on Embedded networked sensor systems (SenSys)*, (pp. 126–137). doi:10.1145/958491.958506

Levis, P., Patel, N., Culler, D., & Shenker, S. (2004). Trickle: A self-regulating algorithm for code propagation and maintenance in wireless sensor networks. In *Proceedings of the First USENIX/ACM Symposium on Networked Systems Design and Implementation (NSDI)*, (pp. 15–28).

Li, Y., & Parker, L. (2008). Detecting and monitoring time-related abnormal events using a wireless sensor network and mobile robot. In *Proceedings of the IEEE/RSJ International Conference on Intelligent Robots and Systems (IROS)*, (pp. 3292–3298).

Lindhorst, T., Lukas, G., & Nett, E. (2010). Modeling fast link failure detection for dependable wireless mesh networks. In *Proceedings of the 2010 Ninth IEEE International Symposium on Network Computing and Applications (NCA)*, (pp. 44–51). doi:10.1109/NCA.2010.13

Liu, K., Ma, Q., Zhao, X., & Liu, Y. (2011a). Self-diagnosis for large scale wireless sensor networks. In *Proceedings of IEEE INFOCOM* (pp. 1539–1547). doi:10.1109/INFCOM.2011.5934944

Liu, Y., He, Y., Li, M., Wang, J., Liu, K., & Mo, L., … Li, X.-Y. (2011b). Does wireless sensor network scale? A measurement study on greenorbs. In *Proceedings of IEEE INFOCOM*, (pp. 873–881).

Loo, C. E., Ng, M. Y., Leckie, C., & Palaniswami, M. (2006). Intrusion detection for routing attacks in sensor networks. *International Journal of Distributed Sensor Networks*, *2*(4), 313–332. doi:10.1080/15501320600692044

Madden, S., Franklin, M. J., Hellerstein, J. M., & Hong, W. (2002). Tag: a tiny aggregation service for ad-hoc sensor networks. In *Proceedings of the 5th symposium on operating systems design and implementation (OSDI)*. doi:10.1145/1060289.1060303

Madden, S. R., Franklin, M. J., Hellerstein, J. M., & Hong, W. (2005). Tinydb: An acquisitional query processing system for sensor networks. *ACM Transactions on Database Systems*, *30*(1), 122–173. doi:10.1145/1061318.1061322

Maróti, M., Kusy, B., Simon, G., & Lédeczi, A. (2004). The flooding time synchronization protocol. In *Proceedings of the 2nd international conference on Embedded networked sensor systems (SenSys)* (pp. 39–49). doi:10.1145/1031495.1031501

Miao, X., Liu, K., He, Y., Liu, Y., & Papadias, D. (2011). Agnostic diagnosis: Discovering silent failures in wireless sensor networks. In *Proceedings of IEEE INFOCOM*, (pp. 1548–1556). doi:10.1109/INFCOM.2011.5934945

Mo, L., He, Y., Liu, Y., Zhao, J., Tang, S.-J., Li, X.-Y., & Dai, G. (2009). Canopy closure estimates with greenorbs: sustainable sensing in the forest. In *Proceedings of the 7th ACM Conference on Embedded Networked Sensor Systems (SenSys)*, (pp. 99–112). doi:10.1145/1644038.1644049

Mpitziopoulos, A., Gavalas, D., Konstantopoulos, C., & Pantziou, G. (2009). A survey on jamming attacks and countermeasures in wsns. *IEEE Communications Surveys and Tutorials*, *11*(4), 42–56. doi:10.1109/SURV.2009.090404

Namos (2012). *Namos deployment -james reserve (lake fulmor) august 28th september 1st 2006*. Retrieved June 3, 2012, from http://robotics.usc.edu/~namos/data/jr aug 06/

Newsome, J., Shi, E., Song, D., & Perrig, A. (2004). The sybil attack in sensor networks: analysis & defenses. In *Proceedings of the 3rd international symposium on Information processing in sensor networks (IPSN)* (pp. 259–268). doi:10.1145/984622.984660

Ngai, E., Liu, J., & Lyu, M. (2006). On the intruder detection for sinkhole attack in wireless sensor networks. In *Proceedings of the IEEE International Conference on Communications (ICC)* (volume 8, pp. 3383–3389).

Ni, K., Ramanathan, N., Chehade, M. N. H., Balzano, L., Nair, S., Zahedi, S., … Srivastava, M. (2009). Sensor network data fault types. *ACM Transactions on Sensor Networks, 5*(3), 25:1–25:29.

Ni, S.-Y., Tseng, Y.-C., Chen, Y.-S., & Sheu, J.-P. (1999). The broadcast storm problem in a mobile ad hoc network. In *Proceedings of the 5th annual ACM/IEEE international conference on Mobile computing and networking (MobiCom)* (pp. 151–162). doi:10.1145/313451.313525

Pei, D., Zhao, X., Massey, D., & Zhang, L. (2004). A study of BGP path vector route looping behavior. In *Proceedings of the 24th International Conference on Distributed Computing Systems (ICDCS)*, (pp. 720–729). doi:10.1109/ICDCS.2004.1281640

Perrig, A., Szewczyk, R., Tygar, J. D., Wen, V., & Culler, D. E. (2002). Spins: Security protocols for sensor networks. *Wireless Networks*, *8*(5), 521–534. doi:10.1023/A:1016598314198

Pires, W. R., J., de Paula Figueiredo, T., Wong, H., & Loureiro, A. (2004). Malicious node detection in wireless sensor networks. In *Proceedings of the 18th International Parallel and Distributed Processing Symposium (IPDPS)*. doi:10.1109/IPDPS.2004.1302934

Rajasegarar, S., Bezdek, J. C., Leckie, C., & Palaniswami, M. (2007). Analysis of anomalies in ibrl data from a wireless sensor network deployment. In *Proceedings of the 2007 International Conference on Sensor Technologies and Applications (SENSORCOMM)* (pp. 158–163). Washington, DC, USA. doi:10.1109/SENSORCOMM.2007.4394914

Ramanathan, N., Balzano, L., Burt, M., Estrin, D., Harmon, T., Harvey, C., … Srivastava, M. (2006). *Rapid deployment with confidence: Calibration and fault detection in environmental sensor networks* (Tech. Rep.). Center for Embedded Networked Sensing, UCLA and Department of Civil and Environmental Engineering, MIT.

Ramanathan, N., Chang, K., Kapur, R., Girod, L., Kohler, E., & Estrin, D. (2005). Sympathy for the sensor network debugger. In *Proceedings of the 3rd international conference on Embedded networked sensor systems (SenSys)*, (pp. 255–267). doi:10.1145/1098918.1098946

Rodrigues, A., Camilo, T., Silva, J., & Boavida, F. (2012). Diagnostic tools for wireless sensor networks: A comparative survey. *Journal of Network and Systems Management*, 1–45.

Sharma, A., Golubchik, L., & Govindan, R. (2007). On the prevalence of sensor faults in real-world deployments. In *Proceedings of the 4th Annual IEEE Communications Society Conference on Sensor, Mesh and Ad Hoc Communications and Networks (SECON)* (pp. 213–222). doi:10.1109/SAHCN.2007.4292833

Sharma, A. B., Golubchik, L., & Govindan, R. (2010). Sensor faults: Detection methods and prevalence in real-world datasets. *ACM Transactions on Sensor Networks*, *6*(3), 23:1–23:39.

Shi, L., Liao, Q., He, Y., Li, R., Striegel, A., & Su, Z. (2011). Save: Sensor anomaly visualization engine. In *Proceedings of the IEEE Conference on Visual Analytics Science and Technology (VAST)*, (pp. 201–210). doi:10.1109/VAST.2011.6102458

Shrivastava, N., Suri, S., & T'oth, C. D. (2005). Detecting cuts in sensor networks. In *Proceedings of the 4th international symposium on Information processing in sensor networks (IPSN)*.

Singh, S., Woo, M., & Raghavendra, C. S. (1998). Power-aware routing in mobile ad hoc networks. In *Proceedings of the 4th annual ACM/IEEE international conference on Mobile computing and networking (MobiCom)* (pp. 181–190). doi:10.1145/288235.288286

Staddon, J., Balfanz, D., & Durfee, G. (2002). Efficient tracing of failed nodes in sensor networks. In *Proceedings of the 1st ACM international workshop on Wireless sensor networks and applications (WSNA)*, (pp. 122–130). doi:10.1145/570738.570756

Szewczyk, R., Mainwaring, A., Polastre, J., Anderson, J., & Culler, D. (2004). An analysis of a large scale habitat monitoring application. In *Proceedings of the 2nd international conference on Embedded networked sensor systems (SenSys)* (pp. 214–226). doi:10.1145/1031495.1031521

The MultihopLQI Protocol. (2012). *The MultihopLQI Protocol*. Retrieved June 16, 2012, from http://www.tinyos.net/tinyos-2.x/tos/lib/net/lqi

Tolle, G., Polastre, J., Szewczyk, R., Culler, D., Turner, N., Tu, K., & Hong, W. (2005). A macroscope in the redwoods. In *Proceedings of the 3rd international conference on Embedded networked sensor systems (SenSys)*, (pp. 51–63). doi:10.1145/1098918.1098925

Tsiftes, N., & Dunkels, A. (2011). A database in every sensor. In *Proceedings of the 9th ACM Conference on Embedded Networked Sensor Systems (SenSys)* (pp. 316–332).

Van Phuong, T., Hung, L. X., Cho, S. J., Lee, Y.-K., & Lee, S. (2006). An anomaly detection algorithm for detecting attacks in wireless sensor networks. In *Proceedings of the 4th IEEE international conference on Intelligence and Security Informatics (ISI)* (pp. 735–736). doi:10.1007/11760146_111

Wälchli, M., & Braun, T. (2009). Efficient signal processing and anomaly detection in wireless sensor networks. In *Proceedings of the EvoWorkshops on Applications of Evolutionary Computing* (pp. 81–86).

Wang, Y., Attebury, G., & Ramamurthy, B. (2006). A survey of security issues in wireless sensor networks. *IEEE Communications Surveys and Tutorials*, *8*(2), 2–23. doi:10.1109/COMST.2006.315852

Wittenburg, G., Dziengel, N., Wartenburger, C., & Schiller, J. (2010). A system for distributed event detection in wireless sensor networks. In *Proceedings of the 9th ACM/IEEE International Conference on Information Processing in Sensor Networks (IPSN)*, (pp. 94–104). doi:10.1145/1791212.1791225

Woo, A., Tong, T., & Culler, D. (2003). Taming the underlying challenges of reliable multihop routing in sensor networks. In *Proceedings of the 1st international conference on Embedded networked sensor systems (SenSys)*, (pp. 14–27). doi:10.1145/958491.958494

Wood, A., & Stankovic, J. (2002). Denial of service in sensor networks. *Computer*, *35*(10), 54–62. doi:10.1109/MC.2002.1039518

Xie, M., Han, S., Tian, B., & Parvin, S. (2011). Anomaly detection in wireless sensor networks: A survey. *Journal of Network and Computer Applications*, *34*(4), 1302–1325. doi:10.1016/j.jnca.2011.03.004

Xing, K., Srinivasan, S. S. R., Rivera, M. J. M., Li, J., & Cheng, X. (2010). Network Security, chapter Attacks and Countermeasures in Sensor Networks. *Survey (London, England)*, 251–272.

Yang, J., Soffa, M. L., Selavo, L., & Whitehouse, K. (2007). Clairvoyant: a comprehensive source-level debugger for wireless sensor networks. In *Proceedings of the 5th International Conference on Embedded Networked Sensor Systems (SenSys)* (pp. 189–203).

Yu, B., & Xiao, B. (2006). Detecting selective forwarding attacks in wireless sensor networks. In *Proceedings of the 20th International Parallel and Distributed Processing Symposium (IPDPS)*.

Yu, Z., & Tsai, J. (2008). A framework of machine learning based intrusion detection for wireless sensor networks. In *IEEE International Conference on Sensor Networks, Ubiquitous and Trustworthy Computing (SUTC)* (pp. 272–279). doi:10.1109/SUTC.2008.39

Zhang, Y., Meratnia, N., & Havinga, P. (2010). Outlier detection techniques for wireless sensor networks: A survey. *IEEE Communications Surveys and Tutorials*, *12*(2), 159–170. doi:10.1109/SURV.2010.021510.00088

ADDITIONAL READING

Abuaitah, G. R., & Wang, B. (2014). Online Data-Centric Anomaly Detection Framework For Sensor Network Deployments. The 2014 IEEE International Conference on Computing, Networking and Communications (ICNC), pages 599-604. doi:10.1109/ICCNC.2014.6785404

Chen, Y., Trappe, W., & Martin, R. (2007). Detecting and localizing wireless spoofing attacks. In *Proceedings of the 4th Annual IEEE Communications Society Conference on Sensor, Mesh and Ad Hoc Communications and Networks (SECON)*, (pp. 193–202).

da Silva, A. P. R., Martins, M. H. T., Rocha, B. P. S., Loureiro, A. A. F., Ruiz, L. B., & Wong, H. C. (2005). Decentralized intrusion detection in wireless sensor networks. In *Proceedings of the 1st ACM international workshop on Quality of service & security in wireless and mobile networks* (Q2SWinet) (pp. 16–23).

Farooqi, A. H., & Khan, F. A. (2009). Intrusion detection systems for wireless sensor networks: A survey. In Communication and Networking, (vol. 56, pp. 234–241). doi:10.1007/978-3-642-10844-0_29

Gnawali, O., Fonseca, R., Jamieson, K., Moss, D., & Levis, P. (2009). Collection tree protocol. In *Proceedings of the 7th ACM Conference on Embedded Networked Sensor Systems (SenSys)*, (pp. 1–14). doi:10.1145/1644038.1644040

Jurdak, R., Ruzzelli, A. G., Barbirato, A., & Boivineau, S. (2011a). Octopus: Monitoring, visualization, and control of sensor networks. *Wireless Communication & Mobile Computing*, *11*(8), 1073–1091. doi:10.1002/wcm.826

Jurdak, R., Wang, X., Obst, O., & Valencia, P. (2011b). Wireless sensor network anomalies: Diagnosis and detection strategies. In Intelligence-Based. *Systems Engineering*, *10*, 309–325.

Khan, M. M. H., Le, H. K., LeMay, M., Moinzadeh, P., Wang, L., Yang, Y., & Jin, X. et al. (2010). Diagnostic powertracing for sensor node failure analysis. In *Proceedings of the 9th ACM/IEEE International Conference on Information Processing in Sensor Networks (IPSN)*, (pp. 117–128). doi:10.1145/1791212.1791227

Liu, K., Ma, Q., Zhao, X., & Liu, Y. (2011). Self-diagnosis for large scale wireless sensor networks. In *Proceedings of IEEE INFOCOM*, (pp. 1539 –1547). doi:10.1109/INFCOM.2011.5934944

Miao, X., Liu, K., He, Y., Liu, Y., & Papadias, D. (2011). Agnostic diagnosis: Discovering silent failures in wireless sensor networks. In *Proceedings of IEEE INFOCOM* (pp. 1548–1556). doi:10.1109/INFCOM.2011.5934945

Mpitziopoulos, A., Gavalas, D., Konstantopoulos, C., & Pantziou, G. (2009). A survey on jamming attacks and countermeasures in wsns. *IEEE Communications Surveys and Tutorials*, *11*(4), 42–56. doi:10.1109/SURV.2009.090404

Newsome, J., Shi, E., Song, D., & Perrig, A. (2004). The sybil attack in sensor networks: analysis & defenses. In *Proceedings of the 3rd international symposium on Information processing in sensor networks (IPSN)*, (pp. 259–268). doi:10.1145/984622.984660

Ngai, E., Liu, J., & Lyu, M. (2006). On the intruder detection for sinkhole attack in wireless sensor networks. In *Proceedings of the IEEE International Conference on Communications (ICC)*, (volume 8, pp. 3383–3389).

Ni, K., Ramanathan, N., Chehade, M. N. H., Balzano, L., Nair, S., Zahedi, S., … Srivastava, M. (2009). Sensor network data fault types. *ACM Transactions on Sensor Networks, 5*(3), 25:1–25:29.

Pires, W. R., J., de Paula Figueiredo, T., Wong, H., & Loureiro, A. (2004). Malicious node detection in wireless sensor networks. In *Proceedings of the 18th International Parallel and Distributed Processing Symposium (IPDPS)*. doi:10.1109/IPDPS.2004.1302934

Ramanathan, N., Chang, K., Kapur, R., Girod, L., Kohler, E., & Estrin, D. (2005). Sympathy for the sensor network debugger. In *Proceedings of the 3rd international conference on Embedded networked sensor systems (SenSys)*, (pp. 255–267). doi:10.1145/1098918.1098946

Rodrigues, A., Camilo, T., Silva, J., & Boavida, F. (2012). Diagnostic tools for wireless sensor networks: A comparative survey. *Journal of Network and Systems Management*, 1–45.

Sharma, A., Golubchik, L., & Govindan, R. (2007). On the prevalence of sensor faults in real-world deployments. In *Proceedings of the 4th Annual IEEE Communications Society Conference on Sensor, Mesh and Ad Hoc Communications and Networks (SECON)*, (pp. 213–222). doi:10.1109/SAHCN.2007.4292833

Shrivastava, N., Suri, S., & T'oth, C. D. (2005). Detecting cuts in sensor networks. In *Proceedings of the 4th international symposium on Information processing in sensor networks (IPSN)*.

Wang, Y., Attebury, G., & Ramamurthy, B. (2006). A survey of security issues in wireless sensor networks. *IEEE Communications Surveys and Tutorials*, *8*(2), 2–23. doi:10.1109/COMST.2006.315852

Xie, M., Han, S., Tian, B., & Parvin, S. (2011). Anomaly detection in wireless sensor networks: A survey. *Journal of Network and Computer Applications*, *34*(4), 1302–1325. doi:10.1016/j.jnca.2011.03.004

Xing, K., Srinivasan, S. S. R., Rivera, M. J. M., Li, J., & Cheng, X. (2010). Network Security, chapter Attacks and Countermeasures in Sensor Networks. *Survey (London, England)*, 251–272.

Yang, J., Soffa, M. L., Selavo, L., & Whitehouse, K. (2007). Clairvoyant: a comprehensive source-level debugger for wireless sensor networks. In *Proceedings of the 5th international conference on Embedded networked sensor systems (SenSys)* (pp. 189–203). doi:10.1145/1322263.1322282

Yu, B., & Xiao, B. (2006). Detecting selective forwarding attacks in wireless sensor networks. In *Proceedings of the 20th International Parallel and Distributed Processing Symposium (IPDPS)*.

Zhang, Y., Meratnia, N., & Havinga, P. (2010). Outlier detection techniques for wireless sensor networks: A survey. *IEEE Communications Surveys and Tutorials*, *12*(2), 159–170. doi:10.1109/SURV.2010.021510.00088

KEY TERMS AND DEFINITIONS

Anomaly: A behavior that is inconsistent with the expected normal behavior. In sensor networks, network and node failures, measurement faults, and malicious network or data behaviors are all considered anomalies. Sensor literature also refers to anomalous sensor readings as outliers, abnormal, faulty, or malicious data.

Broadcast Storms: Serious duplicates, collisions, and contentions caused by simple flooding and redundant rebroadcasts.

Clipping Fault: A series of sensor readings that maintain steadiness over a period of time during which those readings exceed the maximum value allowed by the Analog to Digital converter (ADC).

Data-Centric Anomalies: Those anomalies that only influence the collected data samples (hence data-centric) including measurement faults, malicious data behaviors, and events.

Node Compromise: A compromise node is a node which is captured by an adversary and is re-programmed to inject falsified readings or cause an insider attack.

Outlier (SHORT Fault): A single isolated sensor reading that is significantly far from the expected reading.

Sensor Network Deployment: An installment of a network of sensors at a certain area in a single or multi hop fashion.

Silent Failures: A type of network failures which can go undetected if we solely rely on a single metric. Examples are ingress drops, frequent changes of parents in the routing tree, and routing loops.

Spike Fault: A reading highly deviated from the expected sensor reading which spans over more than one reading consecutively and may or may not return to the normal behavior afterwards.

Stuck-At (CONSTANT) Fault: A series of sensor readings that maintain steadiness for the rest of the node's lifetime or may return back to normal after a significant amount of time.

ENDNOTES

1 Previous research refers to inconsistent sensor readings that are far from the true value as outliers, anomalous, abnormal, faulty or malicious data. We will use all these terms interchangeably unless clearly stated in this chapter.

Chapter 9
Fault-Diagnosis and Decision Making Algorithm for Determining Faulty Nodes in Malicious Networks

A. Khosravi
Shahid Chamran University of Ahvaz, Iran

Yousef S. Kavian
Shahid Chamran University of Ahvaz, Iran

ABSTRACT

This chapter addresses fault diagnosis agreement problem in a network with malicious members. The authors provide a new algorithm to reach an agreement among fault-free members about the faulty ones. The algorithm is designed for fully-connected networks and also can be easily adapted to partially connected networks. The authors contribution is to reduce bit complexity of Byzantine agreement process by detecting the same list of faulty units in all fault-free members. Therefore, the malicious units can be removed from other consensus processes. Also, each healthy unit detects a list of malicious units locally which results in lower packet transmission in the network. The authors provided algorithm solves fault diagnosis agreement problem in 2t+1 rounds and $O(n^{t+1})$ bit complexity for each member.

INTRODUCTION

Decision making and consensus in an unreliable environment is an important issue that has received significant amount of attention in the literature (Maggs et al. 2012; Pasqualetti et al. 2012; Olfati-Saber et al. 2007). Recent interests in research in the area of unmanned vehicles have led to design of automatic control systems. On the other hand, the problem of consensus in a distributed system is vital for members in order to have a reliable system. Researchers have considered the consensus problem from two different approaches. From the first point of view, researchers consider all network members as a healthy unit (Maggs et al. 2012; Alekeish & Ezhilchelvan

DOI: 10.4018/978-1-4666-8251-1.ch009

2012; Kar & Moura 2007). Therefore, there is no adverse condition affecting the consensus process. Maggs et al. (Maggs et al. 2012) proposed an average consensus protocol for the clock synchronization problem in wireless sensor networks. They proposed a consensus approach to compensate clock function in each unit. However, there are many systems subject to failures. Therefore, the consensus problem also has been considered in systems with faulty units. Byzantine Agreement (BA) (Lamport et al. 1982; Cheng et al. 2008; Lee & Ewe 2007; Okun 2009; Wang et al. 2010) problem is one of the well-known problems in the area of distributed consensus introduced by Lamport et al. (Lamport et al. 1982) and since then various algorithms have been proposed for different network models (Cheng et al. 2008; Khosravi et al. 2011; Chiang et al. 2010; Wang & Yan 2000; Lamport et al. 1982). While Byzantine agreement has been studied extensively over the years, the original model has been mostly continued to be used for different topologies (Wang et al. 2010; Lee & Ewe 2007). In (Lamport et al. 1982), the authors presented an algorithm to solve BA problem in $t+1$ rounds for the first time, where t is the number of arbitrary faulty processors in network N, $|N| = n$, and $n > 3t$. The main goal of different studies is to develop an optimum algorithm in terms of number of message exchange rounds and number of communication bits and also to increase maximum number of allowable faulty units in the system. However, many protocols have been designed for solving the BA problem in different network structures like generalized connected and ad hoc networks (Cheng et al. 2008; Chiang et al. 2010; Yan & Wang 2005) and different fault models (Meyer & Pradhan 1991; Siu et al. 1998; Siu et al. 1996). In general, all of the protocols designed for BA problem must satisfy the following conditions:

- All fault-free processors agree on a common value.
- If the source processor is fault-free then all fault-free processors find the source initial value as the agreement value.

Fischer and Lynch (Fischer & Lynch 1981) showed that $t+1$ rounds are necessary for any Byzantine agreement protocol in general. In fact, local fault diagnosis would not help the Byzantine agreement to be solved in fewer rounds. When a faulty unit is detected by the network, there is no agreement between correct units about the detection of this fault or in another word; a correct processor does not know whether a specific faulty unit is detected by other correct units or not.

Many fault diagnosis protocols have been proposed to achieve a reliable distributed system. The main aim of fault diagnosis is to isolate faulty processors from correct system to have a better management on the network resources. There are two well-known fault diagnosis approaches for different systems, namely test-based approach (Yan & Wang 2005; Mallela & Masson 1978; Elhadef et al. 2007) and evidence-based approach (Buskens & Bianchini 1993; Hsiao et al. 2000; Ayeb & Farhat 2004). Test-based approaches consist of a test phase in which a tester processor performs some test methods on the target units to check whether they are faulty or not. However, this approach is not well applicable in malicious systems in which the processors are probable to have arbitrary faults. On the other hand, the evidence-based approaches use the information collected from the network usually in consensus processes like BA to detect faulty processors (Khosravi et al. 2011).

Fault diagnosis models also can be divided to non-agreement (Ayeb & Farhat 2004; Khosravi et al. 2011; Adams & Ramarao 1989; Vaidya & Pradhan 1994) and agreement (Yan & Wang 2005; Hsiao et al. 2000) models. In non-agreement models, the list of detected faulty units may be different from one unit to another. On the other hand, in the agreement model, all correct processors find the same set of faulty units and malicious behavior of faulty units does not affect the process. In fact, by solving fault diagnosis agreement (FDA) problem, faulty units could be removed from message transmission and this will help BA problem to be solved in fewer rounds in the future. All FDA algorithms must satisfy the following conditions:

- **Consensus:** All the fault-free processors detect same set of faulty processors.
- **Fairness:** No fault-free processor is falsely detected as a faulty unit by any fault-free processor.

FDAMIX proposed in (Hsiao et al. 2000) uses BA framework to solve FDA problem. In (Cheng et al. 2008), also authors have implemented similar idea to solve FDA and BA problem in dynamic ad hoc environment. FDAMIX examines information collected from the BA problem to locate faulty processors. At the end of the message exchange phase in a network with n processors, the BA process must be performed n times on the stored message sets in each processor to find a common set of faulty units in all processors. In fact, FDAMIX needs huge number of messages to be exchanged between processors to solve FDA problem. The proposed algorithm reduces number of message exchange rounds and on the other hand, increases number of detected faulty units. Each unit has two set of faulty units. The first set stores locally detected faulty units and the second set stores the detected faulty units that are common between correct processors. Therefore, the number of detected faults in the proposed algorithm is increased in compare of FDAMIX. In (Cheng et al. 2008), authors proposed a protocol to improve FDAMIX model of fault diagnosis. Their method claims to solve FDA problem in 3 message exchange rounds. However, it can be easily shown by an example that their methods cannot satisfy fairness and consensus conditions. Also, based on (Fischer & Lynch 1981), BA cannot be reached in less than t+1 message exchange rounds.

The contributions provided via the proposed algorithm are as follows:

- Reduce bit complexity and transmission complexity of FDA problem.
- Reduce bit complexity and transmission complexity of BA problem by finding local list of faults.

The rest of the chapter is organized as follows: Section 2 describes the provided algorithm. In section 3 an example is provided to illustrate the consensus algorithm. Proofs for correctness of the proposed are given in section 4. Finally, section 5 concludes the chapter.

THE PROPOSED ALGORITHM

In the protocol proposed by Hsiao et al. in (Hsiao et al. 2000), all arbitrary faulty units are detected at the end of agreement process. However, the arbitrary faulty nodes are detected during the BA process in the proposed algorithm. Therefore, as a faulty unit is detected locally by some fault-free processors, they can stop transmitting messages to this faulty processor. However, network's members in the proposed make an agreement about diagnosis of faulty units at the end of the FDA phase and the detected faulty nodes can be removed from next consensus processes. We assume a fully-connected network with reliable communication links and arbitrary fallible processors in a synchronous environment. The definitions and terms used in this chapter are shown as follows:

- S: the source processor of the BA process.
- V: set of all possible values for BA problem, $V = \{0, 1, \lambda\}$.
- λ: a default value such as NIL (Chiang et al. 2010), $\lambda \in V$, for BA process.

- t: maximum number of arbitrary faulty units.
- p_a: set of locally detected faulty units by a fault-free processor.
- p_c: list of common detected faulty units in all fault-free processors.
- $val_i^k(P_j)$: value that has been sent from P_j to P_i in level k.
- $FU(P_j)$: indicates that processor P_j is a faulty unit.

Basic Fault Identification Concepts

Consider a network N, $|N| = n$, with $\lfloor (n-1)/3 \rfloor$ arbitrary faulty processors as presented in Figure 1. At first, the source processor S sends its initial value to rest of the network. Therefore, $val_i^1(S)$ would be the value received by processor P_i from the source processor in the first level of message exchange. At the second round, all receiver processors retransmit the received value from the source processor S to each other. All processors store the received values in an Evidence Gathering tree, called an EG-tree. The EG-tree for a fault-free processor contains the received messages from processors. In a similar way, at the third round all non-source processors transmit the stored messages at the second round to other units and the received messages are stored in the third level of EG-tree. In general, the source processor might also have arbitrary fault and send different values to different processors. In this case, all non-faulty processors must agree on a common value, either the majority of transmitted values or default value λ.

Assume fault-free processor P_i has a ($t+1$)-level EG-tree as depicted in Figure 2. In general, processors in level k of message exchange phase send the stored messages in level $k-1$ to processor P_i. Let $val_i^{k-1}(P_j) = val(P_jP_i)$ be the values that P_i receives directly from processor P_j in level $k-1$ and also $val^k_i(val_v^{k-1}(P_j)) = val(P_jP_vP_i)$ be the same values received by P_i from P_j via P_v in level k. Therefore, the following statement is true:

If $val(P_jP_i) \neq val(P_jP_vP_i)$ then

$$FU(P_j) \vee FU(P_v) \quad (1)$$

Since there is a disagreement between the values transmitted by P_v and P_j and also P_i is a fault-free receiver then, at least one of processors P_v and P_j is faulty. This fact is shown by statement $FU(P_j) \vee FU(P_v)$ and we call it a probabilistic statement as is depicted in (1). By comparing the stored messages at the j'th column of the matrices at the k'th level of the EG-tree, P_i generates a set of probabilistic statements as follows:

$$\sum P_i = \{FU(P_{v1}) \vee FU(P_j), FU(P_{v2}) \vee FU(P_j), \cdots\}, P_v, P_j \in N \quad (2)$$

$\sum P_i$ is set of probabilistic statements in processor P_i related to the j'th column of the matrices at the EG-tree and $\{P_{v1}, P_{v2}, \ldots\}$ are the processors that have sent different values from $val_i^{k-1}(P_j) = val(P_jP_i)$ in level k. $\sum P_i$ always leads to a prime argument as follows:

$$\Delta = \{FU(P_j) \vee (FU(P_{v1}) \wedge FU(P_{v2}) \wedge \cdots)\}, P_v, P_j \in N \quad (3)$$

Based on (1), at least one of the statements $FU(P_j)$ and $(FU(P_{v1}) \wedge FU(P_{v2}) \wedge \cdots)$ must be true. The statement $\left(FU\left(P_{v1}\right) \wedge FU\left(P_{v2}\right) \wedge \cdots\right)$ implies that all of the processors $v = \{P_{v1}, P_{v2}, \cdots\}$ must be faulty. On the other hand, there are at most t faulty units in the network.

$$if\ |\nu| > t \text{ then } FU(P_j) \quad (4)$$

Figure 1. A fully-connected network with n processor and $\lfloor (n-1)/3 \rfloor$ faulty processors

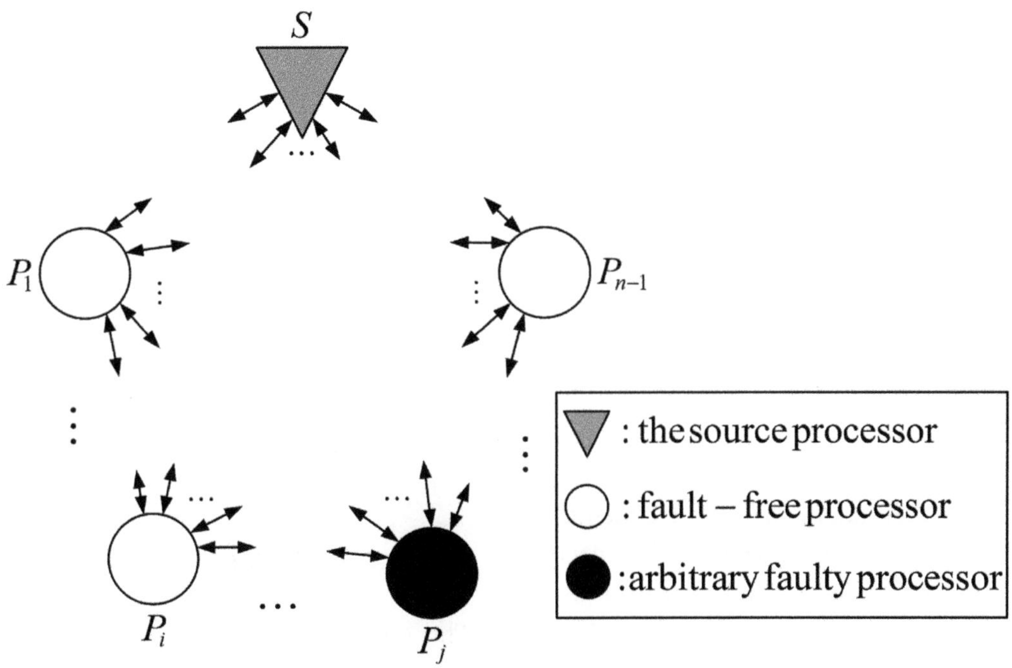

Figure 2. A t-level EG-tree of processor P_i in Figure 1

Processor P_i

Level 1	Level 2		Level t+1
	$val_i^2(P_1) = \left[val_1^1(S) \right]$		$val_i^{t+1}(P_1) = \left[val_1^t(P_1) \quad val_1^t(P_2) \quad \ldots \quad val_1^t(P_{n-1}) \right]$
$val_i^1(S)$	$val_i^2(P_2) = \left[val_2^1(S) \right]$	$\cdots$	$val_i^{t+1}(P_2) = \left[val_2^t(P_1) \quad val_2^t(P_2) \quad \ldots \quad val_2^t(P_{n-1}) \right]$
	$\vdots$		$\vdots \qquad \vdots \qquad \vdots$
	$val_i^2(P_{n-1}) = \left[val_{n-1}^1(S) \right]$		$val_i^{t+1}(P_{n-1}) = \left[val_{n-1}^t(P_1) \quad val_{n-1}^t(P_2) \quad \ldots \quad val_{n-1}^t(P_{n-1}) \right]$

$val^{t+1}{}_i(P_j)$: P_i's received value from P_j in level $t+1$

It implies from (4) that if number of processors in the statement $\{P_{v1}, P_{v2}, \cdots\}$ is greater than t then the statement $\left(FU(P_{v1}) \wedge FU(P_{v2}) \wedge \cdots\right)$ cannot be a diagnosis of faults and therefore $FU(P_j)$ holds.

Let $p_a = \{P_{j1}, P_{j2}, \cdots\}$, $\left|p_a\right| = \tau$, be the set of detected faulty processors by fault-free processor P_i. All probabilistic statements containing at least one of units of p_a must be removed from the set of probabilistic statements since they have no more information about undetected faulty units. Also, removing the unacceptable statements from a prime argument as was used in (3) would be based on $t-\tau$ faulty processors.

Let

$$val_i^k(P_j) = \left[val_j^{k-1}(P_1) \ \ldots \ val_j^{k-1}(P_j) \ \ldots \ val_j^{k-1}(P_{n-1}) \right]$$

be the j'th matrix at the k'th level of the EG-tree of processor P_i and also suppose P_i be a correct unit. Based on a similar argument as before, P_i compares each column of $val_i^k(P_j)$ with corresponding directly received value from $P_1, \ldots, P_{n-1}$ in level k-1 and prepares a new set of probabilistic statements. For example, if $val_j^{k-1}(P_{v1})$ is different from $val_i^{k-1}(P_{v1})$ then new probabilistic statement $FU(P_{v1}) \vee FU(P_j)$ is resulted.

Three basic rules are implemented to identify faulty processors based on the mentioned concepts in this section. Let P_i be a healthy receiver processor and the matrices depicted in Figure 3 be the messages stored in level k of the EG-tree. Also suppose that P_j is a processor that P_i wants to verify its faulty status. Processor P_i identifies a set of different processors that have accused processor P_j. For this purpose, processor P_i compares the gray shaded section of the EG-tree in Figure 3 with the directly received values from corresponding processors in level k-1 and if there is a mismatch between the received messages then the accuser processor is added to a list of processors that accuse P_j of being faulty, called accuser list, AL_j. Let t and $|p_a| = \tau$ be maximum number of faulty processors in the network and number of detected faulty processors by P_i, respectively. Also, let AL_j be the set of processors that accuse processor P_j of being faulty. When a new processor is going to be added to AL_j, it has to be checked that it does not belong to p_a. Based on mentioned above, the following rule is used to identify faulty units in P_i:

If

$$|AL_j| > t - \tau$$

then $FU(P_j)$. (R.1)

If P_j could not be detected faulty using R1, adding new accuser processors to AL_j would be continued in next rounds. As soon as P_j is detected as a faulty processor, it is added to p_a and the related matrices and parameters of P_j for detection of other faulty units will be removed from the list of accusers.

Byzantine Agreement Phase

This section illustrates the Byzantine agreement phase for the proposed algorithm. This is the first stage of the consensus process. This phase consists of two parallel sub phases, message exchange phase and local fault diagnosis phase. The message exchange phase starts by the source processor transmitting its initial value to all other processors and since then the message exchange process is used to build a $(t+1)$-level EG-tree in each fault-free processor. The proposed protocol benefits from the stored messages to identify faulty processors.

Using R1 processors detect faulty units in each round. p_a for each fault-free processor is the set of locally detected faulty units. Therefore, at the end of message exchange phase all processors detect $|p_a|$ faulty units, locally. Figure 4 shows the procedure for finding the agreement value in fault-free processor P_i. Now, all processors are ready to move to the fault diagnosis agreement phase.

Figure 3. Level K of the EG-tree stored in processor P_i. The gray shaded area is used to check faulty status of P_j

$$\text{Processor } P_i$$
$$\text{Level k}$$
$$\begin{aligned}
val_i^k(P_1) &= \left[val_1^{k-1}(P_1) \ \ldots \ val_1^{k-1}(P_j) \ \ldots \ val_1^{k-1}(P_{n-1}) \right] \\
val_i^k(P_2) &= \left[val_2^{k-1}(P_1) \ \ldots \ val_2^{k-1}(P_j) \ \ldots \ val_1^{k-1}(P_{n-1}) \right] \\
&\quad \vdots \qquad\qquad \vdots \qquad\qquad \vdots \\
val_i^k(P_j) &= \left[val_j^{k-1}(P_1) \ \ldots \ val_j^{k-1}(P_j) \ \ldots \ val_j^{k-1}(P_{n-1}) \right] \\
&\quad \vdots \qquad\qquad \vdots \qquad\qquad \vdots \\
val_i^k(P_{n-1}) &= \left[val_{n-1}^{k-1}(P_1) \ \ldots \ val_{n-1}^{k-1}(P_j) \ \ldots \ val_{n-1}^{k-1}(P_{n-1}) \right]
\end{aligned}$$

Fault Diagnosis Agreement Phase

In order to detect the same set of faulty units in all fault-free processors, the processors move to the FDA phase. Using this phase, a general decision could be made on detection of faulty processors. In this phase, there is no initial source processor as we know from original BA problem. However, processors use their list of detected faulty processors p_a as the initial value received from a virtual source. This phase consists of two sub phases, namely fault list exchange sub phase and globally detected faulty units sub phase.

Fault list exchange sub phase is similar to message exchange phase mentioned in section 2.2. In this phase, each processor sends the list of detected faulty units p_a to other units. To provide a better vision to exchange of lists of faulty units, each list is shown in a binary format. For example, if there are seven processors in network N and $p_a = \{S, P_5\}$ for processor P_2 then, p_a is shown as (5).

$$p_a = [1 \ \ 0 \ \ 0 \ \ 0 \ \ 0 \ \ 1 \ \ 0]. \qquad (5)$$

Value 1 and 0 in (5) indicate faulty and fault-free status of a unit, respectively.

A similar tree structure to the EG-tree is used here to store detected faulty sets received from processors, called FG-tree. Figure 5 shows the FG-tree of P_i for two rounds. The FDA phase needs $t+1$ message exchange rounds for the agreement. If we consider p_a as the set of detected faulty units in P_i then $val_i^1(P_i) = p_a$ is considered as the value that is sent from the virtual source as is presented in Figure 5. Each processor sends its list of detected faulty units in second round to other processors. In next rounds, processors send last stored messages in their FG-tree to rest of processors. However, vertices with repeated index are excluded from message transmission. For example, if $val\left(P_j P_i\right)$ be the value received by P_i from P_j in current round then, P_i would not send it to P_j at the next round to save message transmission bits. When processors prepare a $t+1$ level FG-tree, they are ready to move to globally detected faulty units sub phase and determine which of the detected faulty units are also detected by all fault-free processors.

Figure 4. Finding the Byzantine agreement value in fault-free processor P_i

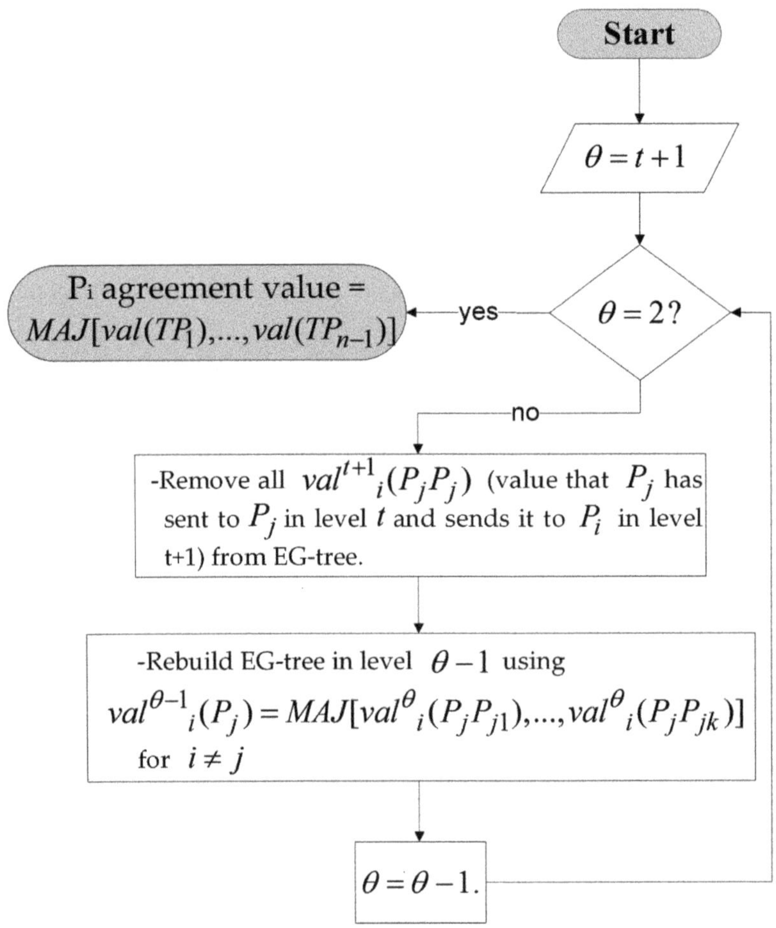

Function MAJ[V]

Take the majority value of $V = [val_i^{\theta}(P_iP_{j1}),...,val_i^{\theta}(P_jP_{jk})]$.
If the majority does not exist, choose default value, λ.

In the globally detected faulty units sub-phase, processors make a common decision about faulty status of each processor in the network and put the common detected faulty units in a list showing by p_c. For this purpose, a decision making strategy is implemented on each column of received lists of faulty units. Figure 6 shows the globally detected faulty units sub phase flowchart. Fault-free processor P_i uses the messages stored in level $\theta = k$ to prepare the voted messages for level $\theta = k-1$. Therefore, when P_i prepares the final matrix of voted messages in level $\theta = 2$, it can be used to make an agreement about detecting of faulty units. Let p_a be the list of detected faulty units in processor P_i and also let $[val(P_1),...,val(P_{n-1})]$ be the prepared matrix of voted messages in second level of FG-tree. If number of value 1 in

Figure 5. The FG-tree stored in processor P_i for two levels

Processor P_i		
Level 0	Level 1	
$val_i^1(P_i)$	$val_i^2(T) = \left[val_T^1(T)\right]$	...
	$val_i^2(P_1) = \left[val_1^1(P_1)\right]$	
	$val_i^2(P_2) = \left[val_2^1(P_2)\right]$	
	$\vdots$	
	$val_i^2(P_{n-1}) = \left[val_{n-1}^1(P_{n-1})\right]$	

$val_j^1(P_j)$: The list of detected faulty units p_a in P_j

$[val(P_1),...,val(P_{n-1})]$ correspond to processor P_{jk} ($P_{jk} \notin p_c$ and $P_{jk} \neq P_j$) is more than $t - |p_c|$ then processor P_i adds P_j to p_c.

The depicted flowchart in Figure 6 is repeated for all columns of p_a to determine the status of all processors. An example is provided in following section to illustrate different aspects of the proposed algorithm.

This section shows how the proposed algorithm acts using an example. A network $N = \{S, P_1, P_2, ..., P_6\}$ with seven processors and two faulty processors $\{S, P_6\}$ are considered for the example. The worst case in BA problem is when the source processor is faulty. Therefore, this situation is considered for the example.

The process starts by the source processor transmitting its initial value to other processors. Since the source processor supposed to be faulty, it might transmit arbitrary values to rest of processors. According to number of faulty units in the network, the BA problem needs $\lfloor (7-1)/3 \rfloor + 1 = 3$ rounds of message exchange. Therefore, the message exchange continues for three rounds to prepare a three level EG-tree in each processor. Consider the faulty source sends $\{1,1,1,1,0,0\}$ to $\{P_1, P_2, P_3, P_4, P_5, P_6\}$ in the first round of message exchange, respectively. At the next round processors send the received value from the source processor to each other. Figure 7 shows the

An Example of Execution The Algorithm

EG-tree saved in fault-free processor P_1 for three rounds. Based on the EG-tree and using R1, processor P_1 locates faulty units in the network. From data stored at the second level of the EG-tree, P_1 adds $\{P_5, P_6\}$ to AL_S. Also, it can be seen that the source processor accuses both P_5 and P_6. Therefore, S is added to AL_{P5} and AL_{P6}. From the received messages at the third level, P_6 is added to AL_{P4} and AL_{P5} and $\{S, P_4, P_5\}$ is added to AL_{P6}. Figure 8 shows the detailed fault detection procedure from the data stored at the EG-tree in Figure 7. Processor P_1 has $p_a = \{\}$, $p_c = \{\}$ and $t = 2$ so far. Since $|AL_{P6}| = 3 > t - |p_a|$, P_6 is detected as a faulty unit and is added to p_a by P_1. Now, a new faulty unit is detected and P_1 must reconsider the evidence received from other processors. Therefore, P_6 is removed from AL_S, AL_{P4} and AL_{P5} and also maximum number of undetected faulty units decreases to $t - |p_a| = 1$.

Figure 6. Flowchart of globally detected faulty units sub phase in processor P_i to check status of P_j

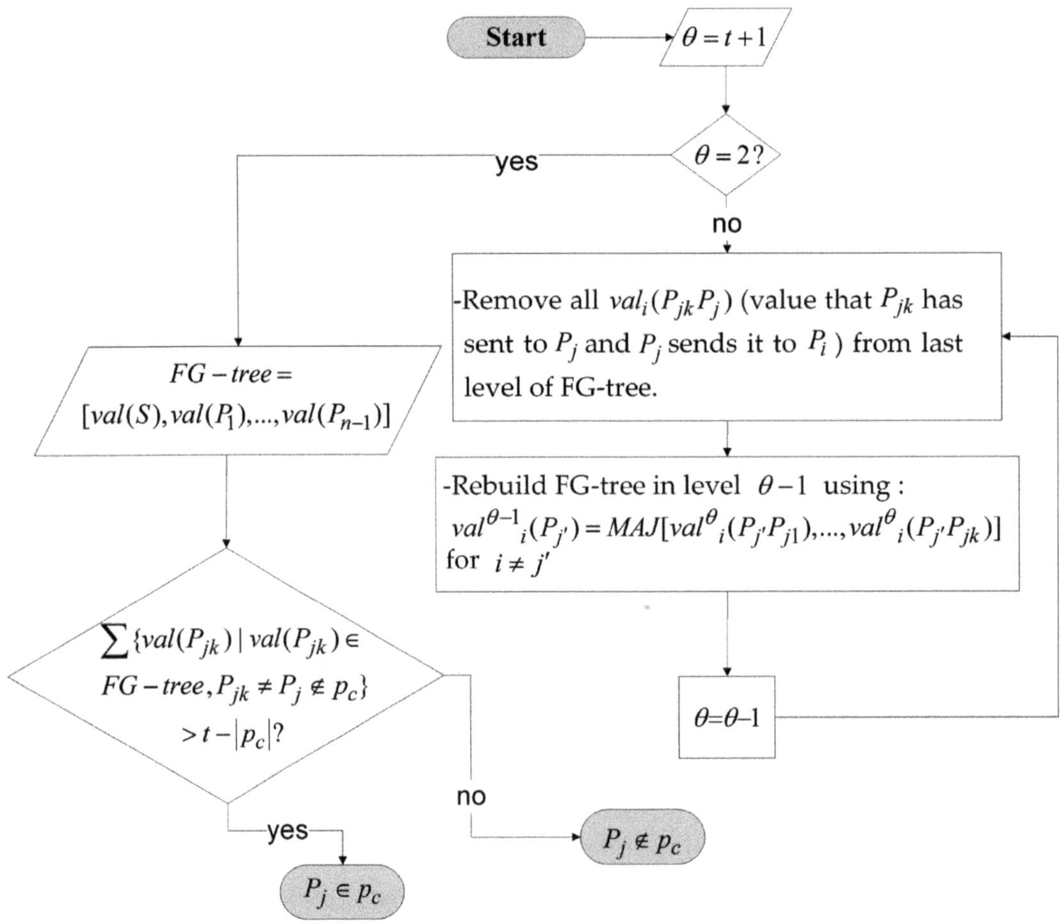

Figure 7. A three level EG-tree for processor in example of section 3

$\Pr ocessor\ P_1$

$Level 1$	$Level 2$	$Level 3$
	$val_1^2(P_1)=1$	$val_1^3(P_1)=[1\ \ 1\ \ 1\ \ 1\ \ 0\ \ 0]$
	$val_1^2(P_2)=1$	$val_1^3(P_2)=[1\ \ 1\ \ 1\ \ 1\ \ 0\ \ 0]$
$val_1^1(S)=1$	$val_1^2(P_3)=1$	$val_1^3(P_3)=[1\ \ 1\ \ 1\ \ 1\ \ 0\ \ 0]$
	$val_1^2(P_4)=1$	$val_1^3(P_4)=[1\ \ 1\ \ 1\ \ 1\ \ 0\ \ 0]$
	$val_1^2(P_5)=0$	$val_1^3(P_5)=[1\ \ 1\ \ 1\ \ 1\ \ 0\ \ 1]$
	$val_1^2(P_6)=0$	$val_1^3(P_6)=[1\ \ 1\ \ 1\ \ 0\ \ 0\ \ 0]$

Figure 8. Local fault diagnosis in processor P_1 during the Byzantine agreement process

$$\text{Processor } P_1$$

Level 2

$$val_1^2(P_1)=1$$
$$val_1^2(P_2)=1$$
$$val_1^2(P_3)=1$$
$$val_1^2(P_4)=1$$
$$val_1^2(P_5)=0$$
$$val_1^2(P_6)=0$$

$$AL_S=\{P_5,P_6\}$$
$$AL_5=\{S\}$$
$$AL_6=\{S\}$$

Level 3

$$val_1^3(P_1)=[1\ \ 1\ \ 1\ \ 1\ \ 0\ \ 0]$$
$$val_1^3(P_2)=[1\ \ 1\ \ 1\ \ 1\ \ 0\ \ 0]$$
$$val_1^3(P_3)=[1\ \ 1\ \ 1\ \ 1\ \ 0\ \ 0]$$
$$val_1^3(P_4)=[1\ \ 1\ \ 1\ \ 1\ \ 0\ \ 0]$$
$$val_1^3(P_5)=[1\ \ 1\ \ 1\ \ 1\ \ 0\ \ 1]$$
$$val_1^3(P_6)=[1\ \ 1\ \ 1\ \ 0\ \ 0\ \ 0]$$
$$AL_4=\{P_6\}$$

$$val_1^3(P_1)=[1\ \ 1\ \ 1\ \ 1\ \ 0\ \ 0]$$
$$val_1^3(P_2)=[1\ \ 1\ \ 1\ \ 1\ \ 0\ \ 0]$$
$$val_1^3(P_3)=[1\ \ 1\ \ 1\ \ 1\ \ 0\ \ 0]$$
$$val_1^3(P_4)=[1\ \ 1\ \ 1\ \ 1\ \ 0\ \ 0]$$
$$val_1^3(P_5)=[1\ \ 1\ \ 1\ \ 1\ \ 0\ \ 1]$$
$$val_1^3(P_6)=[1\ \ 1\ \ 1\ \ 0\ \ 0\ \ 0]$$
$$AL_5=\{P_6\}$$

$$val_1^3(P_1)=[1\ \ 1\ \ 1\ \ 1\ \ 0\ \ 0]$$
$$val_1^3(P_2)=[1\ \ 1\ \ 1\ \ 1\ \ 0\ \ 0]$$
$$val_1^3(P_3)=[1\ \ 1\ \ 1\ \ 1\ \ 0\ \ 0]$$
$$val_1^3(P_4)=[1\ \ 1\ \ 1\ \ 1\ \ 0\ \ 0]$$
$$val_1^3(P_5)=[1\ \ 1\ \ 1\ \ 1\ \ 0\ \ 1]$$
$$val_1^3(P_6)=[1\ \ 1\ \ 1\ \ 0\ \ 0\ \ 0]$$
$$AL_6=\{S,P_4,P_5\}$$

However, $AL_S = \{P_5\}$ and $AL_{P4} = \{\}$ and therefore, no new faulty processor is detected using R1.

At the end of the message exchange phase, processors move to the decision making phase in order to find the agreement value. To find the agreement value, the last level of the EG-tree should be modified by removing redundant values. The majority function is performed on each column of the EG-tree to reconstruct the EG-tree in previous round. Figure 9 shows the procedure of finding the agreement value from the modified EG-tree. Since P_1 is a fault-free receiver, the majority function is only used to find the value of other processors and the directly received value from the source by P_1 helps to find the final agreement in the second level as depicted in Figure 9.

Next, all processors proceed to the next phase to make an agreement about detection of faulty units. At the beginning of FDA phase, P_1 has $p_a = \{P_6\}$ and is ready to send it to other processors. However, in the BA phase, processors P_2, P_3, P_4 and P_5 have detected $p_a = \{P_6\}$, $p_a = \{P_6\}$, $p_a = \{P_6\}$ and $p_a = \{S, P_6\}$, respectively and $p_a = \{S, P_5\}$, a random value, for P_6 is send to P_1. In the fault list exchange sub phase, processors send their detected faulty units to each other. Figure 10 shows the stored FG-tree in P_1 for three rounds.

When all processors prepare a three level FG-tree, they are ready to find the common detected faulty processors from it. For this purpose, two set of values should be removed from the last level of FG-trees. First, if P_1 is verifying the status of P_j then, the majority value indicating P_j`s opinion about its own faulty status should be removed from the FG-tree. Second, processor P_1 does not need to identify its own faulty status. Therefore, the majority function is not performed on the related values. Figure 11 illustrates globally detected faulty units sub phase in processor P_1. The gray shaded area in Figure 11 indicates the vertexes that have performed the majority function on them and the resulted values at the bottom are the values to be replaced data in the second level. However, since P_1 is the receiver processor, it is not necessary to find the majority values for P_1 at the third level and $val_1^1(P_1)$ would be used for this purpose. As soon as the FG-tree is reconstructed at the second level, an agreement is made about faulty processors. For this purpose, P_1 counts number of value 1 at the columns of the second level to check the condition of the corresponding processor. At first, P_1 has $p_c = \{\}$ and if number of 1s in all j^{th} columns is greater than $t - |p_c| = 2$ then P_{j-1} is faulty (considering $P_0 = S$). Based on this, P_1 adds P_6 to p_c and this means that all fault-free processors also have detected $p_c = \{P_6\}$. Next, P_1 rechecks messages at the second level, this time using $t - |p_c| = 1$ bound and noticing that $val_1^2(P_6)$ must be removed from the messages first. As is shown in Figure 11, no new faulty unit is added to p_c and processor P_1 finishes the algorithm by $p_a = \{\}$ as a set of locally detected faulty units and $p_c = \{P_6\}$ as FDA list. By detecting a common set of faults, processors can make a decision to remove faulty processors from the network or rearrange network structure based on this agreement.

Next section provides proofs for correctness of the proposed algorithm. Also, the complexity of the proposed algorithm is compared with previous algorithms in terms of number of message exchange rounds and bit complexity.

Correctness of the Algorithm

This section provides the proofs for correctness of the proposed algorithm. At first, we show in theorem 1 that no fault-free processor is detected as a faulty unit by rest of correct processors at the FDA phase. In other word, the FDA phase only makes agreement about detecting faulty units and satisfies the fairness condition.

Theorem 1. Let N, $|N| = n$, be a network and P_0, P_1, ..., P_{n-1} are its members and P_{j1}, P_{j2}, ..., P_{jt}, $t = \lfloor (n-1)/3 \rfloor$, are maximum number of faulty units in network N. If P_i is a fault-free processor in network N then, there will be no agreement about detection of P_i as a faulty unit among correct processors using the proposed algorithm.

Proof. Since there are at max t faulty units in the network and $|AL_{Pi}| \le t$ for all fault-free units therefore, no fault-free processor sends $p_a = \{P_i\}$

Figure 9. Decision making phase to find BA agreement value in processor P_1

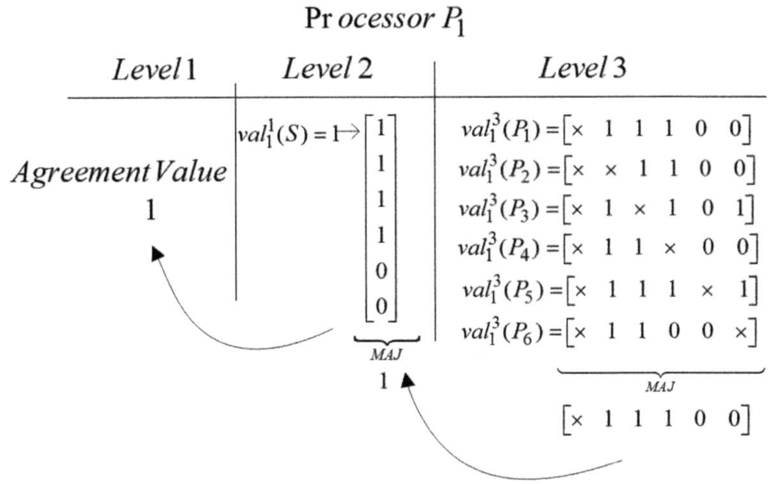

Figure 10. The FG-tree Stored in processor P_1 in FDA globally detected faulty units sub phase

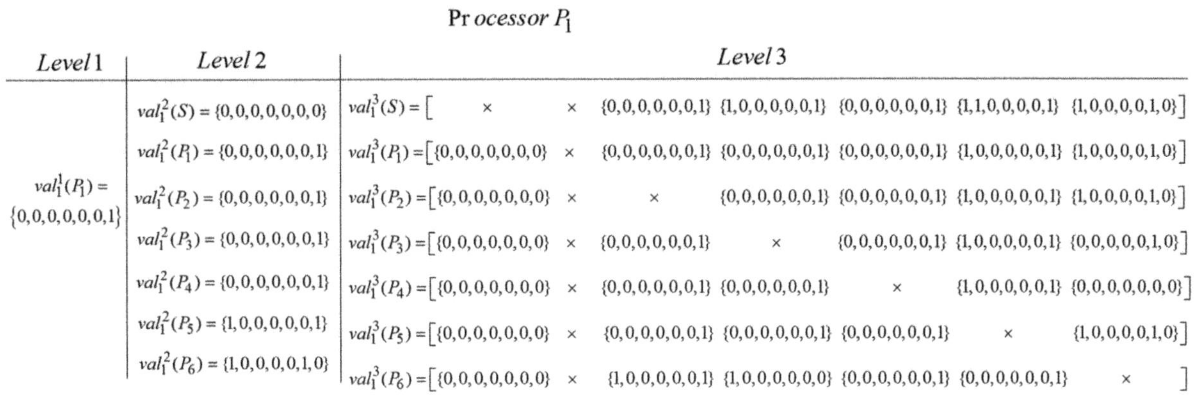

Figure 11. Globally detected faulty units sub phase in processor P_1

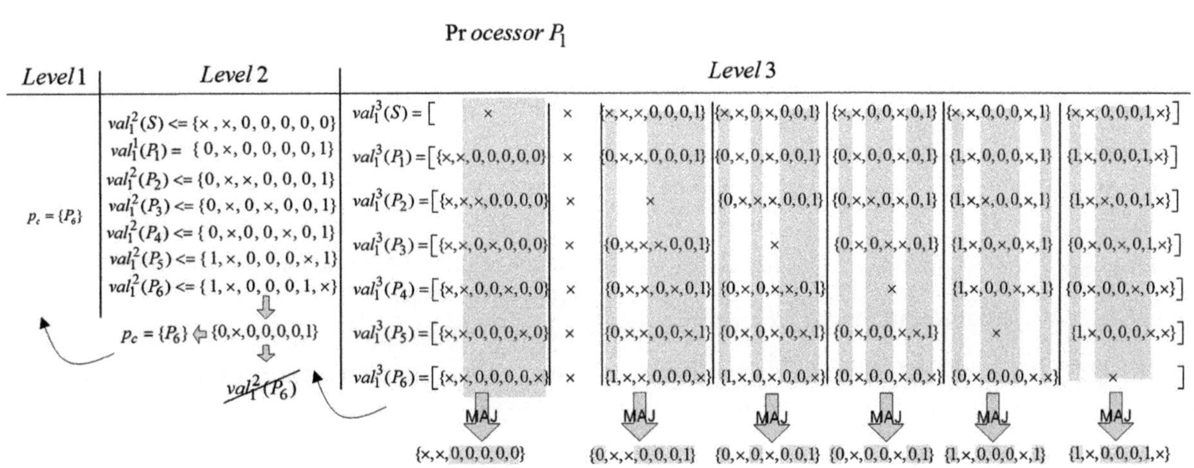

in FDA message exchange phase. Based on this fact, there are at most $t = \lfloor (n-1)/3 \rfloor$ faulty units that may send $p_a = \{P_i\}$ to fault-free processors in this phase. This situation can be rephrased to a situation that a virtual fault-free source sends $p_a = \{0\}$ to network N and only the existing t faulty processors may change this value. However, by considering the virtual source as a member of N, there will be $n+1$ processors in the network and the virtual source acts as a fault-free processor. By removing the values of P_i from the message exchange process in FDA phase, there will be n processors in the network and $n > 3t+1$. Therefore, by implementing the decision making strategy on the FG-tree and preparing the majority values at the second level, there will be at most $t = \lfloor (n-1)/3 \rfloor$ values from P_{j1}, P_{j2}, ..., P_{jt} containing $p_a = \{1\}$, indicate P_i as a faulty unit, and when $p_c = \{\}$ in all processors then $\lfloor (n-1)/3 \rfloor \not> t - |p_c|$ and no processor adds P_i to p_c. On the other hand, if $p_c \neq \{\}$ then, there is an agreement about diagnosis of some faulty units among correct processors and therefore, the related values to the detected faulty unit are removed from the FG-trees. Based on this, in general case, the number of faulty units that have sent $p_a = \{P_i\}$ to the network and their value are not removed from FG-trees is not greater than $\lfloor (n-1)/3 \rfloor - |p_c|$ and the term $\lfloor (n-1)/3 \rfloor - |p_c| \not> t - |p_c|$ is true in all correct processors.

Based on the next theorem, whenever the FDA phase checks the status of a faulty processor, we can suppose that a virtual faulty source has sent different values to different processors. Also, it shows that this situation does not affect the FDA phase.

Theorem 2. Let N, $|N| = n$, *be a network and* P_0, P_1, ..., P_{n-1} *are its members and* P_{j1}, P_{j2}, ..., P_{jt}, $t = \lfloor (n-1)/3 \rfloor$, *are maximum number of faulty units in network* N. *If* P_j *is a faulty unit then, the reconstructed FG-tree at the second level in globally detected faulty units sub phase is common in all correct processors.*

Proof. According to (Yan & Wang 2005), network N, $|N| = n$, with t faulty processor needs $t+1$ rounds of message exchange to reach the agreement among its processors. In the case of the faulty source in BA problem, the reconstructed values in the decision making phase at the second level of all tree structures are common among all correct processors. To illustrate the case, suppose that the source processor has sent value 0 to n' processors and value 1 to n' processors in a network with total $n = 2n'+1$ processors. Since BA results in a common value in all fault-free processors and also the resulted majority values at the second level related to fault-free processors are the same as their received values from the source processor therefore, the values related to faulty processors must be common at the reconstructed second level of all tree structures to have a common agreement. At the FDA phase and at the globally detected faulty units sub phase, when processors want to verify the faulty status of processor P_j, the exchanged values of P_j are removed from the FG-trees. Therefore, number of remaining faulty units decreases to $t-1$. On the other hand, P_j is detected as a faulty unit by some of the processors, i.e. $P_j \in p_a$ for some processors and $P_j \notin p_a$ for others. This situation is similar to the case of the virtual faulty source sending $p_a = 1$ and $p_a = 0$ to different processors. However, by removing the values of faulty processor P_j and considering the effects of the virtual faulty source, number of allowable faulty units does not exceed t bound. Therefore, the number of value 1 and value 0 are the same at the second level of all FG-trees after globally detected faulty units sub phase.

It has been shown in theorem 2 that if processors check the faulty status of a faulty processor in FDA phase then the reconstructed matrices at the second level of all FG-trees are the same. Theorem 3 shows how to find the faulty status of a unit from the stored data at the second level.

Theorem 3. Let N, $|N| = n$, be a network and P_0, P_1, …, P_{n-1} are its members and P_{j1}, P_{j2}, …, P_{jt}, $t = \lfloor (n-1)/3 \rfloor$, are maximum number of faulty units in network N. Also, let $P_{j'1}$, $P_{j'2}$, …, $P_{j'k}$ be the members of p_c and $val^2(P_{j'1})$, $val^2(P_{j'2})$, …, $val^2(P_{j'k})$ are removed from all FG-trees in globally detected faulty units sub phase. If at the FDA phase for checking the status of P_j, there are more than $t - |p_c|$ values equal 1 at the second level of the reconstructed FG-tree of fault-free processor P_i then, P_j is a faulty unit.

Proof. Suppose that fault-free processor P_i has sent $val_i^1 = 1$ to rest the of processors at the message exchange phase of BA. Since there are not more than $|(n-1)/3|$ values from faulty processors affecting the MAJ function at the decision making phase and reconstructing the tree structures in processors then, $val_i^1 = 1$ stores at the second level of all tree structures after the reconstruction phase. First, suppose that $p_c = \{\}$. If the number of 1s at the second level of the reconstructed FG-trees equals τ and $\tau > t$, since there are at most t faulty processors in the network then, the receiver processor is sure that at least one of the accuser transmitter processors are fault-free and based on theorem 1 and (Ayeb & Farhat 2004), P_j is detected as a faulty unit. When $p_c \neq \{\}$, data from processors that are members of p_c are removed from the second level of the FG-trees. Therefore, maximum number of remaining faulty units in the network is $t - |p_c|$ and if $\tau > t - |p_c|$ then, it results in detection of faulty P_j.

Corollary 1. From theorem 1, 2 and 3, it is resulted that the proposed algorithm satisfies the FDA conditions and p_c is the common list of detected faults in fault-free processors and also p_a is the local list of detected faults.

Corollary 2. The proposed algorithm needs $2t+1$ rounds of message exchange to solve BA and FDA problem simultaneously.

Corollary 3. The bit complexity for BA problem in the proposed algorithm is $O(n^t)$ and for FDA problem is $O(n^{t+1})$.

CONCLUSION AND DISCUSSION

The proposed algorithm in this study aims to solve the fault diagnosis agreement problem in a distributed environment. The algorithm has

Table 1. Comparison of FDA protocol

Protocol	L. Fault Diagnosis	Bit Comp.	N. of Rounds	Consensus	
				BA	FDA
Hsiao et al. (Hsiao et al. 2000)	-	$O(n^{2t+1})$	$(n+1)(t+1)$	√	√
Ayeb and Farhat (Ayeb & Farhat 2004)	√	$O(n^{t-\rho})^*$	$(t+1-\rho)$	√	-
Chaing et al. (Chiang et al. 2010)	√	$O(n)$	3	-	-
Our method	√	$O(n^{t+1})$	$2t+1$	√	√

*the protocol can stop message transmission based on number of detected faulty units ρ

three properties: First, it can locally detect high percentage of malicious faulty units during Byzantine agreement process. Second, it implements a new agreement framework to make an agreement among fault-free processors about diagnosis of faulty units. Third, the common detected faulty units are removed from next consensus processes to decrease the number of message exchange rounds to reach the agreement.

The proposed algorithm needs $2t + 1$ message exchange rounds to solve BA and FDA problem, simultaneously. That is $(n+1)(t+1)$ in the algorithm proposed in (Hsiao et al. 2000). Also, the bit complexity is improved from $O(n^{2t+1})$ in (Hsiao et al. 2000) to $O(n^{t+1})$ in the proposed algorithm to solve FDA problem. Table 1 compares the proposed method in this chapter with some known fault diagnosis algorithms. The proposed algorithm in this chapter is able to detect high percentage of faulty units with minimum message overheads.

REFERENCES

Adams, J. C., & Ramarao, K. V. S. (1989). Distributed diagnosis of Byzantine processors and links. *Proceedings. The 9th International Conference on Distributed Computing Systems*. doi:10.1109/ICDCS.1989.37989

Alekeish, K., & Ezhilchelvan, P. (2012). Consensus in Sparse, Mobile Ad Hoc Networks. *IEEE Transactions on Parallel and Distributed Systems*, *23*(3), 467–474. doi:10.1109/TPDS.2011.182

Ayeb, B. & Farhat, A. (2004). A flexible formal framework for masking/demasking faults. *Information Sciences*, *159*, 29–52.

Buskens, R. W., & Bianchini, R. P. J. (1993). Distributed on-line diagnosis in the presence of arbitrary faults. *FTCS-23 The Twenty-Third International Symposium on Fault-Tolerant Computing*.

Cheng, C.-F., Wang, S.-C., & Liang, T. (2008). Byzantine Agreement & Fault Diagnosis Agreement in Dynamic Ad-Hoc Environment. *Fundamenta Informaticae*, *89*, 161–187.

Chiang, M. L., Wang, S. C., & Tseng, L. Y. (2010). The anatomy study of consensus agreement in MANETs. *Computers & Electrical Engineering*, *36*(3), 234–253. doi:10.1016/j.compeleceng.2009.09.001

Elhadef, M., Boukerche, A., & Elkadiki, H. (2007). An adaptive fault identification protocol for an emergency/rescue-based wireless and mobile ad-hoc network. In *Proceedings of the 21st International Parallel and Distributed Processing Symposium, IPDPS 2007*.

Fischer, M. J., & Lynch, N. A. (1981). A Lower Bound for the Time to Assure Interactive Consistency. *Information Processing Letters*, *14*(4), 183–186. doi:10.1016/0020-0190(82)90033-3

Hsiao, H.-S., Chin, Y.-H., & Yang, W.-P. (2000). Reaching Fault Diagnosis Agreement Under a Hybrid Fault Model. *IEEE Transactions on Computers*, *49*(9), 980–986.

Kar, S., & Moura, J. M. F. (2007). Distributed average consensus in sensor networks with random link failures and communication channel noise. In *Conference Record: Asilomar Conference on Signals, Systems and Computers*. 676–680. doi:10.1109/ACSSC.2007.4487299

Khosravi, A., Mohammadi, K., & Shiroie, M. (2011). Locating malicious links in fully-connected networks using a formal framework. In *Proceedings of the ICSEng 2011: International Conference on Systems Engineering*. 247–250. doi:10.1109/ICSEng.2011.51

Lamport, L., Shostak, R., & Pease, M. (1982). The Byzantine Generals Problem. *ACM Transactions on Programming Languages and Systems*, *4*(3), 382–401. doi:10.1145/357172.357176

Lee, K. W., & Ewe, H. T. (2007). Performance study of Byzantine Agreement Protocol with artificial neural network. *Information Sciences*, *177*(21), 4785–4798. doi:10.1016/j.ins.2007.04.011

Maggs, M. K., O'Keefe, S. G., & Thiel, D. V. (2012). Consensus clock synchronization for wireless sensor networks. *IEEE Sensors Journal*, *12*(6), 2269–2277. doi:10.1109/JSEN.2011.2182045

Mallela, S., & Masson, G. M. (1978). Diagnosable Systems for Intermittent Faults. *IEEE Transactions on Computers*, *27*(6), 560–566. doi:10.1109/TC.1978.1675147

Meyer, F. J., & Pradhan, D. K. (1991). Consensus with dual failure modes. *IEEE Transactions on Parallel and Distributed Systems*, *2*(2), 214–222. doi:10.1109/71.89066

Okun, M. (2009). On the round complexity of Byzantine agreement without initial set-up. *Information and Computation*, *207*(12), 1351–1368. doi:10.1016/j.ic.2009.07.002

Olfati-Saber, R., Fax, J. A., & Murray, R. M. (2007). Consensus and Cooperation in Networked Multi-Agent Systems. *Proceedings of the IEEE*, *95*(1), 215–233. doi:10.1109/JPROC.2006.887293

Pasqualetti, F., Bicchi, A., & Bullo, F. (2012). Consensus computation in unreliable networks: A system theoretic approach. *IEEE Transactions on Automatic Control*, *57*(1), 90–104. doi:10.1109/TAC.2011.2158130

Siu, H. S., Chin, Y. H., & Yang, W. P. (1996). A note on consensus on dual failure modes. *IEEE Transactions on Parallel and Distributed Systems*, *7*(3), 225–230. doi:10.1109/71.491575

Siu, H. S., Chin, Y. H., & Yang, W. P. (1998). Byzantine agreement in the presence of mixed faults on processors and links. *IEEE Transactions on Parallel and Distributed Systems*, *9*(4), 335–345. doi:10.1109/71.667895

Vaidya, N. H., & Pradhan, D. K. (1994). Safe system level diagnosis. *IEEE Transactions on Computers*, *43*(3), 367–370. doi:10.1109/12.272443

Wang, S. C., & Yan, K. Q. (2000). Reaching fault diagnosis agreement on dual link failure mode. *Proceedings Seventh International Conference on Parallel and Distributed Systems*. doi:10.1109/ICPADS.2000.857710

Wang, S.-S., Yan, K.-Q., & Wang, S.-C. (2010). An optimal solution for byzantine agreement under a hierarchical cluster-oriented mobile ad hoc network. *Computers & Electrical Engineering*, *36*(1), 100–113. doi:10.1016/j.compeleceng.2009.06.002

Yan, K. Q., & Wang, S. C. (2005). Reaching fault diagnosis agreement on an unreliable general network. *Information Sciences*, *170*(2-4), 397–407. doi:10.1016/j.ins.2004.03.011

KEY TERMS AND DEFINITIONS

Byzantine Agreement: Decision making in wireless sensor networks of malicious members, when some intruder or Byzantine faulty sensors prevent fault-free members from reaching an agreement.

Byzantine Node (Malicious Node): A network sensor is known to be Byzantine, when its behavior is completely unpredictable.

Consensus: The problem of reaching an agreement in distributed environment of Wireless Sensor Networks.

Fault Diagnosis: The process of diagnosing the faulty behavior of a malicious node in a distributed manner is the fault diagnosis process.

Fault Diagnosis Agreement: Reaching an agreement about a set fault nodes is known as the fault diagnosis agreement problem.

Fault Tolerant Network: A network known to be t-fault tolerant when it can work perfectly in presence of maximally t faulty units.

Reliability: The probability of a network to be healthy up to time t is known as the reliability of the system.

Chapter 10
RSS-Based Outdoor Localization with Wireless Sensor Networks in Practice

Tsenka Stoyanova
University of Patras, Greece

Fotis Kerasiotis
University of Patras, Greece

George Papadopoulos
University of Patras, Greece

ABSTRACT

In this chapter the authors discuss the feasibility of sensor node localization by exploiting the inherent resources of WSN technology, such as the received signal strength (RSS) of the exchanged messages. The authors also present a brief overview of various factors influencing the RSS, including the RF-signal propagation and other topology parameters which influence the localization process and accuracy. Moreover, the RSS variability due to internal factors, related to the hardware implementation of a sensor node, is investigated in order to be considered in simulations of RSS-based outdoor localization scenarios. Localization considerations referring to techniques, topology parameters and factors influencing the localization accuracy are combined in simulation examples to evaluate their significance concerning target positioning performance. Finally, the RF propagation model and the topology parameters being identified are validated in real outdoor localization scenario.

INTRODUCTION

The most distinctive features of Wireless Sensor Networks (WSNs), in relation to the other wireless networks, are the large number of the sensor nodes involved (often in the range of hundreds and thousands), their small size and relatively lower price. Furthermore, due to the battery-based power supply of the sensor nodes, there are critical constraints for their operation and energy consumption, which impose the requirement for optimizing the use of their hardware modules and resources.

DOI: 10.4018/978-1-4666-8251-1.ch010

The spatial localization of the sensor nodes is a fundamental problem in the area of WSNs. An attractive way for estimating the location of a wireless object is by using the Received Signal Strength (RSS) attenuation with the distance, as this does not require any additional hardware. This is possible due to the fact that in most sensor node radios, the Received Signal Strength Indicator (RSSI) is a standard feature, which allows the RSS value to be obtained automatically together with the received message.

As we consider outdoor environments, spatial localization via Geo-Positioning System (GPS) is also feasible option. The main disadvantages of the GPS-based localization methods are the incurred cost, the size and the energy consumption of sensor nodes equipped with GPS receiver, the limited accuracy of localization in some locations and environments, as well as the dependence on services controlled by foreign governments. In particular, equipping every WSN node with a GPS receiver increases both the size and cost of the nodes and in consequence, the overall application cost. Furthermore, adding extra hardware increases the energy consumption of the sensor nodes, which in the context of WSNs shortens the operational life-time of the network. In certain potential areas of interest, the localization accuracy offered by geo-positioning systems could be insufficient due to local government restrictions.

On the other hand, the RSS is not bound by these restrictions, but it is known for being noisy, unstable, variable and difficult to use in practice. To achieve better understanding of the nature of these difficulties and limitations, and to identify the range of applicability of the RSS in localization scenarios, a thorough study on the RSS behavior and its dependence on various factors is essential. In connection to this, the following problems are considered of primary importance:

- The RSS behavior in outdoor environment has not been researched thoroughly at medium distances, when a variety of factors change, resulting in poor understanding of the limitations and the feasibility of performing RSS-based localization. For that reason, a detailed analysis of the RSS behavior under various conditions is required.
- The log-normal path loss model (LPLM), which is the most frequently used propagation model for RSS-based localization, is rather general and does not take into account some important factors that influence the RSS, such as the height of the nodes from the ground. Therefore, alternative propagation models need to be considered and evaluated experimentally.
- The existing deployment algorithms provide solutions for the coverage and connectivity problems, with respect to the number of neighbor nodes, without taking into consideration the factors influencing the communication among the network nodes. Most assumptions and simplifications being made towards such solution contribute to the common belief that the communication between two sensor nodes is good until a certain distance and worsens after it, which is not always correct. A communication link should be considered as guaranteed when the RSS is sufficient, so that the sent packets will be received with packet loss rate below a certain threshold.
- In most of the published research on RSS-based localization, the reported results are based more on simulations rather than on real experimentations. Authors commonly assume that the variability of RSS is more or less a statistical quantity, instead of measuring and modeling it. To contribute to the better understanding of the RSS variability, a detailed experimental study is essential.

Dealing with these problems is of primary importance for the successful use of the RSS as an estimator of distance in applications requiring network nodes localization.

BACKGROUND AND RELATED RESEARCH

Outdoor Localization Methods

Every localization system is based on certain mechanism used to determine the position of an object by measuring the physical distances in an indirect way. Different methods, which provide a relationship between distance and some measurable parameter of the transmitted signal, have been used when the radio signal is concerned. These methods can be divided into the following categories: (1) Connectivity based; (2) Timing based; (3) Directionality based; and (4) Signal strength based.

- **Connectivity Based Methods:** It is the simplest radio frequency (RF) location method. The position of unknown source is determined by *Cell of Origin* (CoO) technique, which considers the location of the beacon node to be the location of the unknown node. Therefore, the accuracy of CoO depends on the density of the beacon points' deployment and the signal range. For this reason, when precision is important, CoO is often used in conjunction with some other technology, such as measuring the time of arrival or the RSS. The biggest advantage of the CoO technique is that it is cheap and does not involve complex location algorithms.
- **Timing Based Methods:** These methods compute the time taken by a signal to travel from a transmitter to a receiver. Two parameters can be used: the absolute time of arrival, known as *Time of Arrival* (ToA) and the differential time of arrival named also as *Time Difference of Arrival* (TDoA). The ToA is based on absolute time measurements. It measures the time it takes for a signal to travel from an unknown device to at least three reference points. Each receiver calculates the time taken by the signal to travel from the sender and calculates the distance to it. Three receivers are needed to perform trilateration and find the sender's location. The Global Positioning System (GPS) is typical examples of ToA implementation. The difference is that the senders are the reference points, i.e. the satellites, which send a signal to a GPS receiver. The receiver calculates its own position based on the satellites' positions. The ToA requires synchronized clocks between the reference points and the unknown node, so that each signal can be time-stamped and its travel time measured precisely. The applicability in WSN with radio ranges as much as 100m - 300m is questionable, due to the fact that 1-microsecond inaccuracy in the synchronization results in a 300-meter shift in the node's location. Moreover, the radio module of sensor node has to support such highly accurate clock synchronization, which will influence the cost and the power consumption of the nodes. A variation from ToA is TDoA. TDoA does not require precise synchronized clock between the reference nodes and the unknown node. All TDoA receivers' clocks, i.e. the beacons' clocks, are precisely synchronized together and receive the transmission from an unknown node recording the instant at which it was received. This information is sent to a base station for further processing. Each difference of arrival time produces a hyperbolic curve in the localization space. The base station compares the information provided by all beacons and calculates the position of the unknown node. This method performs well better outdoors than indoor.
- **Directionality Based Methods:** These methods compute the angle of each of the reference points with respect to the node with unknown location. This technique is

referred to as *Angle of Arrival* (AoA) and determines the angle of arrival of the signal coming from the node with unknown location at the reference nodes. For finding the signal direction, it is required highly directional antennas or antenna arrays as additional antennas to the WSN nodes. Using such a solution in WSN will increase the cost and the size of the nodes and in overall the cost to whole set up.

- **Signal Strength Based Methods:** These methods are based on measuring the RSS of the exchanged messages. Two RSS-based techniques are commonly used in WSNs, RSS lateration and RSS fingerprinting. Using RSS does not require specialized receivers, clocks, or time sources. It can be implemented with almost all radio modules, which has RSSI as a feature. *RSS lateration* needs the RSS measured from messages exchanged between the node with unknown location and at least three reference points. The corresponding distance to a beacon node can be calculated based on the measured RSS and the chosen propagation model. After having the distances to the three beacons, the location of the unknown node can be calculated. In general, using RSS for distance prediction suffers a major drawback of that it is difficult to calculate a distance from a RSS value without knowing precisely the antenna type, the transmission power and the environment variables, which influence the propagation model. *RSS fingerprinting* is based on creating radio map for the localization area that contains the fingerprints for certain predefined locations. The radio map is constructed by creating a grid of points. To each grid point is associated a fingerprint vector, consisting the RSS measurements from each of the beacons. When new unknown node is added to the network, it receives the RSS from each of the beacons and creates a test fingerprint vector, which is compared with all vectors in the radio map. The location of the grid point with the closet vector becomes a location of the unknown node. The first limitation about the accuracy of RSS fingerprinting method is due to the density of the radio map, i.e. the distance between the grid points. The fingerprinting technique is applicable for indoor and outdoor applications.

In the following, we focus our discussion on the RSS-based localization as it is not bound by any requirements about additional hardware, antennas or accurate clock synchronization and is often seen as the preferable technology in large scale applications.

Factors Influencing the RSS

Considerable research efforts have been devoted to empirical studies on the RSS characteristics for the low-power radios Chipcon CC1000 and Chipcon CC2420, used in most of the WSN nodes. In particular, Ma, Chen, Zhang & Ni (2006) and Lee, Wang & Xiao (2006) presented a thorough empirical study of the RSS using Mica2 sensor nodes in indoor environment and considering parameters such as operating frequency, transmitters' and receivers' inequality, antenna orientation, battery voltage, temporal and spatial properties of the environment, and the environmental dynamics. The authors studied the radio signal behavior in complex environments, considering the effects of the reflections and the obstructions. They have collected the RSS data in both indoor and outdoor environments to evaluate the influence of the environment over the RSS. Likewise, Lymberopoulos, Lindsey & Savvides (2006) investigated the RSS variability in 3-D indoor environment using sensor nodes equipped with Chipcon CC2420 radio and monopole antennas. Their study mainly focuses on the impact of the antenna orientation on the RSS.

The aforementioned studies report a strong dependency of the RSS on the antenna orientation and the environment specifics, especially for indoor scenarios due to the reflection of the signal from the floor, walls, ceilings and furniture. However, the RSS behavior at long distances in outdoor environment has not been researched thoroughly, resulting in poor understanding of the limitations and feasibility of RSS-based localization.

In the following subsection the most important factors that impact the RSS are indentified based on outdoor experiments (Stoyanova, Kerasiotis, Prayati, Papadopoulos, 2009) using Tmote Sky (Moteiv Corporation, 2006) and TelosB nodes (Crossbow Technology, 2009).

Factors Related to the Sensor Node Hardware

The sensor nodes hardware features, such as the working radio frequency, the internal antenna design and implementation and the tolerances in terms of RSSI accuracy range and linearity are among the factors influencing the RSS behavior as follows:

- **Antenna Orientation:** All real antennas, i.e. non-isotropic, have non-uniform radiation pattern, meaning that the orientation position of the antenna influences the amplitude value of RSS.
- **Radio Frequency:** Tmote Sky and TelosB sensor nodes operate in the band of 2400 MHz, where 16 channels are available, with first central frequency at 2405 MHz and last one at 2480 MHz. Different frequencies may result in different RSS at the receiver due to the unequal frequency response of the antenna.
- **Variation of the Transceivers:** Due to manufacturing tolerance of the hardware components, different transmitters behave differently even when they are initialized uniformly. Moreover, the sensitivity of the receivers of different radios may also vary in the range [-90dBm, -95dBm]. Thus, the RSS may have significant variance when measured with different set of transceivers.
- **RSS Value:** It is observed that when the RSS weaken its uncertainty increases. Possible sources of uncertainty in the RSS are the internal noise from the system hardware and the incomplete description of the RF communication link. Furthermore, the inherent accuracy of producing the RSSI is also playing role to the uncertainty of the RSS measurements. This variation is most likely due to rounding errors in the on-chip RSSI calculation.
- **Transmission Power:** Intuitively one would expect that different transmission (Tx) powers correspond to different RSS averages. Such changes in the Tx power, however, may have an impact on the value and variance of RSS in a non-linear way and when transmitting at a lower Tx power may not achieve the required reliability.

Factors Related to the Environment

The main factors related to the specifics of the operational environment in terms of obstructions (e.g. trees) and the sensor nodes positioning can be summarized as follows:

- **Specifics of the Environment:** Obstruction in the environment may have a significant effect on the RSS variance due to the reflection, diffraction, and scattering phenomena inherent to the radio wave propagation.
- **Height from the Ground:** The height of the sensor node from the ground affects the RSS mainly due to the ground reflection of the RF waves. In this case the direct and reflected waves interact and create fluctuations, i.e. sharp increase and decrease in the measured propagation curve at certain distances.

RSS- Based Localization Techniques

Typically, WSNs are heterogeneous, consisting of sensing nodes deployed at unknown locations and beacons at known locations. The number of the sensor nodes in a particular WSN varies depending on the application scenario, but in general it is expected to be large. Such large-scale deployment usually requires knowledge of the geographical location of the individual sensor nodes, so that the WSN can answers critical questions such as, where an object of interest is located, where a particular event took place, etc. For that purpose a number of RSS-based localization algorithms have been developed in the past years. These algorithms can be categorized as based on:

1. Connectivity information, where the location of a unknown node is retrieved by using only the information about its neighbors, beacons or regular nodes (He, Huang, Blum, Stankovic & Abdelzaher, 2003; Reichenbach, Blumenthal & Timmermann, 2006; Niculescu & Nath, 2003).
2. Ranging through actual RSS measurements and distance computation (Yedavalli, Krishnamachari, Ravula & Srinivasan, 2005; Stoleru & Stankovic, 2004; Savvides, Han & Srivastava, 2001; Savarese, Rabaey & Langendoen, 2002; Sichitiu, Ramadurai & Peddabachagari, 2003).
3. Profiling or mapping the deployment area by collecting the RSS signatures for a number of area points (Lorincz & Welsh, 2006; Bahl, Balachandran & Padmanabhan, 2000).

In addition, numerous techniques for RSS-based localization have been developed employing different position estimation techniques, such as the Maximum-Likelihood Estimator (MLE) and Multi-Dimensional Scaling (MDS). These techniques are more suitable for complex localization application scenarios, where a large-scale network with low-density deployment of reference nodes is considered (Li, 2007). Based on these techniques, new approaches like Semi Definite Programming (SDP) relaxation (Ouyang, Wong & Lea 2010), and Weighted Least Squares (WLS) have been proposed (Wang, Yang, 2011). These contributions established a solid theoretical foundation and RSS-based localization simulation framework. However, to the best of our knowledge there is a lack of respective implementations concerning real outdoor application scenarios, as well as real-world proof-of-concept.

Range-Free Techniques

The range-free techniques rely on connectivity information only. An example is the DV-hop technique proposed in Niculescu & Nath (2003). According to it, the beacons broadcast their location throughout the network. Each sensing node forwards the message until it reaches all other beacons, while executing a hop-count mechanism. Through the later, all nodes in the network, including beacons, are aware of the smallest number of hops to every beacon. In order to convert hop-count into physical distance, the algorithm estimates the average distance per hop. Assuming that the distance between each pair of beacons and the minimum hops between them are known, a straightforward average distance per hop, *HopDist*, is computed as follows:

$$HopDist = \frac{1}{N}\frac{\sum_{i=1,j\neq i}^{N}\left(\sqrt{(x_i - x_j)^2 + (y_i - y_j)^2}\right)}{h_{ij}} \qquad (1)$$

where N is the number of beacons, (x_i, y_i) is the location of the i-th beacon, (x_j, y_j) is the location of the j-th beacon and h_{ij} is the number of hops between the i-th and j-th beacon.

Thus, the WSN nodes can calculate their position based on the minimum hop-count to each beacon and the average-distance per hop, after performing multilateration. Obviously, this method is based on the assumption that the WSN nodes are deployed relatively uniformly in the entire target area.

Range-Based Techniques

The range-based techniques require knowledge of the relation between the distance of the communicating nodes and some measurable parameter of the transmitted signal, such RSS. The RSS has been used in many projects (Bahl & Padmanabhan, 2000; Kloos, Guivant, Nebot & Masson, 2006) and studies (Whitehouse, Karlof & Culler, 2007; Stoep, 2009).

The most popular way of mapping the RSS-to-distance is through the LPLM, given as:

$$RSS\left[dBm\right] = P_t - PL(d) \quad (2)$$

$$PL(d)[dBm] = PL(d_0) + 10n\log_{10}\left(\frac{d}{d_0}\right) + X(\sigma) \quad (3)$$

$$RSS\left[dBm\right] = P_t - PL(d_0) - 10n\log_{10}\left(\frac{d}{d_0}\right) - X(\sigma) \quad (4)$$

where the *RSS* is in decibels (dBm), d is the distance between the sender and the receiver, n is the path-loss exponent, P_t is the transmit power of the sender in dBm, $PL(d_0)$ is the power loss in dBm at a reference distance d_0 (usually d_0 is 1m) and $X(\sigma)$ is a zero-mean Gaussian random variable with standard deviation σ, measured in dBm.

The distance d as a function of RSS can be obtained by solving Equation (4), which results in:

$$d = k10^{\left(\frac{P_t - RSS - X(\sigma)}{10n}\right)} \quad (5)$$

where k is a constant incorporating both $\log_{10}(d_0)$ and $PL(d_0)$ as follows:

$$k = 10^{\left(\log_{10}(d_0) - \frac{PL(d_0)}{10n}\right)} \quad (6)$$

LPLM is used in many deployments, localization or tracking tasks, mainly because of its simplicity due to the log-normal relation between distance and RSS (Whitehouse, Karlof & Culler, 2007; Stoep, 2009). However, this is valid only under certain specific conditions. A more complex, and thus, a more accurate propagation model, which is based on the two-ray ground reflection effect of the propagated signal is proposed in (Suh, Joung & Ko 2007). However, this model does not account for important factors, such as the type of the ground on which the signal is reflected and the antenna radiation irregularity, as well as the RSS variability and uncertainty. Likewise, the model proposed in (Kloos, Guivant, Nebot & Masson, 2006) is also based on the two-ray ground reflection technique and still factors such as the antenna radiation irregularity and hardware manufacturing tolerance effects are not studied.

In (Stoep, 2009) and (Whitehouse, Karlof & Culler, 2007) attempts to model the dependence of the RSS uncertainty are presented. Unfortunately, important factors, such as the influence of height from the ground and the hardware manufacturing tolerances of the radios on the RSS-distance relation, and consequently on RSS uncertainty, have not been widely (or not at all) investigated, thus the reported results are not conclusive concerning the RSS uncertainty sources and dependency.

Hybrid Algorithms

Considering the complex relation between RSS and distance, other hybrid localization algorithms which are based on two-phase refinement positioning, take advantage of both range-free and range-based algorithms in order to reduce the localization error (Savvides, Park & Srivastava, 2002; Langendoen & Reijers, 2003). The idea behind these aproach is that nodes with unknown positions firstly perform an initial estimation of their positions based on techniques such as sum-dist, DV-hop count, etc. Afterwards making use of a certain optimization algorithm they iteratively recalculate and refine their positions based on the RSS measurements being produced during the message exchange with neighbor nodes.

Fingerprinting Technique

The techniques that employ database comparison are fingerprinting, pattern recognition or pattern matching (Bensky, 2007; Munoz, Lara, Vargas & Enriquez-Caldera, 2009). The fingerprinting is based on creating radio map, i.e. data base, for the coverage area of the network. The radio map is constructed by measuring the RSS in predefined locations and creating a fingerprint vectors or signature for each location. All fingerprint vectors compose the database. The position of an unknown node is identified by matching its RSS signature vector with the prerecorded database. This vector must be compared to the database vectors according to a given algorithm, such as the minimum Euclidean distance (Bensky, 2007). The radio map is valid only for the area where it was created. Any changes in the environment and the topology parameters that affect the radio propagation will require creating a new radio map database.

Tracking Methods

In essence, tracking of mobile objects is another aspect of the sensor nodes localization. A WSN can be used to track and secure inventory in large storehouses, locate personnel around an office, track people with special needs or elderly ones in a health care facility, locate enemy troops in the battlefield or for a temporal analysis of animal behavior in natural habitats. The applications which include such services belong to the general category of target tracking. The main aim in these applications is position estimation for moving sensor nodes as accurately as possible.

The target tracking applications are based on the exchange of localization information among the WSN nodes, which require collaborative sensing, communication and computation among multiple sensors that observe the target objects. In general, the target tracking application scenario is composed of four functional components, namely deployment, localization of the fixed nodes, localization and target tracking of the moving target and information exchange. The target-tracking scenario needs at least three reference nodes, i.e. beacons, relatively to which the target will be localized. Since the main goal of tracking is the localization of moving objects, some of the above mentioned localization methods and mostly those referring to the range measuring algorithms could be also utilized. The position estimation of mobile nodes is typically combined with a filtering algorithm, for instance Kalman filtering (Murphy, 1998; Welch & Bishop, 2006), in order to improve the tracking accuracy.

Deployment Aspects

The deployment of sensor nodes reflects two main aspects of a WSN, namely sensing and communication. When localization is concerned, defining deployment constraints and topology parameters is also part of the preceding deployment phase. The deployment process can be considered as consisting of three phases: pre-deployment, actual deployment and post-deployment. The pre-deployment phase includes analysis of the application requirements and the target environment followed by deployment simulation and planning in order to ensure the requirements for coverage, connectivity, optimal energy budget and low packet loss in the physical layer.

Several deployment algorithms can be used for dealing with the coverage problems, when minimizing the number of deployed nodes such as in (Shakkottai, Srikant & Shroff, 2005; Zhang & Hou, 2005) and the connectivity problems, when determining the optimal number of neighbors as in (Xue & Kumar, 2004; Cardei & Wu, 2006). In some studies involving WSNs, the deployment and the RF propagation are not discussed at all as a factor, which influences the communication link between two sensor nodes, but it is just mentioned that the communication link is good within a certain distance. In other studies, such as (Barboni & Valle, 2008; Otero, Kostanic & Otero, 2009) and (Fanimokun & Frolik, 2003), the received power is modeled with LPLM, which does not take in to account the influence of the height from the ground and the ground reflection over the RF signal propagation. Such simplifications provoke the widespread belief that the communication between two sensor nodes is good until a certain distance and worsens after it, which is not always correct. In general, a communication link should be considered as guaranteed when the RSS is sufficient so that the sent packets can be received with packet loss below a certain threshold (Stoyanova, Kerasiotis & Papadopoulos, 2010). Thus, the proper understanding and modeling of the radio wave propagation is essential for deploying a WSN in a way, which guarantees reliable communication among the network nodes.

RSS-BASED OUDOR LOCALIZATION ISSUES: MODELLING, SIMULATION, ACTUAL EXPERIMENTS

Classical Sources of RSS Variability

When the radio module reports the RSS of a received packet, the RSS value represents the sum of the noise power, P_N, and the signal power, P_S. Thus the received power, P_R, can be expressed in dBm as:

$$P_R = P_S + P_N, \tag{7}$$

The noise power, P_N, expressed in dBm is given according to (Vasilescu, 2005) as:

$$P_N = 10\log_{10}(P) + 10\log_{10}(\Delta f) + 10\log_{10}\left(10^{\frac{NF}{10}} - 1\right) \tag{8}$$

where P is the thermal noise power in dBm, Δf is the bandwidth in Hz over which the noise is measured, and NF is the noise figure of the receiver. NF is a measure of degradation of the signal-to-noise ratio (SNR), caused by the components in the radio frequency signal chain, which means that it is specific for a given receiver hardware.

The thermal noise power, P, which is the first term in the right side in Equation (8), is approximately -174dBm/Hz as determined in (Vasilescu, 2005). In the 2.4GHz IEEE 802.15.4 standard the bandwidth of transmission in a single band is 2MHz, and thus, the second term in Equation (8) has value of approximately 63dBm. The third term in Equation (8) is the noise figure NF, which for most of the CC2420-like receivers has a value in the range 5-7dBm. Since precise vendor information about the noise figure value is not available for the radio chip CC2420, we approximate NF=7dBm. Hence, the estimated value for P_N [dBm] is approximately -104dBm. This value of P_N is low to affect strong signals, however it could become a considerable factor in very low-power signal measurements.

In the case of low-power signal measurements and lack of external interferences, the receiver noise power defines the noise floor. In (Hauer, Handziski & Wolisz, 2009) the noise floor of the CC2420 is measured as -99dBm, which is close to the value estimated with Equation (8). The noise floor limits the smallest measurement that can be taken with certainty since any measured amplitude can on average be no less than the noise floor. In the radio communication and electronics, the noise floor may depend on the thermal noise and on other interfering signals, such as other wireless networks. This results in high percentages of packet loss when the RSS is low.

In addition, the above mentioned factors also contribute to the variability of the RSS as well. Although these factors are static in nature, due to the large number of sensors and the great range of parameter selection in the modeling process they are treated as statistical effects. The variability of RSS in this category is caused by the manufacturing tolerances as well as the rounding errors when computing the RSSI. It is often the case that the production tolerance of the hardware components causes different transmitters to behave differently even when they are initialized uniformly. Moreover the sensitivity of the receivers of different sensor nodes may also vary in the range -90dBm to -95dBm. Thus, the actual RSS value might have significant variance when measured with different set of transceivers operating in the same setup (Stoyanova, Kerasiotis, Prayati & Papadopoulos, 2009). Furthermore, the accuracy of producing the RSSI also contributes to the uncertainty of the RSS measurements. The inaccuracy in the RSSI reading is most likely due to rounding errors in the on-chip RSSI calculation. According to the datasheet of the radio chip CC2420 (Zigbee, 2003), the RSSI accuracy is within the range of ±6dBm and the RSSI linearity is within the range of ±3dBm

It is of great importance for the localization applications to identify and characterize the RSS variability sources as well as to model and include them in the localization simulation process for more realistic localization simulations.

Modeling RSS Variability

In the present section we focus on the problems whether and how the RSS measurements are influenced by P_N and the accuracy of producing the RSSI. For the purpose of this investigation, we performed series of measurements in an outdoor environment, and specifically in a free space area of 100m x 100m. During these measurements we made every effort to minimize the presence of external interferences. In particular, the area of measurements was selected far away from obstacles such as buildings, trees or walking people, and thus the only unavoidable obstacle was the ground surface. This area was also free from other wireless networks, such as IEEE 802.11 or Bluetooth, which share the same frequency band of 2.4GHz, as shown in Figure 1 for the selected channel 26. Therefore, in the absence of external interferences, and assuming that the parameters of the environment do not change noticeably during one experiment, we can consider that the uncertainty in RSS measurements is mostly due to the internal hardware noise and the accuracy of the estimation of the RSSI itself.

In our experiments, the process of measuring the RSS consists of the following steps: Every 100msec the transmitter sends packets to the receiver. In total 150 packets were sent for each selected transmitter-to-receiver distance (T-R). The receiver acquires the RSSI of the received message and forwards it to the base station, which is connected through an USB port to a laptop computer. The real-field measurements have been conducted using TelosB nodes (Crossbow Technology, 2009), equipped with CC2420 radio chip (Zigbee, 2003), and operating at 2.48 GHz with internal antenna and Tx power at the maximum of 0dBm. These measurements were performed for different heights from the ground and distances in order to have more distinct RSS levels. For twenty-five of the distance-height combinations, the mean value μ and the standard deviation σ are calculated. Since the noise is considered resultant from multiple factors we assume a Gaussian distribution. The probability density function (PDF) for a Gaussian distribution is given as:

$$PDF_{\sigma,\mu}(x) = \frac{1}{\sigma\sqrt{2\pi}} \exp\left(-\frac{\left(x-\mu\right)^2}{2\sigma^2}\right) \tag{9}$$

Figure 1. RSS, channels and bandwidth for IEEE 802.11 WLANs in the area of measurements

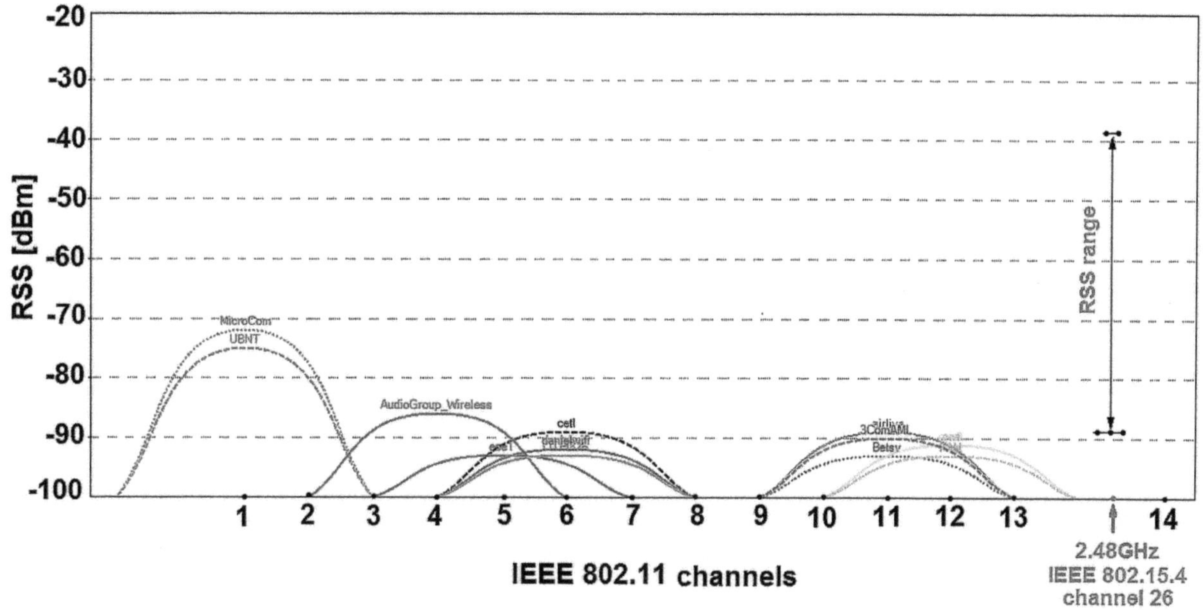

where μ is the RSS mean value, σ is the standard deviation and x is the RSS variable, measured or simulated.

Figure 2 presents RSS measurements for four mean values, μ. The mean values, μ, and the corresponding standard deviations, σ, in dBm are given in Table 1. As expected, the strongest signal, i.e., highest μ, has the smallest deviation, while the weakest signal has the highest deviation, i.e. the measured values are more dispersed around the mean value.

Figure 3 presents the distributions for four sets of RSS measurements, where the mean value μ and the standard deviation σ are calculated over 150 samples. It is noticeable that with the decrease of the signal strength the standard deviation increases. This increase is especially remarkable for signals with RSS below -80dBm.

The distribution of the RSS measurements is used to investigate the corresponding RSS uncertainty when both the receiver and the transmitter are static. The data were analyzed with respect to: (a) the distribution type, (b) the RSS deviation dependency and (c) the deviation modeling. For that purpose, twenty-five measurements were chosen in order to cover RSS values from -40 dBm to -87 dBm. For all the twenty-five the mean value μ and the standard deviation σ were calculated. Figure 4 shows σ with respect to μ for all the twenty-five measurements, marked with black dots.

In conjunction with the measurements presented in Figure 3, it can be concluded that as RSS is decreased the standard deviation increases. Therefore, of paramount importance for the modeling process is to find the relation $\sigma = f(\mu)$. For this purpose, a data-to-curve-fitting approximation is performed and the results are presented in Figure 4. The two most appropriate approximations are given with Equations (10) and (11). Equation (10), the modeling parameters of which are shown in Table 2, approximates the data with curve 1, while Equation (11) approximates the data with curve 2. Both equations are to be used in the RSS uncertainty generation process depending on the mean value for fixed sensor nodes in outdoor unobstructed environment, when localization simulation is considered (Stoyanova, Kerasiotis, Efstathiou & Papadopoulos, 2010).

Figure 2. RSS measurements for four different combinations of μ and σ

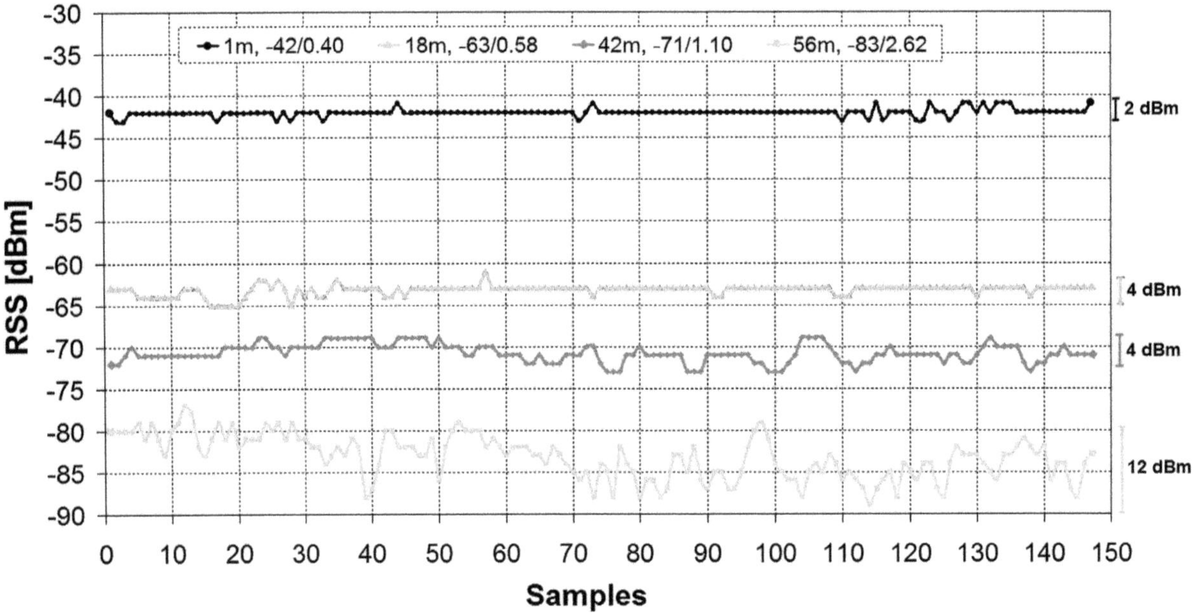

Figure 3. RSS deviation from the mean. Actual measurements for 4 different μ/σ cases

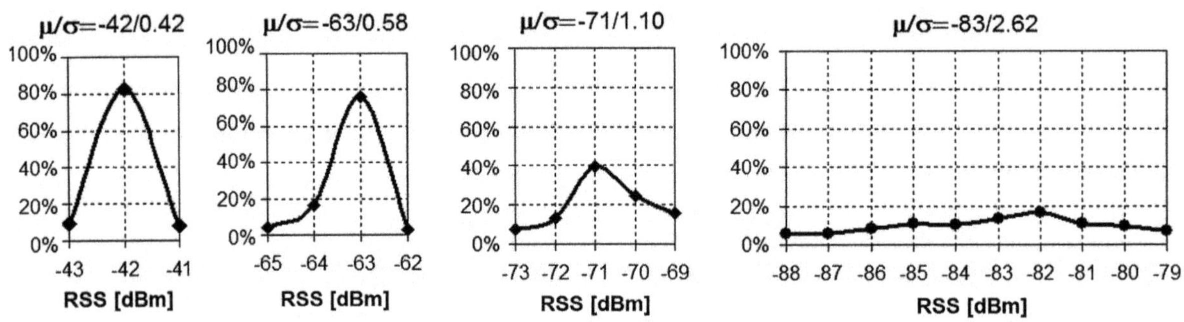

Table 1. μ and the corresponding σ in dBm

μ	-42	-63	-71	-83
σ	0.40	0.58	1.10	2.62

Table 2. Modeling parameters for Eqiation (10)

$p_1 = 1.35$	$q_1 = 174$
$p_2 = 160$	$q_2 = 8560$
$p_3 = 5960$	$q_3 = -200$
$p_4 = -300$	

Figure 4. Twenty-five μ/σ data and the approximation curves

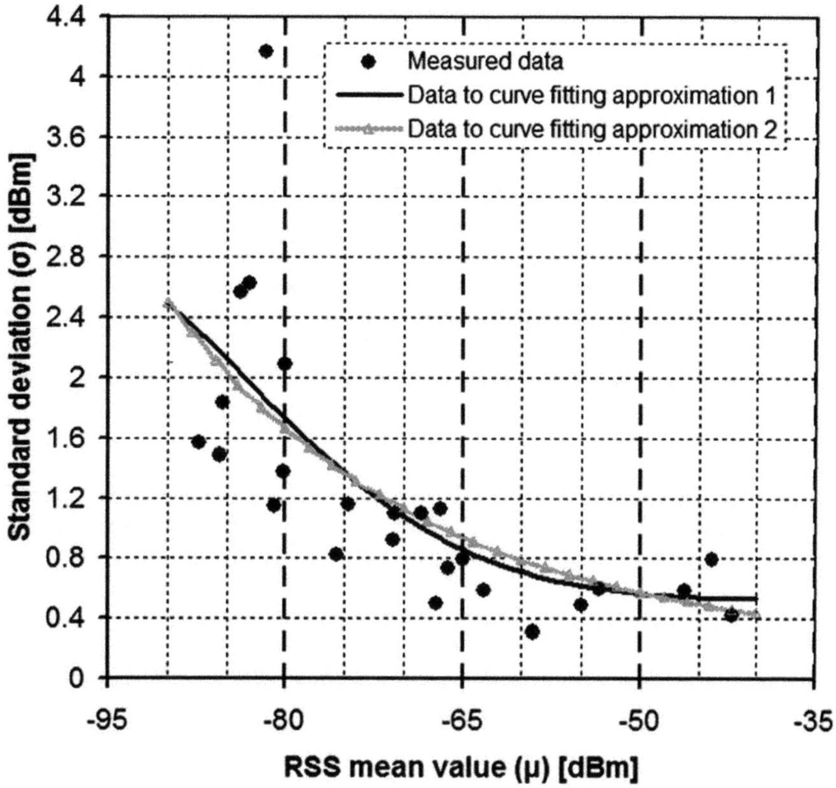

$$\sigma(\mu) = \frac{p_1\mu^3 + p_2\mu^2 + p_3\mu + p_4}{\mu^3 + q_1\mu^2 + q_2\mu + q_3} \quad (10)$$

where the parameters p and q are calculated as follows:

$$\sigma(\mu) = a.\exp(-b.\mu) + c \quad (11)$$

where $a = 0.04$, $b= -0.045$, $c = 0.2$.

The comparison of the simulated deviation with the real measured deviation for the same μ is presented in Figure 5. The measured and generated RSS data deviation are consistent and indicate that the described modeling technique, expressed with Equations (10) and (11), is suitable for modeling the RSS uncertainty when the sensor nodes are static.

Propagation Models and Deployment Considerations

Propagation Models

In localization tasks, the LPLM is the most widely used RF propagation model for estimating distances from RSS measurements. This model is convenient due to the simple logarithmic relation between RSS and distance. However, when the RF signal propagates in obstacle-free outdoor environment, it follows the 2-ray grown reflection model, while only at very small heights from the ground the RF signal propagation can be also represented with the LPLM. Figure 6 depicts a comparison of real measured data with simulated ones, using LPLM and Free space Outdoor Model (FOM) described in (Stoyanova, Kerasiotis, Prayati &

Figure 5. Data deviation from the mean: simulated and measured

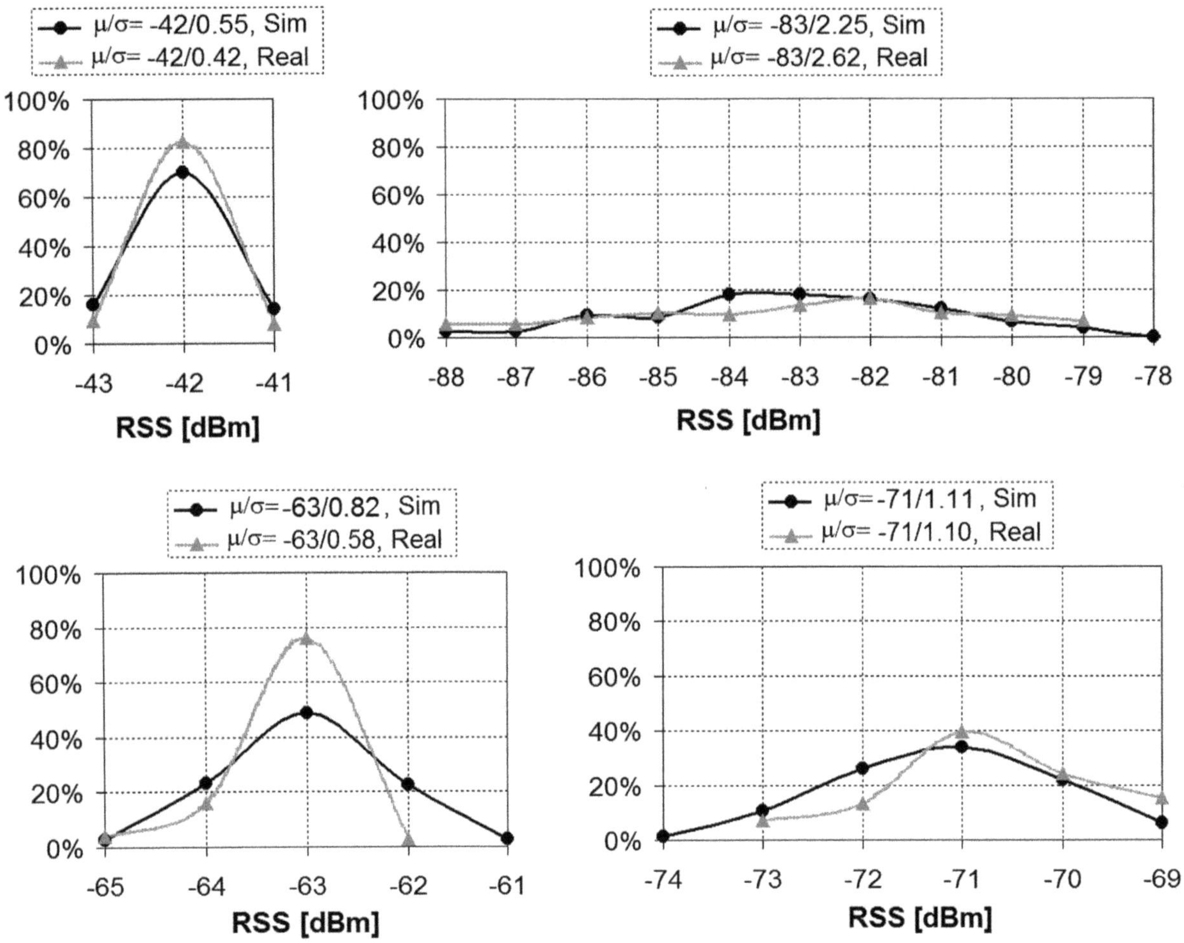

Papadopoulos, 2009), which is based on the 2-ray ground model.

In outdoor obstacle-free environment, the direct and reflected from the ground signals interact and create fluctuations of the RSS, which in turn cause unreliability during the distance estimation process. This means that a certain RSS value could correspond to more than one distance. For example, in Figure 6 (d), distances of 21m, 31m, 33m, and 54m correspond to RSS of -80dBm. When the height from the ground increases, the fluctuations also increase as it is shown in Figure 6 (a), (b), (c) and (d). For that reason, prerequisite for successful localization is such fluctuations to be avoided via network topologies where the nodes are places at lower heights from the ground (if possible). On the other hand, if the height is too low, for example up to 0.10m (see Figure 6 (a)), the propagation of the RF signal even with maximum Tx power cannot reach distance greater than 20-25m. Therefore, medium heights in the range 0.40m-0.70m (see Figure 6 (b)) can guarantee propagation distance of 50m or above, while at the same time, the fluctuations are located in the first 5m-10m and can be avoided during the design of the localization scenario.

RSS Threshold

A major communication consideration is the correct message reception. It is assumed that a packet sent by a transmitter can be received by a receiver with certain confidence, only if the intensity of the received signal is above a given threshold. The relation between the intensity of the received signal or the received signal strength and the packet loss is experimentally investigated to find whether such a threshold exists.

The results for 65 measurements are presented in Figure 7, including the RSS values and the corresponding packet loss. For each measurement the packet loss is calculated over 150 sent packets.

Based on the results presented in Figure 7, the following conclusions are drawn:

- For RSS above -80dBm, packet loss is less than 5%
- For RSS between -80dBm and -87dBm, packet loss rises to 10%.
- For RSS lower than -87dBm and close to approximately -90dBm, which corresponds to the sensitivity threshold of the CC2420 radio, packet loss can reach 100%.
- The RSS value of -80dBm is an acceptable threshold level. It indicates that signals with RSS above -80dBm will be received with less than 5% packet loss. Similar results have been reported in (Srinivasan & Levis, 2006).

RSS-Based Localization Simulations and Experiments

Localization can be defined as a process of discovering the spatial position of all sensor nodes within a particular sensor network via finding the relationships between them. In the general case, most of the nodes are placed at unknown locations, while only few of them, not less than three, are aware of their spatial location. These nodes are referred to as beacons. Thus, a localization algorithm presumes that the unknown nodes can position themselves with respect to their neighbors, including other unknown nodes and beacons. The main idea is that the unknown nodes first have to find their initial position estimations with respect to the beacon nodes, using a certain method, such as sum-dist, DV-hop count, etc. (Niculescu & Nath, 2003; Langendoen & Reijers, 2003), and then using an optimization algorithms to iteratively recalculate and refine the estimation of their positions based on RSS measurements between them (Stoyanova, Ts., Kerasiotis, K., Antonopoulos & Ch., Papadopoulos, G., 2014).

The localization procedure, known as Collaborative Localization Technique (CLT), utilizes distance measurements between pairs of location-unaware sensor nodes and exploits the ad-hoc networking capabilities of all the sensor nodes for location estimation (Koneru, 2006). CLTs allow location-unaware sensor nodes that are not in the range of any reference nodes to be located when they have at least one neighboring sensor node. Such algorithms do not demand an overlapping coverage of reference nodes and work well in sparse reference node configuration.

CLT Description

Even though many varieties of CLT exist, they share a common structure, where three phases can be identified (Langendoen & Reijers, 2003): (1) Estimating the distance between the unknown and the beacon nodes, (2) Deriving for each node an initial position based on its distance to the beacons, (3) Refining the nodes position using information about the measured distance to the neighboring nodes.

In the following we detail on these purpose of these three phases:

Figure 6. Comparison of real measured data and simulation with FOM and LPLM

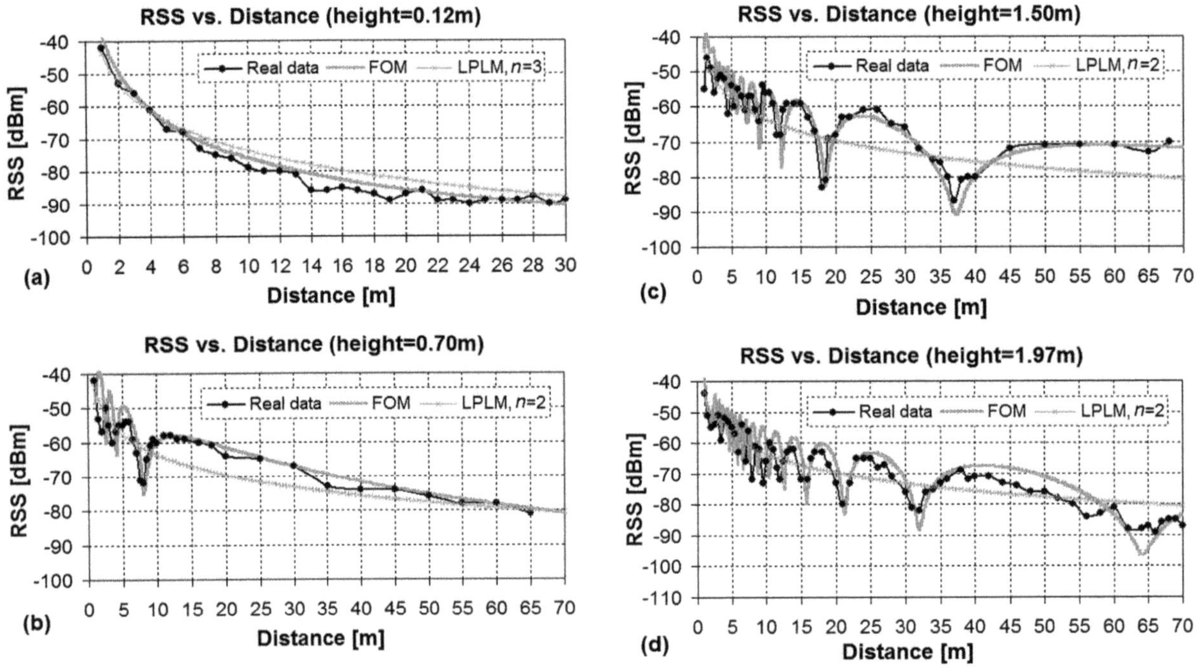

Figure 7. Packet loss versus RSS

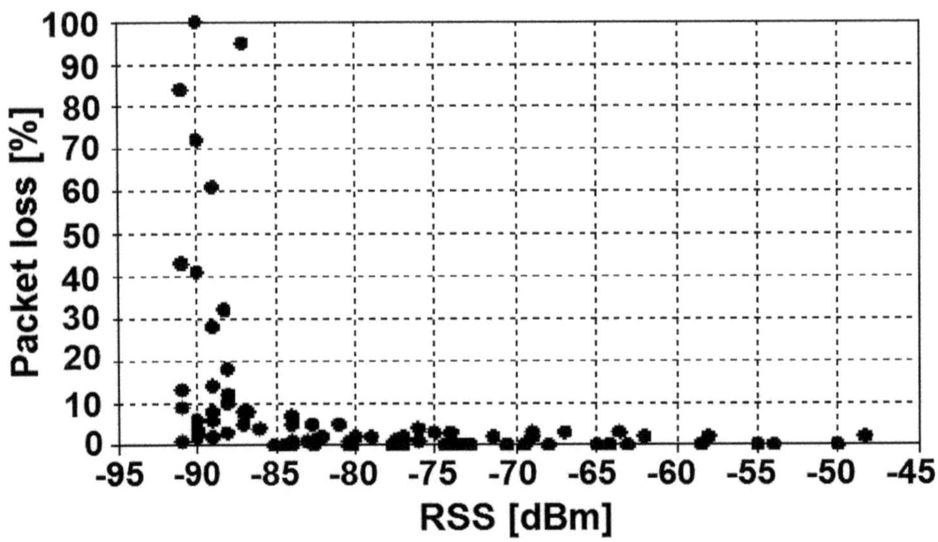

Phase 1: Estimating the distance between unknown and beacon nodes

In this phase the nodes share information to collectively determine the distances between individual nodes and the beacons, so that an initial position can be calculated in Phase 2. Three approaches for estimating the distance to the beacons are presented in (Langendoen & Reijers, 2003), namely Sum-dist, DV-hop and Euclidean. Although they differ in the distance calculation method, these approaches use similar mechanisms to deliver the needed information. The position information of each of the beacon nodes is flooded through the network to all other beacons. The flooding information exchange is not the subject of this work, but it is described in (Langendoen & Reijers, 2003). From the three approaches associated to this phase, the DV-hop is applied in the simulation examples and the real localization tests, presented in this particular work. The description of the DV-hop approach follows.

The DV-hop essentially consists in sending a message from one beacon to all others. Along the path, the message is forwarded from the regular nodes in order to calculate the number of hops between every pair of beacons, taking into account the shortest paths. In a real application, a routing algorithm is necessary to accomplish this step. The final goal is to calculate the average one-hop distance. The calculation of the one-hop distance, d_{1hop}, for four beacons A_i, i=1,…, 4, is giving by:

$$d_{1hop} = \frac{d_{12} + d_{13} + d_{14} + d_{23} + d_{24} + d_{34}}{h_{12} + h_{13} + h_{14} + h_{23} + h_{24} + h_{34}}, \quad (12)$$

where d_{12} is the known distance between the beacons A_1 and A_2, and h_{12} is the number of hops between them.

After the calculation of the average one-hop distance, the distance between any unknown node, j, and a beacon, A_i, can be calculated as follows:

$$d_{Ai,j} = h_{Ai,j} d_{1hop} \quad (13)$$

where $h_{Ai,j}$ is the number of hops for the shortest path between beacon A_i and node j.

At the end of this phase the distance between every unknown node and every beacon will be roughly estimated.

Phase 2: Deriving for each node an initial position based on its distance to the beacons

In the second phase, based on the estimated distances between an unknown node and the three closest beacon nodes, the initial position of the unknown nodes can be calculated.

The most common method for deriving a position is multilateration. In 2-D dimensional space three beacons are quite sufficient and the method is named trilateration. From the estimated distances, $d_{Ai,j}$ the known beacon positions, (x_{Ai}, y_{Ai}), the initial position (x_j, y_j) of any unknown node, j, can be derived forming the following system of equations:

$$\begin{aligned} (x_{A1} - x_j)^2 + (y_{A1} - y_j)^2 &= d_{A1,j}^2, \\ (x_{A2} - x_j)^2 + (y_{A2} - y_j)^2 &= d_{A2,j}^2, \\ (x_{A3} - x_j)^2 + (y_{A3} - y_j)^2 &= d_{A3,j}^2. \end{aligned} \quad (14)$$

The system can be linearized by subtracting the last equation from the first and the second, which gives a system of linear equations, expressed in matrix form as **Au=b**, where:

$$\mathbf{A} = \begin{bmatrix} 2(x_{A1} - x_{A3}) & 2(y_{A1} - y_{A3}) \\ 2(x_{A2} - x_{A3}) & 2(y_{A2} - y_{A3}) \end{bmatrix} \quad (15)$$

$$\mathbf{b} = \begin{bmatrix} x_{A1}^2 - x_{A3}^2 + y_{A1}^2 - y_{A3}^2 + d_{A3,j}^2 - d_{A1,j}^2 \\ x_{A2}^2 - x_{A3}^2 + y_{A2}^2 - y_{A3}^2 + d_{A3,j}^2 - d_{A2,j}^2 \end{bmatrix} \quad (16)$$

$$\mathbf{u} = \begin{bmatrix} x_j \\ y_j \end{bmatrix} \qquad (17)$$

The system can be solved using a standard least-squares approach (Langendoen & Reijers, 2003):

$$\mathbf{u} = (\mathbf{A}^T\mathbf{A})^{-1}\mathbf{A}^T\mathbf{b}$$

When Phase 2 of the algorithm is completed, every node would have an initial position. Based on this we can estimate the distance d_{ij}^{est} between any two nodes.

Phase 3: Refining the estimated positions of nodes

The last phase in the CLT is the refinement phase. During this phase each node measures the RSS for every received message sent by its neighbors and sends the respective RSS values to the base station. Based on these measurements, the approximate distance to every neighbor can be extracted from the employed RF propagation model. Using the initial nodes position and the measured distance to their neighbors, some optimization algorithms, based on Kalman Filter (KF) (Murphy, 1998; Welch & Bishop, 2006), Particle Swarm Optimization (PSO) (Kulkarni & Venayagamoorthy, 2010), Genetic Algorithms (GA) (Tam, Cheng & Lui, 2006), or other method can be applied to refine the node positions.

The refinement can be formulated as minimizing the difference between the estimated distance after Phase2 and the distance calculated via the RSS measurements between each two neighbor nodes, which can be expressed as:

$$\min \left\{ \sum_{i=1}^{Nnode} \left(\frac{1}{N_{neighb}} \sum_{j=1}^{Nneighb} \left(d_{ij}^{est} - d_{ij}^{RSS} \right)^2 \right) \right\} \qquad (18)$$

where *Nnode* is the number of the nodes participating in the WSN, *Nneighb* is the number of the neighbors for each node, d_{ij}^{est} is the distance estimated in Phase 2 between nodes i and j, d_{ij}^{RSS} is the distance corresponding to the measured RSS.

Let's consider PSO as the optimization algorithm used in this study. PSO is a stochastic optimization algorithm introduced by Kennedy and Eberhart (1995), which can be successfully applied for localization in WSNs as presented in (Kulkarni & Venayagamoorthy, 2010). In brief, in PSO-based methods, a number of agents, called particles, explore the solution space of the problem, sampling the objective function at the discrete points they reach. Each particle is characterized with a position and a velocity in the solution space, where the position of a particle represents a candidate solution to the problem. The velocity towards the location of the local or global best-performing particle depends also on the past success of that particle. Thus, relying on knowledge of their own best solution found so far and on the best-found solution for the entire swarm (or a local neighborhood of particles), which are weighed stochastically, each particle strives iteratively to reach better and better solutions. The search process continues until the value of the objective function converges to the target value, or until the maximum allowed number of iterations is reached. Under certain conditions the PSO methods are guaranteed to reach the global optimum, or some point very close to it. Further details about the various PSO methods are available in (Kennedy & Eberhart, 2001). In a real application the refinement can be done in the sensor nodes only if they have sufficient computational power, otherwise in the base station.

Localization Simulation

The simulation process aims at simulating a sensor nodes localization scenario, where the nodes can be randomly placed. The simulation is performed in Matlab environment. Two scenarios are simulated: (a) small area 40mx40m, 4 beacon nodes, 20 nodes with unknown location, T-R height=0.4m and Tx=-25dBm, and (b) large area 500mx500m, 4 beacon nodes, 100 nodes with unknown location, T-R height=0.4m and Tx=0dBm.

The simulation procedure consists of four steps:

Step 1: Generation of sensor nodes location - The positions of the sensor nodes are randomly generated within the area. The beacons can be also randomly generated or with preselected positions, i.e., in the corners of the area, and if more beacons are needed the rest are randomly located within the area.

Step 2: Calculation of distance from node to anchors - The number of beacons is not essential in the CLT, since they are used only to calculate the average one-hop distance in meters using the DV-hop technique. Among different simulations the average one-hop distance does not differ due to the number of nodes, but due to different network topologies, produced by the random positioning of the sensor nodes. The distance from a node to each beacon is calculated through Equation (13).

Step 3: Least-squares approximation is used to find the initial position (x, y) for each unknown node - The distances to the three closest beacons are used in Equations (14) - (17) to calculate the initial coordinates for all unknown nodes.

Step 4: Refinement of the estimated positions of the unknown nodes using multiple simulated RSS measurements so that a close to estimation of variance or uncertainty of RSS to be calculated - During this step an optimization algorithm, such as PSO, is used for refinement of the initial positions of the unknown nodes. The refinement is based on simulated RSS measurements, which are produced in the following way: for each specific distance, using FOM, a RSS is generated and a uncertainty component calculated from Equation (11) is added.

As it was mentioned above two localization scenarios were simulated:

Small Area: 40m x 40m, 4 beacon nodes, 20 nodes with unknown locations, T-R height=0.4m and Tx=-25dBm. The network topology with the true nodes' positions and the localized positions after Step 3 and Step 4 are shown in Figure 8. The characteristics of the network and the localization results expressed as RMSE are summarized in Table 3.

The radio range, for the chosen height from the ground and Tx power, is approximately 13m - 14m, because at the distance of 15m the RSS is close to the sensitivity threshold of the radio, which is around -90 dBm. The calculated one-hop-distance varies from one deployment to another around 7.5m - 8.5m. When the initial nodes positions are calculated, the RMSE after Step 3 is 4.6m for the topology presented in Figure 8. For different topologies (deployments), the RMSE after Step 3 may vary between 4.5m and 5.5m. After the refinement with PSO, there is an improvement of the estimated positions accuracy for the unknown nodes. For the topology presented in Figure 8, the final RMSE after Step 4 is 2m. This error may vary within 2m-3m. This variance is observed for different random deployments, and for repeated localization run for the same topology, due to the uncertainty added in the RSS generation process. The calculated RMSE is about 15-16% of the radio range and 5% of the linear size of the area, which is relatively good result.

Large Area: 500m x 500m, 4 beacon nodes, 100 nodes with unknown location, T-R height=0.4m and Tx=0dBm. The network topology with the true positions of the nodes and the localized positions after Step 3 and Step 4 are shown in Figure 9. The characteristics of the network and the localization results expressed as RMSE are summarized in Table 3.

The radio range for the chosen height from the ground and Tx power is approximately 75m. The calculated one-hop-distance varies from deployment to deployment in the range 48m - 51m.

Figure 8. Localization simulation for 20 nodes with unknown locations

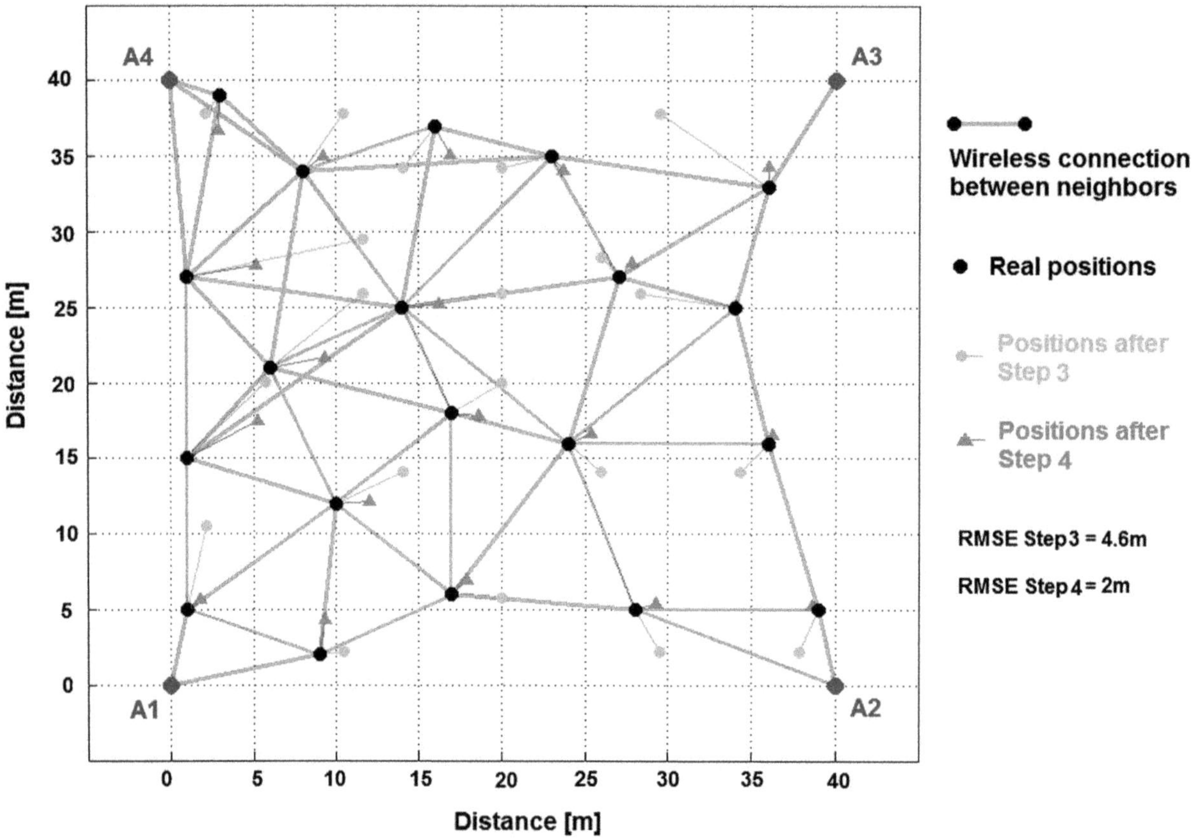

Table 3. Characteristics of the network and the localization results

Area size [m]	Sensor Nodes	Tx Rower [dBm]	Radio Range [m]	RMSE Step 3 [m]	RMSE Step 4 [m]
40 x 40	20	-25	13	4.6	2
500 x 500	100	0	75	44	27

When the initial positions of nodes are calculated, the RMSE after Step 3 is 44m for the presented topology but it may vary between 41m and 55m from one topology to another. When the refinement with PSO is performed, there is an improvement on the estimated node positions. The final RMSE after Step 4 is 27m for the topology presented in Figure 9. This error also may vary within the range 26m - 33m. Due to the uncertainty added in the RSS generation process, this variability is observed for different random deployments but also for repeated runs of the localization simulation for the same topology. The calculated RMSE is about 36% from the radio range, and 5.4% from the linear size of the area.

The results for these two simulation scenarios are summarized in Table 3. It can be seen that the results for the large area are worse than those for the small area, which is most likely related to the increased number of hops between the beacons, which are 4-7 hops for the scenario (a) and 9-15 for the scenario (b).

The generated RSS data (with the uncertainty included) over which the localization is performed are presented together with the propagation model curve. There are 46 wireless connections for the results of the small area shown in Figure 10(a) and 275 wireless connections for the large area reported in Figure 10(b).

Practical Localization Experiment

The goal here is to localize the positions of all unknown nodes using only the RSS taken with the exchanged messages between them, when the position of the beacon is known. The localization procedure is centralized, i.e. all calculations are performed in the BS (in our case a laptop computer).

We report on the results of a real-field experiment and compare to simulation results obtained for the same setup and deployment parameters, including height from the ground of the sensor nodes and beacons and the Tx power.

The deployed WSN consists of 22 sensor nodes, and four of them are beacons with known locations. The topology of the network is shown in Figure 11, where nodes A1-A4 are the beacons with known locations and the nodes with IDs 5 to 22 are with unknown locations. The base station (BS), which collects all messages sent by the sensor nodes, including the beacons, is located approximately 50m-55m away from the most remote sensor node and is connected to a laptop

Figure 9. Localization simulation for 100 nodes with unknown locations

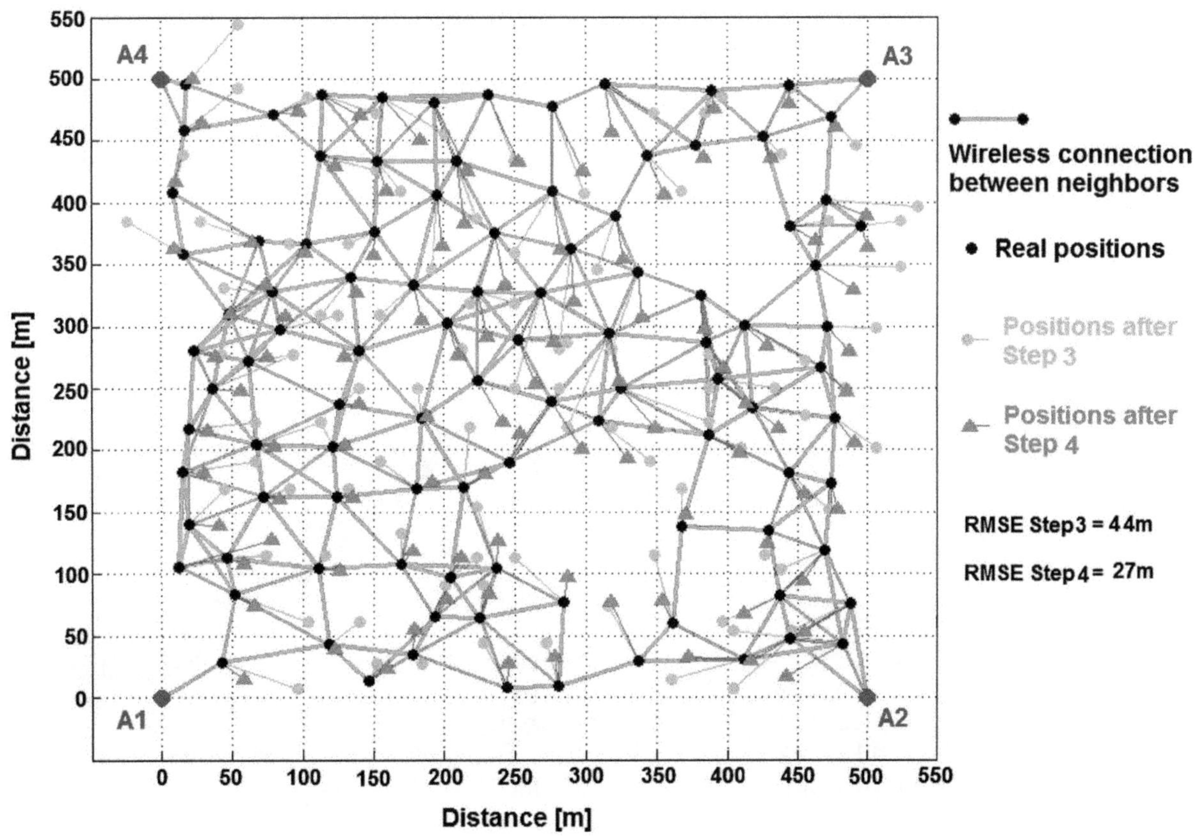

computer. The nodes are deployed in an area of 40m x 40m. The height from the ground is 0.40m and the transmission power is set at -25dBm, which guarantee a maximum communication radius of 15m, i.e. at distance of 15th meter the RSS is close to -90dBm. The beacons are deployed in a way so that they cannot communicate with each other and have only one unknown neighbor node. The solid line in Figure 11 expresses the established connection between the neighboring nodes.

As previously described, for the localization the CLT is used, which consists of the following steps:

1. Estimate the distance between the unknown and beacon nodes,
2. Derive for each node a position based on its distances to beacons,
3. Refine the estimated position of the nodes, making use of information about the measured distance to the neighboring nodes.

The process of obtaining the RSS consists in the following:

- Each node, including the beacons, sends a packet with Tx power of -25dBm. All other nodes listen and if they hear the message, record the sender's ID and the measured RSS of the massage.
- After that, the same node sends a message only for the BS with Tx power of 0dBm. The message includes information for all registered neighbor nodes IDs and the corresponding RSS.
- The time interval between the messages is 100ms. This process ends when all nodes send 150 messages, which takes approximately 15 minutes.
- Two different sets of measurements were performed.

The location estimation is centralized. As many factors influence the RSS value, an average of the RSS measurements between a pair of nodes is used as a simple compensation mechanism. The measured RSS values and the corresponding actual distance are shown in Figure 12(b). The relation between the distance and the measured RSS is rather ambiguous, most likely due to the difference in the orientation and the inequality of hardware among the nodes.

In order to adjust the propagation model to cover the RSS/distance predisposition a data-to-curve fitting approximation (DCF) is applied, and

Figure 10. Generated RSS data for: (a) small area with RSS data for 46 connections, and (b) big area with RSS data for 275 connections

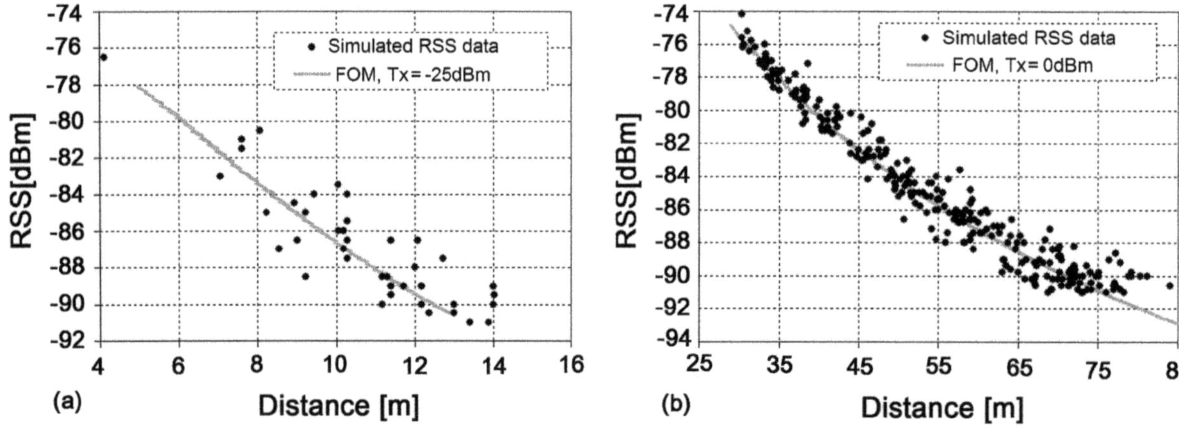

Figure 11. Sensor network topology

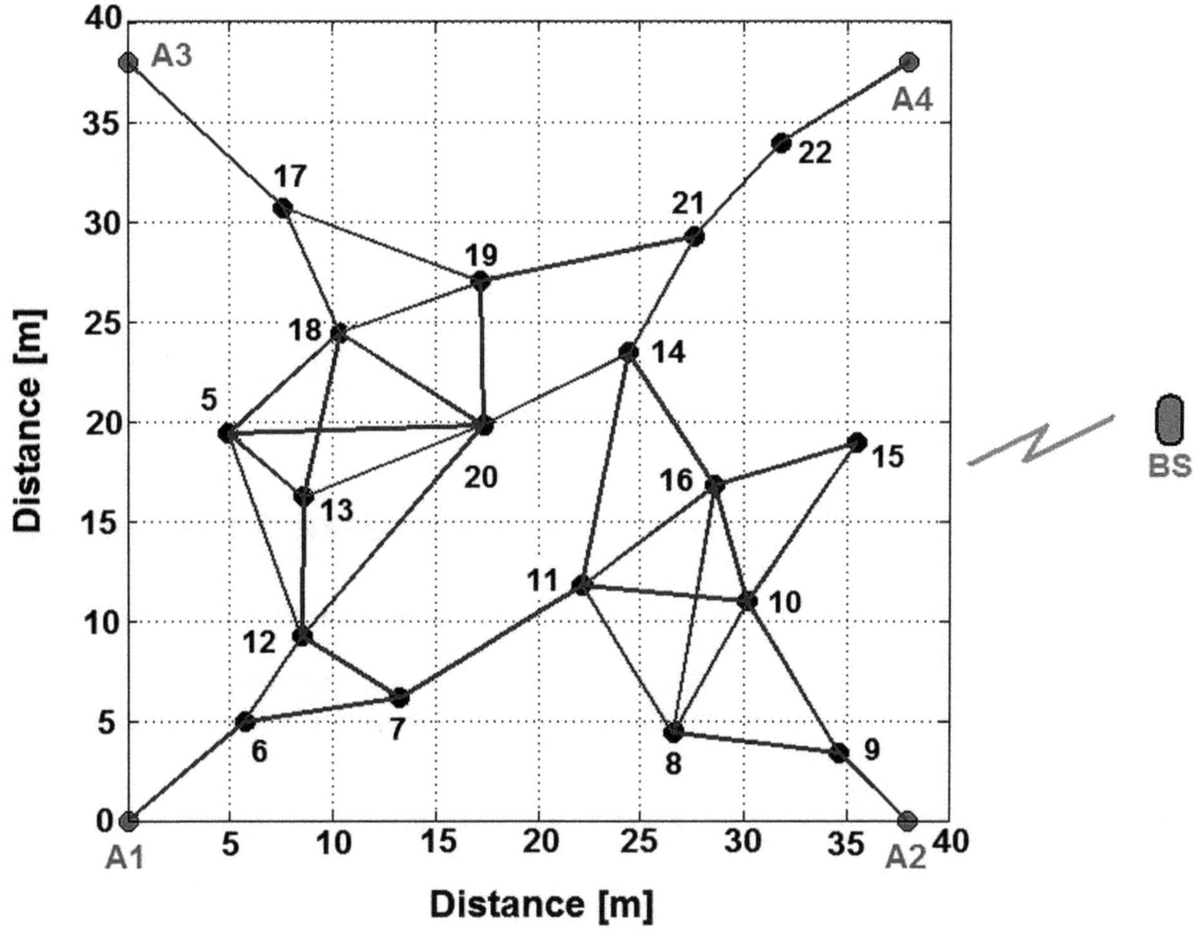

the result is shown with a solid pink line (with squares) in Figure 12(b). The dark dots are the 37 measured RSS values vs. the corresponding distance. The FOM and LPLM propagation models were tested during the localization procedure.

The important parameters for FOM is the Tx power and for LPLM are P_0, measured at distance of 1m, and the path loss factor n. As presented in Figure 12(b), the closest match to DCF are produced by FOM with Tx=-25dBm and by LPLM with P_0=-65dBm and n=2.2. The two models are also shown in Figure 12(a) for the entire radio communication distance of 15m.

The localization procedure described above consists of three steps. The first step, i.e., determine the distance between unknown and beacon nodes, is internal and does not provide visible localization information. Step 2, i.e., derive the initial nodes positions, provides preliminary localization information, based only on the information about the neighbors and the locations of the beacons. The results from Step 2 can be visualized for comparison. The final Step 3, i.e., refinement of position estimations, utilizes the RSS measurements to improve the localization position after Step 2. The results for the RMSE after Step 2 and Step 3, for both mentioned models,

Figure 12. The measured RSS vs. Distance

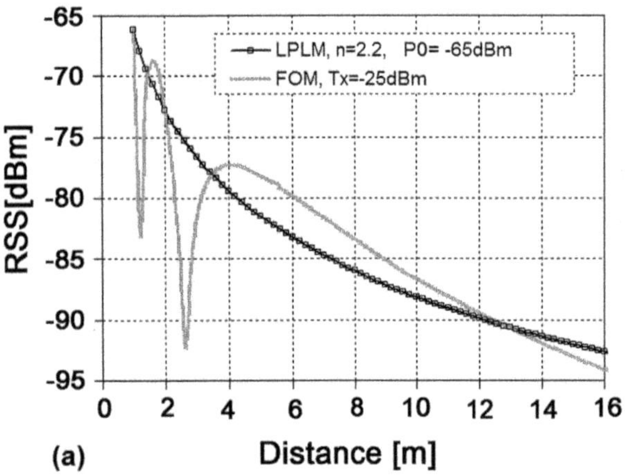

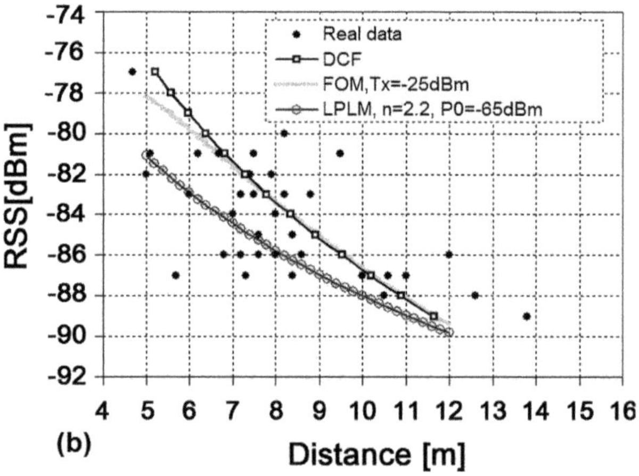

are given in Table 4. In Figure 13 are shown the localization results when using FOM and Figure 14 when using LPLM.

The RMSE results in Table 4 are averaged for 10 runs of the localization process. The RMSE calculated after Step 2 and Step 3 for the two models is similar and confirms that both models can be used when the RF propagation is preliminary evaluated. The results also present that the RMSE can be reduced up to 50% during the refinement phase, when using the RSS measurements taken from the neighboring nodes.

We do not claim that the techniques utilized in Step 2 and Step 3, i.e., DV-hop and PSO, are the most successful ones. Here they are used in their simplistic versions to illustrate the feasibility of performing localization of network nodes using only the RSS. Furthermore, Step 2 does not use the RSS data but only the information about the number of neighbors. The actual RSS measurements are used at Step 3 for refinement and as the results show, refining of localization results is feasible and the results are quite promising.

FUTURE RESEARCH DIRECTIONS

In this chapter we discussed the feasibility of RSS-based localization when performed with the resources of a WSN. Despite of the extensive study of the factors influencing this process some open issues still remain. Testing more advanced localization algorithms as well as optimization algorithms are among the important issues to be addressed. Further investigation on the influence of the number of beacons over the localization performance, and determining the optimal number and positions of the beacons, may improve the localization accuracy. Finally, the WSNs that make use of the 2.4GHz band may take advantage of the difference in the RF propagation between the first and the last channels in order to improve the localization positioning.

CONCLUSION

The present chapter investigated the feasibility of localizing wireless objects to be performed with the resources of the WSN technology, when using

Table 4. RMSE for Step 2 and Step 3 of the localization positioning

RMSE (FOM)		RMSE (LPLM)	
after Step 2	after Step 3	after Step 2	after Step 3
4.36m	2.2m	4.42m	2.2m

Figure 13. Localization positioning when using FOM

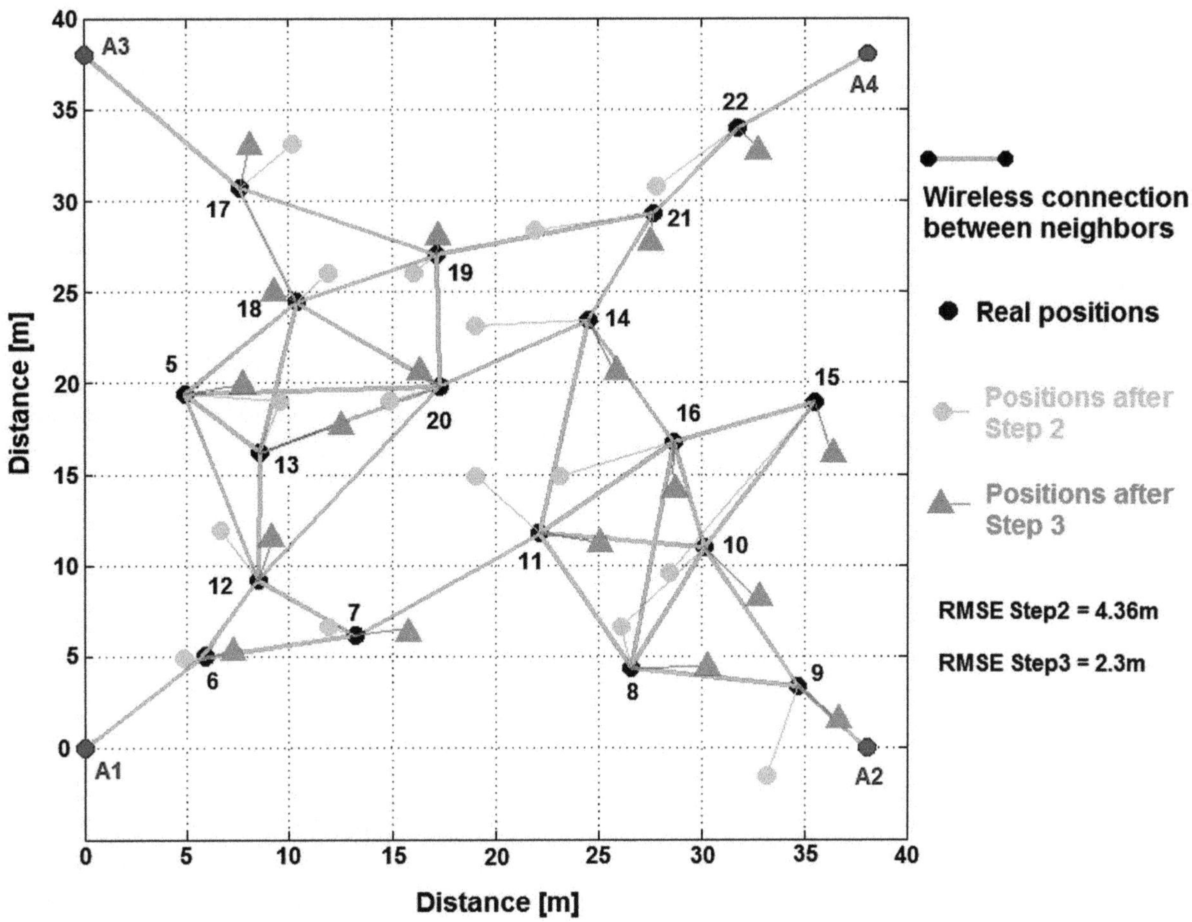

only the RSS of the exchanged messages. In order to evaluate the feasibility of such a solution, the present study was focused on identifying the most crucial RSS-based localization problems and to determine and evaluate the topology parameters that can guarantee successful localization. More specifically we aimed at:

1. Evaluation of the behavior of the propagated RF signal in order to identify the important problems of RSS-based localization,
2. Selection of the topology related parameters, when taking into consideration the localization requirements and the deployment constraints,

Figure 14. Localization positioning when using LPLM

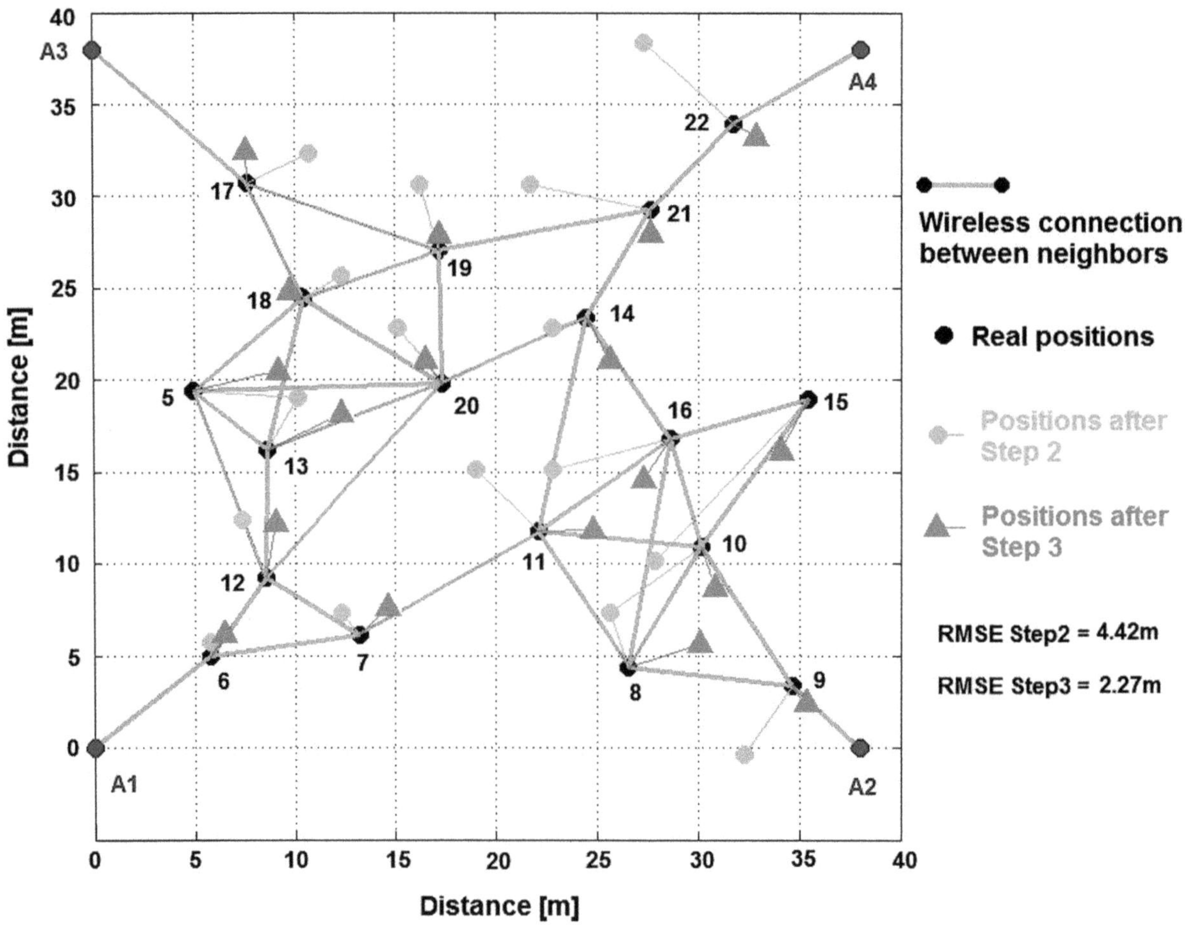

3. Use of the selected propagation model and topology parameters to simulate a localization scenario, and
4. Use of the proposed topology parameters to perform real outdoor localization experiment.

In brief, the main results obtained during the simulations and the practical experiments are:

- When the height from the ground increases, the variability of the RSS also increases, and consequently more than one distances correspond to a certain RSS value. In order to avoid this, the relation RSS vs. distance has to decrease smoothly with the distance as much as possible. This can be achieved by judicious selection of the height from the ground for the sensor nodes. The tendency is to avoid the higher variability by selecting lower heights from the ground.
- If the T-R height is too low, for example 0.10m, the RF signal even with maximum Tx power of 0 dBm cannot propagate at distance greater than 20-25m. Therefore, heights of 0.40m-0.70m can guarantee radio range of 50m or above, while at the same time, the higher RSS variations are located in the first 5m -10m and can be avoided during the design of the setup.

- The inequality of hardware imposes the need of a calibration procedure before the localization process. Based on that, the RSS measurements were averaged between a pair of nodes for minimizing the influence of the inequality of hardware on the RSS value.
- For the CLT localization technique the number of beacons is not essential, since they are used only to calculate the average one-hop distance using the DV-hop technique. For subsequent simulations the average one-hope distance does not differ with the number of beacons, but differs due to the different network topologies resulting from the random locations of nodes.
- The localization error, expressed in terms of RMSE, may vary slightly from one deployment to another and from one localization run to another for the same topology. The calculated RMSE is around 15-20% of the radio range when the area is small and the number of sensor nodes is small, and goes up to 30-40% of the radio range when the area is larger with larger number of sensor nodes with double number of hops. In both cases this error corresponds to 5-6% from the linear size of the area.
- The localization accuracy of the performed practical test, when both LPLM and FOM are applied, presents similar results and confirms that both models can be used when the network topology parameters are preliminary evaluated so that fewer variations in the propagation curve are present.
- We do not claim that the DV-hop and PSO techniques used in the initial estimation step and in the refinement step are the best choices. Here they were used to illustrate the feasibility of performing the node localization process using only the RSS.
- The simulations and the real practical experiments show that the RSS-based localization is feasible when takes in consideration some RSS specifics, such as: the proper RF propagation model, RSS dependence on the height of the nodes, RSS variability due to internal and external factors, etc. The results are quite promising but the applicability of the RSS-based localization depends mainly on the application scenario and the accuracy requirements.

REFERENCES

Bahl, P., Balachandran, A., & Padmanabhan, V. (2000). *Enhancements to the RADAR User Location and Tracking System* (Tech. Rep., MSR, 2000).

Bahl, P., & Padmanabhan, V. (2000). RADAR: An In-Building RF-Based User Location and Tracking System. In *Proceedings of the IEEE INFOCOM '00, March 2000.* doi:10.1109/INFCOM.2000.832252

Barboni, L. & Valle, M. (2008). Wireless Sensor Networks Power-Aware Deployment. *SENSORCOMM 2008, August 2008, Cap Esterel, France* (pp. 252-257), DOI: .10.1109/SENSORCOMM.2008.48

Bensky, A. (2007). *Wireless positioning technologies and applications.* Artech House.

Cardei, M., & Wu, J. (2006). Energy-efficient coverage problems in wireless ad-hoc sensor networks. *Computer Communications, 29*(4), 413–420. doi:10.1016/j.comcom.2004.12.025

Crossbow Technology. (2009). *TelosB mote platform.*

Fanimokun, A., & Frolik, J. (2003). Effects of natural propagation environments on wireless sensor network coverage area. In *Proceedings of the 35th Southeastern Symposium System Theory (SSST03).* 16-18 March 2003.

Hauer, J.-H., Handziski, V., & Wolisz, A. (2009). Experimental Study of the Impact of WLAN Interference on IEEE 802.15.4 Body Area Networks. In *Proceedings of the 6th European Conference on Wireless Sensor Networks, Cork, Ireland, 2009* (LNCS 5432, pp: 17 – 32), doi:10.1007/978-3-642-00224-3_2

He, T., Huang, C., Blum, B. M., Stankovic, J. A., & Abdelzaher, T. (2003). Range-free localization schemes for large scale sensor networks. In *9th annual international conference on MobiCom 2003, San Diego, CA, USA 2003* (pp. 81–95), doi:10.1145/938985.938995

Kennedy, J., & Eberhart, R. (1995). Particle Swarm Optimization. In *Proceedings of IEEE International Conference on Neural Networks. IV* (pp. 1942–1948).

Kennedy, J., & Eberhart, R. (2001). *Swarm Intelligence.* Morgan Kaufmann.

Kloos, G., Guivant, J. E., Nebot, E. M., & Masson, F. (2006). Range Based Localisation Using RF and the Application to Mining Safety. In *Proceedings of International Conference on Intelligent Robots and Systems*, October 2006, Beijing, China (pp: 1304-1310), doi:10.1109/IROS.2006.281894

Koneru, A. (2006, December). *Comparative Study of RSS-based Collaborative Localization Methods in Wireless Sensor Networks.* Master's Thesis, University of North Texas.

Kulkarni, R. V. & Venayagamoorthy, G. K. (2010). Particle Swarm Optimization in Wireless Sensor Networks: A Brief Survey. *IEEE Transaction on systems, Man, and Cybernetics - part C: Applications and reviews.* doi: .0.1109/TSMCC.2010.2054080

Langendoen, K., & Reijers, N. (2003). Distributed localization in wireless sensor networks: A quantitative comparison. *Computer Networks, 43*(4), 499–518. doi:10.1016/S1389-1286(03)00356-6

Lee, E. F., Wang, C., & Xiao, L. (2006). *A Study of Radio Signal Behaviors in Complex Environments* (Tech. Rep. MSU-CSE-06-6). Computer Science Department. Michigan State University.

Li, X. (2007, November). Collaborative localization with received-signal strength in wireless sensor networks. *IEEE Transactions on Vehicular Technology, 56*(6), 3807–3817. doi:10.1109/TVT.2007.904535

Lorincz, K., & Welsh, M.,(2006). MoteTrack: A Robust, Decentralized Approach to RF-Based Location Tracking. In *Proceedings of Personal and Ubiquitous Computing.*

Lymberopoulos, D., Lindsey, Q., & Savvides, A. (2006). *An Empirical Analysis of Radio Signal Strength Variability in IEEE 802.15.4 Networks Using Monopole Antennas.* (ENALAB Tech. Rep. 050501).

Ma, J., Chen, Q., Zhang, D., & Ni, L. M. (2006). *An Empirical Study of Radio Signal Strength in Sensor Networks Using MICA2 Nodes* (HKUST, Tech. Rep.).

Moteiv Corporation Inc (2006). *Ultra low power IEEE 802.15.4 compliant wireless sensor module* (Tmote Sky datasheet).

Munoz, D., Lara, F., Vargas, C., & Enriquez-Caldera, R. (2009). *Position Location Techniques and Applications*. Academic Press.

Murphy, K. (1998). *Kalman filter toolbox for Matlab*. Retrieved from http://www.cs.ubc.ca/~murphyk/Software/Kalman/kalman.html

Niculescu, D., & Nath, B. (2003, January). DV Based Positioning in Ad hoc Networks. *Journal of Telecommunication Systems*, *22*(1), 267–280. doi:10.1023/A:1023403323460

Otero, C. E., Kostanic, I., & Otero, L. D. (2009). Development of a simulator for stochastic deployment of wireless sensor networks. *Journal of Networks*, *4*(8), 754–762. doi:10.4304/jnw.4.8.754-762

Ouyang, R. W., Wong, A. K. S., & Lea, C.-T. (2010, March). Received signal strength based wireless localization via semidefinite programming: Noncooperative and cooperative schemes. *IEEE Transactions on Vehicular Technology*, *59*(3), 1307–1318. doi:10.1109/TVT.2010.2040096

Reichenbach, F., Blumenthal, J., & Timmermann, D. (2006). Improved Precision of Coarse Grained Localization in Wireless Sensor Networks. *9th Conference on DSD 2006, Dubrovnik, Croatia, 2006* (pp. 630-637), doi:10.1109/DSD.2006.61

Savarese, C., Rabaey, J., & Langendoen, K. (2002). Robust positioning algorithms for distributed ad-hoc wireless sensor networks. In *USENIX technical annual conference, vol. 2. Monterey, CA, 2002.*

Savvides, A., Han, C., & Srivastava, M. B. (2001). Dynamic fine-grained localization in ad-hoc networks of sensors. In *7th annual international conference on MobiCom 2001, Rome, Italy* (pp. 166–179), doi:10.1145/381677.381693

Savvides, A., Park, H., & Srivastava, M. (2002). The Bits and Flops of the N-Hop Multilateration Primitive for Node Localization Problems. *First ACM International Workshop on Wireless Sensor Networks and Application*, Atlanta, GA, September 2002. doi:10.1145/570738.570755

Shakkottai, S., Srikant, R., & Shroff, N. (2005). Unreliable sensor grids: Coverage, connectivity and diameter. *Ad Hoc Networks*, *3*(6), 702–716. doi:10.1016/j.adhoc.2004.02.001

Sichitiu, M. L., Ramadurai, V., & Peddabachagari, P. (2003). Simple algorithm for outdoor localization of wireless sensor networks with inaccurate range measurements. *International Conference on Wireless Networks*, 2003 (pp. 300-305).

Srinivasan, K., & Levis, P. (2006). RSSI is Under Appreciated. *Third Workshop on EmNets, 2006.*

Stoep, J. V. (2009). *Design and Implementation of Reliable Localization Algorithms using Received Signal Strength.* Ms Thesis report, University of Washington, 2009.

Stoleru, R., & Stankovic, J. A. (2004). Probability Grid: A Location Estimation Scheme for Wireless Sensor Networks. *IEEE International Conference on Sensor and Ad-Hoc Communications and Networks (SECON)*, Santa Clara, CA, 2004. doi:10.1109/SAHCN.2004.1381945

Stoyanova, T., Kerasiotis, F., Efstathiou, K., & Papadopoulos, G. (2010). *Modeling of the RSS Uncertainty for RSS-Based Outdoor Localization and Tracking Applications in Wireless Sensor Networks. SENSORCOMM 2010, July, 2010* (pp. 45–50). Italy: Venice/Mestre.

Stoyanova, T., Kerasiotis, F., Prayati, A., & Papadopoulos, G. (2009). A Practical RF Propagation Model for Wireless Network Sensors. SENSORCOMM 2009, Athens-Vouliagmeni, Greece (pp. 194-199).

Stoyanova, T., Kerasiotis, F., Prayati, A., & Papadopoulos, G. (2009, December). Evaluation of Impact Factors on RSS Accuracy for Localization and Tracking Applications in Sensor Networks. *Telecommunications Systems Journal*, *42*, 235–248.

Stoyanova, Ts., Kerasiotis, K., Antonopoulos, C., & Papadopoulos, G. (2014). RSS-based Localization for Wireless Sensor Networks in Practice. 9th IEEE/IET CSNDSP 2014, Manchester, UK, 23-25/July 2014

Stoyanova, T., Kerasiotis, K., & Papadopoulos, G. (2010). Deployment Considerations for Reliable Communication in Wireless Sensor Networks. *International Journal on Advances in Networks and Services*, *3*(1&2), 151–162.

Suh, C., Joung, J. E., & Ko, Y. B. (2007). New RF Models of TinyOS Simulator for IEEE 802.15.4 Standard. *Proc. of WCNC 2007* (pp: 2238-2242). doi:10.1109/WCNC.2007.418

Tam, V., Cheng, K.-Y., & Lui, K.-S. (2006, July). Using Micro-Genetic lgorithms to Improve Localization in Wireless Sensor Networks. *Journal of Communication*, *1*(4), 1–10.

Vasilescu, G. (2005). *Electronic Noise and Interfering Signals: Principles and Applications*. Springer-Verlag.

Wang, G., & Yang, K. (2011, May). A New Approach to Sensor Node Localization Using RSS Measurements in Wireless Sensor Networks. *IEEE Transactions on Wireless Communications*, *10*(5), 1389–1395. doi:10.1109/TWC.2011.031611.101585

Welch, G., & Bishop, G. (2006). *An Introduction to the Kalman Filter* (Tech. Rep. 95-041), UNC-Chapel Hill. Retrieved July 24, 2006, from: http://www.cs.unc.edu/~welch/kalman/

Whitehouse, K., Karlof, C., & Culler, D. (2007, January). A Practical Evaluation of Radio Signal Strength for Ranging-based Localization. *Mobile Computing and Communications Review*, *11*(1), 41–52. doi:10.1145/1234822.1234829

Xue, F., & Kumar, P. R. (2004). The number of neighbours needed for connectivity of wireless networks. *Wireless Networks*, *10*(2), 169–181. doi:10.1023/B:WINE.0000013081.09837.c0

Yedavalli, K., Krishnamachari, B., Ravula, S., & Srinivasan, B. (2005). Ecolocation: A Technique for RF Based Localization in Wireless Sensor Networks. IPSN 2005, Los Angeles, CA, April 2005.

Zhang, H., & Hou, J. C. (2005, March). Maintaining Sensing Coverage and Connectivity in Large Sensor Networks. *Ad Hoc & Sensor Wireless Networks*, *1*, 89–124.

Zigbee, (2003). *CC2420: 2.4 GHz IEEE 802.15.4 / Zigbee RF Transceiver.* Chipcon AS SmartRF® CC2420 Preliminary Datasheet (rev 1.0), 2003-11-17.

ADDITIONAL READING

Blake, R., & Chartrand, L. (2000). Wireless communication technology. Delmar Cengage Learning, 2000, ISBN: 0-7668-1266-9.

Chang, K. (2000). *RF and Microwave Wireless Systems*. John Wiley & Sons, Inc.

Chuang, P.-J., & Wu, C.-P. (2008). An Effective PSO-based Node Localization Scheme for Wireless Sensor Networks. *Ninth International Conference on Parallel and Distributed Computing, Applications and Technologies 2008* (pp. 187-194). doi:10.1109/PDCAT.2008.73

Clerc, M., & Kennedy, J. (2002).The particle swarm: explosion, stability, and convergence in a multidimensional complex space. IEEE Transactions on Evolutionary Computation, 6, 58–73.

Eberhart, R. C., & Shi, Y. (2000). Comparing inertia weights and constriction factors in particle swarm optimization. In *Proceedings of Congress on Evolutionary Computing 2000* (pp. 84–88). doi:10.1109/CEC.2000.870279

Ekberg, P. (2009, August). Swarm-Intelligent Localization. (Tech. Rep.). Department of Information Technology, Uppsala University.

Gau, Y.-H., Chu, H.-C., & Jan, R.-H. (2007). A weighted multilateration positioning method for wireless sensor networks. *International Journal of Pervasive Computing and Communications*, *3*(3), 289–303. doi:10.1108/17427370710856246

Goldsmith, A. (2005). Wireless communication. Cambridge University Press. doi:10.1017/CBO9780511841224

Hauer, J. H., Handziski, V. & Wolisz, A. (2009). Experimental Study of the Impact of WLAN Interference on IEEE 802.15.4 Body Area Networks. In *Proceedings of the 6th European Conference on Wireless Sensor Networks*, (LNCS 5432, pp. 17 – 32). *Cork, Ireland, 2009*. doi:10.1007/978-3-642-00224-3_2

Haykin, S., & Moher, M. (2005). *Modern Wireless Communications*. Pearson Education Inc., Pearson Prentice Hall.

Kuo, S. P., Tseng, Y.-C., Wu, F. J., & Lin, C.-Y. (2005). A Probabilistic Signal Strength-based Evaluation Methodology for Sensor Network Deployment. In *Proceedings of the 19th International Conference on AINA'05* (pp: 319 – 324). doi:10.1504/IJAHUC.2005.008015

Patwari, N., & Hero, A. O. (2003). Using Proximity and Quantized RSS for Sensor Localization in Wireless Networks. In *WSNA'03* (pp. 20-29) San Diego, CA, USA. doi:.10.1145/941350.941354

Rappaport, T. S. (2001). Wireless Communications: Principles and Practice. Prentice Hall.

Smith, A., Balakrishman, H., Goraczko, M., & Priyanha, N. (2004). Tracking moving devices with cricket location system. In *Proceedings of the 2th ACM MobiSys* (pp. 190-202). Boston, MA, 2004 doi:10.1145/990064.990088

Trelea, I. C. (2003). The particle swarm optimization algorithm: Convergence analysis and parameter selection. *Information Processing Letters*, *85*(6), 317–325.

Vasilescu, G. (2005). *Electronic Noise and Interfering Signals: Principles and Applications*. Springer-Verlag.

Yu-feng, W., Yan, W., & Chao-Yi, M. (2009). Nodes Localization in Sensor Networks based on Vectors and Particle Swarm Optimization. *GEC'09, June 12-14, 2009, Shanghai, China, 2009*, ACM 978-1-60558-326-6/09/06.

KEY TERMS AND DEFINITIONS

Localization Methods: Methods aiming at finding the position of a target object in a system of coordinates.

Modeling: Process of defining a description or representation of something (e.g. system, phenomenon) based on its properties, which can be used for further study of its characteristics.

Node Deployment: Process of deploying the nodes of a network on such positions so that some particular requirements are satisfied (e.g. connectivity).

Received Signal Strength (RSS): The power of a transmitted signal when it is received by a receiver.

RF Signal Propagation Models: Models which describe the behavior of radio-frequency (RF) electromagnetic waves when they are transmitted or propagated from one point to another.

Simulation: Process of imitating the behavior of something (e.g. phenomenon, process) by means of something analogous.

Wireless communication: Data communication which is performed wirelessly, using electromagnetic waves.

Wireless Sensor Networks (WSNs): Network infrastructures typically composed of a large number of nodes, which cooperatively monitor physical and environmental conditions, such as motion, temperature, humidity, sound etc., using dedicated sensors, and the respective information they collect is transferred to one or more central locations, called base stations.

Chapter 11
The Development of Mobile Wireless Sensor Networks:
A Survey

Yuenong Zhu
Lawrence Technological University, USA

Kun Hua
Lawrence Technological University, USA

ABSTRACT

In this chapter, the authors mainly discuss mobile wireless sensor networks (MWSNs). First, the authors are introduce the evolution of MWSNs from sensor networks to wireless sensor networks, and finally to mobile networks. Second, to provide a general context of MWSNs the authors then compare the peer work of MWSNs in chronological order. The third section discusses typical issues including localization, deployment, resource/energy efficiency and coverage issues. Cross-layer design is considered as one of the most useful ways to improve MWSNs in the future.

INTRODUCTION

Nowadays, sensors have been deployed onto many mobile platforms, such as cars, bikes, planes, animals, and even human body. A general scenario of MWSN is shown in Figure 1. And the organization structure of the whole book chapter is shown in Figure 2.

Sensor Networks

A sensor network consists in a group of specialized transducers with a communication infrastructure. These specialized transducers intended to monitor and record conditions at diverse locations. Commonly, monitored parameters are temperature, humidity, pressure, wind direction and speed, illumination intensity, vibration intensity, sound intensity, power-line voltage, chemical concentrations, pollutant levels and vital body functions, etc.

A sensor network consists of multiple sensor nodes. Sensor nodes are kind of detection stations, each of them are very tiny, low-cost, light weight and portable. However, sensor nodes have various energy and computational constraints because of their inexpensive nature and ad hoc method

DOI: 10.4018/978-1-4666-8251-1.ch011

Figure 1. The schematic diagram of MWSNs

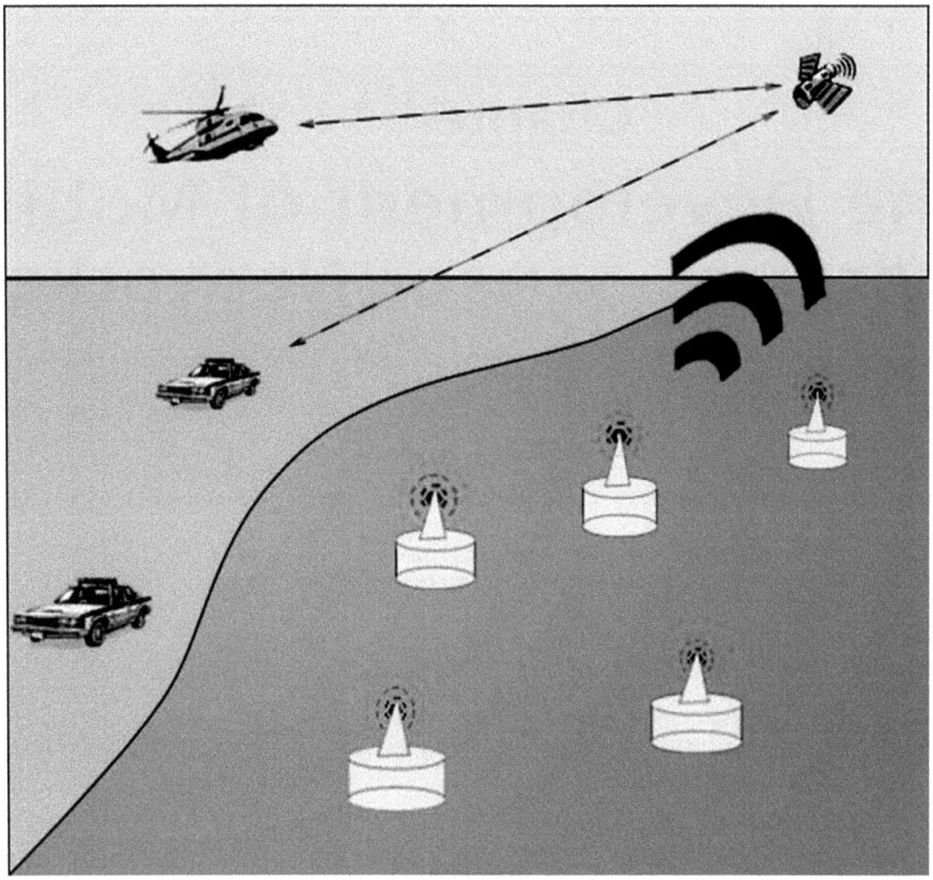

of deployment (Bharathidasan, A., & Ponduru, V.A.S., 2003). Every sensor node is equipped with a transducer, microcomputer, transceiver and power source. The transducer generates electrical signals based on sensed physical effects and phenomena. The microcomputer processes and stores the sensor output. The transceiver, which can be hard-wired or wireless, receives commands from a central computer and transmits data to that computer. The power for each sensor node is derived from the electric utility or from a battery.

Sensor networks are applied in a variety of fields including industrial automation, automated and smart homes, video surveillance, traffic monitoring, medical device monitoring, monitoring of weather conditions, air traffic control, robot control and so on.

Wireless Sensor Networks

Typically, a sensor network node consists of the following several parts: a radio transceiver with an internal antenna or a connection to an external antenna, a microcontroller, an electronic circuit which can work for sensors and an energy source, a battery or an energy harvester.

Sensor nodes have the same composition but different size. The cost of sensor nodes is ranged from a few to hundreds of dollars, depending on the complexity of the individual sensor nodes. Size and cost constraint on sensor nodes result in corresponding constraint on resources, such as energy, memory, computational speed and communications bandwidth. The topology of the WSNs can vary from a simple star network to an

Figure 2. Book chapter organization structure

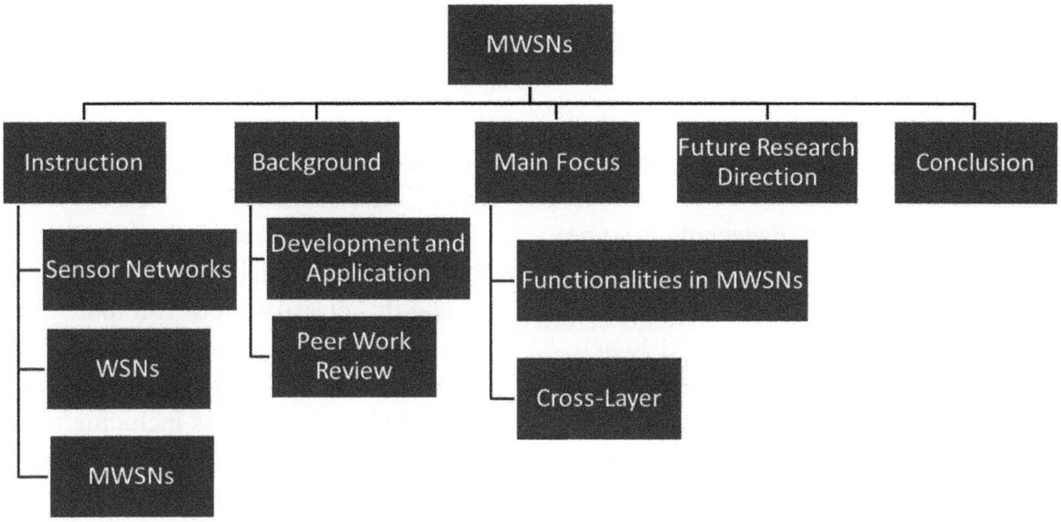

advanced multi-hop wireless mesh network. The propagation technique between the hops of the network can be routing or flooding (Amundson, I., Koutsoukos, X.D., 2009; Sohraby, K., Minoli, D., & Znati, T., 2007).

There are four major differences between MWSNs and WSNs, which are localization, dynamic network topology, power consumption and network sink.

Localization

Mobile nodes must continuously obtain their positions when they are passing away the sensing region, which needs additional time, energy, the availability of a rapid localization service (Al-Jemeli, M., & Hussin, F., 2014). In WSNs, however, locations of static nodes can only be determined during initialization.

Dynamic Network Topology

In dynamic topology, table data and route discovery must update frequently. And it costs in terms of power, time, and bandwidth (Al-Jemeli, M., & Hussin, F., 2014). However, MWSNs can borrow the route of mobile ad hoc networks (MANETs) (Abolhasan, M., Wysocki, T., & Dutkiewicz, E., 2004). Typically, traditional WSN routing protocols rely on routing tables or recent route histories (Akyildiz, I.F., Su, W., Sankarasubramaniam, Y., & Cayirci, E., 2002).

Power Consumption

WSNs can employ collaborative signal enhancement to achieve energy efficiency. Mobile entities need much more additional power for mobility (Ma, T., Hempel, M., Hua, K., Peng, D., & Sharif, H., 2010). Many of them have some energy reservation or they can be charged by being plugged into the power grid to recharge their batteries. Static sensors have a big difference from mobile entities, which use quite a lot of energy and expect energy efficiency (Wang, Q., Hempstead, M., & Yang, W., 2006).

Network Sink

In MWSNs, to have a minimum transmission hops for sensor nodes, mobile base stations hinder the sensing region to collect data or position by themselves (Gandham, S. R., Dawande, M., Prakash, R., & Venkatesan, S., 2003). In WSNs, however, base station can process sensor data by using resource-intensive methods.

The following features make WSNs better than others: First, in many cases, we need to know an unknown environment by using sensors to collect and process information. Especially in some environments, like wilderness and deep sea, it is hard to rearrange nodes. However, people can use redeployment when nodes are mobile (Al-Jemeli, M., & Hussin, F., 2014). Second, mobile nodes can connect the lost or weak communication pathways in networks. There are two conditions. One is these communication pathways may be sparse or disjoint. Another one is stationary nodes die. This is not available with static WSNs in which form dead or disconnected nodes would lost (Al-Jemeli, M., & Hussin, F., 2014). Third, when network sinks are stationary, the nodes closer to the base station will die sooner, since the closed nodes must forward more data messages than the further nodes. By using mobile base stations, this problem is eliminated. At the same time, the lifetime of the network is extended (Gandham, S. R., Dawande, M., Prakash, R., & Venkatesan, S., 2003). Forth, by creating multiple communication pathways and reducing the number of hops, messages must travel before reaching their destination. Thus, mobility characteristic can increase channel capacity and maintain data integrity (Kansal, A., Somasundara, A.A., Jea, D.D., Srivastava, M.B., & Estrin, D., 2004).

WSNs may consist of many different types of sensors, such as seismic, low sampling rate magnetic, thermal, visual, infrared, acoustic and radar (Akyildiz, I. F., Su, W., Sankarasubramaniam, Y., & Cayirci, E., 2002). Different sensors are able to monitor a wide variety of ambient conditions. The different characteristics are temperature, humidity, vehicular movement, lightning condition, pressure, soil makeup, noise levels, the presence or absence of certain kinds of objects, mechanical stress levels on attached objects.

In addition, sensor nodes can be used for continuous sensing, event detection, event ID, location sensing, and local control of actuators. The concept of micro-sensing and wireless connection of these nodes promise many new application areas (Akyildiz, I. F., Su, W., Sankarasubramaniam, Y., & Cayirci, E., 2002). We can use the applications into military, environment, health, home and other commercial areas. It is also possible to expand this classification with more categories, such as space exploration, chemical processing and disaster relief. Furthermore, WSNs will be spread to broader aspects including multimedia wireless sensor networks and ad hoc networks.

Due to the flexible and low cost infrastructure of multimedia wireless sensor networks, wireless Local Area Networks (LANs) are poised to enable a variety of delay-sensitive multimedia, many challenging and interesting topics will be involved in research of mobile multimedia networks, (e.g., energy-efficient multimedia source coding, MIMO techniques for multimedia delivery, cross-layer optimization, secure multimedia streaming, network localization, coverage and deployment, and cooperative transmission for multimedia delivery, etc.). Unlike file transmission, real-time multimedia applications do not require complete insulation from packet loss, but rather that the application layer cooperate with the lower layers to select the optimal wireless transmission strategy that maximizes multimedia performance (Schaar, M.V.D., Shankar N.D.S., 2005).

An ad hoc wireless network is a collection of wireless nodes that is self-configured to form a network without the aid of any established infrastructure. Some or possibly all of these nodes are mobile. These networks are extremely compelling for applications where a communications infrastructure is too expensive to deploy, cannot be deployed quickly, or is simply not feasible. There are numerous potential applications for ad hoc wireless networks, ranging from multi-hop wireless broadband Internet access, to sensor networks, to building or highway automation, to voice, image, and video communication for disaster areas. However, ad hoc networks still need to overcome some challenges, for instance, the lack of established infrastructure, the network and channel dynamics, and the nature of the wireless medium offer an unprecedented (Setton, E., Yoo, T., Zhu, X., Goldsmith, A., & Girod, B., 2005).

Mobile Wireless Sensor Networks

Simply, MWSNs is a kind of Wireless Sensor Networks in which sensor nodes are mobile. Mobility may apply to all nodes or only to subsets of nodes. Commonly, sensor nodes in the WSNs consist of a radio transceiver and a microcontroller powered by batteries. As well known, these sensors detect light, height, temperature, etc. Among these functions, MWSNs also discover neighboring nodes that pass away. As mobile sensors are smaller and much more versatile, they have broader application fields like mobile robots, vehicles, bikes, airplanes, boats, etc. Comparing to WSNs, MWSNs will meet more challenges and serious problems because of its mobile characteristic. Generally speaking, MWSNs have two types of challenges, which are hardware and environment. Hardware constrains battery power and devices price. Environment factors concern the shared medium and varying topology. In this chapter, we will mainly discuss four key issues of MWSNs, which are localization, deployment, resource/energy efficiency and coverage. The remainder of this chapter will unfold through above four key issues. At the same time, we will need to discuss another technology—cross-layer design, which is very important to mobile wireless sensor networks.

BACKGROUND

Development and Application

Over the past decade, we have witnessed the rapid development of WSNs, with advancements in network communication protocols, system design, energy efficiency, and other aspects. With the growth of technology, however, it is far away to meet human needs. We need to create something new to make our life better. Therefore, MWSNs emerged. Unlike WSNs, they work with mobile sensor nodes. So many people developed an interest in it recently. MWSNs consist of one or many remote sinks and many sensor nodes. Each equipped with some actuators, sensing devices, and a wireless transceiver (Al-Jemeli, M., & Hussin, F., 2014). Sensor nodes are mounted onto mobile robots, cars (Eriksson, J., Girod, L., Hull, B., Newton, R., Madden, S., & Balakrishnan, H., 2008), bikes (Eisenman, S.B., Miluzzo, E., Lane, N.D., Peterson, R.A., Ahn, G-S., & Campbell, A. T., 2007), airplanes (Allred, J., Hasan, A.B., Panichsakul, S., Pisano, W., Gray, P., & Huang, J., 2007), boats, even carried by people or animals (Liu, T., Sadler, C. M., Zhang, P., & Martonosi, M. R., 2004) and move to desired areas (Navarro-Serment, L.E., Grabowski, R., Paredis, C.J., & Khosla, P.K., 2002; Bergbreiter, S., & Pister, K., 2003; Sibley, G.T., Rahimi, M.H., & Sukhatme, G.S., 2002; Batalin, M.A., & Sukhatme, G.S., 2004). Such mobile sensor networks are extremely valuable in situations where traditional deployment mechanisms environment, may move with the surrounding air or water currents. Research of robotics, controls, communications, signal processing and networking techniques will be involved in MWSNs. Sometimes, it fail or just not suitable for it, and where sensors cannot be manually deployed or air-dropped. For example, airborne sensors for atmosphere monitoring and underwater sensors for underwater, etc.

Peer Work Review

The developing process of MWSNs will be introduced in chronological order.

2002

In the initial stage of the research in MWSNs, people considered the basic issues, including miniature size, low power, low cost, simple fabrication, and a sensor/actuator suite that facilitates navigation and localization (Sibley, G.T., Rahimi, M.H., & Sukhatme, G.S., 2002). However, Sibley T., Rahimi M.H., and Sukhatme G.S. also focused on developing to explore problems in large-scale distributed robotics and sensor networks, and extending longevity.

It is well known that the coordination of individual with similar or disparate abilities can build up a team with abilities greater than the summation of each part. Navarro-Serment L.E., Grabowski R., Paredis C.J., and Khosla P.K. took advantage of this idea to robots (Navarro-Serment, L.E., Grabowski, R., Paredis, C.J., & Khosla, P.K., 2002). They made a group of robots coordinating as a team to share information and resources. In this robots team, a task will be solved by a team of collaborating robots. Furthermore, they could collect information from multiple locations at the same time.

2003

The paper (Bergbreiter, S., & Pister, K., 2003) named as "CotsBots: An Off-the-Shelf Platform for Distributed Robotics" is a classic example for it. CotsBots are inexpensive and modular robots built entirely from commercial off-the-shelf components. Each robot costs approximately $200 in quantity. And these robots provide a convenient platform. Each is equipped with on-board processing, radio communication, and a base platform for mobility. It means that everyone can get and use it. Thus, the most significant breakthrough in this period is MWSNs into an extremely useful technology for multi-robot research.

2004

Batalin M.A. and Sukhatme G.S. (2004) want to keep mobile robots working, although they are in an unknown dynamic environment. In addition, there is no global information, for example, map and GPS (Global Positioning System).

With continued application, people find nodes have critical resource constraints such as processing speed, memory size, and energy supply in many sensor or wireless systems. Nodes also face special hardware issues, such as sensing device sample time, data storage/access restrictions, and wireless transceiver capabilities (e.g., Liu, T., Sadler, C. M., Zhang, P., & Martonosi, M. R., 2004). So they try to use ZebraNet's system design decisions in the face of a range of real-world constraints.

2007

In this year, MWSNs have been applied from military and professions to daily life. For example, Australian has broader interest in fitness by cycling. Thus, there is substantial interest in the mainstream recreational cycling community in collecting data quantifying various aspects of the cycling experience. Eisenman S. B., Miluzzo E., Lane N. D., Peterson R. A., Ahn G-S., and Campbell A. T. (2007) study BikeNet, which describe a multifaceted sensing system and explores personal, bicycle, and environmental sensing by using dynamically role-assigned bike area networking.

Every country spends a lot of money to build and repair their road ways. But the result is few people are satisfied with the quality of the road where they work and live. Thus, Eriksson J., Girod L., Hull B., Newton R., Madden S., and Balakrishnan H. (2008) propose a pothole patrol of using a mobile sensor network for road surface monitoring. They can use their method to keep roadways in good condition and inform drivers of hazardous road conditions especially at night or when lighting is poor.

Allred J., Hasan A. B., Panichsakul S., Pisano W., Gray P., and Huang J. (2007) implemented SensorFlock airborne WSN, spanned the development of Micro Aerial Vehicle (MAV) airplane, its avionics, semi-autonomous flight control software, launch system, flock control algorithm, and wireless communication networking between MAVs.

2012

Alnawaiseh, A. and Abdelghany, K (2012) proposed TeamSense system for object-tracking applications of autonomous mobile wireless sensor networks. In their design, each sensor node is allowed for peer communication and collaboration to track several objects concurrently while reducing the network's overall energy consumption. Sensor nodes in dynamic, decentralized teams continuously fuse gathered data to enhance the estimation and prediction of the objects' trajectories and optimally reconfigure the formed teams. The optimal reconfiguration of the team is achieved by solving a kind of mixed-integer problem. The TeamSense algorithms are considered to achieve an energy efficient tracking performance by using limited sensing resources.

2014

Recently, people still focus on the development of the MWSNs. And energy efficiency is still a major issue. Al-Jemeli, M. and Hussin, F. (2014) proposed a cross-layer operation model that can improve the energy consumption and system throughput of IEEE 802.15.4MWSNs. They want to use control packet overhead and delivery ratio degradation to increases the network's energy consumption. In contrast to the previous journey, Wang C., Chen W. and Li W. (2014) proposed a Distributed Kalman Filter (DKF) with an average algorithm which reduces the sensor computation load and significantly reduces the data exchange between the adjacent nodes to save energy.

From above mentioned research activities, we can see that the application of MWSNs is wide. And in the future of MWSNs are vibrant and deserves our attention.

MAIN FOCUS OF THE CHAPTER

Functionalities in MWSNs

MWSNs include different functionalities from WSNs like network lifetime extension by resource (e.g. energy) optimization, localization for mobile platforms, coverage area optimization, deployment (placement) and relocation (Lee, J., Dharne, A., & Jayasuria, S., 2007). For these functionalities, positioning is more general than others. Refer to (Marks, M., 2010), they propose a taxonomy for positioning of MWSNs. Figure 3 can illustrate their relationship. Positioning is recognized as a center in taxonomy. It consists of three parts, which are location, deployment and relocation, respectively. The goal of location is to locate where the nodes are placed. The aim of Deployment is to determine where the nodes should be placed. Because of the mobility of MWSNs, we need to reconsider location. And the function of relocation is exactly the same as location. We need to focus on resource optimization because we need to collect sensing data effectively and it is hard to distribute sensors again. For coverage area optimization, we need to think about it. But it is not necessary. And it depends on the applications.

It is normal in MWSNs for sensor nodes to change locations after initial deployment. Mobility applies either all nodes or some part of subsets of nodes. And its mobility could be either passive or active. Active sensors are intelligent enough to move with their path while passive sensors have two situations. One of the problems is that sensors need to be moved to the destination by human or environmental assistance. Another issue is that humans cannot control sensors' movement. Their movement is passive and relies on environment.

We witnessed the evolution of WSNs in past decades. Recently, we gradually turned to MWSNs. Although MWSNs can solve several issues relating to sensor network coverage and connectivity and many challenges remain depend on mobility. Besides above mentioned issues, MWSNs still have additional challenges waiting for people to solve. They are localization, deployment, and energy/resource efficiency and coverage issues. Later, we will introduce these four issues separately.

Localization

MWSN localization can be described as a group of nodes which work together to initiate localization (Al-Jemeli, M., & Hussin, F., 2014). One or more than one nodes emit signals. Then, some signal properties, such as arrival time, phase, and signal strength, are watched by one or many receivers. Node position is then determined by transforming signal measurements into position estimates by means of a localization algorithm. For the purpose of determining position, it is necessary to ask help from cooperating sensor nodes. These sensor nodes have been deployed into the environment at known positions a priori. These devices go by the name of anchor, infrastructure, or seed nodes. The localization will be introduced as Figure 4 and Figure 5.

Problem Statement

MWSNs still have some challenges. One of the challenges is the need of localization. Because mobile sensors must estimate their positions frequently, they will spend a lot of time and energy. They can also consume other resources needed by the sensing application. Furthermore, localization schemes provide high-accuracy positioning information in WSNs. They cannot be employed by mobile sensors, because they require centralized processing typically. They take too long to

Figure 3. The taxonomy of MWSNs

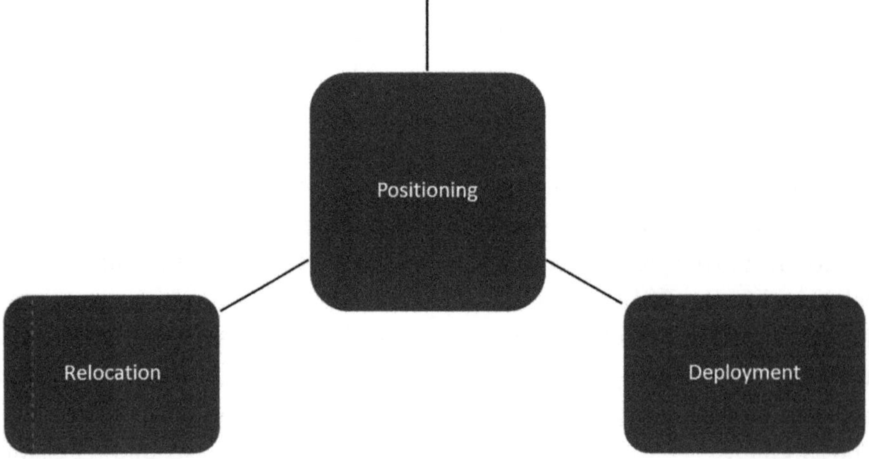

Figure 4. The procedure of introducing localization

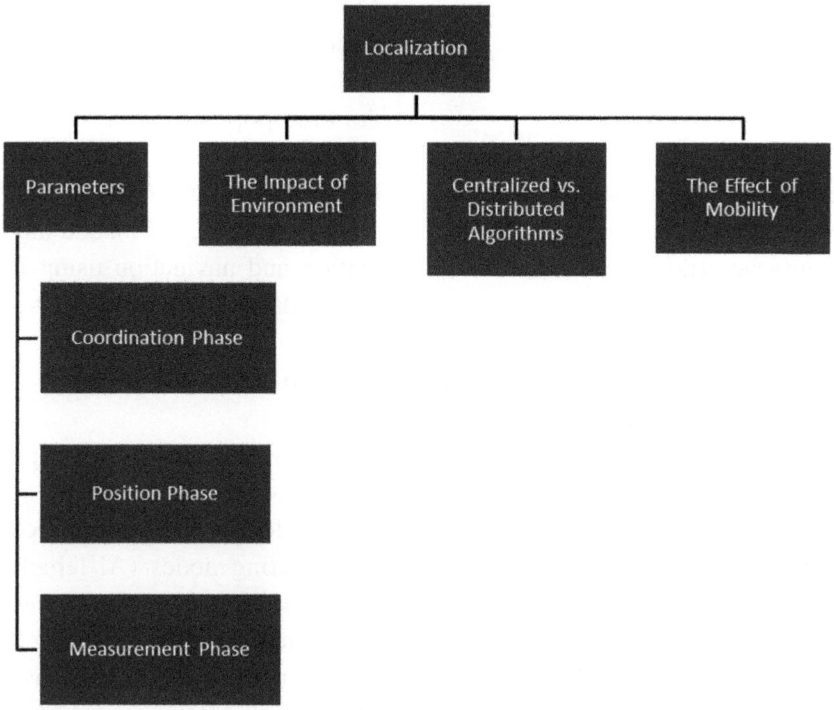

Figure 5. The procedure of introducing parameters

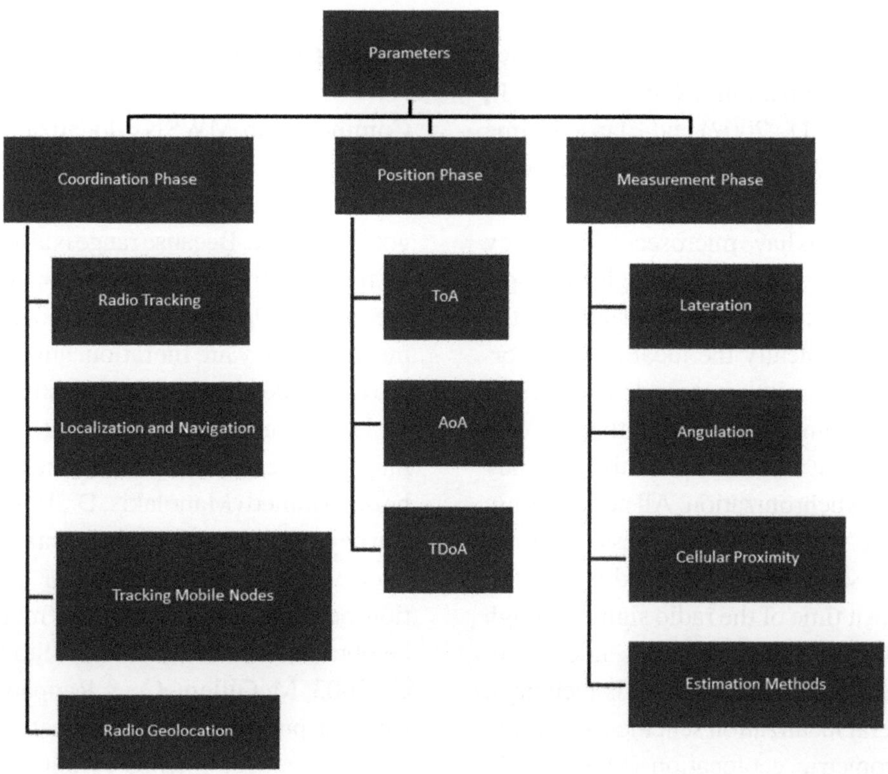

run, or make assumptions about the environment or network topology. It does not apply to dynamic networks.

Solution

Diverse hardware, algorithms and signal modalities are used in localization techniques, which can be categorized along several different dimensions. There are three phases typically used in localization (Brooks, R.R., Griffin, C., & Friedlander, D.S., 2002; Moore, D., Leonard, J., Rus, D., & Teller, S., 2004; Girod, L., Lukac, M., Trifa, V., & Estrin, D., 2006). They are coordination phase, measurement phase, and position phase.

Coordination Phase

Before transmitting signal, nodes work together to participate in the localization (Al-Jemeli, M., & Hussin, F., 2014). Typically, such coordination can include notification that the localization process is about to begin, and clock synchronization, which enables received signal data to be analyzed within a common timeframe. Such as Reference Broadcast Synchronization (RBS) (Elson, J., Girod, L., & Estrin, D., 2002) and Elapsed Time on Arrival (ETA) (Kus´y, B., Dutta, P., Levis, P., Mar´oti, M., L´edeczi, A., & Culler, D., 2006), such coordination methods have microsecond accuracy and require transmission of only a single message. In the message, encoded is the timestamp of the message sender, typically the localization coordinator. Before transmission, encoded is inserted into the message immediately. This is the reason of reducing the amount of non-deterministic latency involved in the synchronization. All nodes within broadcast range will receive the message at approximately the same time instant, and assuming a negligible transit time of the radio signal through air, will be able to transform the sender timestamp into their local timescale. This technique is used in several localization schemes, including radio interferometric geolocation (Mar´oti, M., Kus´y, B., Balogh, G., V¨olgyesi, P., N´adas, A., Moln´ar, K., D´ora, S., & L´edeczi, A., 2005), radio interferometric tracking of mobile wireless nodes (Kus´y, B., Sallai, J., Balogh, G., L´edeczi, A., Protopopescu, V., Tolliver, J., DeNap, F., & Parang, M., 2007), tracking mobile nodes using RF doppler shifts (Kus´y, B., L´edeczi, A., & Koutsoukos, X., 2007), and mobile sensor localization and navigation using RF doppler shifts (Amundson, I., Koutsoukos, X., Sallai, J., 2008).

Measurement Phase

Typically, the measurement phase includes the transmission of a signal by at least one node, which is followed by signal processing on the other participating nodes (Al-Jemeli, M., & Hussin, F., 2014). Based on signal measurement, some techniques exist for obtaining bearing, range, or proximity information, such as Angle-of-Arrival (AoA) method, Time-of-Arrival (ToA) method, Time-Difference-of-Arrval (TDoA) method and others.

Position Phase

Commonly, in MWSNs, localization techniques are based on range (Al-Jemeli, M., & Hussin, F., 2014). Then, distance or angle approximations are got from range. Because range is always influenced by noise, optimization methods are used to filter the noise. Optimization methods include five methods. They are literation, angulation, cellular proximity, dead reckoning and estimation methods.

Literation is used to estimate position, when ranges between landmarks and the mobile node can be determined (Manolakis, D., 1996). Angulation can be used to determine the position of the mobile node when anchor bearings or angular separation between anchors and the mobile node can be obtained (Esteves, J., Carvalho, A., & Couto, C., 2003; McGillem, C., & Rappaport, T., 1989). Cellular proximity is an alternative approach. It is the range-free method (Want, R., Hopper, A.,

Falcao, V., & Gibbons, J., 1992; Shang, Y., Ruml, W., Zhang, Y., & Fromherz, M., 2004; Niculescu, D., Nath, B., 2003), whereby a node is localized to the region in which it is detected. Dead reckoning is a widely used localization technique for mobile robots (Allred, J., Hasan, A.B., Panichsakul, S., Pisano, W., Gray, P., & Huang, J., 2007; Zou, Y., & Chakrabarty, K., 2003; Chen, J., Li, S., & Sun, Y., 2007; Sibley, G.T., Rahimi, M.H., & Sukhatme, G.S., 2002). The major drawback of dead reckoning is that the position estimation accrues error over time, primarily because of noisy encoder data due to uneven surfaces, wheel slippage, dust, and other factors. State estimation methods can be used, when measurement data is noisy, or the system is under defined (Al-Jemeli, M., & Hussin, F., 2014).

Next, we will discuss other aspects of MWSN-based localization, e.g., effects of mobility, centralized vs. distributed processing, and environment.

Effect of Mobility

Localization latency is a main challenge of MWSNs. If the time to complete the localization takes too long, the sensor will significantly change its position since the measurement takes place. For example, in order to derive the proper control outputs for wheel angular velocity, robot navigation requires periodic position estimates. If the robot is moving at 1 meter per second and the localization algorithm takes 5 seconds, the robot may be 5 meters off from its intended position (Al-Jemeli, M., & Hussin, F., 2014).

Centralized vs. Distributed Algorithms

The work of a centralized localization algorithm is based on a base station, and all working nodes must transmit their measurement data to the base station (Al-Jemeli, M., & Hussin, F., 2014). Centralized localization algorithm has a major advantage and a major disadvantage. The advantage of centralized localization algorithm is that this algorithm has more accuracy, precision, and can process greater amounts of data. The disadvantage of centralized localization algorithm is that base station processing suffers from the common pitfalls of centralization, such as poor scalability, single point of failure, data routing complexity, and greater power consumption. The closer nodes to the base station, the more significant disadvantage.

Because of the mobility characteristic, MWSNs need continuous localization. At the same time, they need to update localization frequently. Therefore, system needs a faster approach for certain applications to meet these above requirements. Distributed algorithm is a good idea for that. However, mobile sensors still use centralized algorithms, such as tracking mobile nodes using RF Doppler shifts (Kus´y, B., L´edeczi, A., & Koutsoukos, X., 2007).

The Impact of Environment

Environment can influence localization accuracy greatly. So we cannot ignore environment of the effectiveness of a localization method (Amundson, I., Koutsoukos, X., Sallai, J., 2008). There are many parameters can affect localization accuracy, such as ambient temperature, pressure, humidity and so on, because these directly impact the crystal oscillator in the transceiver. Radio wave propagation is affected by precipitation, including moisture in the air. Therefore, localization technique that uses RF measurements can be impaired under these conditions (Sizun, H., 2004). In addition, many of the existing localization techniques that provide good accuracy outdoors will not work indoors. Thus, in door environment is a kind of independent environment for localization accuracy. Because indoor environment has many obstacles, such as walls, electric equipments, furniture, people, etc. The obstacles will cause signal reflections that interfere with each other. Thus, most propagation methods and measurement techniques will suffer from these multipath effects (Hashemi, H., 1993).

Conclusion of MWSN Localization

After analyzing key features of MWSNs, localization entails entirely new challenges that result from integrating resource (especially energy) constrained wireless sensors on a mobile platform. The localization methods and algorithms could not provide greater accuracy on larger-footprint mobile entities with fewer resource restrictions as before. Similarly, centralized and high-latency localization techniques for traditional static WSNs will no longer be desirable for future MWSNs applications (Al-Jemeli, M., & Hussin, F., 2014).

Future of MWSN Localization

There are several directions for future work in MWSNs localization (Al-Jemeli, M., & Hussin, F., 2014). They are:

1. Reducing localization latency. A tradeoff exists between the rapid execution of an algorithm and its accuracy.
2. Reducing run-time latency, while maintaining positioning accuracy.
3. More distributed localization techniques to replace current centralized localization algorithms, due to the additional latency and energy costs incurred.
4. New methods in localization in urban and indoor areas where obstacles such as vehicles, walls, people, and furniture cause multipath propagation and loss of line of sight.
5. Changes of MWSNs localization by motorized form factors. Advanced mobile sensor navigation is expected in the near future.

Deployment

MWSNs consist of a large number of tiny, inexpensive, mobile and distributed autonomous sensor nodes that are generally deployed in large areas. They have some additional functions to WSNs, such as self-deployment and relocation. In this section, we will discuss deployment of MWSNs. An efficient deployment technique is necessary to handle severely constrained resources, e.g., communication bandwidth and power, storage and computational capabilities, which is closely relevant to network lifespan, robustness and efficiency. An efficient deployment technique is a critical issue which affects the total network cost and MWSN detection capability (Akyildiz, I. F., Su, W., Sankarasubramaniam, Y., & Cayirci, E., 2002).

Mobility is kind of essential for MWSNs self-deployment and relocation. It can be active or passive. (Batalin, M.A., & Sukhatme, G.S., 2004; Lee, J., Dharne, A., & Jayasuria, S., 2007; Marks, M., 2010; Azlina, N., Aziz, B. A., Mohemmed, A.W., & Alias, M.Y., 2009; Li, J., Zhang, B., & Cui, L., 2010; Wu, C., Lee, K. & Chung, Y., 2006; Garetto, M., Gribaudo, M., Chiasserini, C., & Leonardi, E., 2007) that is to find the position and move to deploy themselves. Therefore, this type of mobile sensors is active. They are smart enough to find their path and move to the target area. Another type of mobility is passive. The passive mobile sensors need to be moved by assistance. The assistance can be from human, animals, machines or environment. A typical mobility is introduced in (Luo, J., Wang, D., & Zhang, Q., 2009), which are C-mobility and U-mobility respectively (also called as double mobility). C-mobility is controllable while U-mobility is uncontrollable. The U-Mobility should not only be considered as a curse that the C-Mobility should always countereffect, but also a fortune (Luo, J., Wang, D., & Zhang, Q., 2009). Two mobile sensor platforms Racemote (Song, G., Zhuang, W., Wei, Z., & Song, A., 2006) and (Robomote Dargie, W., & Poellabauer, C., 2010) are applied for MWSNs deployment. Variant approaches have been proposed in previous research work for the deployment of MWSNs for different issues. Different types of deployment patterns have been proposed, like grid, triangle, polygon (Navarro-Serment, L. E., Grabowski, R., Paredis, C. J., & Khosla, P. K., 2002), tillings, and diamond for achieving both coverage and connectivity benefits.

Key Issues

After initial distribution, mobile sensors need to find their positions and place themselves over the destinations. According to different approaches, mobile sensors are required to address different issues, for instance, energy efficient movements of sensors, obstacle adaptability, lifetime of network, fault tolerance etc. From (Akyildiz, I. F., Su, W., Sankarasubramaniam, Y., & Cayirci, E., 2002), we can find that coverage and connectivity are the main issues of deployment. Next, many key issues of mobile sensor network deployment will be introduced, such as localization, connectivity, coverage, communication and sensing range, energy, node density, obstacle adaptability, lifetime, sensor relocation, movement of sensors and fault tolerance.

To solve these problems, we need to find the target area to place the mobile sensors. Because of the mobility, localization is more difficult. There are many methods to find the target area. For the complex area, map of the area is needed to get the position or location information. When the position of sensors is required at each point in time, we can use GPS. In mobile sensor nodes, GPS is so expensive that it can't be used in anywhere. Thus, we must propose different gathering schemes for location information. In localization for mobile sensor networks (Hu, L., & Evans, D., 2004), partially mobile sensor networks (Dyo, V., & Mascolo, C., 2007), coverage algorithms in GPS-less wireless mobile sensor networks (Yu, C. W., Wang, C. H., Hsu, L. C., & Cheng, K. J., 2008) various localization techniques with or without GPS are discussed.

Connectivity is one of the key issues. The topology is very important for sensors. It affects many network characteristics. These network characteristics include latency, robustness, capacity, complexity of data routing and processing (Romer, K. & Mattern, F., 2004). Because the sensor node can wake up joining the network, or go to sleep leaving WSN, topology in WSN is dynamic. If there is a network connection (possibly over multiple hops) between any two nodes, the network is said to be connected (Romer, K. & Mattern, F., 2004).

The coverage area of a sensor node can be defined as the effective range of the sensors attached to a sensor node. The degree of coverage influences information processing algorithms. High coverage is very important to robust systems. We can use high coverage to extend the network lifetime by switching redundant nodes to power-saving sleep modes. Simultaneously, the observation and redundancy required can determine the actual degree of coverage. (Romer, K. & Mattern, F., 2004)

Coverage may be different when it crosses a network. For example, nodes may be deployed more densely at interesting physical locations. There are three kinds of coverage. They are sparse coverage, dense coverage and redundant coverage. Only parts of interesting physical locations are covered with sensor nodes is spares coverage. All the interesting physical locations can be most covered by sensor nodes. This is dense coverage. While for redundant coverage, multiple sensors are deployed in the same physical locations. (Romer, K. & Mattern, F., 2004)

Communication range (R_c) and sensing range (R_s) are very important parameters for MWSNs because R_c and Rs can affect the performance of the network and deployment need to consider the R_c and R_s ratios. However, range can be classified by the type of sensors. There are two types of ranges. One is R_c and another one is (R_s).

$\frac{R_c}{R_s}$ ratio is considered in most of the pattern based approaches. $R_c > R_s$ means we have the problem of coverage and $R_s > R_c$ means we have the problem of connectivity. Both coverage and connectivity should be guaranteed.

It is essential to manage energy resources for MWSNs. At the same time, we can not ignore balancing energy consumption by optimal placement of sensors. Thus, there are four reasons for managing energy resources as following.

1. Sensors need battery for energy because they are autonomous device that they will work without any human assistance.
2. It is infeasible to recharge all battery since WSN consists of thousands of sensors.
3. In MWSNs, mobility itself may consume most of energy available.
4. Managing energy resources can increase the life of network.

Areas with high density increase network cost and computation overhead. Areas with low density may have the problem of coverage holes or network partitions. Node density is based on the node distribution. And node distribution can be classified into two types. They are uniform and non-uniform, respectively. If sensors in the network are uniformly distributed, sensors close to sink are more likely used in data transmission. The node density near sink must be more. However, non-uniform sensor densities using centralized algorithm when sensors can move by flipping (Cardei, M., Yang, Y., & Wu, J., 2008). In other area, the node density near sink would be less called non-uniform density node as discussed in (Cardei, M., Yang, Y., & Wu, J., 2008; Liu, C-H., & Ssu, K-F., 2008).

Obstacle adaptability is one of the very important issues of MWSNs in real time applications. Real-world environments, such as a metropolitan area with buildings, or a rural environment with plants and water etc., those naturally have obstacles or holes. In most of the applications the target field is not always a plane surface, so we should consider the obstacles during deployment. From (Tan, G., Stephen, A., Jarvis, Member, & Kermarrec, A. M., 2009), they proposed two new schemes. The major difference of their schemes from previous work is previous adaptability to arbitrary network densities or communication ranges, and to obstacles. Their first scheme is an enhanced form of the traditional virtual force based method, and latterly a floor-based scheme is considered. (Tan, G., Stephen, A., Jarvis, Member, & Kermarrec, A. M., 2009)

The necessary lifetime has a high impact on the required degree of energy efficiency and robustness of the nodes (Romer, K. & Mattern, F., 2004). Depending on the application, the required lifetime of a sensor network may range from some hours to several years.

Many nodes with communication, sensing and processing abilities have restricted energy and the lifetime of the system is affect by the energy on the nodes (Howard, A., Matari´c, M.J., & Sukhatme, G.S., 2002). There are different ways to increase network lifetime. For instance, incremental deployment (Howard, A., Matari´c, M. J., & Sukhatme, G. S., 2002) is to add new nodes after detecting failed nodes, sensor relocation, load balancing and balancing energy consumption by placing sensors in different densities which varies with the distance of sinks (Akyildiz, I. F., Su, W., Sankarasubramaniam, Y., & Cayirci, E., 2002).

Sensor relocation means that, after detecting the coverage hole, the redundant nodes are relocated to the whole region (Akyildiz, I. F., Su, W., Sankarasubramaniam, Y., & Cayirci, E., 2002). Compared with sensor deployment, sensor relocation has four special difficulties (Wang, G., Cao, G., Porta, T. L. & Zhang, W., 2005):

1. Sensor relocation may have a strict response time requirement. For example, if the sensor monitoring a security-sensitive area dies, another sensor should move to replace it as soon as possible.

2. Relocation should not affect the application currently using the sensor network, which means that the relocation should minimize its effect on the current sensing topology.
3. Since movement may be much more expensive in terms of energy than computation and communication, every algorithm must balance energy costs with response time.
4. Care must be taken to balance the energy costs of an individual node with the overall network energy cost to ensure maximum network lifetime.

Mobility is very necessary to mobile WSNs. It gives the network many new functions, such as locomotion, relocation and coverage optimization property. Because mobility requires power, managing movement of sensors is very important. If one can manage the sensor movement, that is whether a sensor moves or not, so that proper energy allocation could be done and used in other functionalities like sensing and communication. (Akyildiz, I. F., Su, W., Sankarasubramaniam, Y., & Cayirci, E., 2002)

Fault tolerance is the property that enables a system to operate properly in the event of the failure of (or one or more faults within) some of its components. If its operating quality decreases in general, such decrease is proportional to the severity of the failure, as compared to a naïvely-designed system in which even a small failure can cause total breakdown. Fault tolerance is particularly prevalent in high-availability or life-critical systems.

RESOURCE (ENERGY) EFFICIENCY

Disadvantages and Advantages in MWSNs

Since limited number of mobile sensors, poor coverage and heterogeneous distribution of the mobiles, sensing data may not be collected effectively. This is the shortcoming of mobile WSNs. For overcoming this shortcoming, people propose two approaches. Some people want to increase the transmission power of sensors to reach the mobile nodes. However, it will result in more energy and the lifetime of sensors will be shortened. Others want to direct the mobiles to visit specified areas. However, this approach is not practical. Moreover, both of these positions only consider a certain area to improve data collection instead of whole network. Hence, this is an aspect we need to study in the future.

Comparing to the WSNs, mobile WSNs also have many advantages, such as low cost, high throughput, large scale sensing coverage (Fu, Q., Banitalebi, B., Zhang, L., & Beigl, M., 2013). And the most important thing is it provides mobiles. Thus, we prefer using mobile WSNs to using WSNs in specified environment.

Improving Methods

Collision of transmitting signal, nodes process unnecessary information (overhearing), excessive idle listening, improper transition and data redundancy are main reasons for the waste of energy (Peng, C., Guo, J., Pei, L., & Wen, L., 2010). Energy efficiency is a part of effective collection. In this paper, we will discuss how to improve energy efficiency.

Because communication energy efficiency is much more important in Wireless Multimedia Sensor Networks (WMSNs) than traditional multimedia transmission in general wireless environments, the study of communication energy efficiency is based on WMSNs. In WMSNs, communication can ensure the effective data collection by using Cooperative Diversity (CD). The energy efficiency can also be achieved by using Unequal Error Protection (UEP) for cross-layer resource allocations. In addition, network resources (energy) are optimally allocated for differently important levels among sensing data cross PHY, MAC and APP layers regarding inter-segment dependency, and energy efficiency is assured while the data transmission quality is optimized (Wang, W., Peng, D., Wang, H., Sharif, H., & Chen, H. H., 2008). This technology is called cross-layer approach.

The traditional layered approach is separating communication network into seven layers which are physical layer, data link layer, network layer, transport layer, session layer, presentation layer and application layer. Each layer has its special function, and each layer provides service for upper layer. The running order of traditional layered approach doesn't work arbitrarily instead of following some specific rules. However, cross-layer approach transports dynamic feedback via different layers to reduce latency, improve Quality of Service (QoS) and others.

There are three layers for cross-layer energy efficiency approach in MWSNs:

PHY layer is a physical media to transfer actual data. In this layer, we need to consider two problems. One problem is transmitter power consumption. Another problem is extra energy waste, and includes reducing circuit complexity and radio transmitter power. From (Zhou, Y., & Yu, L., 2007), we know FSK is a good modulation. And Direct Sequence-Code Division Multiple Access (DS-CDMA) is a suitable data coding. We can use these two above reduce the collision of multiple access to cut down energy consumption via minimum energy coding algorithm. (Zhou, Y., & Yu, L., 2007)

Desirable BER for BPSK and QPSK modulation schemes can express optimal transmission power value Pt as follows (Wang, W., Peng, D., Wang, H., Sharif, H., & Chen, H.H., 2008):

$$P_t = \frac{N_0 * BW * b}{A}\left[\mathrm{erfc}^{-1}(2e)\right]^2 \qquad (1)$$

$$\begin{cases} N_0 = \text{noise power density} \\ BW = \text{frequnecy bandwidth} \end{cases}$$

where,

b=constellation size

A= channel state information factor

e=desirable BER

Assuming BPSK modulation scheme is used for overhead packet transmission, the transmission data rate of overhead packets can be expressed as:

$$R = BW * b \qquad (2)$$

Desirable BER for M-QAM modulation schemes can express optimal transmission power value Pt as follows (Wang, W., Peng, D., Wang, H., Sharif, H., & Chen, H.H., 2008):

$$P_t = \frac{2N_0 * BW * \left(2^b - 1\right)}{3A}\left[\mathrm{erfc}^{-1}\left(\frac{be}{2\left(1 - \frac{1}{2^{\frac{b}{2}}}\right)}\right)\right]^2 \qquad (3)$$

where, $b \in [2, 4, 6, 8]$.

A low transmission data rate produces a higher robustness leading to a much better power efficiency. A high transmission data rate has an opposite effect. Thus, multirate power versatility can effectively improve energy efficiency.

Because nodes are mobile in the MWSNs, we cannot know the accurate node arrival time. We need to keep a node always awake if it wants to discover all the nodes passing away at any time. However, this idea is not energy efficient at all. Thus, MAC layer optimizations for listening times, which offers some form of optimization of the power consumption; however, not at the level of granularity which could be achieved with patterns (Dyo, V., & Mascolo, C., 2008).

In APP layer, we need to solve two main problems. First, every node must keep awake to detect other nodes because of duty cycling. It is energy inefficient. Second, the topology in these networks changes rapidly because of mobility of MWSNs. These two problems can be solved if nodes are equipped with specialized sensors, such as motion detectors or accelerometers. However, it will lead to some other questions, such as devices' cost, size and installation (Dyo, V., & Mascolo, C., 2008).

If above three layer can work together, we can get the maximum energy efficiency for MWSNs. By the above describes, we can find optimal strategy to get the highest energy efficiency via cross-layer approach.

Coverage Issues

Sensor coverage is measured by the general monitoring area of the MWSNs. Sensor coverage is closely related to the network Quality of Service (QoS), and undesirable sensor deployment and sensor failures will influence the coverage severely. Extremely critical application scenarios, natural limitations and external harsh environments will make the initial deployment obviously far from having the desirable features of full coverage (Zhu, C., Shu, L., Hara, T., Wang, L., & Nishio, S., 2010). The coverage of a mobile sensor networks now depends not only on the initial network configurations, but also on the mobility behavior of the sensors.

Because sensor coverage is one of the basic and significant issues in MWSNs, we need to maintain sensor coverage. As Figure 6 shows, self-deployment and relocation are two major ways to achieve it.

Self-Deployment

Through self-deployment, mobile sensors can relocate themselves for autonomous coverage of critical areas (Tariq, M., Zhou, Z., Park, Y. J, & Sato, T., 2010). In MWSNs, self-deployment includes three methods, which are movement-assisted, potential fields and virtual force.

There are some requirements in movement-assisted. One of the requirements is placing sensors in a target field to maximize the sensing coverage. (Wang, G., Cao, G., & Porta, T. L., 2004) designed three distributed protocols to move mobile sensors from densely deployed areas to sparsely deployed areas in an iterative way based on Voronoi diagrams. The three distributed protocols are The VECtor-based Algorithm (VEC), VORonoi-based Algorithm (VOR) and minimal algorithm.

- **VEC (Wang, G., Cao, G., & Porta, T. L., 2004):** VEC tries to make sensors distributed evenly.

Figure 6. The way to preserve coverage

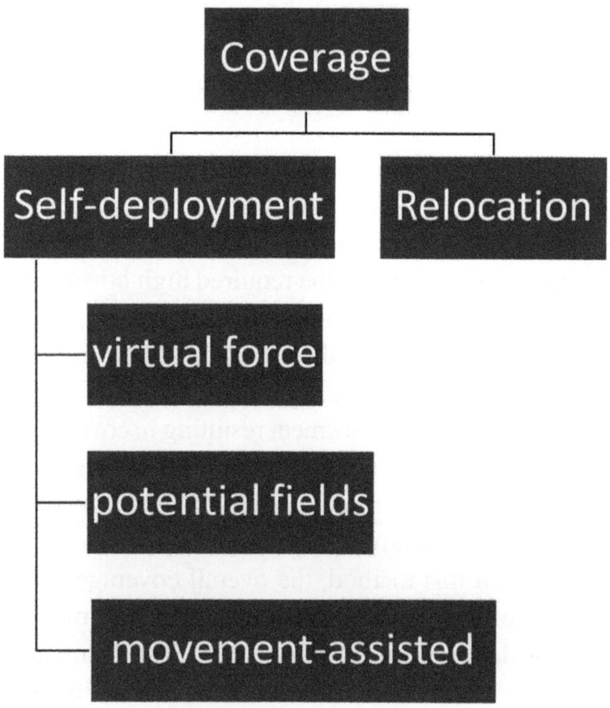

Note: If one sensor is in the Voronoi polygon, force should another sensor move $2 * d_{move}$.

- **VOR (Wang, G., Cao, G., & Porta, T. L., 2004):** VOR is a kind of pull-based algorithm. It pulls sensors to the farthest Voronoi vertex.
- **Minimal Algorithm (Wang, G., Cao, G., & Porta, T. L., 2004):**

$$maxd \in dp\left\{d\left(O_{on}, d\right)\right\} < r$$

$$\begin{cases} d = \text{distance between point} \\ \text{to the Voronoi vertex} \\ \text{C(O,r)=a circle centered} \\ \text{at point O with radius r} \\ \text{O}_{on}\text{=the point on line from O to the} \\ \text{furthest distance between} \\ \text{point to Voronoi vertext} \end{cases}$$

Minimal algorithm uses the above formula to achieve its aim.

Another requirement is that the system can operate well in a severe environment. Therefore, (Wang, G., Cao, G., & LaPorta, T., 2003) concerned to provide the required high coverage in some severe environment. Because in such environments, manually deploying sensors is not possible. In addition, alternative methods may lead to imprecise placement resulting in coverage holes. Thus, (Wang, G., Cao, G., & LaPorta, T., 2003) proposed deploying sensor networks composed of a mixture of mobile and static sensors. Based on this method, the overall coverage can be improved through mobile sensors can move from dense areas to sparse areas. The same as (Wang, G., Cao, G., & Porta, T. L., 2004), in the protocol, static sensors detect coverage holes locally by using Voronoi diagrams.

Potential fields are commonly used in mobile robotics. In this aspect, they are typically applied to tasks, such as local navigation, obstacle avoidance and so on. They apply potential fields to the deployment problem. The fields are constructed in such a way that each node is repelled by both obstacles and by other nodes, thereby forcing the network to spread itself throughout the environment (Howard, A., Matari´c, M.J., & Sukhatme, G.S., 2002).

Virtual Force Algorithm (VFA) is an important method to solve the area of coverage issue. Zou, Y., & Chakrabarty, K., (2003) proposed a sensor deployment. They use Virtual Force Algorithm to enhance the coverage after an initial random placement of sensors. Because a large extent on the coverage provided by the sensor deployment can affect the effectiveness of cluster-based distributed sensor networks, the VFA algorithm attempts to maximize the sensor field coverage for a number of sensors. Virtual motion paths and the rate of movement for the randomly-placed sensors can be determined by the combination of attractive and repulsive forces. A one-time movement with energy consideration incorporated is carried out when the effective sensor positions are decided. At the same time, Zou, Y., & Chakrabarty, K., (2003) proposes a novel probabilistic target localization algorithm that is executed by the cluster head.

Zou, Y., & Chakrabarty, K., (2003) used VFA to solve coverage issue. Next, Chen, J., Li, S., & Sun, Y., (2007) introduced methods to improve VFA. After improvement, convergence, the boundary in Region of Interest (RoI), effective distance of acting force, useless moving and expression of exponential function for the relationship of virtual force will be better.

Relocation

Wang, G., Cao, G., Porta, T.L. & Zhang, W., (2005) proposed a two-phase sensor relocation solution, which can be used to deal with sensor failure or response to new events. To effectively relocate sensors and minimize the effect on the application, they proposed a Grid-Quorum solution to quickly locate the closest redundant sensor with low message complexity. The sensors exchange communication messages first, and ask all relevant sensors to physically move at the same time. And they propose to use cascaded movement to relocate the redundant sensor in a timely, efficient and balanced way. It means that the cascaded movement solution can significantly reduce the relocation time. Further, a distributed protocol has been proposed to find the best cascading schedule to minimize the difference between the total energy consumption and the minimum remaining power.

Cross-Layer

Cross-layer is becoming an important studying field for wireless communications. Cross-layer design breaks away from traditional network design where each layer of the protocol stack operates independently (Setton, E., Yoo, T., Zhu, X., Goldsmith, A., & Girod, B., 2005).

- **Problem Statement:** The traditional layered approach brings three main problems to us in MWSN.
    - It cannot share different information among different layers.
    - It means that traditional layered approach leads to each layer not having complete information. The traditional layered approach cannot guarantee the optimization of the entire network.
    - The traditional layered approach does not have the ability to adapt to the environmental changes.
    - Because of the interference between different users, access confliction, fading, and the change of environment in the multimedia wireless sensor networks, traditional layered approach for wired networks is not applicable to wireless networks.
- **Solution:** To solve the abovementioned problems in the cross-layer, each layer is not designed separately instead of jointly with other layers (Kun, H., Zhu, Y., Wang, W., & Wang, H., 2014). It allows multi-jointly adjust of key parameters to achieve the general goals. Thus, it will result in the higher efficiency of resource usage, which is expected to improve the transmission performance, such as data rate, energy efficiency, QoS (Quality of Service) and so on.

Although Open Systems Interconnection model (OSI) has seven layers, we don't use all of their seven layers in the cross-layer. We always use application layer, transport layer, network layer, link layer and physical layer. We need to pay attention to some specialized parameters in some specialized layers. Based on this cooperation, we can gain the optimal scheme.

For example, PHY and MAC can cooperate to gain the optimal capacity based on Shannon formula as follows:

In the multimedia WSNs, the information of bearing signal $d(t)$ is given as

$$d(t) = \sum_{i=-\infty}^{\infty} d_i P_{T_b}(t - iT_b) \quad (4)$$

where T_b is bit duration, and $P_{T_b}(t)$ is rectangular signal.

$$P_{T_b}(t) = \begin{cases} 1, 0 \le t \le T_b \\ 0, otherwise \end{cases} \quad (5)$$

Spreading sequence ($c(t)$) is given as

$$c(t) = \sum_{j=-\infty}^{\infty} c_j PT_c(t - jT_c) \quad (6)$$

Where T_c is chip duration, C_j is spreading sequence with values of ±1 and $P_{T_c}(t)$ is rectangular signal.

$$P_{T_c}(t) = \begin{cases} 1, 0 \le t \le T_c \\ 0, otherwise \end{cases} \quad (7)$$

$$r(t) = s(t) + n(t)$$

Assuming that four sensors are picked to transmit signals simultaneously.

Thereby,

$$y_1(t) = f_1 * A_1 * c(t) + n_1$$

$$y_2(t) = f_2 * A_2 * c(t) + n_2 \quad (8)$$

$$y_3(t) = f_3 * A_3 * c(t) + n_3$$

$$y_4(t) = f_4 * A_4 * c(t) + n_4$$

In general,

$$y_k(t) = f_k * A_k * c(t) + n_k k = 1,2,3,4$$

where n_k is the AWGN (Additive White Gaussian Noise) noise at receiver, f_k is the fading coefficient at channels (we assume that the same noise power at both channel and receiver) and A_k is amplification factors to maintain constant average output power of the channels.

$$A_k = \sqrt{\frac{\frac{E_b}{N_0}}{f_k^2 \frac{E_b}{N_0} + 1}} \quad (9)$$

$$SINR(T_s, R_s, M_s) = \sum_{k=1}^{M=4} \frac{A_k f_k}{N_k B + \sum_{t_x \neq t_j} A_k f_k} \quad (10)$$

$$BER = \frac{c(M)(k+1)^{M+1}}{M_S^{N+2}} * \frac{A_k f_k}{N_0} * \prod_{m=1}^{M} \frac{A_m f_m}{N_0} \quad (11)$$

Where K denotes the factor in non-central Chi-squared distribution, and $K = 0$ for exponential distribution, constant M_s represents the modulation type, e.g. $M_s = 2$ for BPSK (Binary Phase Shift Keying).

Kun, H., Zhu, Y., Wang, W., and Wang, H. (2014) simulated the result as Figure 7 and Figure 8. From these two figures, we can observe that throughput performance is strongly impacted by modulation types N and transmission set selection m. The highest throughput

$$\left\lceil \frac{T}{t} \right\rceil$$

is at modulation N is the highest value. This result can support the theoretical results. It means that the joint-optimization of transmitter number, receiver number, modulation type and channel state has the most significant effect on the throughput enhancement.

- **Application:** Nowadays, because of some prominent aspects of low power, low cost,

Figure 7. Throughput Performance with various transmitters/sink numbers and modulations

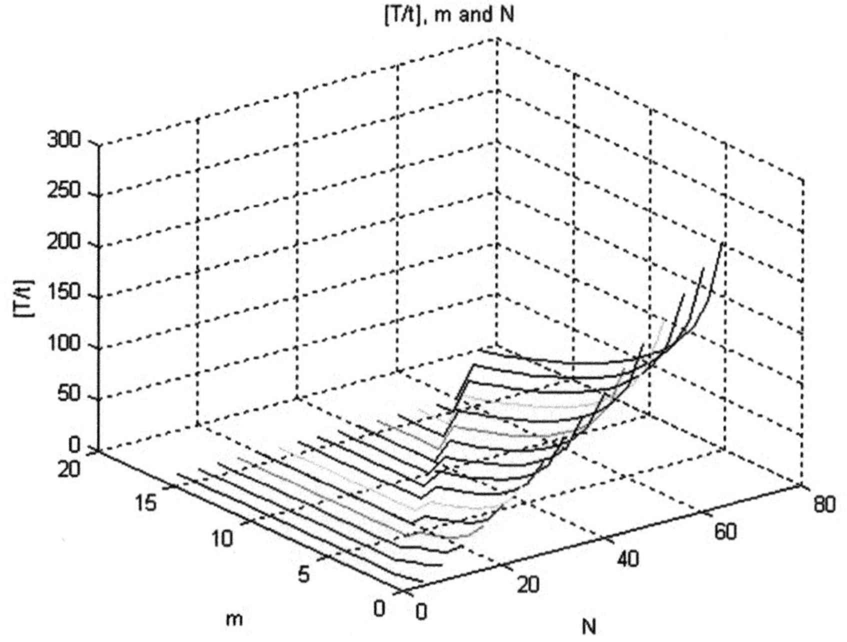

Figure 8. Throughput under various Modulation types and transmission set selection

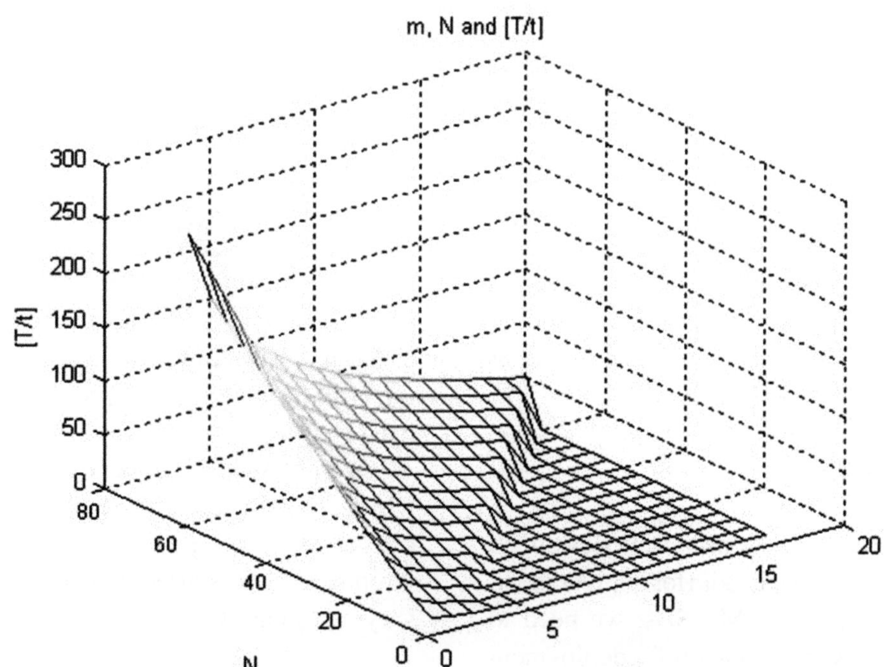

distributed and self-organization, wireless sensor networks are used in various fields, including video surveillance, traffic monitoring, air traffic control, robotics, automotive, home health monitoring, and industrial automation. However, WSNs have the same problem as the traditional layered approach. The wireless channel is easily affected by various elements, such as multipath fading, interferences, mobility issues and others (Su, W., & Lim, T. L., 2006).

It can be used for many applications. From other research, we find that there is a strong need to gain an effective data rate by using cross-layer design in the multimedia wireless sensor networks. There are many challenges for cross-layer design, such as how to choose the transmitters and receivers, how to select the optimal modulation, how to figure out the channel environment problems and so on.

In addition, professions explore the potential synergies of exchanging information between different layers to support real-time video streaming. In the approaches, information is exchanged between different layers of the protocol stack, and end-to-end performance is optimized by adapting to these information at each protocol layer (Setton, E., Yoo, T., Zhu, X., Goldsmith, A., & Girod, B., 2005).They have discussed key parameters being used in the cross-layer information exchange along with the associated cross-layer adaptation. Substantial performance gains through this cross-layer design are demonstrated for video streaming (Setton, E., Yoo, T., Zhu, X., Goldsmith, A., & Girod, B., 2005).

Future Trends

Although current MWSNs have been developed successfully, there is much room for improvement. For the future trends on MWSNs, we need to consider the real world problems for deployment (Akyildiz, I.F., Su, W., Sankarasubramaniam, Y., & Cayirci, E., 2002). The soft computing approaches and bio inspired algorithm should also be studied in the future work (Akyildiz, I.F., Su, W., Sankarasubramaniam, Y., & Cayirci, E., 2002). A higher degree of coverage/connectivity in dense networks will also be considered (Tan, G., Stephen, A., Jarvis, Member, & Kermarrec, A. M., 2009). Reducing the delay in sensor random deployment, and designing fault tolerance strategies by providing sensor redundancy in the unstable sensor networks environments (Tan, G., Stephen, A., Jarvis, Member, & Kermarrec, A. M., 2009) are problems for us, either. By using the existing research, we can improve some researches; for instance, extend existing schemes (Tan, G., Stephen, A., Jarvis, Member, & Kermarrec, A. M., 2009), line-of-sight relationship (Howard, A., Matari´c, M.J., & Sukhatme, G.S., 2002) and increase lifetime of the entire network (Ma, T., Hempel, M., Hua, K., Peng, D., & Sharif, H., 2010).

CONCLUSION

Mobile Wireless Sensor network is a kind of wireless sensor networks in which sensor nodes are mobile. In this chapter, we have introduced the development of MWSNs from original wireless sensor networks. We also compared the peer work about MWSNs bi using chronological order. Various MWSNs functions and issues have been discussed, including localization, deployment, and energy/resource efficiency and coverage issues. Cross-layer design is specially considered as an important solution to improve MWSN performance in the future.

Overall, MWSNs have two major challenges, which are hardware and environment. Hardware constrains battery power and devices price. Environment factors affect the shared medium and varying topology.

REFERENCES

Abolhasan, M., Wysocki, T., & Dutkiewicz, E. (2004). A Review of Routing Protocols for Mobile Ad Hoc Networks. *Ad Hoc Networks*, *2*(1), 1–22. doi:10.1016/S1570-8705(03)00043-X

Akyildiz, I. F., Su, W., Sankarasubramaniam, Y., & Cayirci, E. (2002). Wireless Sensor Networks: A Survey. *Computer Networks*, *38*(4), 393–422. doi:10.1016/S1389-1286(01)00302-4

Al-Jemeli, M., & Hussin, F. (2014). An Energy Efficient Cross-Layer Network Operation Model for IEEE 802.15.4-Based Mobile Wireless Sensor Networks. *IEEE Sensors Journal*, *99*, 1–9.

Allred, J., Hasan, A. B., Panichsakul, S., Pisano, W., Gray, P., & Huang, J. (2007, November). *Sensor Flock: An Airborne Wireless Sensor Network of Micro-Air Vehicles*. Paper Presented at the 5th ACM International Conference on Embedded Networked Sensor Systems. Sydney, Australia.

Alnawaiseh, A., & Abdelghany, K. (2012, August). *Team Sense: Energy-Efficient Autonomous Mobile Wireless Sensor Networks for Object Tracking*. Paper presented at the 8th International Wireless Communications and Mobile Computing Conference (IWCMC), Limassol.

Amundson, I., Koutsoukos, X., & Sallai, J. (2008, September). *Mobile Sensor Localization and Navigation Using RF Doppler Shifts*. Paper Presented at the 14th Annual International Conference on Mobile Computing and Networking. San Francisco, CA. doi:10.1145/1410012.1410034

Amundson, I., & Koutsoukos, X. D. (2009, September). *A Survey on Localization for Mobile Wireless Sensor Networks*. Paper Presented at the 15th Annual International Conference on Mobile Computing and Networking. Beijing, China.

Azlina, N., Aziz, B. A., Mohemmed, A. W., & Alias, M. Y. (2009, March). *A Wireless Sensor Network Coverage Optimization Algorithm Based on Particle Swarm Optimization and Voronoi Diagram*. Paper Presented at IEEE International Conference on Networking, Sensing and Control. Okayama, Japan.

Batalin, M. A., & Sukhatme, G. S. (2004). Coverage, Exploration and Deployment by a Mobile Robot and Communication Network. *Telecommunication Systems*, *26*(2-4), 181–196. doi:10.1023/B:TELS.0000029038.31947.d1

Bergbreiter, S., & Pister, K. (2003, October). *Cotsbots: An Off-the-Shelf Platform for Distributed Robotics*. Paper Presented at the IEEE/RSJ International Conference on Intelligent Robots and Systems. Las Vegas, Nevada. doi:10.1109/IROS.2003.1248878

Bharathidasan, A., & Ponduru, V. A. S. (2003). *Sensor Networks: An Overview. Unpublished overview*. Davis: University of California.

Brooks, R. R., Griffin, C., & Friedlander, D. S. (2002). Self-Organized Distributed Sensor Network Entity Tracking. *International Journal of High Performance Computing Applications*, *16*(3), 207–219. doi:10.1177/10943420020160030201

Cardei, M., Yang, Y., & Wu, J. (2008, June). *Non-Uniform Sensor Deployment in Mobile Wireless Sensor Networks*. Paper Presented at the International Symposium on a World of Wireless, Mobile and Multimedia Networks. Newport Beach, CA. doi:10.1109/WOWMOM.2008.4594823

Chen, J., Li, S., & Sun, Y. (2007). Novel Deployment Schemes for Mobile Sensor Networks. *Sensors (Basel, Switzerland)*, *7*(11), 2907–2919. doi:10.3390/S7112907

Dargie, W., & Poellabauer, C. (2010). *Fundamentals of Wireless Sensor Networks: Theory and Practice* (1st edition). Wiley.

Dyo, V., & Mascolo, C. (2007, November). *A Node Discovery Service for Partially Mobile Sensor Networks*. Paper Presented at the 2nd International Workshop on Middleware for Sensor Networks. Newport Beach, CA. doi:10.1145/1376860.1376863

Dyo, V., & Mascolo, C. (2008, June). *Efficient Node Discovery in Mobile Wireless Sensor Networks*. Paper Presented at the 4th IEEE International Conference on Distributed Computing in Sensor Systems. Santorini Island, Greece. doi:10.1007/978-3-540-69170-9_33

Eisenman, S. B., Miluzzo, E., Lane, N. D., Peterson, R. A., Ahn, G.-S., & Campbell, A. T. (2007, November). *The BikeNet Mobile Sensing System for Cyclist Experience Mapping*. Paper Presented at the 5th International Conference on Embedded Networked Sensor Systems. New York, NY. doi:10.1145/1322263.1322273

Elson, J., Girod, L., & Estrin, D. (2002, December). *Fine-Grained Network Time Synchronization Using Reference Broadcasts*. Paper Presented at the 5th Symposium on Operating Systems Design and Implementation. Boston, MA. doi:10.1145/1060289.1060304

Eriksson, J., Girod, L., Hull, B., Newton, R., Madden, S., & Balakrishnan, H. (2008, June). *The Pothole Patrol: Using a Mobile Sensor Network for Road Surface Monitoring*. Paper Presented at the 6th International Conference on Mobile Systems, Applications, and Services. Breckenridge, CO. doi:10.1145/1378600.1378605

Esteves, J., Carvalho, A., & Couto, C. (2003, August). *Generalized Geometric Triangulation Algorithm for Mobile Robot Absolute Self-Localization*. Paper Presented at the IEEE International Symposium on Industrial Electronics. doi:10.1109/ISIE.2003.1267272

Fu, Q., Banitalebi, B., Zhang, L., & Beigl, M. (2013, March). *Energy-Efficient Collaborative Data Collection in Mobile Wireless Sensor Networks*. Paper Presented at the 47th Annual Conference on Information Sciences and Systems, Baltimore, MD.

Gandham, S. R., Dawande, M., Prakash, R., & Venkatesan, S. (2003, December). *Energy Efficient Schemes for Wireless Sensor Networks with Multiple Mobile Base Stations*. Paper Presented at the Global Telecommunications Conference, San Francisco, CA. doi:10.1109/GLOCOM.2003.1258265

Garetto, M., Gribaudo, M., Chiasserini, C., & Leonardi, E. (2007, October). *A Distributed Sensor Relocation Scheme for Environmental Control*. Paper Presented at the IEEE International Conference on Mobile Ad Hoc and Sensor Systems, Pisa, Italy.

Girod, L., Lukac, M., Trifa, V., & Estrin, D. (2006, October). *The Design and Implementation of a Self-Calibrating Acoustic Sensing Platform*. Paper Presented at the 4th International Conference on Embedded Networked Sensor Systems, Boulder, CO. doi:10.1145/1182807.1182815

Hashemi, H. (1993, July). The Indoor Radio Propagation Channel. *Institute of Electrical and Electronics Engineers*, *81*(7), 943–968. doi:10.1109/5.231342

Howard, A., Matari'c, M. J., & Sukhatme, G. S. (2002). *Efficient Incremental Sensor Network Deployment Algorithm*. Unpublished dissertation, Federal University, Brazil.

Howard, A., Matari'c, M. J., & Sukhatme, G. S. (2002). Mobile Sensor Network Deployment Using Potential Fields: A Distributed, Scalable Solution to the Area Coverage Problem. In A. Howard (Ed.), *Distributed Autonomous Robotic Systems 5* (pp. 299–308). Springer Japan Publishing. doi:10.1007/978-4-431-65941-9_30

Hu, L., & Evans, D. (2004, September). *Localization for Mobile Sensor Networks*. Paper Presented at the 10th Annual International Conference on Mobile Computing and Networking, Philadelphia, PA.

Kansal, A., Somasundara, A. A., Jea, D. D., Srivastava, M. B., & Estrin, D. (2004, June). *Intelligent Fluid Infrastructure for Embedded Networks*. Paper Presented at the 2nd International Conference on Mobile Systems, Applications, and Services, Boston, MA. doi:10.1145/990064.990080

Kun, H., Zhu, Y., Wang, W., & Wang, H. (2014, October). *Cross-Layer Design of MAC-PHY for Optimal Capacity of Multimedia Wireless Sensor Networks*. Paper Presented at the ACM Research in Adaptive and Convergent Systems (RACS), Towson, MD.

Kusy, B., Dutta, P., Levis, P., Maroti, M., Ledeczi, A., & Culler, D. (2006). Elapsed Time on Arrival: A Simple and Versatile Primitive for Canonical Time Synchronization Services. *International Journal of Ad Hoc and Ubiquitous Computing*, *1*(4), 239–251. doi:10.1504/IJAHUC.2006.010505

Kusy, B., Ledeczi, A., & Koutsoukos, X. (2007, November). *Tracking Mobile Nodes Using RF Doppler Shifts*. Paper presented at the 5th International Conference on Embedded Networked Sensor Systems, Sydney, Australia. doi:10.1145/1322263.1322267

Kusy, B., Sallai, J., Balogh, G., Ledeczi, A., Protopopescu, V., Tolliver, J., . . . Parang, M. (2007, June). *Radio Interferometric Tracking of Mobile Wireless Nodes*. Paper Presented at the 5th International Conference on Mobile Systems, Applications and Services. San Juan, Puerto Rico.

Lee, J., Dharne, A., & Jayasuria, S. (2007, July). *Potential Field Based Hierarchical Structure for Mobile Sensor Network Deployment*. Paper Presented at the American Control Conference. New York, NY. doi:10.1109/ACC.2007.4282565

Li, J., Zhang, B., & Cui, L. (2010). A Restricted Delaunay Tringulation Graph Based Algorithm for Self-Deployment in Mobile Sensor Networks. *Journal of Computer Information Systems*, *6*(10), 3155–3162.

Liu, C.-H., & Ssu, K.-F. (2008, September). *A Moving Algorithm for Non-Uniform Deployment in Mobile Sensor Networks*. Paper Presented at the 6th International Conference on Mobile Systems, Applications, and Services. Breckenridge, CO. doi:10.1145/1506270.1506316

Liu, T., Sadler, C. M., Zhang, P., & Martonosi, M. R. (2004, June). *Implementing Software on Resource-Constrained Mobile Sensors: Experimences with Impala and ZebraNet*. Paper Presented at the 2nd International Conference on Mobile Systems, Applications, and Services. Boston, MA. doi:10.1145/990064.990095

Luo, J., Wang, D., & Zhang, Q. (2009, January). *Double Mobility: Coverage of the Sea Surface with Mobile Sensor Networks*. Paper Presented at ACM SIGMOBILE Mobile Computing and Communications Review. New York, NY. doi:10.1145/1558590.1558601

Ma, T., Hempel, M., Hua, K., Peng, D., & Sharif, H. (2010, October). *A Novel Cooperative Image Transmission Scheme in Wireless Sensor Networks*. Paper Presented at the 35th IEEE Conference on Local Computer Networks. Denver, CO. doi:10.1109/LCN.2010.5735711

Manolakis, D. (1996). Efficient Solution and Performance Analysis of 3-D Position Estimation by Trilateration. *IEEE Transactions on Aerospace and Electronic Systems*, *32*(4), 1239–1248. doi:10.1109/7.543845

Marks, M. (2010). A Survey of multi-objective deployment in wireless sensor networks. *Journal of Telecommunications & Information Technology, 36*(3).

Maroti, M., Kusy, B., Balogh, G., Volgyesi, P., Nadas, A., Molnar, K., ... Ledeczi, A. (2005, November). *Radio Interferometric Geolocation*. Paper presented at the 3rd International Conference on Embedded Networked Sensor Systems, Sydney, Australia. doi:10.1145/1098918.1098920

McGillem, C., & Rappaport, T. (1989). A Beacon Navigation Method for Autonomous Vehicles. *IEEE Transactions on Vehicular Technology, 38*(3), 132–139. doi:10.1109/25.45466

Moore, D., Leonard, J., Rus, D., & Teller, S. (2004, November). *Robust Distributed Network Localization with Noisy Range Measurements.* Paper presented at the 2nd International Conference on Embedded Networked Sensor Systems. Baltimore, MD. doi:10.1145/1031495.1031502

Navarro-Serment, L. E., Grabowski, R., Paredis, C. J., & Khosla, P. K. (2002). Millibots: The Development of a Framework and Algorithms for a Distributed Heterogeneous Robot Team. *IEEE Robotics and Automation Magazine, 9*(4), 31–40. doi:10.1109/MRA.2002.1160069

Niculescu, D., & Nath, B. (2003). Dv Based Positioning in Ad Hoc Networks. *Journal of Telecommunication Systems, 22*(1/4), 267–280. doi:10.1023/A:1023403323460

Peng, C., Guo, J., Pei, L., & Wen, L. (2010). Energy-Efficiency Strategies in Wireless Sensor Networks. *Control Engineering of China, 17*(3).

Romer, K., & Mattern, F. (2004). The Design Space of Wireless Sensor Networks. *IEEE Wireless Communications, 11*(6), 54–61. doi:10.1109/MWC.2004.1368897

Schaar, M. V. D., & Shankar, N. D. S. (2005). Cross-Layer Wireless Multimedia Transmission: Challenges, Principles, and New Paradigms. *IEEE Wireless Communications, 12*(4), 50–58. doi:10.1109/MWC.2005.1497858

Setton, E., Yoo, T., Zhu, X., Goldsmith, A., & Girod, B. (2005). Cross-layer Design of Ad Hoc Networks for Real-time Video Streaming. *IEEE Wireless Communications, 12*(4), 59–65. doi:10.1109/MWC.2005.1497859

Shang, Y., Ruml, W., Zhang, Y., & Fromherz, M. (2004). Localization from Connectivity in Sensor Networks. *IEEE Transactions on Parallel and Distributed Systems, 15*(11), 961–974. doi:10.1109/TPDS.2004.67

Sibley, G. T., Rahimi, M. H., & Sukhatme, G. S. (2002, May). *Robomote: A Tiny Mobile Robot Platform for Large-Scale Sensor Networks.* Paper Presented at the IEEE International Conference on Robotics and Automation. Washington, DC. doi:10.1109/ROBOT.2002.1014697

Sizun, H. (2004). *Radio Wave Propagation for Telecommunication Applications*. Berlin, Heidelberg: Springer.

Sohraby, K., Minoli, D., & Znati, T. (2007). *Wireless Sensor Networks: Technology, Protocols, and Applications*. Wiley-Interscience. doi:10.1002/047011276X

Song, G., Zhuang, W., Wei, Z., & Song, A. (2006, October). *Racemote: A Mobile Node for Wireless Sensor Networks.* Paper presented at the 5th IEEE Conference on Sensors. Daegu.

Su, W., & Lim, T. L. (2006, June). *Cross-Layer Design and Optimization for Wireless Sensor Networks*. Paper presented at the IEEE 7th Software Engineering, Artificial Intelligence, Networking, and Parallel/Distributed Computing. Las Vegas, NV.

Tan, G., Stephen, A., Jarvis, Member, & Kermarrec, A. M. (2009). Diffusion Based self-deployment algorithm for mobile sensor networks deployment schemes for mobile sensor networks. *IEEE Transactions on Mobile Computing*, *8*(6).

Tariq, M., Zhou, Z., Park, Y. J., & Sato, T. (2010, September). *Diffusion Based Self-Deployment Algorithm for Mobile Sensor Networks*. Paper presented at the 72nd IEEE Vehicular Technology Conference, Ottawa, ON. doi:10.1109/VETECF.2010.5594514

Wang, C., Chen, W., & Li, W. (2014, April). *Energy Efficient Distributed Kalman Filter for Wireless Sensor Networks.* Paper presented at the 11th IEEE International Conference on Networking, Sensing and Control, Miami, FL. doi:10.1109/ICNSC.2014.6819618

Wang, G., Cao, G., & LaPorta, T. (2003, November*). A Bidding Protocol for Deploying Mobile Sensors*. Paper presented at the 11th IEEE International Conference on Network Protocols, Atlanta, GA.

Wang, G., Cao, G., & Porta, T. L. (2004). Movement-Assisted Sensor Deployment. *IEEE Transactions on Mobile Computing*, *5*(6), 640–652. doi:10.1109/TMC.2006.80

Wang, G., Cao, G., Porta, T. L., & Zhang, W. (2005, March). *Sensor Relocation in Mobile Sensor Networks.* Paper presented at the 24th Annual Joint Conference of the IEEE Computer and Communications Societies, Miami, FL.

Wang, Q., Hempstead, M., & Yang, W. (2006, September). *A Realistic Power Consumption Model for Wireless Sensor Network Devices.* Paper presented at the 3rd Annual IEEE Communications Society on Sensor and Ad Hoc Communications and Networks. Reston, VA. doi:10.1109/SAHCN.2006.288433

Wang, W., Peng, D., Wang, H., Sharif, H., & Chen, H. H. (2008). Energy-Constrained Distortion Reduction Optimization for Wavelet-Based Coded Image Transmission in Wireless Sensor Networks. *IEEE Transactions on Multimedia*, *10*(6), 1169–1180. doi:10.1109/TMM.2008.2001354

Want, R., Hopper, A., Falcao, V., & Gibbons, J. (1992). The Active Badge Location System. *ACM Transactions on Information Systems*, *10*(1), 91–102. doi:10.1145/128756.128759

Wu, C., Lee, K., & Chung, Y. (2006, July). *A Delaunay Triangulation Based Method for Wireless Sensor Network Deployment.* Paper Presented at the 12th International Conference on Parallel and Distributed Systems. Minneapolis, MN.

Yu, C. W., Wang, C. H., Hsu, L. C., & Cheng, K. J. (2008, September). *Coverage Algorithms in GPS-Less Wireless Mobile Sensor Networks.* Paper Presented at the 5th International Conference on Mobile Technology, Applications, and Systems, Mobility Conference. Yilan, Taiwan. doi:10.1145/1506270.1506368

Zhou, Y., & Yu, L. (2007). The Current Status of Energy-Efficiency for Wireless Sensor Networks. *Network Communications and Security,* 128-130.

Zhu, C., Shu, L., Hara, T., Wang, L., & Nishio, S. (2010, August). *Research Issues on Mobile Sensor Networks.* Paper presented at the 5th International Conference on Software Testing on Communications and Networking. Beijing, China.

Zou, Y., & Chakrabarty, K. (2003, April). *Sensor Deployment and Target Localization Based on Virtual Forces.* Paper presented at the 23nd Annual Joint Conference of the IEEE Computer and Communications. San Francisco, CA. doi:10.1109/INFCOM.2003.1208965

ADDITIONAL READING

Akewar, M. C., & Thakur, N. V. (2012). A Study of Wireless Mobile Sensor Network Deployment. *International Journal of Computer Networks and Wireless Communications*, *2*(4), 533–541.

Allred, J., Hasan, A. B., Panichsakul, S., Pisano, W., Gray, P., & Huang, J. (2007, November). *Sensor Flock: An Airborne Wireless Sensor Network of Micro-Air Vehicles*. Paper presented at the 5th ACM International Conference on Embedded Networked Sensor Systems. Sydney, Australia.

Amundson, I., & Koutsoukos, X. D. (2009, September). *A Survey on Localization for Mobile Wireless Sensor Networks.* Paper presented at the 15th Annual International Conference on Mobile Computing and Networking. Beijing, China.

Batalin, M. A., & Sukhatme, G. S. (2004). Coverage, Exploration and Deployment by a Mobile Robot and Communication Network. *Telecommunication Systems*, *26*(2-4), 181–196. doi:10.1023/B:TELS.0000029038.31947.d1

Choi, S., Choi, Y., & Lee, I. (2006). IEEE 802.11 MAC Level FEC Scheme with Retransmission Combining. *Wireless Communications Transfer: European Review of Labour and Research*, *5*(1), 203–211.

Dargie, W., & Poellabauer, C. (2010). *Fundamentals of Wireless Sensor Networks: Theory and Practice* (1st edition). Wiley.

Foschini, G. J., & Salz, J. (1983). Digital Communication over Fading Radio Channels. *The Bell System Technical Journal*, *62*(2), 429–456. doi:10.1002/j.1538-7305.1983.tb03104.x

Hua, K., Wang, H., & Wang, W. (2012, August). *Alamouti Based Cooperative Wireless Networks for Multiplexing-Diversity Balanced Multimedia Transmission.* Paper presented at the 8th International Wireless Communications and Mobile Computing Conference. Limassol, Cyprus. doi:10.1109/IWCMC.2012.6314307

Hua, K., Wang, W., & Alghamdi, A. (2012, May). *Multiplexing-Diversity Balanced Cooperative Wireless Cellular Networks Based on Alamouti Space Time Code for Multimedia Transmission.* Paper presented at the IEEE Global Communications Conference. Anaheim, CA. doi:10.1109/GLOCOM.2012.6503392

Hua, K., Zhu, G. C., & Vasilakos, A. V. (2013, December). *A Secure and Robust Self-Encoded Spread Spectrum Multiple-Access Approach for Multimedia Communication System.* Paper presented at the IEEE Global Communications Conference. Atlanta, GA.

Kun, H., Zhu, Y., Wang, W., & Wang, H. (2014, October). *Cross-Layer Design of MAC-PHY for Optimal Capacity of Multimedia Wireless Sensor Networks.* Paper presented at the ACM Research in Adaptive and Convergent Systems (RACS). Towson, MD.

Li, J., Zhang, B., & Cui, L. (2010). A Restricted Delaunay Tringulation Graph Based Algorithm for Self-Deployment in Mobile Sensor Networks. *Journal of Computer Information Systems*, *6*(10), 3155–3162.

Liu, Q., Zhou, S., & Giannakis, G. B. (2004). Cross-Layer Combining of Adaptive Modulation and Coding With Truncated ARQ over Wireless Links. *Wireless Communications*, *3*(5), 1756–1754.

Marks, M. (2010). A Survey of Multi-Objective Deployment in Wireless Sensor Networks. *Journal of Telecommunications & Information Technology*, *36*(3).

Quazi, T., Xu, H. J., & Takawira, F. (2009). Quality of Service for Multimedia Traffic Using Cross-Layer Design. *Communications*, *3*(1), 83–90.

Romer, K., & Mattern, F. (2004). The Design Space of Wireless Sensor Networks. *IEEE Wireless Communications*, *11*(6), 54–61. doi:10.1109/MWC.2004.1368897

Schaar, M. V. D., & Shankar, N. D. S. (2005). Cross-Layer Wireless Multimedia Transmission: Challenges, Principles, and New Paradigms. *IEEE Wireless Communications*, *12*(4), 50–58. doi:10.1109/MWC.2005.1497858

Setton, E., Yoo, T., Zhu, X., Goldsmith, A., & Girod, B. (2005). Cross-Layer Design of Ad Hoc Networks for Real-Time Video Streaming. *IEEE Wireless Communications* *12*(4), 59–65. doi:10.1109/MWC.2005.1497859

Su, W., & Lim, T. L. (2006, June). *Cross-Layer Design and Optimization for Wireless Sensor Networks*. Paper presented at the 7th IEEE Software Engineering, Artificial Intelligence, Networking, and Parallel/Distributed Computing. Las Vegas, NV.

Tan, G., & Stephen, A., Jarvis, Member, & Kermarrec, A. M. (2009). Diffusion based self-deployment algorithm for mobile sensor networks deployment schemes for mobile sensor networks. *IEEE Transactions on Mobile Computing*, *8*(6).

Tariq, M., Zhou, Z., Park, Y. J., & Sato, T. (2010, September). *Diffusion Based Self-Deployment Algorithm for Mobile Sensor Networks*. Paper presented at IEEE 72nd Vehicular Technology Conference, Ottawa, ON. doi:10.1109/VETECF.2010.5594514

Wang, G., Cao, G., & Porta, T. L. (2004). Movement-Assisted Sensor Deployment. *IEEE Transactions on Mobile Computing*, *5*(6), 640–652. doi:10.1109/TMC.2006.80

Wang, G., Cao, G., Porta, T. L., & Zhang, W. (2005, March). *Sensor Relocation in Mobile Sensor Networks*. Paper presented at the 24th Annual Joint Conference of the IEEE Computer and Communications Societies, Miami, FL.

Wang, W., Peng, D., Wang, H., Sharif, H., & Chen, H. H. (2008). Energy-Constrained Distortion Reduction Optimization for Wavelet-Based Coded Image Transmission in Wireless Sensor Networks. *IEEE Transactions on Multimedia*, *10*(6), 1169–1180. doi:10.1109/TMM.2008.2001354

Wang, W., Peng, D., Wang, H., Yang, Y., Sharif, H., & Chen, H. (2008, December). *Position Based Unequal Error Protection for Image Transmission with Energy Constraint over Multirate XPD MIMO Sensor Networks*. Paper presented at the meeting of the Global Telecommunications Conference. New Orleans, LA. doi:10.1109/GLOCOM.2008.ECP.906

Zhu, C., Shu, L., Hara, T., Wang, L., & Nishio, S. (2010, August). *Research Issues on Mobile Sensor Networks*. Paper presented at the 5th International Conference on Software Testing on Communications and Networking, Beijing, China.

Zou, Y., & Chakrabarty, K. (2003, April). *Sensor Deployment and Target Localization Based on Virtual Forces*. Paper presented at the 23nd Annual Joint Conference of the IEEE Computer and Communications. San Francisco, CA. doi:10.1109/INFCOM.2003.1208965

KEY TERMS AND DEFINITIONS

Ad Hoc Networks: They are a special kind of wireless mobile networks. They don't have central control node. All the nodes are equality.

Energy Efficiency: The utilization of per unit of energy.

Local Area Networks: A computer group, in which many computers interconnect.

Mobile Wireless Sensor Networks (MWSNs): A kind of Wireless Sensor Networks in which sensor nodes are mobile.

Multimedia: It consists in different medias, such as sound, video, etc.

Sensor Network: A sensor network consists in a group of specialized transducers, which intend to monitor and record conditions at diverse locations, with a communication infrastructure.

Virtual Force Algorithm (VFA): An important method to solve the area of coverage issue.

Wireless Sensor Networks (WSNs): They are composed of a large number of sensors. It can collect process and transmit the information.

Chapter 12
Object Tracking Algorithms in Wireless Sensor Networks

Maryam Sadat Mirsadeghi
Shahid Bahonar University of Kerman, Iran

Ali Mahani
Shahid Bahonar University of Kerman, Iran

ABSTRACT

Mobile target tracking is one of the most important applications of wireless sensor networks (WSNs). But, the use of sensor networks for object tracking faces a number of issues in which the limited energy supply is the most important. So in target tracking problem, using methods to decrease the energy consumption as well as high accuracy and quality of tracking is the main goal. Hence, reducing the number of participant nodes in tracking phase, increasing the sleep duration of noninvolved nodes and decreasing the number of transmitted packets to the sink are the most referred methods. In this chapter the authors introduce the most suitable methods for energy efficient mobile object tracking.

INTRODUCTION

Rapid advances in miniaturization in computing and sensor technologies and advent of low-power short-range radios recently have given rise to strong interest in smart sensor networks (Tyagi & Kumar; Safa El-Hajj, & Zoubian; Tadayon, Khoshroo, Askari, Wang, & Michel, 2013; Kuk Liu, & Liu, 2011).Each sensor node has capability of sensing the environment, processing data and communicating with the base station. All these tasks are done without human involvement and this makes such networks have a lot of applications in both military and civilian fields(Rezaii, 2011; Pathan & Hong, 2008).

Target tracking is one of the most important and useful application of WSNs. In the early dawn of the research area, the main driving force was military applications, such as detecting and tracking enemy vehicles and soldiers and estimating the positions, speeds and directions of incoming aircrafts. However, as the field has developed and matured, the methods have spread to disparate domains, such as bio-medicine, finance, wild animal habit monitoring, civilian scenarios, air traffic control and so forth (Pathan et al., 2008).

DOI: 10.4018/978-1-4666-8251-1.ch012

The foundation of tracking is to recursively estimate the position of considered mobile object in each moment. In fact a target tracking system is often required to ensure continuous monitoring and estimate the trajectory (Daniel Svensson., 2007). In WSN thousands of sensor nodes are deployed and used to find mobile objects position which is a complicated scenario and is accompanied with collaborative works between nodes. They should sense the environment, detect intruding objects which are within the sensing range of them, collect data, process data and send information to the base station. So there should always exist nodes that can detect the target along its trajectory, with low detection delay or high coverage level, and base stations can always receive the latest reports about the target in a timely manner.

Sometimes sensor nodes are to identify and track one or more individual objects, such as people, animals, and vehicles. The individual objects usually have very small size comparing with the large area with sensor network deployed. There are some large objects, such as diffused poison gas, biochemical, and chemical liquid, spreading in very large region with sensor network deployed (Shih, Wang, Chen, & Yang, 2008). Detection and tracking this type of objects usually done different from individual ones and we do not consider them in this chapter.

The use of sensor networks for object tracking faces a number of main issues. These issues contain limited energy supply and communication bandwidth, distributed algorithms and control, and handling the fundamental performance limits of sensor nodes, especially as the size of the network becomes large. (Monowar, Alam, Obaidur Rahman, Hong, & Lee, 2010; Martirosyan, Boukerche, Richard, & Pazzi, 2008.)

Tracking of the mobile objects should solve main problems such as object detection, localization, and prediction. As mentioned above, in the localization problem, excessive sensors may join in detection and tracking for only a few objects. And, if all nodes have to always wake up to detect a mobile target, there are a lot of waste of battery power and channel utilization. Also if each sensor node sends packets directly to the sink, it consumes a lot of energy. Actually, power conservation is one of the most critical issues in target tracking since it would be difficult to replace the battery of the sensor nodes that are once deployed in the network area (Pathan et al., 2008). So we can say that energy consumption is the most important design factor for WSN target tracking systems.

In this chapter we consider energy supply limitation in target tracking sensor networks and try to introduce suitable methods to decrease the energy consumption as well as high accuracy and quality of tracking. Reducing the number of participant nodes in detection and tracking phases, increasing the sleep duration of noninvolved nodes and decreasing number of transmitted packets to the sink are the most referred methods. However these methods have negative effect on the accuracy of tracking, e.g., reducing the number of active nodes causes to miss the trajectory of object and so increase the needed to find the lost object which consumes a lot of energy. As we see, there is direct relationship between accuracy and energy and so we should create a trade off between energy consumption and accuracy of tracking in a target tracking sensor network.

Our proposed solution to solve the above problem and implement the methods is prediction and clustering mechanisms which are used a lot in many target tracking papers and researches. By using prediction mechanism, in each time instance only some nodes that are near to the next predicted location of object wake up and others go to low power modes. Accuracy of prediction and also the clustering method, affecting the selection mechanism, the accuracy in tracking and the energy consumption of system. By clustering in each time instance only some head nodes are responsible of gathering data, processing and sending to the sink which consumes more energy. Main advantages of cluster-based WSN in target tracking that help to have lower energy consumption are:

- Collaborative signal processing,
- Data aggregation,
- Small number of transmission/reception,
- Low consumed energy with long lifetime,
- High scalability and robustness.

And the main parameters whose are considered in clustering phase are as follows:

1. Low control overhead to have less energy consumption during clustering.
2. Good distribution of cluster heads to have high coverage, low number of orphan nodes (nodes that are not covered by any cluster head) and low collision between cluster heads as well as reasonable number of CHs.

Since in each time instant head nodes should sense and detect the object and also orphan nodes transmit directly to the sink and do not participate efficiently in tracking of moving object, so coverage of network has direct effect on accuracy and energy consumption of target tracking system. According to our proposed solution in each time instant, only some nodes with the following characteristics will be selected for tracking and other nodes go to the power saving mode:

- Near to the target,
- Less distance to their cluster head.

Also to increase the robustness of target tracking a low power recovery mechanism is introduced to find lost object.

BACKGROUND

As we mentioned in the previous section, energy consumption is the most important design factor for WSN target tracking systems. Based on this factor many research works about object tracking are focus on finding the optimum point of energy consumption, tracking accuracy and calculation time. So many researchers try to reduce the number of active nodes during tracking process in which nodes active all of their components. The tracking methods are classified into two categories (Figure1):

1. Distributed methods.
2. Centralized methods.

In distributed target tracking methods, the nodes are not in any groups or classification. Each node detects the target, communicates with sink or other nodes independently and sends packets to the base station directly. In such a situation there is a lot of waste of energy due to communication, since communication with the base station consumes more than 80% of network's energy. Therefore centralized methods are proposed in which sensor nodes are grouped in categories such as tree or cluster. In Figure 1 classification of target tracking methods in WSNs is shown. Notice that in both methods we can adopt prediction mechanisms to improve the energy consumption by reducing the number of participant nodes.

DISTRIBUTED TARGET TRACKING METHODS

As we mentioned before, in distributed target tracking methods, nodes are not in any groups or classification. Each node that detects the mobile object sends information to the sink directly which causes a lot of waste of battery power and channel utilization. However by using prediction we can improve the energy efficiency of network. For example in (Yang, Feng, Rozenblit, & Qiao, 2006), authors propose PAM (Prediction And Mesh) method in which the node that detects the object remains active and other nodes go to sleep mode. The active node estimates the current location of object based on sensed data and predicts the next location of it by linear predictor. Then informs the node which is near the predicted location to

Figure 1. Classification of Target tracking methods

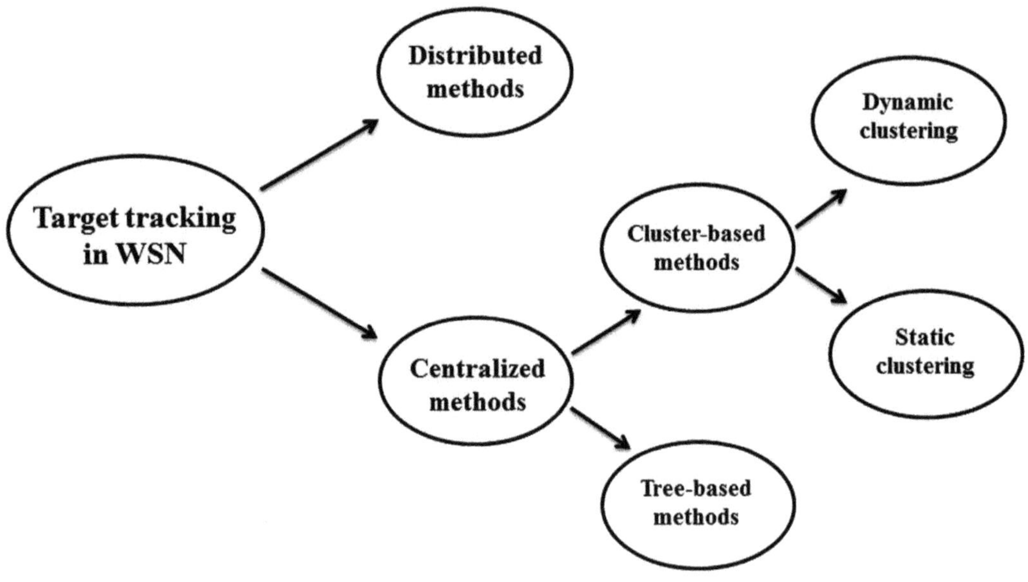

wake up and track the object in the next instant. Although being awake of only one node in each time instance reduces the energy consumption of network, the probability of missing object is high. So here authors propose MESH algorithm as recovery mechanism to find the lost object.

In (Wang, Ma, Wang, & Bi, 2007), method of prediction-based dynamic energy management was proposed. A particle filter was employed to predict next position of object, which was adopted to awaken wireless sensor nodes so that their sleep time was prolonged. Each sensor node can estimate its idle interval and go to sleep itself. And also to select suitable nodes, an optimization approach of distributed genetic algorithm and simulated annealing was proposed.

CENTRALIZED TARGET TRACKING METHODS

In distributed target tracking methods there is a lot of waste of energy due to too much communication. So centralized methods are introduced in which sensor nodes are grouped in categories such as tree or cluster. In such methods only head nodes, roots in tree-based method and cluster heads in clustering-based method, are responsible of gathering data from ordinary nodes (members in clustering and leaves in tree), computing and transmitting packets to the base station which reduces the energy consumption of network. Centralized methods are divided in to two categories:

1. Tree-based target tracking method,
2. Cluster-based target tracking method.

TREE-BASED METHODS

Tree-based tracking methods organize the network as a graph or hierarchical tree. Examples of tree-based methods include DCTC (Dynamic Convoy Tree-based Collaboration), OCO (Optimized Communication and Organization), DOT (Dynamic Object Tracking) and POOT (Prediction-based Optimistic Object Tracking strategy). In such methods in each time instance only a few nodes which are belonged to the tree are active and other nodes can go to low power consumption modes.

For example in Zhang& Cao, (2004), authors propose a dynamic convoy tree-based collaboration (DCTC) framework to detect and track the mobile object. It relies on a tree structure called *convoy tree*, which includes sensor nodes around the moving target. The node which is nearer to the mobile object becomes root and configures its tree, gathers data from its leaves and sends to the base station. And the tree is dynamically configured to add some nodes and prune some nodes as the target moves which cause a lot of power waste.

In Tran, Snoussi, & Richard, (2006), decreasing the energy consumption of tree reconfiguration is the main goal. OCO has four phases: position-collecting, processing, tracking, and maintenance. In the position-collecting phase the base sensor node collecting positions of all the nodes in the network. In the processing phase the base sensor node cleans up the redundant nodes, detects the border nodes, and routs. The tracking phase detects all objects coming from outside of the network area. At the first only border nodes sense the environment. When a border node detects an object, it periodically sends its position information to the base sensor node. When it has lost the object, it sends a message to turn on all its neighbors. If a neighbor detects the object, it will continue sending its position to the base sensor and, right after it has lost the object, it turns all its neighbors on, and so on. If activated neighbors detect nothing, they automatically turn themselves off after a short interval. This way, the objects are tracked as long as they remain within the network area. The maintenance phase is started when the network has a dead node (for example, the node has run out of its energy). In this case, the base sensor node just deletes the dead node from the list, and then reorganizes the network by starting the four-step procedure again.

In Tsai, Chu, & Chen, (2007), a Dynamic Object Tracking is proposed in which the face neighbours are identified by Gabriel graph. First of all source sends request to sensor nodes and the node which is nearer to the target replies back as the beacon node. And to detect moving object continuously, it wakes its spatial neighbours up. Then, source sends query to beacon node (node keeping track information), which reply back target's next location and the source moves towards next beacon node. The process is repeated until the source catches the target. In this method in each time instance the beacon node should wakes all of its spatial neighbours up to track the mobile object continuously. But in (Hsua, Chenb, & Li, 2012), authors propose Prediction-based Optimistic Object Tracking strategy which uses prediction mechanism to reduce the number of active nodes in compared with DOT method. So the energy consumption will be reduced.

CLUSTER-BASED METHODS

Although the tree based approaches track the moving object more accurately, tree organizations result in high energy consumptions. So another low energy consumption network organizing method is proposed. Clustering is the most common technique used for energy aware routing in WSNs. By clustering the network, nodes are classified as either cluster heads or members. In a cluster-based tracking algorithm, the member nodes of the cluster detect the object and send information to the cluster head (CH). The cluster head collects all information from its members; it calculates the location of the target and sends this information to the sink node.

Cluster-based target tracking algorithms are divided into two general groups:

1. Static clustering (Khedr & Osamy; Deldar & Yaghmaee, 2011; Teng, Snoussi, & Richard; Wang et al., 2007; Wang, Lou, Z; Wang, Ma, & Chen, 2010 & 2013).

2. Dynamic clustering (Kuk Liu et al; Sheng, Changfeng, & Zhangsong, 2011).

In dynamic clustering, cluster would be created by a predefined event and then the cluster head is selected. For example in reference (Kuk Liu et al., 2011), an adaptive dynamic clustering scheme based on the number of the sensors in cluster for target tracking was proposed in which the active radius of the cluster was adjusted dynamically. In this method sensor module of all nodes always is ON and after detecting the mobile object, the following conditions are considered in cluster head selection:

1. Distance from the sink.
2. Residual energy.
3. Received signal strength.

Then the cluster head tunes its radio power adaptively and selects its members. So when the target is moving in the monitored area the dynamic cluster is formed around the target. The cluster members put the information to the CH and then the CH tracks the target by a certain filtering algorithm.

In reference (Sheng et al., 2011), the tracking model is the same but authors use prediction mechanism to decrease number of active nodes. The CH predicts next position of object by using particle filter. The closest node to the predicted location is selected as cluster head and its radio power is tuned adaptively based on the number of cluster members. Sensors in (Kuk Liu et al., 2011) and (Sheng et al., 2011) do not statically belong to a cluster, and may support different clusters at different times. Moreover, since only one cluster is active around the target, redundant data is suppressed. Although dynamic clustering offers many benefits and is more favorable for accurate target tracking. But changing the cluster head at the beginning of each tracking process round Consumes more energy and so is not a suitable approach for target tracking purposes.

In static clustering all attributes of each cluster such as the size of a cluster, the coverage area, and the cluster members are static. In some studies (Khedr et al. Deldar et al., 2011. Teng et al., 2007), we have a heterogeneous WSN consisting of two types of wireless devices. So clusters are formed statically at the time of network deployment and doesn't change or update anyway. On the other hand in some studies (Wang, Z et al., 2010. 2013. Wang, X et al., 2007), the WSN is homogenous and the cluster structure change periodically, which is considered in this paper. However in such structures, the clustering period is much larger than tracking period that can reduce the energy consumed for clustering.

The main parameters whose are considered in clustering phase are as follows:

1. Low control overhead to have less energy consumption during clustering.
2. Good distribution of cluster heads to have high coverage, low number of orphan nodes and low collision between cluster heads as well as reasonable number of CHs.

Since orphan nodes transmit directly to the sink and do not participate efficiently in tracking of moving object, so coverage of network has direct effect on missing rate and energy consumption of target tracking system.

Several clustering methods have been proposed so far to achieve such. In (S.H. Lee, S. Lee, Song, & H.S. Lee, 2011), a gradual cluster head election algorithm is proposed that gradually elects cluster heads according to the proximity to neighbor nodes and the residual energy level at each clustering step. Although this method guarantees high network connectivity and high coverage, also the number of cluster heads and their collision in each clustering process round is so high. In (Mahani Askari-Hemmat, & Kavian, 2012), anew duplication avoidance method is proposed. Each node has a random number generator based on logistic map function and by random updating in each clustering process round, decide to be either a cluster head or not without any communication overheads. In (Dahnil, Singh, & Ho, 2012), cluster heads are elected based on their probability and residual energy and authors present a new topology control mechanism in which nodes can control their transmission power level. This technique ensures that any elected cluster head is connected under an optimal degree determined the number of neighbors. In (Tripathy & Chinara, 2012), a staggered clustering protocol is proposed which selects the high energy node with low communication cost as a cluster head and also considers the load balancing. For CH selection authors in (Taheri, Neamatollahi, Younis, Naghibzadeh, & Yaghmaee, 2012) combine fuzzy logic, on demand clustering, non-probabilistic CH election with considering residual energy to reach longer life time.

As we see some methods use prediction mechanism in which the next location of moving object is predicted to decrease the energy consumption of network. Then, during each defined time step, only some nodes near the predicted location are activated and other nodes stay in sleep mode as energy save state.

In most target tracking applications, tracking and prediction mechanisms are based on object motion model. For example (Hsua, Chenb, & Li, 2012; Deldar et al., 2011; Bhuiyan, Wang, Yong, & Zhang, 2010; H. Kim, E. Kim, & Han, 2006; Di, Joo, & Beng, 2008; Yadav, Naik, Ananthasayanam, Gaur, & Singh, 2012; Ribeiro, Schizas, Roumeliotis, & Giannakis, 2010) assume that the target has a linear mobility with constant velocity. In (Hsua et al., 2012) the regression based prediction and Kalman filter are used. In (Deldar et al., 2011; Bhuiyan et al., 2010) the linear predictor is used to predict the next position of object in which the current location and the previous location of object are taking account. Authors in (Kim et al., 2006) assume that the object just sometimes moves non-linear, so they use moving average estimator with a proposed correction mechanism. In (Di et al., 2008. Yadav et al., 2012.Ribeiro et al., 2010), Kalman filter based methods such as Extended-KF and SOI_KF were proposed.

In (Wang et al., 2007; Tripathy et al., 2012) particle filter is proposed to predict the location of the object and Gauss Markove mobility model is considered in (Xu, Winter, & Lee, 2004). Authors propose a simple prediction model in which the object mobility stays constant for a certain period of time and in (Medeiros, Park, & Kak, 2008), authors use extended Kalman filter to track the moving target with nonlinear mobility model.

Preliminaries

A WSN is composed of a large number of stationary sensor nodes each capable of sensing, computing and communication and know their own location by using GPS or other location awareness techniques. Nodes can sense the environment, detect the target and track it. Since transmitting data to the sink consumes more energy, to prolong the network lifetime here we use static clustering based method and also a prediction-based scheme has been designed to reduce the number of activated nodes. Low communication overhead, good distribution of CHs with high coverage, low numbers of orphan nodes and low collision between cluster heads as well as low number of CHs are important metrics in our clustering mechanisms. Orphan nodes are nodes that are not covered by any cluster head and so they should stay active all the time, sense the environment continuously and transmit their data to the sink without any positive impact on tracking task. In the clustering scheme with low coverage, the number of orphan nodes is increased and so the missing rate and energy consumption of target tracking system is increased strongly. Although high coverage reduces the number of orphan nodes but large number of cluster heads increases the collision and the number of direct transmission to the sink and so has negative effect on energy consumption and tracking quality. Also based on the accuracy of the predictor the best cluster is selected to provide more accurate tracking while reducing dramatically the miss rate. So the accuracy of predictor has direct effect on missing rate and energy consumption of tracking system.

None of mentioned clustering methods in previous section satisfy all of the goals at once. Also so far there is no research works about prediction based WSN tracking object with variable velocity. So the previous methods lost the variable speed object during tracking process. It is clear that the missing rate of the object leads to high total energy consumption in the network. In WSN object tracking mechanisms, the energy consumption is dependent on the missing rate, since the nodes must find the lost object via recovery mechanism as a high energy consumption procedure. In an accurate predictor less nodes would be activated for collaborative tracking and leads to negligible consumed energy and transmission overheads.

In this chapter we define a deferent kind of mobility model for moving object with a new prediction based energy efficient and high accurate tracking method in a clustered sensor network. And two new clustering algorithms with low communication overheads are introduced while have good distribution of cluster heads to increase the quality of tracking. This method has capability of tracking more than one object come into considered area simultaneously.

Energy Consumption Model

In this section the mathematical model of consumed energy in target tracking sensor network is introduced. We assume that all nodes work in two modes, sleep mode and active mode. In each time step, just a few numbers of nodes nearby the predicted location of target will be activated as tracker nodes and the others will stay in power saving mode. During sleep mode, nodes just can listen to communication channel to check messages and their power consumption is ignored. For active nodes, all of their components are active and the energy consumption of sensing and communication components is considered.

Each node that wakes up, senses the environment for a certain period of time t_{sense} and consumes energy, E_{sense} then sends its data packets to the cluster head. The energy consumption model for communication is discussed in (Heinzelman, Chandrakasan, & Balakrishnan, 2002), in which the consumed energy of transmitter node is:

$$\begin{aligned} &if \quad d \succ d_0 \quad E_t = lE_{elec} + l\varepsilon_{mp} d^4 + lE_{DA} \\ &if \quad d \preceq d_0 \quad E_t = lE_{elec} + l\varepsilon_{fs} d^2 + lE_{DA} \end{aligned} \tag{1}$$

where E_{DA} is the energy for data aggregation in head nodes, E_{elec} is the dissipated per bit to run the transmitter or receiver circuit, ε_{fs} and ε_{mp} depend on the transmitter amplifier model, d is distance, d_0 is threshold distance and l is the packet length. The consumed energy for receiver nodes E_r is as follows:

$$E_r = lE_{elec} \tag{2}$$

Target Tracking Mechanism

At the initialization step, all nodes are in sleep mode in which just can listen to communication channel to check messages. Once cluster heads receive an alarm message from the sink, wake up and sense the environment. If they cannot find the object, wake their members up to sense the environment and find it. Each cluster head that object is in its cluster, stays active and other nodes go back to sleep mode. For the first tracking round the active cluster head wakes all of its members up to sense the object but for the next periods that in each time step it should find the best sensor nodes for object tracking. The cluster head is responsible of doing necessary computation and finding the best sensor nodes. Then sends a wake up message to them, and if object is going to leave its cluster, it should inform the next cluster head and send a wake up message and necessary current position information of object to that CH. Cluster heads send their information to the sink for the period of T second and clustering is repeated after period of T_0 second in which≫ $T_0 >> T$.

Measurements and Localization Model

It is assumed that each sensor node can produce bearing angle estimates of the target in its sensing range. So the nonlinear measurement model for sensor i∈{1, 2, ..., n} at time instant t is as follow:

$$\theta_i = \arctan(\frac{y_t - s_y^i}{x_t - s_x^i}) + v_t \quad (3)$$

where (S_x^i, S_y^i) is position of the sensor i, (x_t, y_t) is real position of target at time t and v_i is sensing error, which is zero mean, Gaussian distributed with constant standard deviation of σ_θ.

Based on assumption that the node sensing errors are sufficiently small, head node can adopt Least Square algorithm to determine target location, with node sensing collaboratively (Vaghefi et al., 2010). The true value of angle can be written as:

$$\tan\left(\theta_i^0\right) = \left(\frac{\sin\left(\theta_i^0\right)}{\cos\left(\theta_i^0\right)}\right) = \left(\frac{y_t - s_y^i}{x_t - s_x^i}\right) \quad (4)$$

In presence of noise, we can write

$$FX = h \quad (5)$$

where X is location of target and F and h matrices are given by

$$F = \begin{bmatrix} \sin\theta_1 & \cos\theta_1 \\ \vdots & \vdots \\ \sin\theta_n & \cos\theta_n \end{bmatrix},$$

$$h = \begin{bmatrix} S_x^1.\sin\theta_1 - S_y^1.\cos\theta_1 \\ \vdots \\ S_x^n.\sin\theta_n - S_y^n.\cos\theta_n \end{bmatrix},$$

The least square solution of (5) is

$$\hat{X}_{Ls} = (F^T F)^{-1} F^T h \quad (6)$$

Target Motion Model

There are many mobility models for moving objects such as models in (Tsai et al., 2007; Tsai et al., 2007; Deldar et al., 2011; Wang et al., 2007; Xu et al., 2004). In this paper a mobile target is assumed which moves for a random period of time in random direction, speed and acceleration that are in ranges $[0, a_{\max}]$ and $[0, v_{\max}]$.

In fact we use Random way point and Constant Acceleration Model (RCAM) for target motion in which Random way point model uses pause times before changing speed, direction and acceleration. The state equations for the RCAM take into account acceleration component which is time invariant. This model will be essential in the estimation of the vehicle's trajectory moves in the street. The equations of motion are as follows:

$$R(t) = R(t-1) + v(t) * \Delta t \quad (7)$$

$$v\left(t\right) = v\left(t-1\right) + a\left(t\right) * \Delta t \quad (8)$$

$$a\left(t\right) = a\left(t-1\right) \quad (9)$$

where, $R(t), v(t)$ and $a(t)$ are the location, velocity and acceleration of target at time t.

By considering the process noise for target movement, dynamic model of object under tracking can be defined as:

$$X_k = AX_k + Gw_k \quad (10)$$

Where k is time index,

$$X_k = [x(k)\ y(k)\ v_x(k)\ v_y(k)\ a_x(k)\ a_y(k)]^T$$

is the state vector contains relative positions, velocities and accelerations of target in two dimension plan, A is the model state transition matrix, W_k is process noise that is white Gaussian noise with zero mean and variance matrix $Q(k)$ and $G(k)$ is process noise input matrix. The related matrices are given by:

$$A = \begin{bmatrix} 1 & 0 & T & 0 & \frac{T^2}{2} & 0 \\ 0 & 1 & 0 & T & 0 & \frac{T^2}{2} \\ 0 & 0 & 1 & 0 & T & 0 \\ 0 & 0 & 0 & 1 & 0 & T \\ 0 & 0 & 0 & 0 & 1 & 0 \\ 0 & 0 & 0 & 0 & 0 & 1 \end{bmatrix},$$

$$G = \begin{bmatrix} 0 & 0 \\ 0 & 0 \\ 0 & 0 \\ 0 & 0 \\ T & 0 \\ 0 & T \end{bmatrix}, W_k = \begin{bmatrix} w_x \\ w_y \end{bmatrix}$$

where T is tracking period.

Solutions and Recommendations

As we mentioned before, there should be a trade off between energy consumption and accuracy of mobile target tracking, so we introduce suitable methods to decrease the energy consumption as well as high accuracy and quality of tracking which will be dealt here.

Predictor

As we mentioned before, by using prediction mechanism we can reduce the number of active nodes and increase their sleep duration. In a predictor, accuracy and speed are two important factors. The high prediction accuracy leads to lower missing rate of object which reduces the need to recovery mechanism as a high energy consumption procedure. On the other hand, computational capacities of sensor nodes are constraint and so the number of predictor's operation shouldn't be too high. There are different predictors which have been proposed in different papers and here we introduce some of them:

Linear Predictor

This is a simple and rapid method of computing with too low number of operation and only if the object moves linearly, can give us accurate answers. This method is used in many articles such as (Deldar et al., 2011) and (Yang et al., 2006), due to its low computational complexity. The current information of object at time t, its position (x_t, y_t) and velocity (v), are used to predict the next location of moving target after time Δt:

$$x_{t+1} = x_t + v * \Delta t * \cos(\theta) \quad (11)$$

$$y_{t+1} = y_t + v * \Delta t * \sin(\theta) \quad (12)$$

In which Δt is the period of time between t and $t+1$, and θ is the angle of the movement direction. And the velocity v and movement direction θ are obtained as follows:

$$v = \frac{\sqrt{\left(x_t - x_{t-1}\right)^2 + \left(y_t - y_{t-1}\right)^2}}{\Delta t} \quad (13)$$

$$\theta = \tan^{-1}\left(\frac{y_t - y_{t-1}}{x_t - x_{t-1}}\right) \quad (14)$$

Extended Kalman Filter

One other prediction method which has more accurate results is Extended Kalman Filter (EKF) (Barrios, Himberg, Motai, & Sadek, 2006;_Pant & Alkin, 2012) which is used in (Hamouda & Phillips, 2011; Vaidehi, Sandhya, Karthika, & Azhagarasan, 2011; Xiao, Tham, & Das, 2010) as predictor. EKF is improved version of the Kalman filter and can be used in nonlinear system because it linearizes transformation via Taylor Expansion. Like KF, EKF and its variants use observed measurements of sensor nodes containing noise and other inaccuracies to predict next position of target. EKF is used to correct noise and generate better quality prediction. It is a two step process named correct and prediction.

Correct

The equations for correct step are as follows:

$$S = HP_k^- H^T + R \quad (15)$$

$$K_k = \frac{P_K^- H^T}{S} \quad (16)$$

$$\hat{X}_k^- = \hat{X}_k^- + K_k(Z_k - h(\hat{x}_k^-, 0)) \quad (17)$$

$$P_k = (I - K_k H)P_k^- \quad (18)$$

Prediction

Prediction equations are described as follows:

$$\hat{X}_k^- = f(\hat{x}_{k-1}^-, 0) \quad (19)$$

$$P_k^- = AP_{k-1}A^T + WQW^T \quad (20)$$

Definition of variables are as follows:

- A jacobian of system model with respect to state.
- W is jacobian of system model with respect to process noise.
- Q process noise covariance.
- R measurement noise covariance.
- K kalman gain.
- P estimated error covariance.
- H jacobian of measurement model.
- Z measurement data.

The nonlinear system (A) and measurement (H) matrices are linearized according to its corresponding jacobian, which is a first-order partial derivative. However here only measurements are nonlinear, This linearization can sometimes cause instabilities when implementing the EKF.

Unscented Kalman Filter

The UKF is a popular extension of the KF introduced in 1997, and is employed in (Anandhakumar, Priyadarshini, Rajeswari, Srividhya, & Niveditha, 2011) as a predictor. It is typically more accurate than the EKF since it is able to capture a higher order of the nonlinearities and no calculation of the first-order Jacobian is necessary, which can be a challenging task. It doesn't linearize the nonlinear functions but instead propagates statistical information through the nonlinear transform (Julier & Uhlmann, 1997).

If x_k is an N_x dimensional Gaussian random variable with mean $\hat{x}_{k|k}$ and variance $P_{k|k}$, a set of $(2N_x + 1)$ sample points X_k are chosen and by assumption that posterior density of the state is Gaussian, this density is presented using two equations shown in Box 1 (Wood, 2008).

The predicted density and measurement are calculated as follows:

$$X^i_{k+1|k} = f_{k+1}(X^i_k) \tag{23}$$

$$\hat{z}_k = \sum_{i=0}^{2N_x} w^i h_x(X^i_{k+1|k}) \tag{24}$$

Table 1. Parameters for energy consumption model

E_{elec}	50 *nJ/bit*
E_{DA}	5 *nJ/bit/signal*
E_{sense}	106 *μJ/s*
ϵ_{mp}	0.0013 *pJ/bit/m*^{4}
ϵ_{fs}	100 *pJ/bit/m*^{2}
d_0	$100\sqrt{2}$
t_{sense}	4.81 *msec*
Data packet size	4000 *bits*
Signal packet size	64 *bits*

The update step is described by:

$$\hat{X}_{k+1} = \hat{X}_{k+1|k} + K_{k+1}(Z_k - \hat{Z}_k) \tag{25}$$

$$P_{k+1} = P_{k+1|k} - K_{k+1} S_{k+1} K^T_{k+1} \tag{26}$$

where (Box 2)

Although UKF has less number of calculation compared with particle filter and has precise results, but anyway it has a lot of calculations. So here we propose a predictor which has high accuracy and lower computing.

Proposed Predictor (PEP)

The proposed predictor (Mirsadeghi & Mahani, 2014, 2013) is based on prediction error (PE) described by the angle between the actual location and previously predicted, denoted by θ_n. Since it is assumed that moving object has constant acceleration, so after predicting the speed of object in the next time step, we try to find next position of the object based on prediction error. If θ_n is in range $[-\alpha, \alpha]$, the next speed of moving object at time $t+1$ is predicted by:

$$v(t+1) = 2 * v(t) - v(t-1) \tag{29}$$

Table 2. Comparison of speed and missing rate of different predictors

Predictor	Number of Operations	Missing Rate
PEP	28440	0.69%
UKF	702017	0.88%
EKF	107734	3.92%
Linear	14193	4.14%

Box 1.

$$\hat{x}_{k+1|k} = \sum_{i=0}^{2N} w^i f_k(X_k^i) \tag{21}$$

$$P_{k+1|k} = Q_{k+1} + \sum_{i=0}^{2N_x} w^i [f_k(X_k^i) - \hat{X}_{k|k}][f_k(X_k^i) - \hat{X}_{k|k}]^T \tag{22}$$

Box 2.

$$S_{K+1} = R_{K+1} + \sum_{i=0}^{2N_x} w^i [h_{k+1}(X_{k+1|k}^i) - \hat{z}_k][h_{k+1}(X_{k+1|k}^i) - \hat{z}_k]^T \tag{27}$$

$$K_{k+1} = (\sum_{i=0}^{2N_x} w^i [X_{k+1|k}^i - \hat{x}_{k+1|k}][h_{k+1}(X_{k+1|k}^i) - \hat{z}_k]^T) S_{k+1}^{-1} \tag{28}$$

This equation is obtained based on constant acceleration:

$$a(t-1) = a(t) \tag{30}$$

Then the next location of object at time $t+1$ is predicted as follows:

$$x(t+1) = v(t+1) * T * \cos\phi + x(t) \tag{31}$$

$$y(t+1) = v(t+1) * T * \sin\phi + y(t) \tag{32}$$

where

$$\phi = \arctan\left(\frac{y(t+1) - y(t)}{x(t+1) - x(t)}\right) \tag{33}$$

But if θ_n is not in the range $[-\alpha, \alpha]$, ,so we have:

$$v_x(t+1) = 2 * v_x(t) - v_x(t-1) \tag{34}$$

$$v_y(t+1) = 2 * v_y(t) - v_y(t-1) \tag{35}$$

$$x(t+1) = v_x(t+1) * T + x(t) \tag{36}$$

$$y(t+1) = v_y(t+1) * T + y(t) \tag{37}$$

Here α is set by using trial and error method.

Simulation Results

In this section the accuracy of predictors and the effect of their accuracy on missing rate and energy consumption are considered. So the linear predictor (Deldar et al., 2011), modified version of Kalman filter named EKF which introduced in (Pant et al., 2012)and another modified version of Kalman filter named UKF (Anandhakumar et al., 2011)are considered to have a fair comparison. Finally the impact of node selection algorithm on network lifetime is investigated.

Simulation Environment

The network area is $400 * 400$ which is divided into 16 equal size cluster and so 16 head nodes with initial energy of 50 jules, are placed at the center of these clusters. The sink node is located out of the area

$(x = 200, y = 450)$.

300 sensor nodes are uniformly distributed in the network area with sensing range $R_{sensing} = 60m$ and initial energy of 0.5 jules. For each sensor node, the standard deviation of measurement noise is $\sigma_\theta = 1^\circ$.

During target tracking process, the sink receives information about moving object, every T=0.5 sec. For the target we have the maximum acceleration $a_{\max} = 15m / s^2$, the maximum velocity $v_{\max} = 30m / s$ and variance matrix of process noise $Q = 5^2 * I_{2*2}$. For the predictor the bound of the prediction location error is $\alpha = 15^\circ$. The parameters of energy consumption model are shown in Table 1.

Accuracy and Missing Rate

First we compare the accuracy of four mentioned predictors that is determined by square root of the difference between predicted and actual location of object in two dimensions. The prediction error is given by:

$$pre\_error = \sqrt{\left(x_{pre}^t - x_{real}^t\right)^2 + \left(y_{pre}^t - y_{real}^t\right)^2} \tag{38}$$

where the pre_error^t is prediction error of predictor at time t.

As shown in Figure 2 the UKF predictor and PEP have lowest prediction error and so are more accurate than the others. But as we expect the linear current predictor has greatest prediction error and EKF predictor is in middle. EKF predictor has good accuracy most of the time, except in some peak points. The EKF predictor uses a first-order linearization that sometimes leads to instabilities during simulation time.

Figure 2. Prediction error of different predictors

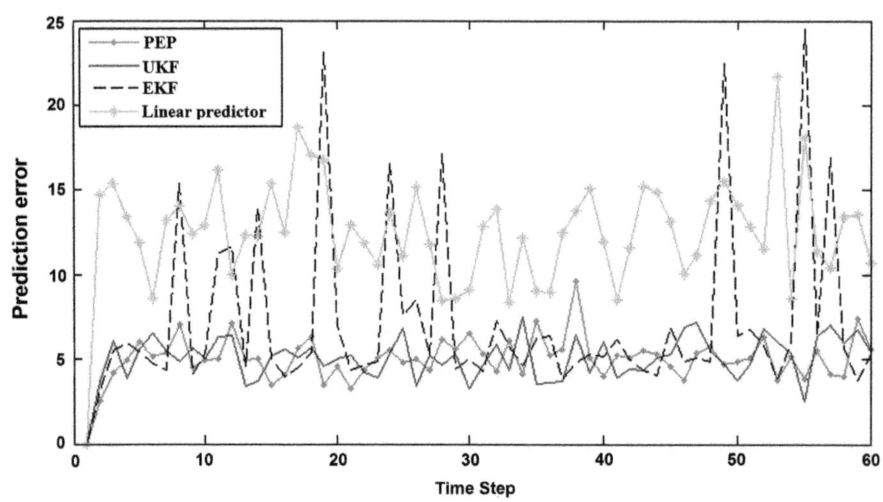

Energy Consumption

When the object is missed, a recovery mechanism is needed to find the lost object, and this leads to excessive energy consumption in the network. After missing the object all nodes in a cluster should wake up, sense the environment and transmit their packet to head nodes. Figure 3 shows total energy consumption of network from 10 min to 60 min. As we expect the predictors with lower missing rate have less energy consumption than the other. We point out that consumed energy of prediction procedure is neglected in Figure 3.

Also another main evaluation metric for target tracking algorithm by WSN is the number of operations. Table 2 shows a comparison between accuracy and number of operations in these four predictors in one hour target tracking. As shown in Table 2 the linear predictor has lower accuracy with fastest run time because of its low number of operation. But UKF predictor with the high accuracy has a large number of operations while computational capacities of sensor nodes are constraint. The PEP gives an optimum point between accuracy and runtime speed. So it has good accuracy that can be equal to the UKF predictor and the number of operations about 24.7 times less than UKF. The number of operations is a man evaluating factor of predictors especially in WSNs with processing resource constraints.

Evaluation Parameter

To have a fair comparison between the predictors the following factor is defined in which the normalized number of operations, predictor error rate and consumed energy are considered (Box 3).

As shown in Figure 4 the PEP gives the best ζ, as defined factor, during simulation time. The linear predictor has lowest number of operation and so leads to great ζ, in comparison with UKF and EKF.

Node Selection Algorithm

The node selection algorithm is the next step after prediction mechanism. This algorithm finds and activates three suitable sensor nodes to guarantee the accuracy of next prediction and also prolong the network lifetime. So far, deferent methods have been proposed for this purpose that are as follows:

Distance_based Selection Algorithm

Some prediction based methods consider nearest nodes to the predicted location such as [34, 38]. In such a situation the cluster head should calculates its member's distance to the predicted location of target and chooses nodes which have lower distance to stay active.

Distance and Remaining Energy Based Selection Algorithm

In this algorithm, the cluster head chooses nodes which have more remaining energy as well as lower distance such as (Deldar et al., 2011). In such a situation all of the members should send their energy to the cluster head and it will calculate the selection parameter for its members as follows:

$$\text{selection}_{\text{i}} = \frac{\text{remaining energy}_{\text{i}}}{\text{distance}_i^2} \quad (40)$$

Then the cluster head chooses three nodes with maximum selection parameter for tracking purpose.

Proposed Selection Algorithm (DESA)

Three parameters are used as selection criterion in our distance and energy based selection algorithm (DESA) [Mirsadeghi, 2014]:

- Distance,

Figure 3. Consumed Energy of different predictors vs. simulation time

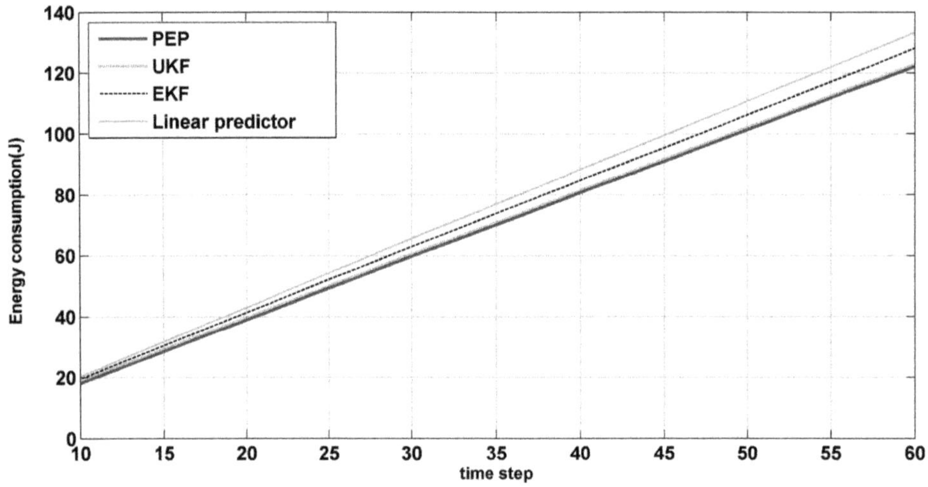

Box 3.

$$\zeta = \frac{1}{(Normalized\ number\ of\ operations) * (Error\ rate) * (Energy\ Consumption\ ratio)} \tag{39}$$

Table 3. Network lifetime comparison for different methods of node selection

Node Selection Method	Lifetime
DESA	11946 seconds
Distance and Remaining energy [19]	10061 seconds
Distance Based selection [38]	5339 seconds

- Remaining energy of nodes,
- Energy consumption for sending a packet to the head node.

The active head node calculates the selection criterion as follows:

$$\text{selection}^{i} = \frac{^{E}\text{remaini}^{-E}\text{sendi}}{d_i^2} \tag{41}$$

where d_i is distance of node i from the object, E_{remain_i} is remaining energy of node i and E_{send_i} is consumed energy of node i for sending a packet to the head node. Then head node chooses three nodes with maximum selection parameter for tracking purpose. Hence in each time step, the closer nodes to the object with more energy and shorter distance from head node will be selected. This selection criterion with three mentioned parameters helps us to prolong the life time of target tracking system.

Effect of Node Selection Algorithm on Network Lifetime

Now we try to evaluate our proposed selection criterion. Table 3 shows the effect of DESA on network lifetime. As depicted in Table 3, when considering three elements selection criterion the

Figure 4. The defined performance factor ζ, vs. simulation time

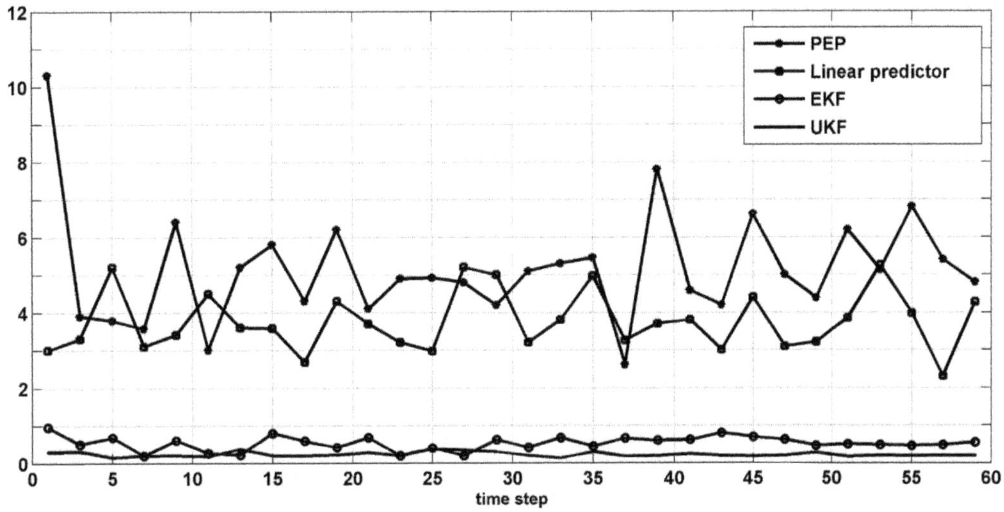

lifetime of network is increased about 10%. Since in each time step, three nodes with highest level of energy and lowest distance to the CH are selected.

Recovery

There is possibility of getting lost object in presence of link failures, node failures, prediction error or incomplete coverage of network. So a recovery procedure is needed to find the lost object. In this procedure when a cluster head cannot find the object and doesn't receive enough information from selected nodes, inform the sink that I have lost the object. Therefore all of the cluster heads are wakeup by the sink node to sense the environment and find the lost object.

Clustering Methods

Here we introduce two clustering methods and three following assumptions are used in proposed clustering mechanism:

- All nodes have the same initial energy and the base station knows this value.
- The base station knows the size of considered environment.
- The base station has information about the initial number of nodes in the network.
- CHs send their information to the base station in single-hop fashion.

In the first method the fuzzy logic is employed to assess the chance of a node to become a CH. Node density, node energy and node's centrality are taken into account to calculate the cluster head's chance. The second method has three steps for CH selection. In the first two steps, CHs are selected without communication overhead and finally some control messages are exchanged between remaining nodes to increase the network coverage.

Duplication Avoidance Clustering (Mahani et al., 2012)

Here assumed that all nodes have an ID number which is unique in the network area and a dynamic ID number that can be changed during cluster head selection. At initial state each node should try to find its neighbors defined as nodes in its communication range. In this method the CHs are selected in four steps. In the first step, the nodes should update their dynamic ID number and also their neighbor's ID using a random number. Each node has a random number generator based on logistic map function.

In the second step, each node should decide whether be the cluster head or not. So a common threshold is assumed between nodes. If the node's ID number was bigger than its neighbor's ID and assumed threshold, the node will decide to be a cluster head otherwise it will stop the decision process. During the third step, between 2 to 8 out of 100 nodes are set as a CHs and nodes that already are CHs and their neighbors shouldn't attend in this step. And in the fourth step, some other CHs will set randomly. For this purpose, the best range of this probability is between 0.2 to 0.4. So in this method CHs are selected in forth steps and although this can help us to increase the coverage area strongly, but random selection of CHs causes high collision and not suitable distribution of CHs in the area which lead to higher energy consumption.

Energy-Aware Distributed Dynamic Clustering Protocol (Taheri et al., 2012)

In this method fuzzy logic is employed in which fuzzy if-then rules and two fuzzy input variables named *Node_degree* and *Node_centrality* are used to evaluate the cost of a node in order to choose a final CH from the set of neighboring tentative CHs. The clustering process is done in three phases. In the first phase, neighbor information are updated, each node computes its cost by using fuzzy logic, sends it to its neighbors and sets its own delay time:

$$delay\ time = \frac{1}{\max(\frac{E_{residual}}{E_{\max}}, d)} \quad (42)$$

Here $E_{residual}$ is remaining energy on node, $E_{\max}$ is initial energy of nodes and d limits the period of time should be elapsed in the clustering process.

In the second phase, every node must wait until the expiration of its delaytime.If a node doesn't hear the *CH_msg* message from any other nodes during the delay time, upon delay time expiration it should introduce itself as atentative_CH to its neighbors. After that if this node has the least cost among the tentative_CHs in its proximity, it will become a final_CH and inform its neighbors. Otherwise if it receives a final_CH message, it can no longer be elected as a CH.In the third step,if a node isn't a final_CH and has received at least one final_CH message, it will elect the final_CH with the least cost to join it.

In this method although there is a good distribution of CHs and high coverage due to using fuzzy logic, the data overhead increased which increases energy consumption of network.

Proposed Method I

Distributed Clustering Method Using Fuzzy Logic: (DCFL) (Mirsadeghi & Mahani, 2013)

At the beginning of the proposed network deployment, the base station broadcasts a beacon signal to all sensor nodes which contains some necessary information such as the initial energy of the sensor nodes in the network, number of nodes in the network and network dimension. The same process is used by the sensor nodes, to determine the number of their neighbor nodes. TheDCFL scheme is implemented during several rounds and in the beginning of each round, the neighbor information is updated and each node calculates its chance of becoming CH by using fuzzy logic.

Fuzzy Logic

Fuzzy logic (FL) is used to model human experience and human decision making behavior. In the FL the input-output relationship is expressed by using a set of linguistic rules or relational expressions. FL basically consists of four important parts including Fuzzification, Defuzzification, inference engine, and a rule base. Here we have employed the most commonly used fuzzy inference technique, called the Mamdani and Assilian (Mamdani & Assilian, 1975) method. For all of the sensor nodes in the network, the following three parameters are collected as fuzzy sets of input variables.

- Remaining energy (represented by $energy$), which is the remaining battery power of the sensor node.
- Local Density (represented by $density$), which is the normalized number of neighbor nodes in a sensor node's neighborhood:

$$node\_density_i = \frac{|node\_neighbers|}{node\_numbers|} \tag{43}$$

Here $node\_neighbers$ is the number of node's neighbers and $node\_numbers$ is the number of all nodes in the environment which is constant.

- Node centrality (represented by $centrality$):
- Shown in Box 4.

The fuzzy sets of input variables and output variable (chance) are depicted in Figure 5. The DCFL calculates a probability which shows the chance of a node to be selected as a CH, using fuzzy if-then rules. Based on the three fuzzy variables, fuzzy if-then rules can be defined which are similar to those presented in Table 4. A greater chance means that the node has a higher priority to be elected as CH.

Clustering Process

After computing the chance, each sensor node sets its own delay_time as follows:

$$Delay - time_i = \frac{1}{node\_chane_i} \tag{45}$$

Each node must wait up to its delay time expiration. If a node does not hear the CH_msgm message from any other sensor node during the delay time, it declare itself as a tentative CH after its delay-time expiration. So it broadcasts a CH_msg message to all the nodes within the cluster range to announce its status. For a node with greater chance, its delay time is lower. So, its delay time would be expired sooner and it has a higher priority of being selected as a tentative CH. If this particular node has the greatest chance among the tentative CHs in its proximity, it will become a final CH and should inform nodes within its cluster range by sending a final_msgmessage within its cluster range.

If a node receives a final_msg message, it can no longer be elected as a CH. Therefore, it must choose to connect to one of the final CHs in its cluster radius based on its distance to selected CHs. On the other hand if a node completes the clustering process and has not yet received anyfinal_msgmessage, it will find itself uncovered and so becomes an orphan node. The main problem of this method is convergence in the cluster head election when two nodes have the same chance and so the same delay time. In this situation, the node with lower ID is selected as CH.

Table 4. Fuzzy rules

Energy	Density	Centrality	Chance
high	high	close	vH(very High)
high	high	adequate	vH
high	high	far	H (high)
high	medium	close	vH
high	medium	adequate	H
high	medium	far	H
⋮	⋮	⋮	⋮
low	low	close	L (Low)
low	low	adequate	vL (very Low)

Box 4.

$$node\ centrality = \frac{\sqrt{\sum_{j\mathit{\hat{I}}node\_neighbors(i)} dist^2(i,j) / \mid node\ neighbors(i) \mid}}{Network\ Dimension} \quad (44)$$

where for an $N * N$ region of space, the $Network\_Dimension$ is defined as N. And $dist(i, j)$ is the distance between nodes i and j.

Proposed Method Π

Three Steps Clustering Method (TSCM): (Mirsadeghi, 2014)

We assume that all nodes have an ID number which is unique in the network area and a dynamic ID number that can be changed during cluster head selection. At initial state each node should try to find its neighbors defined as nodes in its communication range. TheTSCM has three steps for CHs selection and cluster creation which describe as follows:

Step 1: At the beginning of each clustering process round the nodes should update their dynamic ID number and also their neighbor's ID. Each clustering process round is from the beginning of clustering process to the restarting it again. To use dynamic ID number there are two main challenges:

1. The dynamic ID number should be a random number (the chance of CH selection distributes uniformly).
2. Each node should know the dynamic ID number of their neighbors without any negotiation overhead.

So the logistic map function is used to generate random numbers. The function is given by:

$$P(\mathrm{i}+1) = r * p(\mathrm{i}) * (1 - P(\mathrm{i})) \quad (46)$$

where r is constant and $p(i)$ is a random number. Based on logistic map of Chaos Theory, if

Figure 5. Fuzzy logic set. (a)Energy, (b) Density, (c) Chance.

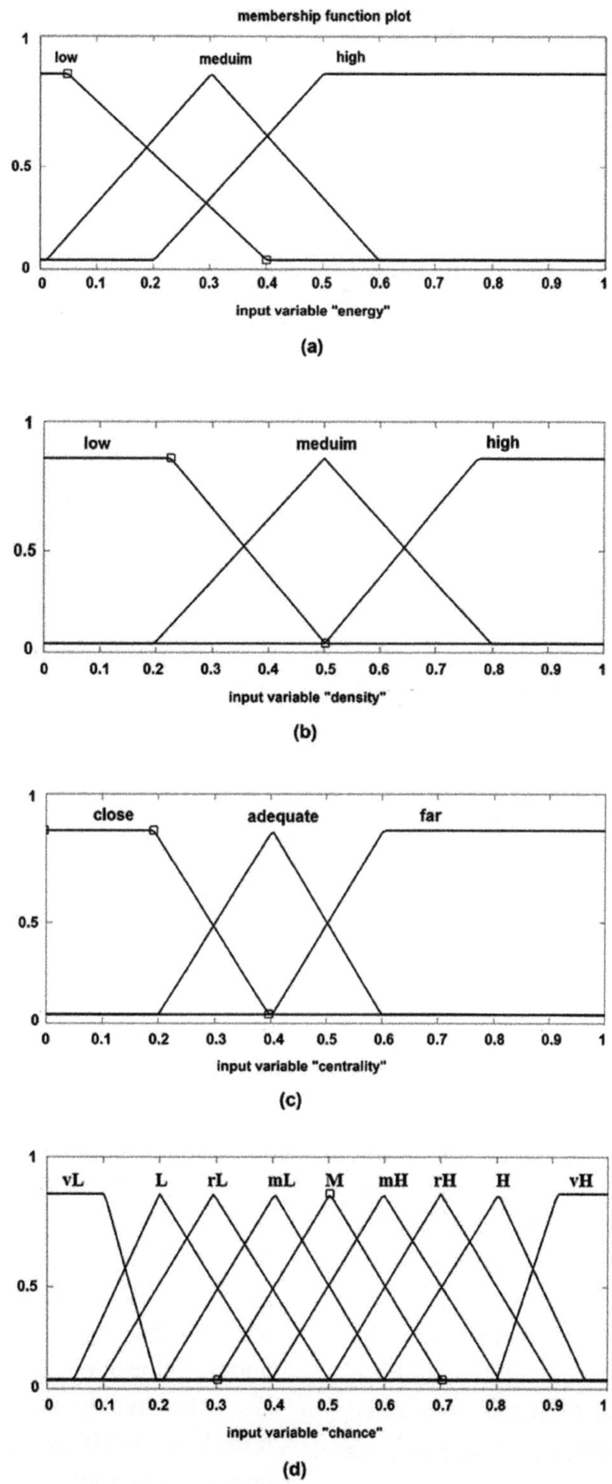

$r > 1+\sqrt{6}$ then the function will generate random numbers. The primary seed is needed to generate the random numbers. In Each clustering process round, nodes should generate a random number to change their dynamic ID. Nodes will update their dynamic ID number and their neighbor's dynamic ID using the following equation:

$$n(m,\mathrm{i}+1) = n(m,\mathrm{i}) + RAND \tag{47}$$

where $n(m,i)$ is dynamic ID Number of node m after i^{th} iteration and RAND is a random number which is generated through the following formula:

$$RAND = ceil(rand(i) * number\ of\ nodes) \tag{48}$$

:*rand(i)*: Random number (between 0 to 1) in iteration i.

:*ceil:*MATLAB round function up

If

$$n(m,i+1) > number\ of\ nodes$$

then:

$$n(m,i+1) = n(m,i+1) - number\ of\ nodes\ .$$

Because of using the same seed at the beginning which come from design stage, each node will generate the same random number like the other nodes. Now each node should decide whether be the cluster head or not. So a threshold is defined and each node compares their dynamic ID number with the defined threshold. If its ID number was bigger than the predefined threshold (CT_1), the node has a chance to be a cluster head otherwise it will stop the decision process in this phase. Then all of the elected cluster heads inform to their neighbors by () CH_msg(ID_i) message which includes their unique ID and nodes that don't receive any message are considered as firs step remaining nodes. So the next two steps are done by these remaining nodes.

Step 2: Between the first step remaining nodes if its number was smaller than the assumed threshold (CT_2) and was the smallest number among its neighbors, the node will decide to be cluster head otherwise it will stop the decision level and wait for next step. Then the CHs inform to their neighbors by CH_msg message and nodes that don't receive any message are considered as second step remaining nodes. So the last step will be done just on these remaining nodes.

Step 3: Despite of previous steps, step 3 needs to exchange some control messages among second step remaining nodes. So each node calculates the weight (W_{e_i}) according to equation 49.

$$W_{e_i} = (\delta \frac{E_{ri}}{E_{initial}} + \beta \frac{1}{avg(dist)}) \tag{49}$$

where, E_{ri} is the remaining energy of node i, $E_{initial}$ is initial energy of node and *avg(dist)* is average distance of node i from all of its neighbors. δ and β are constant values for scaling two terms and are set according to the distribution of nodes within network and their residual energy situation. It can be seen from this equation that node with more remaining energy and lower distance to its neighbor nodes, will have more weight. Then each node sets up a timer and waits for a certain delay (which is inversely proportional to remaining energy of that node). The node whose delay expires first among the neighboring nodes, if doesn't hear the CH_msgmessages from any other sensor nodes during the delay time, introduces itself as a cluster head candidate by broadcasting the:

()candidate_msg(Node ID, candidate_CH, weight),

to all nodes within the cluster range. The node with higher energy, has lower delay time. So that node has a higher chance to be selected as a CH.

Next, if this particular node has the biggest weight among the candidate CHs, it will become a final cluster head and broadcast a CH_msg message within its cluster range. On the other hand, if a node receives aCH_msgmessage, it can no longer be elected as a CH. At the end of this step, CHs inform the sink their location and the clusters formed as follows: each non cluster head node after receivingCH_msgfrom all CHs in its neighborhood, chooses the nearest cluster head and joints to its cluster. Clusters will keep unchanged for a relatively longer period of time until the next round of clustering process begins.

Simulation Results

In this section we have a comparison between the simulation results of different static clustering methods via MATLAB. We run each method for 1000 rounds and evaluate coverage, numbers of orphan nodes (nodes that are not covered by any cluster head), data overhead, lifetime and energy consumption. Network coverage is calculated based on sensing area of nodes and data overhead are the messages that are exchanged during cluster creation. Suppose that in each clustering process round all nodes are active, sense the environment and send their information to the cluster heads. Then all the cluster heads send the data packet to the sink. The results show that DCFL and TSCM have better condition in comparison with other methods. The network area is 200*200 m^2 in which 200 sensor nodeswith initial energy of 1 Joule, are uniformly distributed. Sensing range $R_{sen\sin g} = 35m$ and maximum communication range $R_{communication} = 50m$.

But when CHs wants to send data to the sink, they can increase their communication range. The sink is located out of the area (x = 100, y = 250) and the packets send to the sink every T=0.5 sec. The Table 5 contains simulation parameters.

Here the best values of thresholds CT_1 and CT_2 are obtained by trial and error method. For larger values of CT_1 and smaller values of CT_2, the number of remaining nodes in the third step increased which increases the energy consumption due to the overhead. Also for smaller values of CT_1 and larger values of CT_2 the number of CHs increased greatly which have bad effect on collision and so on quality energy consumption of tracking system.

Coverage, Number of CHs and Energy Consumption

Table 6 compares the coverage ratio, number of orphan nodes, number of CHs and all energy

Table 5. Parameters for energy consumption model

E_{elec}	50 *nJ/bit*
E_{DA}	5 *nJ/bit/signal*
E_{sense}	22 *mW*
ϵ_{mp}	0.0013 *pJ/bit/m*^{4}
ϵ_{fs}	10 *pJ/bit/m*^{2}
d_0	87 *m*
Data packet size	2000 *bits*
δ	0.8
β	5
Permanent counter threshold	75
Temporary counter threshold	2
CT_1	150
CT_2	50

consumption of all the methods in the whole working duration. As shown in Table 6, network coverage influences the number of orphan nodes which increase energy consumption of network because they are not covered by any cluster head and send directly to the sink. So when clustering method has high coverage it means that the number of orphan nodes is reduced and so the energy consumption of network would be reduced. According to the Table 6, almost all nodes in our methods are covered by the selected cluster heads and there aren't any orphan nodes.

Although ECPF and GCA improve the network coverage, but their energy consumption is greater than proposed methods since almost both of them have more cluster heads than DCFL and TSCM in each clustering process round and this high coverage is achieved by large number of CHs. As a result the number of unnecessary transmission to the sink and so the energy consumption of network increased. Also according to the Figure 6, ECPF has relatively higher data overhead. In other methods there are very bad situation. They have high data overhead, also few numbers of cluster heads cannot cover all the area. So there are a lot of orphan nodes and less coverage which increase the energy consumption of network.

Network Lifetime and Evaluation Parameter

According to Figure 7, our second scheme has larger FND, HNA and LND and Figure 8 shows that its lifetime is more than other clustering methods. And although the first method has lower LND than ECPF method, its FND and HNA are larger and has better lifetime than other methods. The lifetime of clustering algorithms are dependent on numbers of cluster heads, coverage and number of orphan nodes, in which all of them are optimized in DCFL and TSCM. So we propose a new parameter, λ, to have fair comparison between the clustering algorithms. The λ is proportion with normalized numbers of orphan nodes ($orph_n$), normalized numbers of cluster heads (CH_n), coverage ratio (*cov*), data overhead ratio and energy consumption ratio (*en*):

$$\lambda = \frac{1}{cov} + CH_n + orph_n + en + overhead \tag{50}$$

In Figure 9 the value of λ in each clustering process round for such methods is illustrated and the average of λ is shown in Table 7. Each method that has the best condition has lower value of λ. So as we see beforeDCFL and TSCM has the best point of λ.

Table 6. Comparison of coverage rate, ratio of orphan nodes, number of cluster heads, and energy consumption

Method	Coverage Ratio (%)	Orphan Nodes	Average Number of CHs	Energy Consumption (%)
DCFL	89	0.009	12	35
TSCM	86	0	13	32
ECPF	88	0.0035	14	38.2
DAC	55	22	7	54.5
GCA	91	0	42	64.2
SCP	43	27	5	72
TCAC	36	40	4	76

Figure 6. Data overhead rate

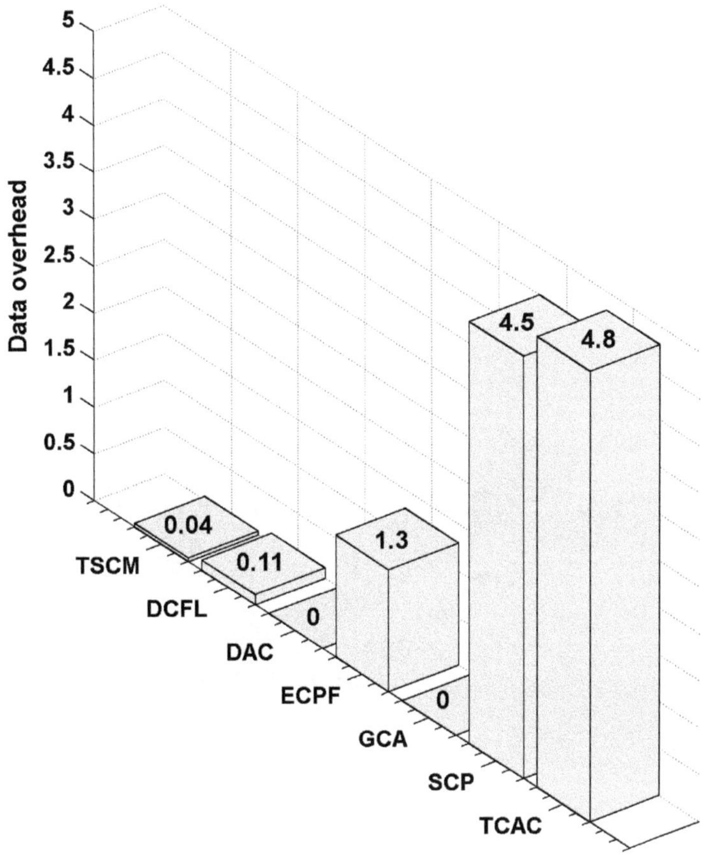

Simulation Results for Tracking Performance

In this section we focus on tracking just one moving object to evaluate our proposed clustering and prediction methods and analyze the effect of clustering on energy efficiency and quality of target tracking via MATLAB, although this method is able to track more than one object. For this purpose, we employ DCFL and TSCM clustering methods in target tracking and compare their performance to three other clustering methods, ECPF, DAC and SCP and evaluate the introduced algorithm for target tracking through simulations. Show that using suitable clustering mechanism can decrease energy consumption of network and has direct effect on accuracy and energy efficiency of tracking system. Also we evaluate the effect of proposed recovery mechanism on energy consumption of network. We assume that all sensor nodes are similar and have standard deviation of measurement noise $\sigma_\theta = 1^\circ$, the simulation environment is like 4.7 and simulation parameters are same as Table 5.

Missing Ratio and Energy Consumption

Our clustering methods reduce the missing rate of object. In target tracking, when the object lost, cluster heads should sense the environment to find the lost object, and orphan nodes are always active Therefore as depicted in Figure 10, network coverage and the numbers of orphan nodes influence

Figure 7. Comparison of network lifetime

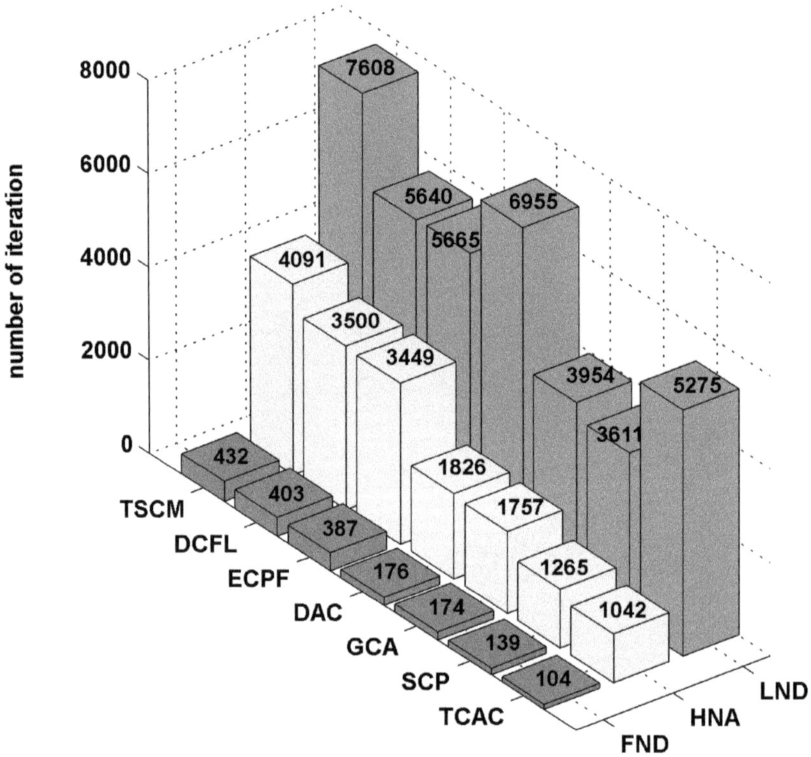

Figure 8. Comparison of network lifetime

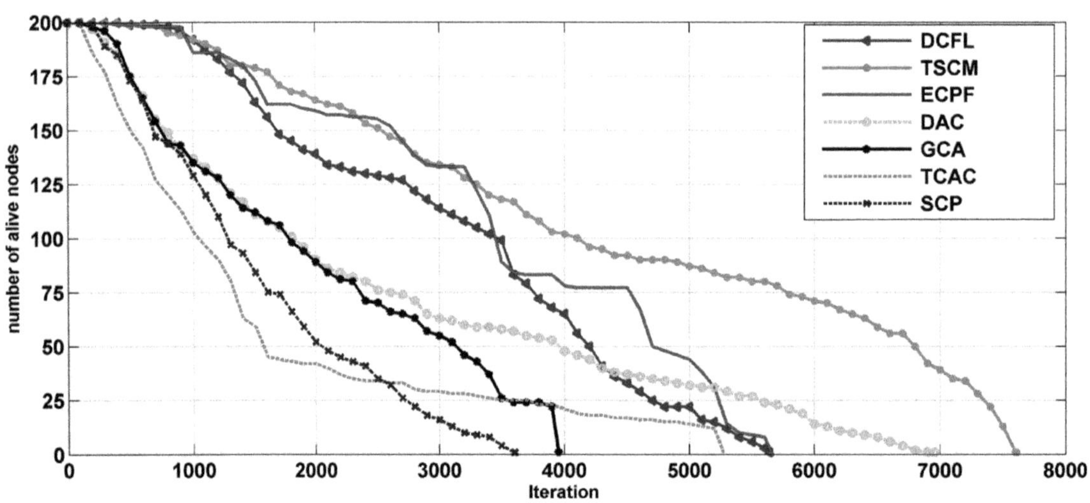

Figure 9. The trace of landa for 1000 rounds

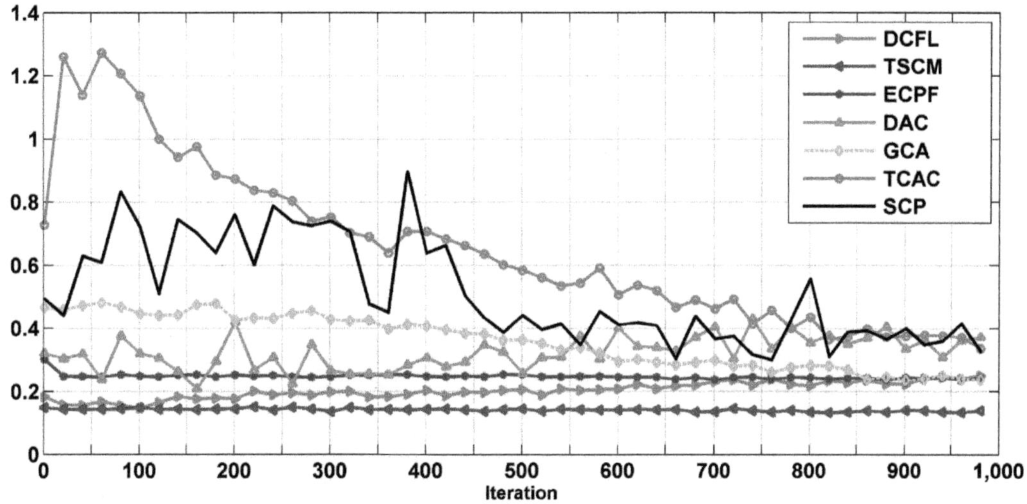

Table 7. Average of λ for clustering methods

Method	λ (Average)
DCFL	0.2016
TSCM	0.1406
ECPF	0.2447
DAC	0.3305
GCA	0.3546
SCP	0.5011
TCAC	0.4729

the missing rate of object. If clustering has a low coverage and some orphan nodes are exist, nodes cannot track the moving object successfully, cluster heads cannot find the lost object and so missing ratio increases. Based on Figure 10, our clustering methods that have high coverage and don't have more orphan nodes, leads to low missing ratio in target tracking.

Figure 11 illustrates all energy consumption of target tracking system in different clustering methods. As shown in Figure 10, network with high coverage and low number of orphan nodes reduces the energy consumption. According to figure11DCFL and TSCM schemes have the lowest energy consumption during target tracking period and their lifetime are increased, because they have a good distribution of CHs in the network and the low missing ratio and numbers of orphan nodes.

Evaluation Parameter (ɣ)

Another new parameter named ɣ is introduced to compare the mentioned tracking methods. Five variables are considered to calculate ɣ: Normalized numbers of orphan nodes ($orph_n$), normalized numbers of cluster heads (CH_n), coverage ratio (*cov*), tracking error ratio (*er*) and energy consumption ratio (*en*):

$$\gamma = \frac{1}{\text{cov}} + CH_n + orph_n + en + er \qquad (51)$$

According to equation 45 the best method has lower value of ɣ. In Figure 12 the value of ɣ in each time step for such methods is drawn. As is clear ɣ in our methods, is lower most of the time and so according to Table 8 its average is lower than others.

Figure 10. Comparison of Missing rate based on network coverage and orphan nodes ratio

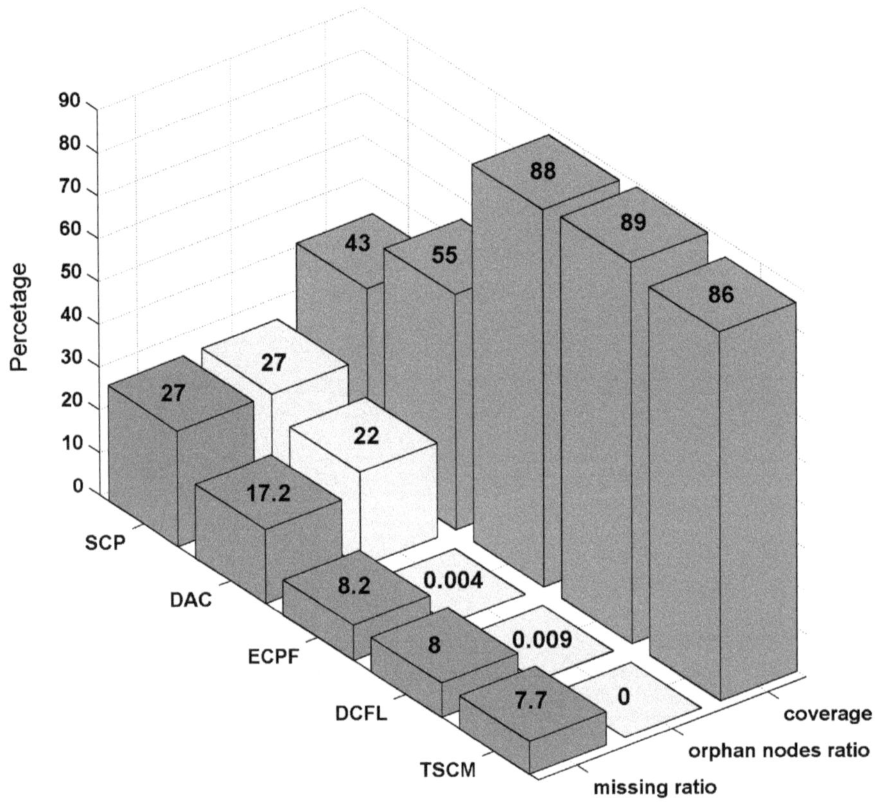

Recovery

Here we evaluate the effect of proposed recovery mechanism on the energy consumption of target tracking system. Figure 13 shows the energy consumed during one hour target tracking using proposed recovery and without that. Here we use the PEP prediction and TSCM clustering scheme. As the proposed clustering scheme has high coverage ratio, so most of the time being awake of cluster heads is enough for finding the lost object. Therefore as we can see from Figure 13, when we use this recovery mechanism, lower number of sensor nodes wake up and sense the environment which consumes less energy than when all nodes wake up.

FUTURE WORK

Up to now we study mobile object tracking in which objects are individual and move far from each other. In such a situation using static clustering and prediction mechanisms are suitable and energy efficient. As we see, predictor accuracy in prediction mechanism andcoverage network in clusteringare two main factors. In the existing studies mentioned above, the current head node at time t predicts the location only for time t+1 and if the target moves in high speed (more than 30m/s), it can pass by a group of nodes very fast without being detected. Therefore, as the target increases its speed, the probability of missing that target increases which doesn't considered in this chapter.Hencefor the future work, we are looking into the different moving patterns of the mobile

Figure 11. Energy consumption of network for target tracking application

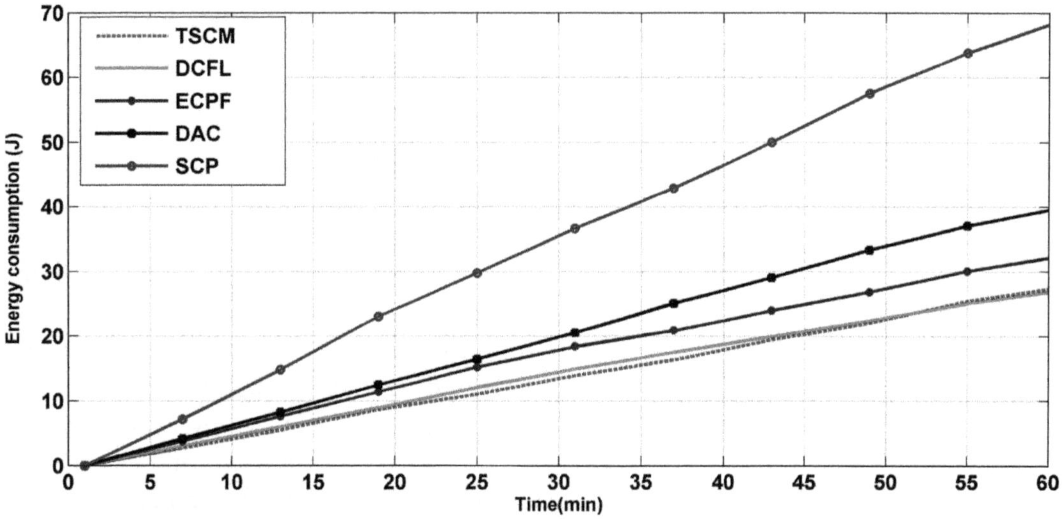

Figure 12. Comparison of gama for 60 minutes

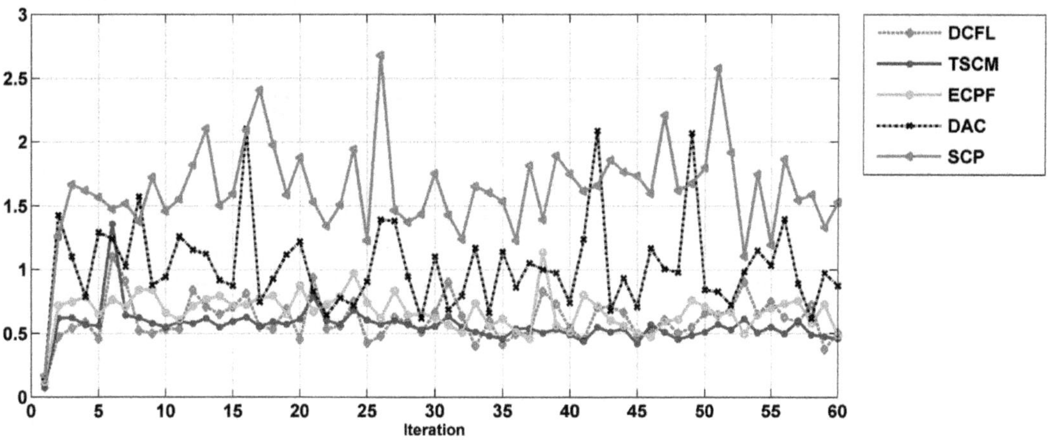

Figure 13. Comparison of energy consumption

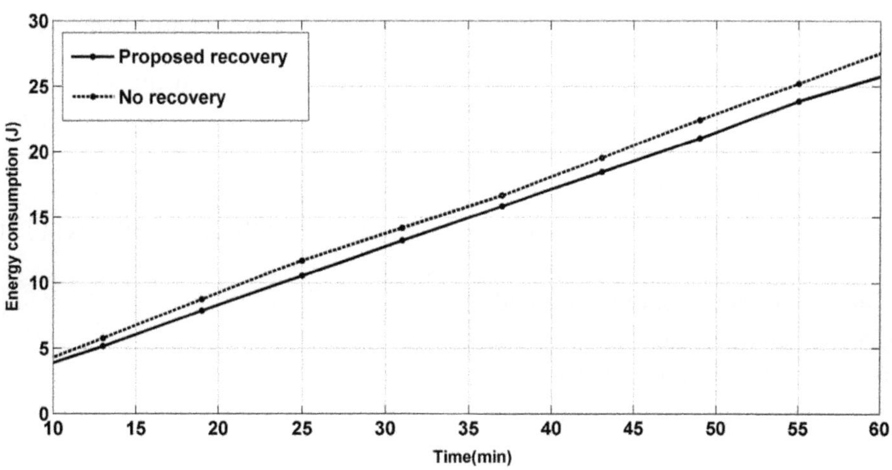

Table 8. Average of gamma for tracking method

Method	ɣ (Average)
DCFL	0.63
TSCM	0.55
ECPF	0.7
DAC	1.02
SCP	1.63

objects, exploring more sophisticated prediction models to improve the prediction accuracy (Hsua et al., 2012; Wang,X et al., 2007; Xu,y et al., 2004),and using proper prediction mechanisms to track fast moving objects accurately. Furthermore, we would like to extend the clustering protocols to meet QoS requirements of WSNs, such as coverage preservation, because complete coverage of the monitored area over long period of time is an outstanding issue (Taheri et al., 2012).

CONCLUSION

In this chapter we introduce a novel prediction based clustering algorithm for target tracking in wireless sensor networks. A static clustering approach and prediction mechanism is used to decrease energy consumption and prolong the lifetime of network. We introduce two new high coverage clustering schemes which increases accuracy and energy efficiency of target tracking system. The first method is based on fuzzy logic the same as ECPF. It has 1% higher coverage than ECPF, 2.5% lower coverage and better FND and HNA. The main idea of the TSCM method is based on our previous work (Tsai et al., 2007) which has zero overhead. Here we have only 0.04% extra overhead while the coverage ratio increases about 31% and the energy consumption of tracking system reduces about 9%. Also in comparison with ECPF and GCA, we have 2% and 5% reduction in coverage respectively but consumed energy reduces about 6.2% and 32.2%. In addition the PEP predictor causes lower missing rate than linear, EKF and UKF predictors. It has 11% reduction in error rates compared to the linear predictor and the number of operations which is the main determinant factor in the speed of the predictor is 24.7 times less than UKF. Also to reduce the energy consumption of finding the lost object, a new low power consumption recovery mechanism and a compatible node selection algorithm are introduced. Simulation results show that our new clustering schemes reduce the energy consumption of tracking system because they have high coverage, low CH selection's overhead and zero number of orphan nodes and the PEP predictor can help the network for more accuracy. And the recovery method helps to have less energy consumption during finding the missing object, too because of the small number of active nodes.

REFERENCES

Anandhakumar, P., Priyadarshini, J., Rajeswari, L., Srividhya, S., & Niveditha, C. K. (2011). *Occlusion detection and object tracking using filter banks.* In IEEE International Conference on Recent Trends in Information Technology (pp. 992-998). doi:10.1109/ICRTIT.2011.5972465

Barrios, C. Himberg, H. Motai, Y. Sadek, A. (2006, September). *Multiple Model Framework of Adaptive Extended Kalman Filtering for Predicting Vehicle Location.* In IEEE Intelligent Transportation Systems Conference. Toronto, Canada.

Bhuiyan, M. Z. A., Wang, G.-, Zhang, L., & Peng, Y. (2010). Prediction-based energy-efficient target tracking protocol in wireless sensor networks. *Journal of Central South University of Technology, 17*(2), 340–348. doi:10.1007/s11771-010-0051-1

Dahnil, D. P., Singh, Y. P., & Ho, C. K. (2012). Topology-controlled adaptive clustering for uniformity and increased lifetime in wireless sensor networks. *The Institution of Engineering and Technology*, *2*(4), 318–327.

Deldar, F., & Yaghmaee, M. H. (2011). Designing a Prediction-based Clustering Algorithm for Target Tracking in Wireless Sensor Networks. In *International Symposium on Computer Networks and Distributed Systems (CNDS)*. doi:10.1109/CNDS.2011.5764572

Di, M., Joo, E. M., & Beng, L. H. (2008). A Comprehensive Study of Kalman Filter and Extended Kalman Filter for Target Tracking in Wireless Sensor Networks. In *IEEE International Conference on Systems, Man and Cybernetics* (pp. 2792-2797).

Hamouda, Y. E. M., & Phillips, C. (2011). Adaptive sampling for energy-efficient collaborative multi-target tracking in wireless sensor networks. *The Institution of Engineering and Technology*, *1*, 15–25.

Heinzelman, W. B., Chandrakasan, A. P., & Balakrishnan, H. (2002). An Application-Specific Protocol Architecture for Wireless Microsensor Networks. *IEEE Transactions on Wireless Communications*, *1*(4), 660–670. doi:10.1109/TWC.2002.804190

Hsua, J. M., Chenb, C. C., & Li, C. C. (2012). POOT: An efficient object tracking strategy based on short-term optimistic predictions for face-structured sensor networks. *Computers & Mathematics with Applications (Oxford, England)*, *63*(2), 391–406. doi:10.1016/j.camwa.2011.07.034

Julier, S. J., & Uhlmann, J. K. (1997). A new extension of the Kalman filter to nonlinear systems. In *Proceedings of the Society of Photo-Optical Instrumentation Engineers*.

Khedr, A. M., & Osamy, W. (2011). Effective target tracking mechanism in a self-organizing wireless sensor network. *Journal of Parallel and Distributed Computing*, *71*(10), 1318–1326. doi:10.1016/j.jpdc.2011.06.001

Kim, H., Kim, E., & Han, K. (2006). An Energy Efficient Tracking method in Wireless Sensor Networks. In Next Generation Teletraffic and Wired/Wireless Advanced Networking (LNCS, 4003, pp. 278-286). doi:10.1007/11759355_27

Kuk Liu, Z., & Liu, Z. (2011). *A Dynamic Clustering Protocol for Wireless Sensor Networks*. In *IEEE International Conference on Computational and Information Sciences*.

Lee, S. H., Lee, S., Song, H., & Lee, H. S. (2011). Gradual Cluster Head Election for High Network Connectivity in Large-Scale Sensor Networks. In *IEEE International Conference on Advanced Communication Technology*, pp. 168-172.

Mahani, A., Askari-Hemmat, M. H., & Kavian, Y. S. (2012). Duplication Avoidance for Energy Efficient Wireless Sensor Networks. In *8th IEEE, IET International Symposium on Communication Systems, Networks and Digital Signal Processing*.

Mamdani, E. H., & Assilian, S. (1975). An experiment in linguistic synthesis with a fuzzy logic controller. *International Journal of Man-Machine Studies*, *7*(1), 1–13. doi:10.1016/S0020-7373(75)80002-2

Martirosyan, A., Boukerche, A., Richard, W., & Pazzi, N. (2008). *Energy-aware and quality of service-based routing in wireless sensor networks and vehicular ad hoc networks* (pp. 669–681). Institut TELECOM and Springer-Verlag France.

Medeiros, H., Park, J., & Kak, A. C. (2008). Distributed Object Tracking Using a Cluster-Based Kalman Filter in Wireless Camera Networks. *IEEE Journal of Selected Topics in Signal Processing*, *2*(4), 448–463. doi:10.1109/JSTSP.2008.2001310

Mirsadeghi, M. (2014). *Energy Efficient Mobile Target Tracking by Wireless Sensor Networks*. A Thesis Submitted as a Partial Fulfillment of the Requirements for the Degree of Master of science in Electronic (M. Sc.).

Mirsadeghi, M., & Mahani, A. (2013). *Low power prediction mechanism for WSN-based object tracking*. In *Conference on Electronics, Telecommunications and Computers (CETC)*.

Mirsadeghi, M., & Mahani, A. (2013). *A novel distributed clustering protocol using fuzzy logic*. In Conference on Electronics, Telecommunications and Computers (CETC).

Mirsadeghi, M. Mahani, A. (2014). Energy efficient fast predictor for WSNbased target tracking. *Springer Annals of Telecommunications*. doi:.10.1007/s12243-014-0430-y

Monowar, M. M., & Alam, M. M. Obaidur Rahman, Md., Hong, C. S., & Lee, S. (2010). A load-aware energy-efficient and throughput-maximized asynchronous duty cycle MAC for wireless sensor networks (pp. 777–794). Institut Télécom and Springer-Verlag.

Pant, B., & Alkin, O. (2012, October). Correlated Movement Mobility Model and Constant Acceleration Model for EKF-Based Tracking Applications. In *IEEE Wireless and Mobile Computing* (pp. 869–874). Networking and Communications.

Pathan, A. K., & Hong, C. S. (2008). *SERP: secure energy-efficient routing protocol for densely deployed wireless sensor networks* (pp. 529–541). Institut TELECOM and Springer-Verlag France.

Rezaii, T. Y., & Tinati, M. A. (2011). *Distributed multi-target tracking using joint probabilistic data association and average consensus filter* (pp. 553–566). Institut Télécom and Springer-Verlag.

Ribeiro, A., Schizas, I. D., Roumeliotis, S. I., & Giannakis, G. B. (2010). Kalman filtering in wireless sensor network. *IEEE Control Systems Magazine*, *30*(2), 66–86. doi:10.1109/MCS.2009.935569

Safa, H. El-Hajj, W., & Zoubian. (2013). A robust topology control solution for the sink placement problem in WSNs. *Journal of Network and Computer Applications*.

Sheng, X., Changfeng, X., & Zhangsong, S. (2011). An Adaptive Target Tracking Scheme for Binary Wireless Sensor Networks. In *IEEE International Conference on Electronics, Communications and Control* (pp. 121-124). doi:10.1109/ICECC.2011.6067926

Shih, K., Wang, S. S., Chen, H. C., & Yang, P. H. (2008). *CollECT: Collaborative event detection and tracking in wireless heterogeneous sensor networks*. Elsevier B.V.

Svensson, D. (2010). *Target Tracking in Complex Scenarios*. Göteborg: Department of Signals and Systems, Chalmers University of Technology.

Tadayon, N., Khoshroo, S., Askari, E., Wang, H., & Michel, H. (2013). Power management in SMAC-based energy-harvesting wireless sensor networks using queuing analysis. *Journal of Network and Computer Applications*, *36*(3), 1008–1017. doi:10.1016/j.jnca.2012.01.009

Taheri, H., Neamatollahi, P., Younis, O. M., Naghibzadeh, S., & Yaghmaee, M. H. (2012). An energy-aware distributed clustering protocol in wireless sensor networks using fuzzy logic. *Elsevier Ad Hoc Networks*, *10*(7), 1469–1481. doi:10.1016/j.adhoc.2012.04.004

Teng, J., Snoussi, H., & Richard, C. (2007). Prediction-based Proactive Cluster Target Tracking Protocol for Binary Sensor Network. In *IEEE International Symposium on Signal Processing and Information Technology* (pp. 234-239). doi:10.1109/ISSPIT.2007.4458178

Tran, S. P. M., & Yang, T. A. (2006). OCO: Optimized Communication and Organization for Target Tracking in Wireless Sensor Networks. In *Proceedings of the IEEEE International Conference on Sensor Networks.*

Tripathy, A. K., & Chinara, S. (2012). Staggered Clustering Protocol: SCP an Efficient Clustering Approach for Wireless Sensor Network. In *IEEE World Congress on Information and Communication Technologies.* doi:10.1109/WICT.2012.6409209

Tsai, H. W., Chu, C. P., & Chen, T. S. (2007). Mobile object tracking in wireless sensor Networks. *Computer Communications*, *30*(8), 1811–1825. doi:10.1016/j.comcom.2007.02.018

Tyagi, S., & Kumar, N. (2013). A systematic review on clustering and routing techniques based upon LEACH protocol for wireless sensor network. *Journal of Network and Computer Applications*, *36*(2), 623–645. doi:10.1016/j.jnca.2012.12.001

Vaghefi, R. M., Gholami, M. R., & Strom, E. G. (2010, November). Bearing-only target localization with uncertainties in observer position. *IEEE PSPB Operations Manual, amended 19.*

Vaidehi, V., Sandhya, M., Karthika, J., & Azhagarasan, S. (2011). Power optimization for object detection and tracking in wireless sensor networks. In *Recent Trends in Information Technology (ICRTIT)*, (pp. 555-560). doi:10.1109/ICRTIT.2011.5972433

Wang, X., Ma, J. J., Wang, S., & Bi, D. W. (2007). Prediction-based Dynamic Energy Management in Wireless Sensor Networks Sensor Networks. *Sensors (Basel, Switzerland)*, 7.

Wang, X., Ma, J. J., Wang, S., & Bi, D. W. (2007). Cluster-based Dynamic Energy Management for Collaborative Target Tracking in Wireless Sensor Networks. *Sensors (Basel, Switzerland)*, *7*(7), 1193–1215. doi:10.3390/s7071193

Wang, Z., Lou, W., Wang, Z., Ma, J., & Chen, H. (2010). A Novel Mobility Management Scheme for Target Tracking in Cluster-Based Sensor Networks. *Springer-Verlag Distributed Computing in Sensor Systems*, *6131*, 172–186. doi:10.1007/978-3-642-13651-1_13

Wang, Z. Lou, W. Wang, Z. Ma, J., & Chen, H. (2013). A Hybrid Cluster-Based Target Tracking Protocol for Wireless Sensor Networks. *International Journal of Distributed Sensor Networks*, Article ID 494863.

Wood, T. (2008). *Mathematical Modelling Of Single Target SONAR and RADAR Contact Tracking*. Thesis submitted for transfer to DPhil status Michaelmas Term.

Xiao, W., Tham, C. K., & Das, S. K. (2010). Collaborative sensing to improve information quality for target tracking in wireless sensor networks. In 8th IEEE International Conference on Pervasive Computing and Communications Workshops (PERCOM Workshops) (pp. 99-104). doi:10.1109/PERCOMW.2010.5470610

Xu, Y., Winter, J., & Lee, W. C. (2004). Prediction-based strategies for energy saving in object tracking sensor networks. In *Proceedings of the Fifth IEEE International Conference on Mobile Data Management (MDM'04)* (pp. 346-357).

Yadav, A., Naik, N., Ananthasayanam, M. R., Gaur, A., & Singh, Y. N. (2012). A Constant Gain Kalman Filter Approach to target tracking in Wireless Sensor Networks. In *IEEE International Conference on Industrial and Information Systems* (pp. 1-7). doi:10.1109/ICIInfS.2012.6304803

Yang, H., & Sikdar, B. A. (2003). Protocol for Tracking Mobile Targets using Sensor Networks. In *Proceedings of IEEE Workshop*.

Yang, L., Feng, C., Rozenblit, J. W., & Qiao, H. (2006). Adaptive Tracking in Distributed Wireless Sensor Networks. *Proceedings of the 13th Annual IEEE International Symposium and Workshop on Engineering of Computer Based Systems*. doi:10.1109/ECBS.2006.20

Yang, W. C., Fu, Z., Kim, J. H., & Park, M. S. (2007). *An Adaptive Dynamic Cluster-Based Protocol for Target Tracking in Wireless Sensor Networks* (pp. 157–167). Advances in Data and Web Management. doi:10.1007/978-3-540-72524-4_19

Zhang, W., & Cao, G. (2004). *DCTC: Dynamic Convoy Tree-Based Collaboration for Target Tracking in Sensor Networks*. Published in the IEEE Transactions on Wireless Communications.

KEY TERMS AND DEFINITIONS

Duplication Avoidance Clustering: In DAC method sensor nodes are grouped in to clusters in order to extend the network coverage with low overlap area and collision probability.

Localization: A technique to determine the location of the sensor nodes.

Missing Ratio: In target tracking, when the object lost, cluster heads should sense the environment to find the lost object. The missing ratio is increased when the heads could not find the lost object.

Motion Prediction: A mechanism to predict the next position of mobile object.

Network Lifetime: Time to live number of sensor nodes, to keep the network topology as a connected graph.

Recovery: Getting lost object in presence of link failures, node failures, prediction error or incomplete coverage of network.

Target Tracking: One of the most important applications of wireless sensor networks in which sensor nodes track the mobile object in the network area.

Chapter 13
Distributed Adaptive Parametric Power Spectral Estimation Using Wireless Sensor Networks

Hamed Nosrati
Sahand University of Technology, Iran

Mousa Shamsi
Sahand University of Technology, Iran

Sayed Mostafa Taheri
The University of Edinburgh, UK

Mohammad H. Sedaaghi
Sahand University of Technology, Iran

ABSTRACT

Spectrum analysis is one of the momentous fields in signal processing. It has a large variety of applications in radar, sonar, speech and image processing. Parametric methods have been proposed and employed for spectrum analysis including power spectral density (PSD) estimation. These methods estimate the parameters of a statistical model and compute the PSD, afterwards. In some circumstances one is obliged to deal with observations of numerous geographically dispersed sensors, to either increase the precision or based on application demands. Having a set of sensors linked together to take the advantages of cooperation and network topology, one obtains a more comprehensive estimation. In this chapter, the authors propose and study four different algorithms capable of facing spatio-temporal variations for parametric modeling and PSD estimation using wireless sensor networks (WSNs). For this purpose, the authors first validate the proposed algorithms using theoretical and mathematical formulations. Thereafter, performing simulation tasks demonstrates and supports the theoretical achievements. The next section of the chapter illustrates the concepts to a greater degree, the authors analyze and compare the performance of these algorithms with each other, as well as with the simple PSD estimation using individual sensors, wherein there is no cooperation among the nodes.

INTRODUCTION

The researches involved in WSNs paradigm are typically divided into three main fields. The first is the circuit and hardware, which predominantly emphasizes on optimal design of the sensor nodes based on two fundamental criteria: having more functionality as well as less resource consumption (Hou & Bergmann, 2012; Papotto, Carrara, Finocchiaro, & Palmisano, 2014; Somov, Baranov, Spirjakin, & Passerone, 2014; Stojcev, Kosanovic, & Golubovic, 2009). Furthermore, the topics related to the processors, proper communication links, and sensors are categorized in this branch

DOI: 10.4018/978-1-4666-8251-1.ch013

(Sizov, Gashinova, Zakaria, & Cherniakov, 2013; Sung & Hsu, 2011; Tolentino, Juson, Tan, & Talampas, 2010). The second field is the network and communications, which investigates the WSNs from topology, data packet, coding, and security points of view (Huang, Xiao, Soltani, Mutka, & Xi, 2013; Liu, Gong, & Zheng, 2014; Ma, Xu, Hempel, Peng, & Sharif, 2014; Naderi, Nintanavongsa, & Chowdhury, 2014). Eventually, the third category is signal processing. Notwithstanding the previous fields which mainly concentrate on data, the signal processing division proposes various algorithms to convert data into information. To extract meaningful information from the raw data, the statistical signal processing employs estimation, classification, and recognition methods (Agarwal & Jagannatham, 2014; Chaudhari & Serpedin, 2008; Koutsopoulos & Halkidi, 2014; Msechu & Giannakis, 2012; Soltanmohammadi, Orooji, & Naraghi-Pour, 2013; Wang & Ssu, 2013). In this chapter, we study WSNs under the third field, and intend to extract the model parameters and power spectral density from raw data in a sensor network.

One of the key issues in signal processing is to estimate the power spectrum of a time series, for the purpose of so many applications. Various approaches have been proposed for PSD estimation over the past few decades. These methods could generally be divided into two main categories. In classical methods, in the first instance, Autocorrelation Function (ACF) is aimed to be estimated for a time series, and afterwards, the power spectrum is obtained by transforming the estimated autocorrelation function into the frequency domain using Fourier transform. Periodogram (Welch, 1967) and Blackman-Tukey (Blackman & Tukey, 1959) approaches can be noted as examples of these methods. This relation between spectral density and autocorrelation function is considered as a nonparametric description of a second-order statistics of a random process. In the second category, parametric description of second-order statistics is achievable by considering the time-series model in a random process. In this designation, spectral density is a function of the model parameters instead of autocorrelation function (Marple Jr, 1987). The Autoregressive (AR), Moving Average (MA) and Autoregressive-Moving Average (ARMA) models well thought-out as some of these models. The fundamental reason for parametric modeling of a random process is to achieve a more comprehensive spectral density than in classical methods. In other words, using the classical methods leads to obtaining the spectral density from a windowed data or an autocorrelation function estimation, while both the unavailable data or unestimated ACFs are considered to be zero. This unrealistic assumption leads to distortions in spectral estimation, nevertheless, model parameters provide more realistic assumptions about the data outside the window rather than assigning zero values for them (Kay, 1988).

Among the aforementioned models, the autoregressive model has received more attentions. Attractiveness of the AR model arises from the fact that the calculation of the model parameters is possible by solving linear equations while, parameters in other models are computed using nonlinear equations. Yule initially introduced the modeling idea of a time series with a p order autoregressive model (AR(p)). In this method, the autocorrelation lag, i.e. p , is calculated based on the available data, in which the known Yule-Walker equation is then solved. Burg (Burg, 1967) reached a similar conclusion by maximizing the process entropy.

The methods for calculating the autoregressive model parameters are divided into two categories. The first group includes the algorithms for calculating the coefficients of a fixed-size block of accumulated data, while in the second category, coefficients are calculated from sequential data. The methods based on data blocks are called recursive-in-order, as they operate on a block of time samples and recursively achieve higher-order parameters based on lower-order estimates of the AR parameter. This method could be beneficial for the situations in which the order of the model is unknown, and thus, different orders have to be tested to achieve the best. Specifically, it is essentially advantegous in cases where the best possible performance is desired while owning the limited data available. Sequential approaches are called fixed order recursive-in-time due to the sequential data processing used to update the model coefficients with a fixed order. These techniques are precisely tailored to situations in which tracking or adapting to a slowly time-varying spectral pattern is required. Sequential parameter estimation methods vary from gradient approaches, such as Least Mean Squares (LMS), to Recursive Least Squares (RLS) algorithms. A hierarchical plot of the discussed methods can be seen in Figure 1.

Adaptive sequential methods can solve the problem of time variations. Nevertheless, spectrum analysis is confronted with another challenge in some cases which is spatial variations. In a variety of circumstances, the process is distributed over a geographical region leading to spatio-temporal variations. In these cases, relying only on a single point's observations can make the estimation to be imprecise. Moreover, a vast variety of applications such as detection, classification, tracking, acoustic arrays, radar, and sonar have to deal with spatial variations in order to make an inference (Hlinka, Hlawatsch, & Djuric, 2013; Neuzil, Kreibich, & Smid, 2014; Zhong & Premkumar, 2014). For spatio-temporal analysis, employment of sensor networks is recommended since involving observations of numerous points, leads to a more comprehensive estimation. Among all relevant methods proposed in wireless sensor networks, adaptive networks proposed by Sayed & Lopes (2006) fulfill the aims properly. Adaptive networks consist of a set of adaptive filters cooperating with each other to solve an optimization problem, using a predefined cost function. In general, adaptive networks work in two different manners, centralized and distributed. In the centralized method, the observations from all nodes are gathered in a single point and the optimization step occurs at the fusion center, while in the distributed manner, it is fully distributed throughout the network. Omitting the fusion center as a potential critical point, distributed schemes optimize the cost function in some inaccuracy expenses, presumably.

The two main distributed adaptive strategies are incremental (Bertsekas, 1997; Lopes & Sayed, 2007; Nedic & Ozdaglar, 2009; Rabbat & Nowak, 2005; Tsitsiklis, Bertsekas, & Athans, 1986) and diffusion (Chen & Sayed, 2011; Lopes & Sayed, 2008). Incremental adaptive networks cycle the network in a Hamiltonian path, where each adaptation step needs a full cycle. Although a full cycle in each adaptation step gives an overall result, finding a Hamiltonian path, susceptibility to node and link failure, and time-consuming adaptation steps are some serious obstacles for this topology. In diffusion topology, every node prepares a local estimation consulting with its adjacent nodes only, though addressing the incremental algorithm's drawbacks in cost of less generality. Using this methodology, we study spatio-temporal parametric modeling, and power spectral density estimation in this research, respectively.

This chapter is organized in the following way. First, the autoregressive model and parameters estimation are discussed and then four adaptive algorithms for estimating the AR model parameters and power spectral density are proposed. Afterwards by defining theoretical error criteria, an evaluation procedure is mathematically conducted for the proposed algorithms. Eventually, the performances of the presented algorithms are compared, both theoretically and numerically.

Notation

In this paper, lower-case letters are used to denote vectors, upper-case letters for matrices, plain letters for deterministic variables, and bold-face letters for random variables. For instance, $\boldsymbol{d}$ represents a random quantity while d is a realization or measurement of that, R is a deterministic

Figure 1. Conventional methods for PSD estimation

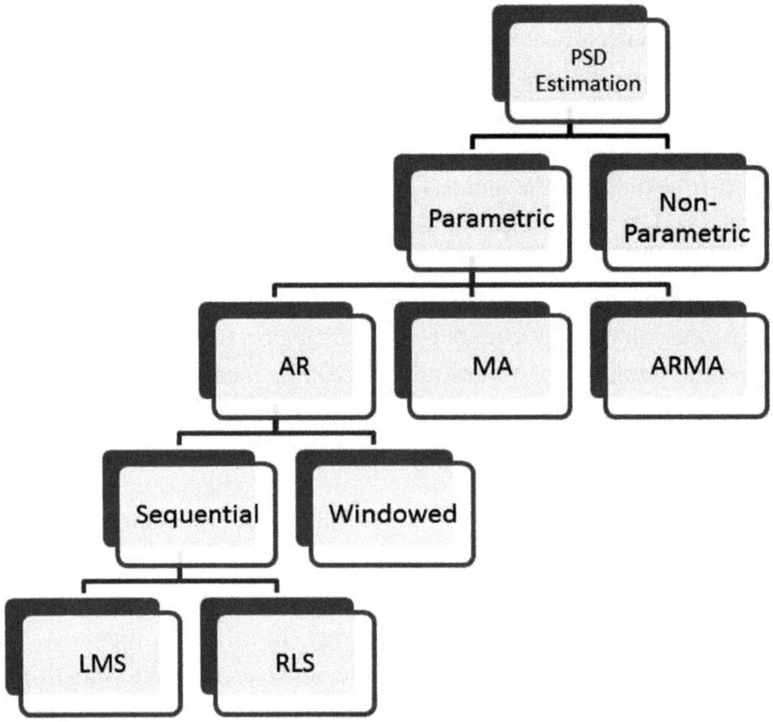

matrix where $\boldsymbol{w}$ denotes a random vector. All the vectors in this paper are column vectors, with the exception to the regression vectors ($\boldsymbol{u}_{k,i}$, $x_{p,i}$), which are taken to be row vectors. The notation $\mathbb{E}$ denotes the expectation operator, Tr(.) shows the trace of its matrix argument, while $(.)^*$ indicates complex conjugation for scalars, and complex conjugate transposition for matrices. col{.} is used to construct a block column vector and diag(.) operator makes a diagonal matrix using the vector operand. The vec(.) operator vectorizes matrices by stacking its columns on top of each other and $\otimes$ is used to represent the Kronecker multiplication.

BACKGROUND

A large number of stochastic processes that are encountered in real world are well approximated by a time series model. In this model, the driving sequence $u(n)$ and the output sequence $x(n)$ that model the data are related to each other via some coefficients, $a(k)$ and $b(k)$, by the following equation:

$$x(n) = -\sum_{k=1}^{p} a(k)\, x(n-k) + \sum_{k=0}^{q} b(k)\, u(n-k) \tag{1}$$

In which, p and q are the order of the corresponding branch. Generally, the driving sequence is considered as a white noise with variance σ^2. The most general model for time series is called moving average autoregressive (ARMA). The block diagram of this model is depicted in Figure 2.

The corresponding transfer function can be expressed in the following manner:

$$\mathcal{H}(\mathcal{Z}) = \frac{\mathcal{B}(\mathcal{Z})}{\mathcal{A}(\mathcal{Z})} \quad (2)$$

where $\mathcal{A}(\mathcal{Z})$ and $\mathcal{B}(\mathcal{Z})$ denote the Z-transform of the first part or AR, and the second part or MA, respectively. Needless to remind that the mentioned system is valid as long as it is stable. In other words, the poles related to the roots of $\mathcal{A}(\mathcal{Z})$ are obliged to reside within the unit circle.

If all $b(k)$s except $b(0)$ are zero in ARMA model we will have:

$$x(n) = -\sum_{k=1}^{p} a(k)x(n-k) + u(n), \quad (3)$$

which represents an AR process of order p. The present value of the process can be expressed as a weighted sum of the previous values and the noise (Figure 3).

Given this model, the power spectral density is calculated from the following equation:

$$P_{\text{AR}}(f) = \frac{\sigma^2}{|A(f)|^2}, \quad (4)$$

in which $A(f)$ is the Discrete Fourier Transform (DFT) of parameters series.

The AR coefficients are related to the autocorrelation functions (ACF) as:

$$r_{xx}(m) = \begin{cases} -\sum_{k-1}^{p} a(k) r_{xx}(m-k) & m > 0 \\ -\sum_{k-1}^{p} a(k) r_{xx}(-k) + \sigma^2 & m = 0 \\ r_{xx}^{*}(-m) & m < 0 \end{cases} \quad (5)$$

or in a matrix form (shown in Box 1).

Figure 2. ARMA model block diagram

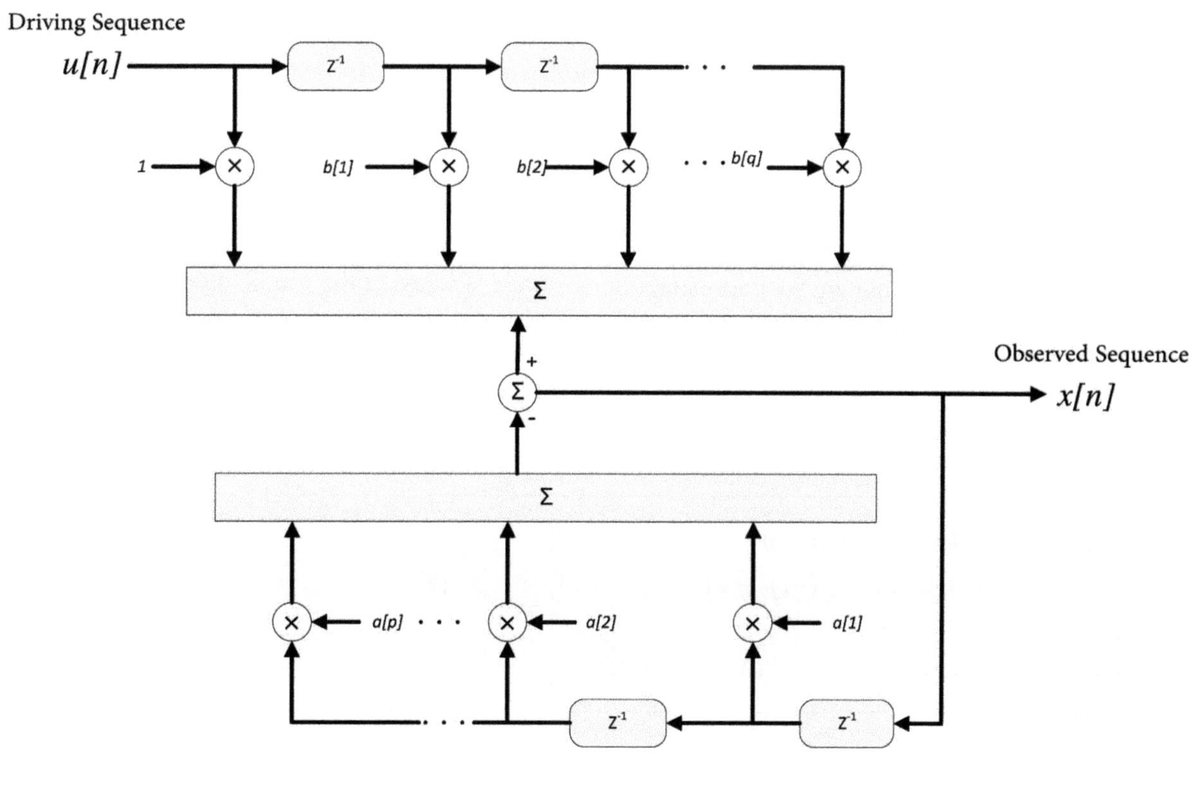

Figure 3. AR model block diagram

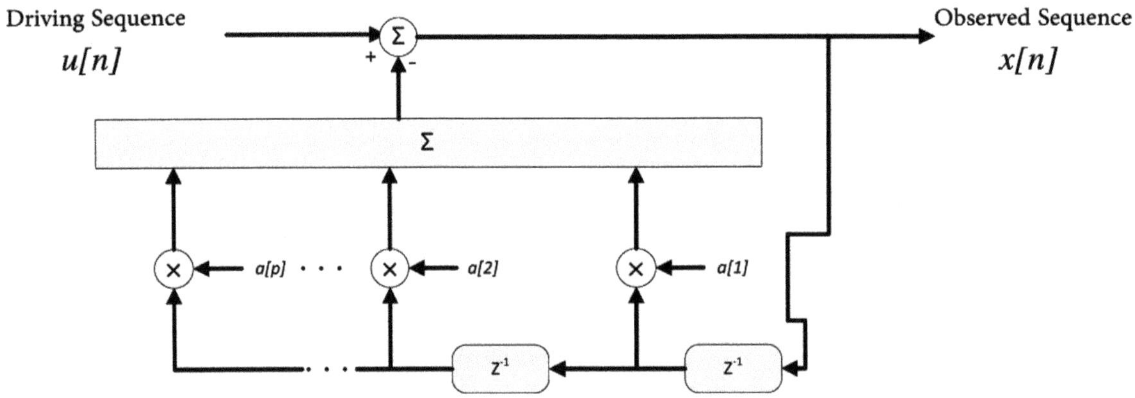

Using the autocorrelation functions, the AR model's parameters can be calculated after solving this set of linear equations called Yule-Walker equations.

Adaptive Estimation of AR Parameters

As mentioned earlier, in cases where the spectral pattern is assumed to have small time variations, adaptive methods are used to estimate the model parameters. In this case, the model coefficients are updated by each new input sample, and a new spectral density curve can be consequently drawn based on the updated coefficients. It prepares a fast and real-time time-frequency analysis (Zheng, 1998). The Doppler tone frequency of a sonar signal recorded from a moving target is an example of a signal source having the parameters that slowly change with time.

Adaptive algorithms used for estimating the AR model parameters are divided into two categories. The most fundamental approach is based on the gradient method. The Least Mean Square (LMS) algorithm, as a well-known gradient-based method, may be an appropriate choice for estimating adaptive coefficients. The second class of adaptive algorithms comprises the RLS-based methods providing better performance but at some additional computation expenses.

Considering the vector of AR model parameters to be:

$$\mathbf{a}_p^T = \left(a_p\left(1\right),\ldots,a_p\left(p\right)\right), \tag{7}$$

p elements of the AR model parameters vector at time i can be calculated based on the previous time $i-1$, using the gradient descent method:

Box 1.

$$\begin{bmatrix} r_{xx}(0) & r_{xx}(-1) & \ldots & r_{xx}(-(p-1)) \\ r_{xx}(1) & r_{xx}(0) & \ldots & r_{xx}(-(p-2)) \\ \vdots & \vdots & \ddots & \vdots \\ r_{xx}(p-1) & r_{xx}(p-2) & \ldots & r_{xx}(0) \end{bmatrix} \times \begin{bmatrix} a(1) \\ a(2) \\ \vdots \\ a(p) \end{bmatrix} = -\begin{bmatrix} r_{xx}(1) \\ r_{xx}(2) \\ \vdots \\ r_{xx}(p) \end{bmatrix} \tag{6}$$

$$\hat{\mathbf{a}}_{p,i} = \hat{\mathbf{a}}_{p,i-1} - \mu\nabla\mathbb{E}\left\{e_{p,i}(i)\right\}, \tag{8}$$

wherein μ is the learning rate, ∇ is gradient, $\mathbb{E}$ is the expected value, and $e_{p,i}\left(i\right)$ is the estimation error expressed as follows:

$$\begin{aligned} e_{p,i}(i) &= x(i) + \mathbf{x}_p^T(i)\hat{\mathbf{a}}_{p,i-1} \\ \mathbf{x}_p^T(i) &= (x(i-1),\ldots,x(i-p))^T. \end{aligned} \tag{9}$$

By defining autocorrelation and cross correlation functions, we acquire:

$$\begin{aligned} \boldsymbol{r}_p &= \mathbb{E}\left\{x(i)\mathbf{x}_p^*(i)\right\} \\ \boldsymbol{R}_p &= \mathbb{E}\left\{\mathbf{x}_p^*(i)\mathbf{x}_p^T(i)\right\}. \end{aligned} \tag{10}$$

Using LMS instantaneous approximations instead of statistical approximations,

$$\begin{aligned} \hat{\boldsymbol{r}}_p &= x(i)\mathbf{x}_p^*(i) \\ \hat{\boldsymbol{R}}_p &= \mathbf{x}_p^*\mathbf{x}_p^T, \end{aligned} \tag{11}$$

Algorithm 1 can be employed to estimate the AR model coefficients.

DISTRIBUTED ADAPTIVE POWER SPECTRAL ESTIMATION

Let consider a general set of N sensor nodes dispersed in a geographical region measuring a particular parameter of a random process. At each time instant i, every node (k) collects a scalar measurement (i.e., $\boldsymbol{d}_k\left(i\right)$) and a regressor vector (i.e., $\boldsymbol{u}_{k,i}$) while these two measured quantities are assumed to be linearly related through an unknown vector, w°. It is also assumed that the observed quantities are corrupted by an additive noise, $\boldsymbol{v}_k\left(i\right)$, through a conventional linear regressor model:

$$\boldsymbol{d}_k(i) = \boldsymbol{u}_{k,i}\, w^\circ + \boldsymbol{v}_k(i), \qquad k = 1,\ldots,N. \tag{12}$$

In this model, w° is the optimum parameter to be estimated. Interpreting this unknown parameters in different ways, we can introduce vast variety of applications. In this chapter we assume that the unknown vector is the parameter of an AR system.

The objective in adaptive networks is to estimate the optimal vector, w°, that minimizes the mean square error:

Algorithm 1. PSD Estimation pseudo-code based on LMS approximation for an AR process

For each time iteration $i \geq 0$:

Start with an initial estimate, $\mathbf{a}_p(-1) = 0$:

1: repeat

2: $\hat{\mathbf{a}}_{p,i} = \hat{\mathbf{a}}_{p,i-1} - \mu\mathbf{x}_{p-1}^* e_{p,i}(i)$

3: until $\left| e_{p,i}(i) \right| \leq \epsilon$

$A(f) = \mathcal{F}\{\mathbf{a}_p\}$

$$P_{AR}(f) = \frac{\sigma^2}{|A(f)|^2}$$

$$w^{\circ} = \underset{w}{\operatorname{argmin}} \ \frac{1}{N}\sum_{k=1}^{N} \mathbb{E} \,|\, \boldsymbol{d}_k - \boldsymbol{u}_k w \,|^2 . \qquad (13)$$

A reasonable way to determine w° in an adaptive network is to aggregate all the measurements in a central node and employ an adaptive filter to estimate w° (Figure 4(a)). Considering $\boldsymbol{d}_k(i)$, $\boldsymbol{u}_{k,i}$, and w_i for this fusion-based strategy, called centralized algorithm, to be

$$\begin{aligned} \boldsymbol{d}_k(i) &= x(i) \\ \boldsymbol{u}_{k,i} &= \mathbf{x}_p^T(i) = (d(i-1),\ldots,d(i-p)) \qquad (14) \\ w_i &= \mathbf{a}_p^T = (a_p(1),\ldots,a_p(p)), \end{aligned}$$

Algorithm 2 is proposed to calculate the spectral density based on the LMS method and centralized cooperation. If we take the observation noise into account, Equation 3 could be rewritten as follows:

$$x(n) = -\sum_{k=1}^{p} a_k x(n-k) + u(n) + u_o(n), \qquad (15)$$

in which $u_o(n)$ is the observed additive noise. In the case that both the driving sequence and the additive noise are considered as normal processes with variances σ^2 and σ_o^2, and are also both generated independently, we can aggregate them in terms of a total normal process as follows:

$$u_v(n) = u(n) + u_o(n)$$

$$\sigma_v^2 = \sigma^2 + \sigma_o^2.$$

The centralized algorithm involves all measurements in every adaptation step. Although this feature increases the generality of the final estimation, it leads to some technical difficulties. The network is not only dependent on a single node, turning it to a single point of failure, but also, is seen to be inefficient from telecommunication and power consumption points of view. An alternative algorithm, explored by Lopes and Sayed (2007), is called distributed incremental strategy. Despite the centralized algorithm, this approach is not reliant on a single node, and as a result, it does not generally have a single point of failure, where the estimation procedure is distributed throughout the network (Figure 4(b)). It starts from the first node and cycles throughout the network sequentially. Instead of applying network parameter w_i, the incremental strategy uses local estimations $\psi_{k,i}$ including the information about nodes 1 through k. Each transition leads to a knowledge increment, hence all measurements about the estimated parameter is observed over the last node. This strategy for PSD estimation is listed in Algorithm 3.

As it can be seen in Algorithm 3, each adaptation step includes N sub-steps to cycle through N sensor nodes. In this condition, the increasing number of network sensors reduces the performance of the whole network as a consequence of

Algorithm 2. Centralized LMS PSD estimation

Start with an initial estimation $\{w_{-1} = 0\}$:

In each iteration $\geq$ 0:

repeat:

1: receive $\{d_{k,i}, u_{k,i}\}$ from all nodes.

2: Adapt the parameter vector $\{w_{i-1}\}$:

$$w_i = w_{i-1} + \mu_k \sum_{k=1}^{N} u_{k,i}^{*}\left(d_k(i) - u_{k,i} w_{i-1}\right).$$

3: Send the adapted vector $\{w_i\}$ to all nodes.

After Convergence:

$$W(f) = \mathcal{F}\{w_i\}$$

$$P_{\mathrm{AR}}(f) = \frac{\sigma_v^2}{|\, W(f) \,|^2}$$

Figure 4. Modes of cooperation.(a): Centralized. (b): Incremental. (c): Diffusion. (d)Non-cooperative.

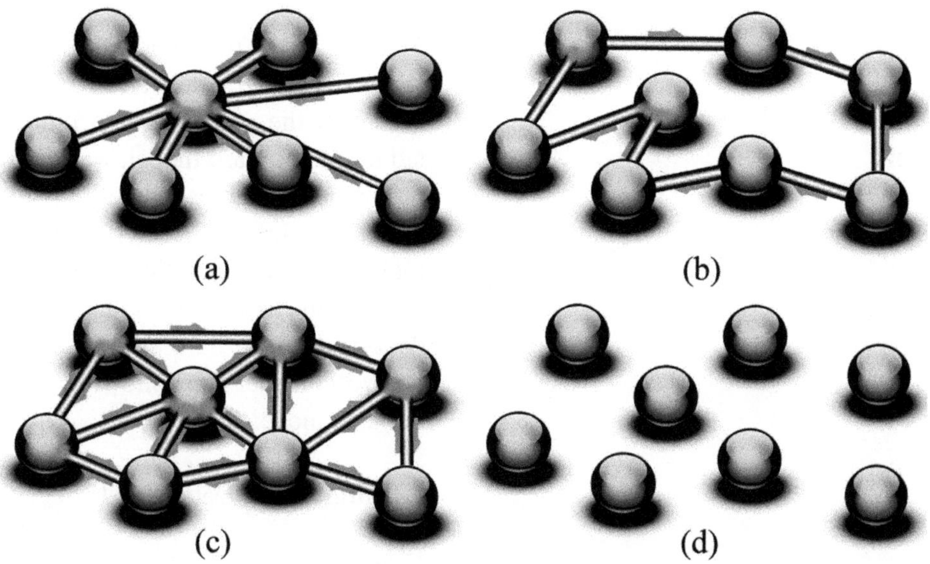

Algorithm 3. Distributed incremental LMS PSD estimation

Start with an initial estimation, $\psi_{N,-1} = 0$:

In each iteration $i \geq 0$:

1: for nodes $k = 1$ to $N - 1$:

$\psi_{k,i} = \psi_{k-1,i} + \mu_k \, u^*_{k,i}(d_k(i) + u_{k,i} \, \psi_{k-1,i})$

2: in node N

$w_i \leftarrow \psi_{N,i}$

After Convergence:

$W_k(f) = \mathcal{F}\{\psi_{k,i}\}$

$P_{\mathrm{AR}_k}(f) = \dfrac{\sigma^2_{v_k}}{| W_k(f) |^2}$

the latency caused by long-path loop (Sayed & Cattivelli, 2009). Furthermore, the network is still susceptible to link and node failure. As a solution, distributed diffusion plan has been proposed in (Lopes & Sayed, 2008). In adaptive networks with diffusion strategy, each node uses the information of all the adjacent nodes instead of just the previous one (Figure 4(b)). For this purpose, node k makes a local combination of data from all the neighboring nodes. This combination can be expressed by the following equation

$$\psi_{k,i-1} = \sum_{\ell \in \mathcal{N}_k} a_{k,l} \, \phi_{l,i}, \tag{16}$$

wherein $\phi_{l,i}$ is an individual estimation in node ℓ and iteration i. $\mathcal{N}_k$ represents the set of neighboring nodes (including node k itself), and $a_{k,l}$ denotes the coefficient used to combine the adjacent individual estimations. PSD estimation using distributed diffusion technique is summarized in Algorithm 4, wherein the adaptation step is followed by the combination step (adapt-then-combine (ATC)).

The combination and adaptation steps can be swapped. In this case, combine-then-adapt (CTA) version of the diffusion strategy works as indicated in Algorithm 5.

A primitive adaptive solution to minimize Equation 13 is employing individual adaptive nodes whereby the network and cooperation are negated (Figure 4(d)). Expectedly, this mode will not prepare a comprehensive estimation

Algorithm 4. Distributed LMS diffusion ATC PSD estimation

Start with an initial estimate, $\phi_{\ell,-1}$:
In each iteration $i \geq 0$:
For each node k:
1: Adapt local estimation:
$$\phi_{k,i} = \psi_{k,i-1} + \mu_k\, u^*_{k,i}(d_k(i) - u_{k,i}\, \psi_{k,i-1})$$
2:Combine local estimations:
$$\psi_{k,i} = \sum_{\ell \in \mathcal{N}_k} a_{k,l}\, \phi_{l,i}$$
After Convergence:
$$W_k(f) = \mathcal{F}\{\psi_{k,i}\}$$
$$P_{\mathrm{AR_k}}(f) = \frac{\sigma^2_{v_k}}{|\, W_k(f)\,|^2}$$

Algorithm 5. Distributed LMS diffusion CTA

Start with an initial estimate, $\psi_{\ell,-1}$:
In each iteration $i \geq 0$:
For each node k:
1: Combine local estimations:
$$\phi_{k,i-1} = \sum_{\ell \in \mathcal{N}_k} a_{k,l}\, \psi_{l,i-1}$$
2: Adapt local estimation:
$$\psi_{k,i} = \phi_{k,i-1} + \mu_k\, u^*_{k,i}(d_k(i) - u_{k,i}\, \phi_{k,i-1})$$
After Convergence:
$$W_k(f) = \mathcal{F}\{\psi_{k,i}\}$$
$$P_{\mathrm{AR_k}}(f) = \frac{\sigma^2_{v_k}}{|\, W_k(f)\,|^2}$$

as the nodes do not consult with each other. If we are aimed at investigating the efficiency of adaptive networks, obviously, the performance of each algorithm should be compared with the non-cooperative algorithm.

PERFORMANCE EVALUATION

To evaluate the proposed algorithms we resort to fitting our model to the standard adaptive networks model. If the proposed algorithms are well fitted to the standard model, we can then calculate the theoretical error indices, and compare them with the simulation results. In adaptive networks realm, we can generally exploit error references under some assumptions that concentrate on the data model.

We consider the following assumptions for the data model:

Assumption 1: Data model.

The unknown vector, $\mathbf{w}^\circ_i$ is related to $\boldsymbol{d}$ and $\boldsymbol{u}$ as follows:

$$\boldsymbol{d}_k = \boldsymbol{u}_k\, w^\circ + \boldsymbol{v}_k. \tag{17}$$

where v_k denotes white noise with variance σ^2 and independent of $\{\, \boldsymbol{d}_l, \boldsymbol{u}_l \,\}$ for all l s.

1. $\boldsymbol{u}_k$ is independent of $\boldsymbol{u}_l$ for $k \neq l$ (spatial independence).
2. For every k, the sequence $\boldsymbol{u}_k$ is independent over time (time independence).

If we want to take the PSD estimation and AR parametric modeling into account, some extra assumptions should be added as follows:

Assumption 2: AR model assumptions.

1. All of the distributed nodes are observing the same process: $R_{u,k} = R_u$.
2. The AR model is driven by a white noise. Hence, the process is driven from a circularly Gaussian distribution, and the following eigendecomposition is valid: $R_u = U \Lambda U^*$.

3. The learning rate, μ, is assumed to be constant for all nodes: $\mu_k = \mu$.

In the incremental strategy, according to the work of Lopes and Sayed (2007), we define the error vectors and signals as follows:

$$\widetilde{\psi}_{k,i} = w^{\circ} - \psi_{k,i} \quad (18)$$

$$\mathbf{e}_{a,k}(i) = \boldsymbol{u}_{k,i}\widetilde{\psi}_{k,i}.$$

Considering these relations, the Mean Square Deviation (MSD) and Excess Mean Square Error (EMSE) at each node can be defined as:

$$\text{MSD}_k^{\text{inc}} = \mathbb{E}\left\|\widetilde{\psi}_k\right\|^2 \quad (19)$$

$$\text{EMSE}_k^{\text{inc}} = \mathbb{E}\,|\,\mathbf{e}_{a,k}\,|^2\,. \quad (20)$$

By averaging the error values over the nodes, the network performance criteria can be defined in each iteration:

$$\text{MSD}_{\text{network}}^{\text{inc}}(i) = \frac{1}{N}\sum_{k=1}^{N}\text{MSD}_k^{\text{inc}}(i) \quad (21)$$

$$\text{EMSE}_{\text{network}}^{\text{inc}}(i) = \frac{1}{N}\sum_{k=1}^{N}\text{EMSE}_k^{\text{inc}}(i). \quad (22)$$

The steady-state error at each node is calculated in the following manner:

$$\text{MSD}_k^{\text{inc}} = \mathbb{E}\left\|\widetilde{\psi}_{k,\infty}\right\|^2 \quad (23)$$

$$\text{EMSE}_k^{\text{inc}} = \mathbb{E}\,|\,\mathbf{e}_{a,k}(\infty)\,|^2\,. \quad (24)$$

Considering the eigen-decomposition of the regressors covariance in each node to be:

$$R_{u,k} = U_k\Lambda_k U_k^*, \quad (25)$$

where U_k is a unitary matrix and Λ_k represents a diagonal matrix containing the eigenvalues of $R_{u,k}$ which is defined in the following way (diag(.) operator builds a diagonal matrix)

$$\Lambda_k = \text{diag}\{\lambda_{k,1}, \lambda_{k,1}, \ldots, \lambda_{k,M}\}, \quad (26)$$

and defining further quantities as:

$$g_k = \mu_k^2\sigma_{v,k}^2\lambda_k^T \quad (27)$$

$$F_k^{\text{inc}} = I - 2\mu_k\Lambda_k + \gamma\mu_k^2\Lambda_k^2 + \mu_k^2\lambda_k\lambda_k^T \quad (28)$$

$$\Pi_{k,l} = F_{k+l-1}^{\text{inc}}F_{k+l}^{\text{inc}}\ldots F_N^{\text{inc}}F_1^{\text{inc}}\ldots F_{k-1}^{\text{inc}} \quad (29)$$

$$a_k = g_k\Pi_{k,2} + g_{k+1}\Pi_{k,3}\ldots + g_{k-2}\Pi_{k,N} + g_{k-1}, \quad (30)$$

then the steady-state MSD and EMSE at each node can be calculated by the following equations (Lopes & Sayed, 2007; Sayed & Cattivelli, 2009):

$$\boxed{\text{MSD}_k^{\text{inc}} = a_k(I - \Pi_{k,1})^{-1}q} \quad (31)$$

$$\boxed{\text{EMSE}_k^{\text{inc}} = a_k(I - \Pi_{k,1})^{-1}\lambda_k}. \quad (32)$$

Assuming the following error vector in the centralized mode:

$$\tilde{\boldsymbol{w}}_i = w^{\circ} - \boldsymbol{w}_i, \tag{33}$$

and considering the following variables:

$$R = \sum_{k=1}^{N} R_{u,k} \tag{34}$$

$$r_k = \text{vec}(R_{u,k}), \quad q = \text{vec}(I) \tag{35}$$

$$\begin{aligned} F_{\text{cnt}} &= I - \mu(I \otimes R) - \mu(R^T \otimes I) \\ &+\mu^2(R^T \otimes R) + \mu^2 \sum_{k=1}^{N} r_k r_k^* \end{aligned}, \tag{36}$$

the steady-state MSD and EMSE can consequently be estimated as:

$$\boxed{\text{MSD}^{\text{cnt}} = \mu^2 \sum_{k=1}^{N} \sigma_{v_k}^2 r_k^* (I - F_{\text{cnt}})^{-1} q} \tag{37}$$

$$\boxed{\text{EMSE}_k^{\text{cnt}} = \mu^2 \sum_{k=1}^{N} \sigma_{v_k}^2 r_k^* (I - F_{\text{cnt}})^{-1} r_k}. \tag{38}$$

To aggregate the analysis of different types of diffusion mode, we had better deal with a general diffusion adaptive solution (Sayed, 2013):

$$\phi_{k,i-1} = \sum_{\ell \in \mathcal{N}_k} a_{1,\ell k} \boldsymbol{w}_{\ell,i-1}$$

$$\boldsymbol{\psi}_{k,i} = \phi_{k,i-1} + \mu_k \boldsymbol{u}_{k,i}^* [\boldsymbol{d}_k(i) - \boldsymbol{u}_{k,i} \phi_{k,i-1}] \tag{39}$$

$$\boldsymbol{w}_{k,i} = \sum_{\ell \in \mathcal{N}_k} a_{2,\ell k} \boldsymbol{\psi}_{\ell,i},$$

where $\{a_{1,\ell k}, a_{2,\ell k}\}$ are real non-negative coefficients corresponding to the (ℓ, k) entries of $N \times N$ combination matrices $\{A_1, A_2\}$. The advantage of using such method is that different choices of $\{A_1, A_2\}$ lead to alternative types of diffusion mode. If we choose $A_1 = I_N$ and $A_2 = A$, diffusion ATC is set up. To reach diffusion CTA, we just ought to choose $A_1 = A$ and $A_2 = I_N$, and eventually selecting $A_1 = I_N$ and $A_2 = I_N$ negates the cooperation.

Error vectors in diffusion mode are defined as follows:

$$\tilde{\phi}_{k,i} \triangleq w^{\circ} - \phi_{k,i} \tag{40a}$$

$$\tilde{\psi}_{k,i} \triangleq w^{\circ} - \psi_{k,i} \tag{40b}$$

$$\tilde{\boldsymbol{w}}_{k,i} \triangleq w^{\circ} - \boldsymbol{w}_{k,i}. \tag{40c}$$

The performance analysis method for diffusion algorithms is based on network matrices, in which the same quantities of the network are collected in a single matrix. Error matrices are consequently defined as:

$$\tilde{\psi}_i = \begin{bmatrix} \tilde{\psi}_{1,i} \\ \tilde{\psi}_{2,i} \\ \vdots \\ \tilde{\psi}_{N,i} \end{bmatrix}, \tilde{\phi}_i = \begin{bmatrix} \tilde{\phi}_{1,i} \\ \tilde{\phi}_{2,i} \\ \vdots \\ \tilde{\phi}_{N,i} \end{bmatrix}, \tilde{\boldsymbol{w}}_i = \begin{bmatrix} \tilde{\boldsymbol{w}}_{1,i} \\ \tilde{\boldsymbol{w}}_{2,i} \\ \vdots \\ \tilde{\boldsymbol{w}}_{N,i} \end{bmatrix}. \tag{41}$$

Let consider the following matrices:

$$\mathcal{A}_1 = A_1 \otimes I_N \tag{42}$$

$$\mathcal{A}_2 = A_2 \otimes I_N \tag{43}$$

$$\mathcal{M} = \text{diag}\left\{\mu_1 I_M, \mu_2 I_M, \ldots, \mu_N I_M\right\} \tag{44}$$

$$\mathcal{R}_i = \text{diag}\left\{\boldsymbol{u}_{1,i}^* \boldsymbol{u}_{1,i}, \boldsymbol{u}_{2,i}^* \boldsymbol{u}_{2,i}, \ldots, \boldsymbol{u}_{N,i}^* \boldsymbol{u}_{N,i}\right\} \tag{45}$$

$$\mathrm{s}_i = \text{col}\left\{\boldsymbol{u}_{1,i}^* \boldsymbol{v}_1(i), \boldsymbol{u}_{2,i}^* \boldsymbol{v}_2(i), \ldots, \boldsymbol{u}_{N,i}^* \boldsymbol{v}_N(i)\right\}, \tag{46}$$

wherein col{.} is used to construct a block column vector, a compact relation is extracted for the evolution of error matrix as follows (Sayed, 2013):

$$\tilde{\boldsymbol{w}}_i = \mathcal{B}_i \tilde{\boldsymbol{w}}_{i-1} - \mathcal{G}\mathrm{s}_i, \qquad i \geq 0, \tag{47}$$

in which:

$$\mathcal{B}_i = \mathcal{A}_2^T \left(I_{NM} - \mathcal{M}\mathcal{R}_i\right) \mathcal{A}_1^T \tag{48}$$

$$\mathcal{G} = \mathcal{A}_2^T \mathcal{M}. \tag{49}$$

Likewise the incremental mode, we need some further quantities to be properly defined:

$$\mathcal{S} = \text{diag}\left\{\sigma_{v,1}^2 R_{u,1}, \sigma_{v,2}^2 R_{u,2}, \ldots, \sigma_{v,N}^2 R_{u,N}\right\} \tag{50}$$

$$\mathcal{Y} = \mathcal{G} \ \mathcal{S} \ \mathcal{G}^T \tag{51}$$

$$\mathcal{F} = \mathcal{B}^T \otimes \mathcal{B}^*. \tag{52}$$

The steady-state MSD and EMSE can be calculated at each node, in terms of defined quantities in diffusion strategy as:

$$\boxed{\text{MSD}_k^{\text{diff}} = \left[\text{vec}\left(\mathcal{Y}^{\mathrm{T}}\right)\right]^{\mathrm{T}} .(\mathrm{I} - \mathcal{F})^{-1}.vec\left(\mathcal{J}_{\mathrm{k}}\right)} \tag{53}$$

$$\boxed{\text{EMSE}_k^{\text{diff}} = \left[\text{vec}\left(\mathcal{Y}^{\mathrm{T}}\right)\right]^{\mathrm{T}} .(\mathrm{I} - \mathcal{F})^{-1}.vec\left(\mathcal{T}_{\mathrm{k}}\right)}, \tag{54}$$

wherein $\mathcal{J}_k$ is a $N \times N$ block diagonal matrix with blocks of size $M \times M$, and all blocks on the diagonal are zero except for the diagonal block of index k whose value is I_M for $\mathcal{J}_k$ and $\mathrm{R}_{u,k}$ for $\mathcal{T}_k$:

$$\mathcal{J}_k = \text{diag}\left\{0_M, \ldots, 0_M, I_M, 0_M, \ldots, 0_M\right\} \tag{55}$$

$$\mathcal{T}_k = \text{diag}\left\{0_M, \ldots, 0_M, \mathrm{R}_{u,k}, 0_M, \ldots, 0_M\right\}. \tag{56}$$

Simulations

Consider an AR(4) process with desired coefficients while guaranteeing the process to be Wide Sense Stationary (WSS). The PSD pattern of the generated process is depicted in Figure 5(a). Pole-Zero plot of the system function is shown in Figure 5(b) in which the poles are all inside the unit circle, thus generate a WSS process. Figure 6(a) shows fifteen scattered nodes observing an AR process arose from white noises with dissimilar variances, as depicted in Figure 6(b). The choice of noise variances at each node is completely random. In order to have a clue to choose a step size guaranteeing convergence and being small enough too, the trace of regressor covariances is shown in Figure 6(c).

Once the data was generated using the predefined coefficients and the aforementioned conditions, we employed different algorithms to calculate the error criteria. The learning rate is considered to be 3 in all modes except the incremental mode, in which this number is divided by the number of nodes to reach comparable results. Running the simulation for 100 times and averaging the results thereafter, assures that the results are sufficiently valid. After calculating the steady-state MSD and EMSE with the last 50 averaged coefficients, simulation measures are available. Based on Equations 31, 32, 37, 38, 53, and 54 we can calculate theoretical quantities.

Simulation quantities versus theoretical results are depicted in Figures 8-11. A holistic glance at the curves convinces us that the simulation results follow the theoretical achievements to an acceptable extent, supporting the concepts. In centralized and incremental modes, simulation results closely resembles the theoretical measures, while this resemblance decreases in diffusion mode. The small observable dissimilarity is due to approximations applied in theoretical MSD and EMSE in diffusion mode.

PERFORMANCE ANALYSIS

In this section, we aim to compare different algorithms considering AR model characteristics. For this purpose, using the MSD criterion described earlier, we try to investigate the best performance for various algorithms. Comparisons in this section will be done in two kinds. For algorithms which are possible to be compared using algebraic formulations, the algebraic comparison will be performed after mathematical simplifications. In this kind, we can theoretically compare centralized with incremental, and diffusion with non-cooperative algorithms. However, for the expressions which are not theoretically comparable, we are compelled to do the task by employing numerical results.

Centralized vs. Incremental

To compare Algorithms 2 and 3 we use the variance relations for the centralized and incremental algorithms, coming respectively (Cattivelli & Sayed, 2011):

$$\boxed{\mathbb{E}\,\|\,\tilde{\boldsymbol{w}}_i\,\|_{\Sigma'}^2 = \mathbb{E}\,\|\,\tilde{\boldsymbol{w}}_{i-1}\,\|_{\Sigma'}^2 + \mu^2 \sum_{k=1}^{N} \sigma_{v_k}^2 \,\mathrm{Tr}\left(R_{u,k}\Sigma\right)} \tag{57}$$

$$\boxed{\mathbb{E}\,\|\,\bar{\boldsymbol{\psi}}_{k,i}\,\|_{\bar{\sigma}_k}^2 = \mathbb{E}\,\|\,\bar{\boldsymbol{\psi}}_{k-1,i}\,\|_{\bar{F}_k\bar{\sigma}_k}^2 + \mu_k^2 \sigma_{v,k}^2 \left(\lambda_k^T \bar{\sigma}_k\right)}. \tag{58}$$

Since all nodes are observing the same process, the covariance matrix of the regressor will be the same at all nodes:

$$R_{u,k} = R_u. \tag{59}$$

In centralized mode, according to Equation 34 we will have:

$$R = NR_u. \tag{60}$$

Additionally, the learning rate assumed to be equal for all nodes ($\mu_k = \mu$). Taking this into consideration, we can simplify the variance relation of the centralized algorithm. Defining the following modified values:

$$\begin{aligned} &\bar{\boldsymbol{w}}_i \triangleq U^* \tilde{\boldsymbol{w}}_i, \bar{\boldsymbol{u}}_k = \boldsymbol{u}_k U, \\ &\bar{\Sigma} \triangleq U^* \Sigma U,\ \bar{\Sigma}' \triangleq U^* \Sigma U, \end{aligned} \tag{61}$$

variance relation can be expressed as follows:

$$\mathbb{E}\,\|\,\tilde{\boldsymbol{w}}_i\,\|_{\bar{\Sigma}}^2 = \mathbb{E}\,\|\,\tilde{\boldsymbol{w}}_{i-1}\,\|_{\bar{\Sigma}'}^2 + \mu^2 \sum_{k=1}^{N} \sigma_{v_k}^2\, Tr\left(\Lambda\bar{\Sigma}\right) \tag{62}$$

Figure 5. (a) Power Spectral Density of the generated process. (b) System function Pole-Zero plot.

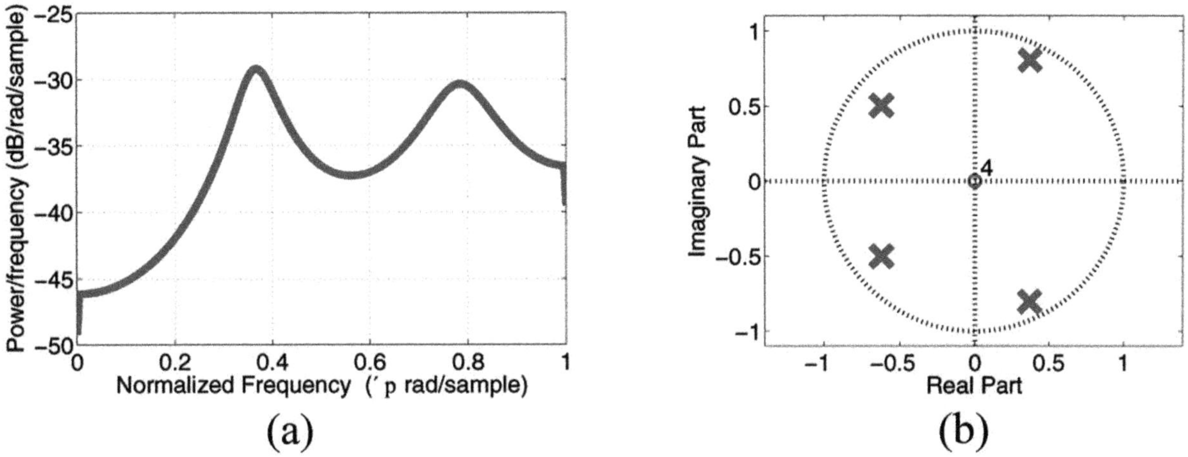

Figure 6. (a) Network topology (b) Noise power, (c) Trace of regressor covariances.

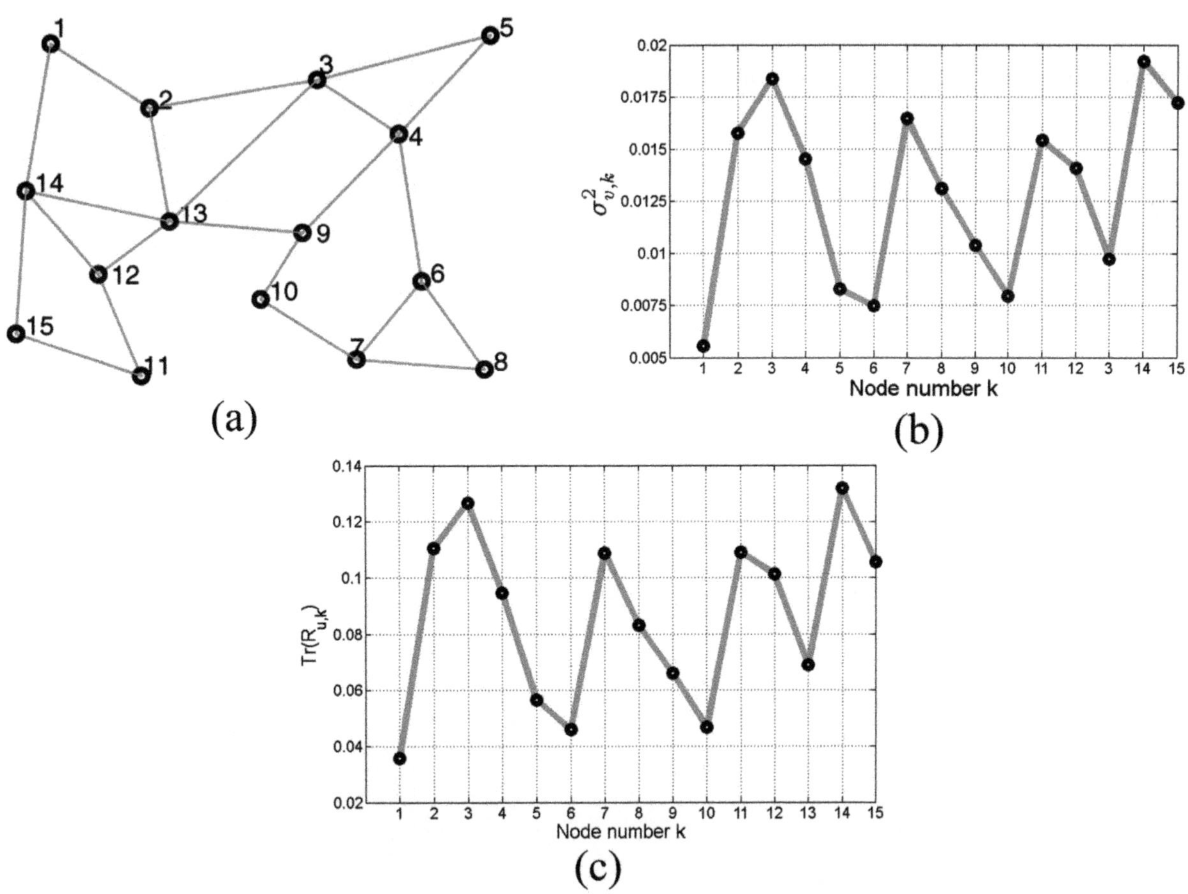

Figure 7. Steady-state error indices for centralized theory vs. simulation.

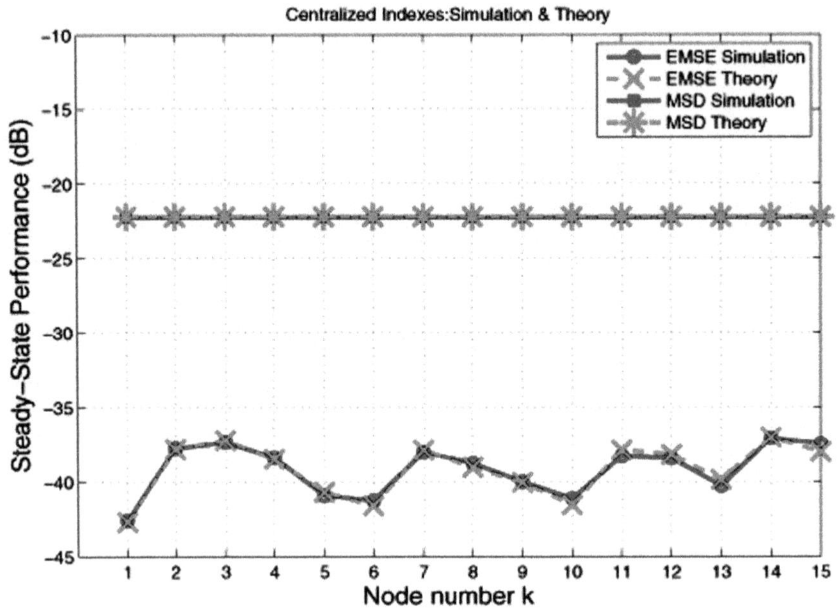

Figure 8. Steady-state error indices for incremental theory vs. simulation

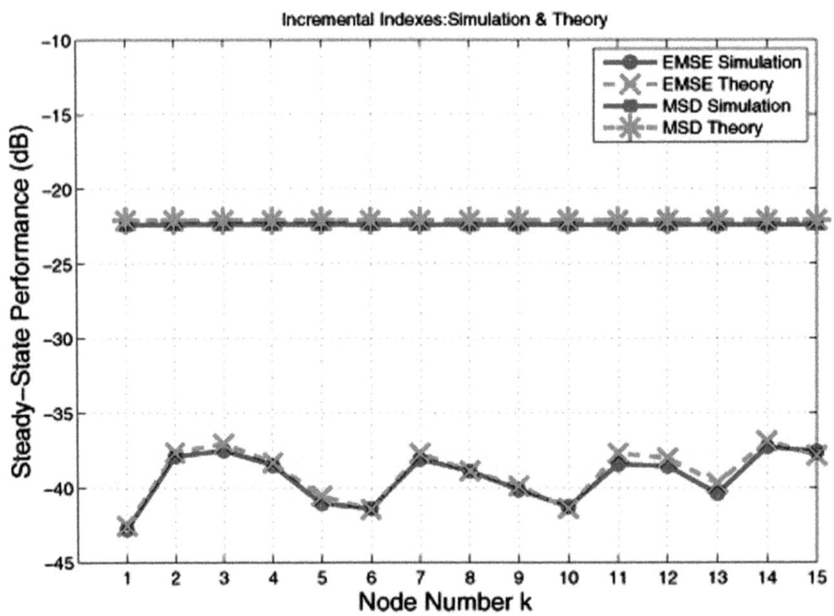

Figure 9. Steady-state error indices for ATC theory vs. simulation

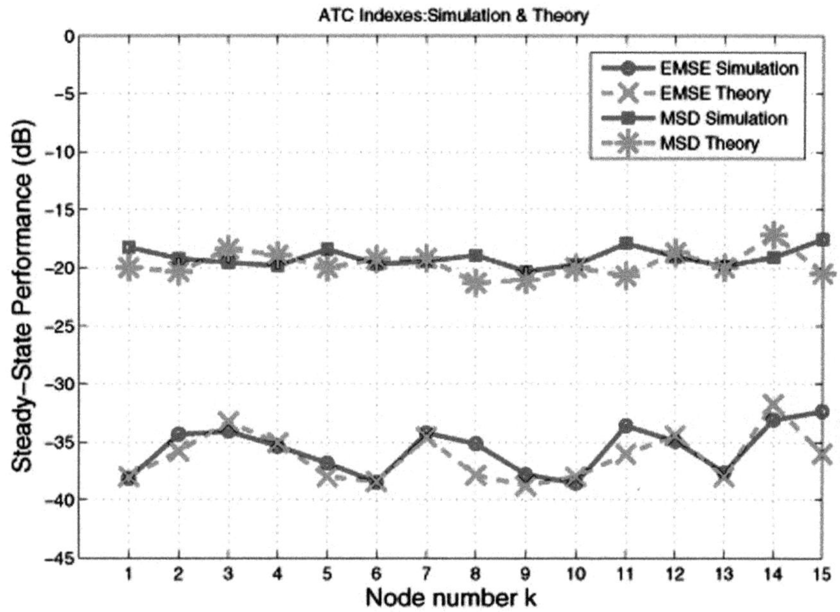

Figure 10. Steady-state error indices for CTA theory vs. simulation

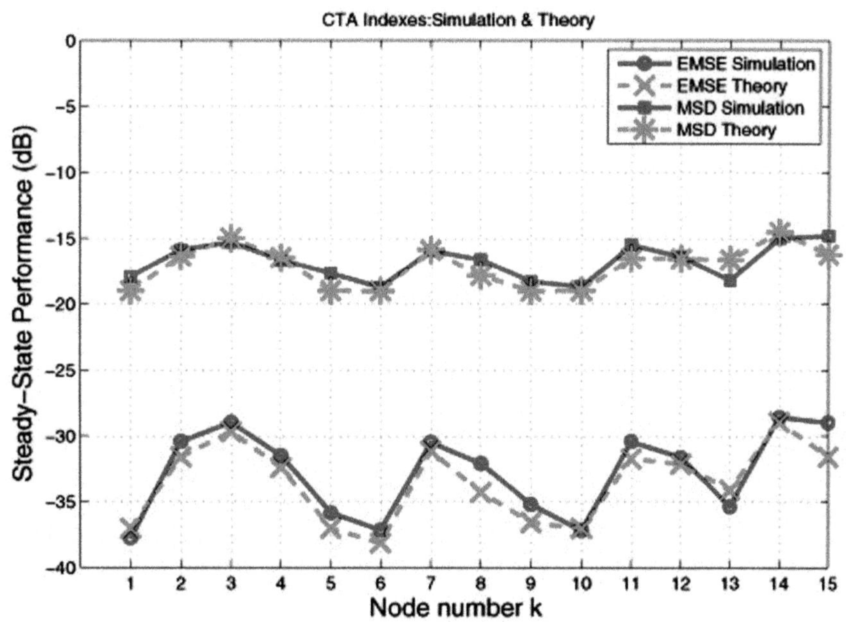

Figure 11. Steady-state error indices for no cooperation theory vs. simulation

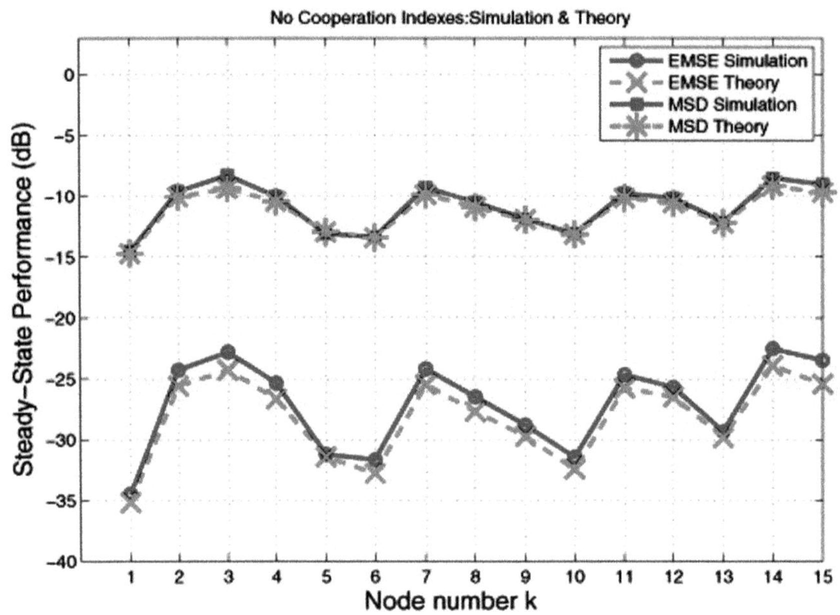

$$\bar{\Sigma}' = \bar{\Sigma} - \mu N \Lambda \bar{\Sigma} - \mu N \bar{\Sigma} \Lambda + \mu^2 \left[N^2 \Lambda \bar{\Sigma} \Lambda + N \Lambda \mathrm{Tr}\left(\bar{\Sigma} \Lambda\right) \right]. \quad (63)$$

Using the diag(.) operator and introducing the following items:

$$\bar{\sigma} = \mathrm{diag}\left(\bar{\Sigma}\right), \overline{\sigma'} = \mathrm{diag}\left(\overline{\Sigma'}\right), \bar{\lambda} = \mathrm{diag}\left(\Lambda\right) \quad (64)$$

$$\bar{F}_{\mathrm{cnt}} \triangleq I - 2\mu N \Lambda + \mu^2 N^2 \Lambda^2 + \mu^2 N \bar{\lambda} \bar{\lambda}^T, \quad (65)$$

and employing the inverse matrix formulas we can conclude to the following relation (proved in Appendix).

$$\bar{\lambda}^T (I - \bar{F}_{\mathrm{cnt}})^{-1} = \frac{\bar{\lambda}^T (2\mu N \Lambda - \mu^2 N^2 \Lambda^2)^{-1}}{1 - \mu^2 N \bar{\lambda}^T (2\mu N \Lambda - \mu^2 N^2 \Lambda^2)^{-1} \bar{\lambda}} \quad (66)$$

$$= \frac{\bar{\lambda}^T (2\mu N \Lambda - \mu^2 N^2 \Lambda^2)^{-1}}{1 - \mu \sum_{m=1}^{M} \frac{\lambda_m}{2 - \mu N \lambda_m}}. \quad (67)$$

Considering the above equation, we can rewrite the network MSD in centralized algorithm for an AR system as:

$$\boxed{\mathrm{MSD}_{\mathrm{AR}}^{\mathrm{cnt}} = \mu \left(\frac{1}{N} \sum_{\ell=1}^{N} \sigma_{v_\ell}^2 \right) \frac{\sum_{m=1}^{M} \frac{1}{2 - \mu N \lambda_m}}{1 - \mu \sum_{m=1}^{M} \frac{\lambda_m}{2 - \mu N \lambda_m}}}. \quad (68)$$

By applying the AR modeling assumptions for the incremental algorithm, we will have:

$$\bar{F}_{\mathrm{inc}} = I - 2\mu \Lambda + \mu^2 \Lambda^2 + \mu^2 \bar{\lambda} \bar{\lambda}^T. \quad (69)$$

Repeating Equation 58 for the whole network and in consecutive iterations, it can be proven that the equation will be reformed as follows:

$$\mathbb{E}\left\|\overline{\psi}_{k,\infty}\right\|^2_{(I-\overline{F}_{\text{inc}}^N)\overline{\sigma}} = \mu^2 \sum_{\ell=1}^{N} \sigma_{v_\ell}^2 \overline{\lambda}^T \overline{F}_{\text{inc}}^{(k-\ell)\text{mod}N} \overline{\sigma}, \tag{70}$$

in which mod(.) operator is used for the remainder calculation of dividing to . Averaging this equation for all nodes in the network and the desired vector , we obtain the following equation:

$$\frac{1}{N}\sum_{k=1}^{N} \mathbb{E}\left\|\overline{\psi}_{k,\infty}\right\|^2_{p} = \mu^2 \frac{1}{N}\sum_{\ell=1}^{N} \sigma_{v_\ell}^2 \overline{\lambda}^T \sum_{k=1}^{N} \overline{F}_{\text{inc}}^{(k-\ell)\text{mod}N} (I - \overline{F}_{\text{inc}}^N)^{-1} p. \tag{71}$$

Considering the following equation:

$$\sum_{k=1}^{N} \overline{F}_{\text{inc}}^{(k-\ell)\text{mod}N} = (I - \overline{F}_{\text{inc}})^{-1}(I - \overline{F}_{\text{inc}}^N), \tag{72}$$

we will have:

$$\frac{1}{N}\sum_{k=1}^{N} \mathbb{E}\left\|\overline{\psi}_{k,\infty}\right\|^2_{p} = \mu^2 \left(\frac{1}{N}\sum_{\ell=1}^{N} \sigma_{v_\ell}^2\right) \overline{\lambda}^T (I - \overline{F}_{\text{inc}})^{-1} p. \tag{73}$$

If the inverse matrix relations are applied, as in Equation 66, we can extract the network MSD for the incremental algorithms as:

$$\boxed{\text{MSD}_{\text{AR}}^{\text{inc}} = \mu\left(\frac{1}{N}\sum_{\ell=1}^{N}\sigma_{v_\ell}^2\right)\frac{\sum_{m=1}^{M}\frac{1}{2-\mu\lambda_m}}{1-\mu\sum_{m=1}^{M}\frac{\lambda_m}{2-\mu\lambda_m}}}. \tag{74}$$

To compare Equations 68 and 74 we ought to compare the following fractions in fact:

$$\frac{\sum_{m=1}^{M}\frac{1}{2-\mu\lambda_m}}{1-\mu\sum_{m=1}^{M}\frac{\lambda_m}{2-\mu\lambda_m}}, \qquad \frac{\sum_{m=1}^{M}\frac{1}{2-\mu N\lambda_m}}{1-\mu\sum_{m=1}^{M}\frac{\lambda_m}{2-\mu N\lambda_m}}. \tag{75}$$

Considering small step iteration coefficient it is obvious that:

$$\frac{1}{2-\mu\lambda_m} \le \frac{1}{2-\mu N\lambda_m}; \tag{76}$$

Therefore, the left fraction has a smaller nominator as well as greater denominator. For that reason, we can conclude the following inequality having equal conditions and small step iteration coefficient:

$$\frac{\sum_{m=1}^{M}\frac{1}{2-\mu\lambda_m}}{1-\mu\sum_{m=1}^{M}\frac{\lambda_m}{2-\mu\lambda_m}} \le \frac{\sum_{m=1}^{M}\frac{1}{2-\mu N\lambda_m}}{1-\mu\sum_{m=1}^{M}\frac{\lambda_m}{2-\mu N\lambda_m}}, \tag{77}$$

hence,

$$\boxed{\text{MSD}_{\text{AR}}^{\text{inc}} \le \text{MSD}_{\text{AR}}^{\text{cnt}}}. \tag{78}$$

We can conclude that under equal conditions in a network and for centralized and incremental algorithms which are observing and modeling an AR process, the incremental algorithm displays smaller MSD which accordingly means a more accurate PSD estimation.

Diffusion and Non-Cooperative Algorithms Comparison

In this section, we aim to compare the ATC and CTA algorithms of the diffusion topology with the non-cooperative mode. Considering the AR model assumptions (Assumption 2), we will have:

$$\mathcal{M} = \mu I_{NM} \tag{79}$$

$$\mathcal{R} = I_N \otimes R_u \tag{80}$$

$$\mathcal{G} = \mu \mathcal{A}_2^T. \tag{81}$$

As a result, we can rewrite Equation 48 in the following manner:

$$\mathcal{B} = \mathcal{A}_2^T \left(I_{NM} - \mathcal{M}\mathcal{R}\right)\mathcal{A}_1^T$$

$$= \left(\mathcal{A}_2^T \otimes I\right)\left(I - \mu\left(I \otimes R_u\right)\right)\left(\mathcal{A}_1^T \otimes I\right)$$

$$= \left(A_2^T A_1^T \otimes I\right) - \mu\left(A_2^T A_1^T \otimes R_u\right)$$

$$= A_2^T A_1^T \otimes \left(I - \mu R_u\right).$$

Moreover, the Equation 50 is also modified as

$$\mathcal{S} = \operatorname{diag}\left\{\sigma_{v,1}^2 R_u, \sigma_{v,2}^2 R_u, \ldots, \sigma_{v,N}^2 R_u\right\} \tag{82}$$

$$= R_v \otimes R_u. \tag{83}$$

Equation 51 can thus be simplified in the following fashion:

$$\mathcal{Y} = \mu^2 \mathcal{A}_2^T \mathcal{S} \mathcal{A}_2 \tag{84}$$

$$= \mu^2 \left(A_2^T \otimes I\right)\left(R_v \otimes R_u\right)\left(A_2 \otimes I\right) \tag{85}$$

$$= \mu^2 \left(A_2^T R_v A_2 \otimes R_u\right). \tag{86}$$

Employing the following theorem which is proven in (Sayed, 2013):

Theory 1: We can write the network MSD by defining the following series representation:

$$\mathrm{MSD}_{\text{network}}^{\text{diff}} = \frac{1}{N}\sum_{j=0}^{\infty} \operatorname{Tr}\left(\mathcal{B}^j \mathcal{Y} \mathcal{B}^{*j}\right) \tag{87}$$

we can now obtain the network MSD to be:

$$\mathrm{MSD}_{\text{AR}}^{\text{diff}} = \frac{\mu^2}{N}\sum_{j=0}^{\infty} \operatorname{Tr}\left(\begin{array}{l}\left[\left(A_2^T A_1^T\right)^j \otimes (I - \mu R_u)^j\right] \times \\ \left(A_2^T R_v A_2 \otimes R_u\right) \times \\ \left[(A_1 A_2)^j \otimes (I - \mu R_u)^j\right]\end{array}\right).$$

By using the following feature of the Kronecker multiplication:

$$\left(X \otimes Y\right)\left(W \otimes Z\right) = \left(XW \otimes YZ\right), \tag{88}$$

we reach the following expression:

$$\mathrm{MSD}_{\text{AR}}^{\text{diff}} = \frac{\mu^2}{N}\sum_{j=0}^{\infty} \operatorname{Tr}\left(\begin{array}{l}\left[\left(A_2^T A_1^T\right)^j \left(A_2^T R_v A_2\right)(A_1 A_2)^j\right] \otimes \\ \left[(I - \mu R_u)^j R_u (I - \mu R_u)^j\right]\end{array}\right). \tag{89}$$

Now, according to the general model expressed in Equation 39 we consider the following three cases:

$$A_1 = A, A_2 = I_N \quad \text{CTA} \tag{90}$$

$$A_1 = I_N, A_2 = A \quad \text{ATC} \tag{91}$$

$$A_1 = I_N, A_2 = I_N \quad \text{No} - \text{Cooperation} \tag{92}$$

in which A is regarded as doubly stochastic. Employing another feature of the Kronecker multiplication:

$$\mathrm{Tr}\left(X \otimes Y\right) = \mathrm{Tr}\left(X\right)\mathrm{Tr}\left(Y\right), \tag{93}$$

we can rewrite Equation 89 for the aforementioned cases in the following way (Box 2).

Looking at the above equations we reckon that we need to discuss about the following values to do the comparison:

$$\mathrm{Tr}\left(AA^T R_v\right) \tag{97}$$

$$\mathrm{Tr}\left(\left(AA^T\right)\left(A^T R_v A\right)\right) \tag{98}$$

$$\mathrm{Tr}\left(R_v\right). \tag{99}$$

The above expressions are written noting that the circular shift is allowed in Tr(.) operator. To continue, we present the following lemma:

Lemma 1: For nonnegative matrix H and doubly stochastic matrix A the following inequality is always valid:

$$\mathrm{Tr}\left(AA^T H\right) \leq \mathrm{Tr}\left(H\right). \tag{100}$$

Proof: As $AA^T \geq 0$ is always true for the doubly stochastic matrix A, and the eigenvalues are always within $\left[0,1\right]$ we can formulate the following eigen-decomposition:

$$AA^T = U\Lambda U^T \tag{101}$$

thus we can write the followings:

$$\mathrm{Tr}\left(A^T AH\right) = \mathrm{Tr}\left(U\Lambda U^T H\right)$$

$$= \mathrm{Tr}\left(\Lambda U^T HU\right)$$

$$\leq \mathrm{Tr}\left(U^T HU\right)$$

H is nonnegative and Λ elements are always in $\left[0,1\right]$, hence

$$\mathrm{Tr}(A_T AH) = \mathrm{Tr}\left(U^T HU\right)$$

$$= \mathrm{Tr}\left(H\right).$$

According to the above lemma we can conclude that the highest MSD belongs to the non-cooperative algorithm. To compare ATC and CTA we go through the following procedure:

$$\mathrm{Tr}\left(\left(AA^T\right)\left(A^T R_v A\right)\right) \leq \mathrm{Tr}\left(A^T R_v A\right) = \mathrm{Tr}\left(AA^T R_v\right). \tag{102}$$

Box 2.

$$\mathrm{MSD}_{\mathrm{AR}}^{\mathrm{cta}} = \frac{\mu^2}{N}\sum_{j=0}^{\infty}\mathrm{Tr}\left(\left(I_N^T A^T\right)^j (I_N^T R_v I_N)(A I_N)^j\right)\mathrm{Tr}\left((I-\mu R_u)^j R_u (I-\mu R_u)^j\right) \quad (94)$$

$$\mathrm{MSD}_{\mathrm{AR}}^{\mathrm{atc}} = \frac{\mu^2}{N}\sum_{j=0}^{\infty}\mathrm{Tr}\left(\left(A^T I_N^T\right)^j (A^T R_v A)(I_N A)^j\right)\mathrm{Tr}\left((I-\mu R_u)^j R_u (I-\mu R_u)^j\right) \quad (95)$$

$$\mathrm{MSD}_{\mathrm{AR}}^{\mathrm{ncp}} = \frac{\mu^2}{N}\sum_{j=0}^{\infty}\mathrm{Tr}\left(\left(I_N^T I_N^T\right)^j (I_N^T R_v I_N)(I_N I_N)^j\right)\mathrm{Tr}\left((I-\mu R_u)^j R_u (I-\mu R_u)^j\right). \quad (96)$$

Consequently, according to the presented reasons we can come to the conclusion that the following inequality is true for the network MSD:

$$\boxed{\mathrm{MSD}_{\mathrm{AR}}^{\mathrm{ncp}} \geq \mathrm{MSD}_{\mathrm{AR}}^{\mathrm{cta}} \geq \mathrm{MSD}_{\mathrm{AR}}^{\mathrm{atc}}} \;. \quad (103)$$

As a result, we can eventually bring to an end that under the AR modeling conditions and equal step iteration coefficient at all nodes, the ATC algorithm presents smaller MSD than the CTA and non-cooperative algorithms, and hence, a more precise PSD estimation.

Simulations

Considering the simulation example presented earlier, after calculating the theoretical quantities, the transient and steady-state criteria for different algorithms are compared as shown in Figures 12-15. The best performance in both the transient and steady-state conditions appertains to the incremental and centralized followed by the diffusion ATC and CTA, respectively. As previously discussed, in incremental mode the estimation procedure starts from one point and evolves through the network to cover all the sensor nodes. In other words, the incremental algorithm takes all the local estimations into account at each step, and hence surpasses the averaging strategy as in centralized. Nonetheless, diffusion estimation is limited by the neighbors' estimations. From another perspective, the incremental mode provides a more precise estimation at some additional time expenses, as it covers all the network nodes at each single step. Nevertheless, the centralized algorithm prepares an equal or less precise estimation as fast as possible. The diffusion methods deliver a faster and less accurate estimation after centralized. Since the adaptation step surpasses the combination step in ATC mode, it obtains more precise results compared to the CTA mode.

Expectedly, when the cooperation is negated, the worst performance occurs. In this mode wherein each node individually estimates the PSD itself, error indices in both the transient and steady-state senses experience quantities far beyond cooperative methods. This comparison clearly points out the effectiveness of the cooperative parametric modeling. The capability of cooperative parametric modeling methods is closely related to the noise power. Accordingly, in situations where the noise power seen to be high, for instance in nodes 3,7 and 14 (Figure 6(b)), the distance between the cooperative-based and non-cooperative methods is at its climax.

Figure 16 demonstrates the similarity of the estimated PSD with its original depicted in Figure 5(a), which was obtained via correlation measurement. In parametric PSD estimation, the resolution is closely tied with the estimated parameters which causes the aforementioned discipline to be repeated in this plot as well. Therefore, we can conclude in essence that the incremental

Figure 12. Network MSD

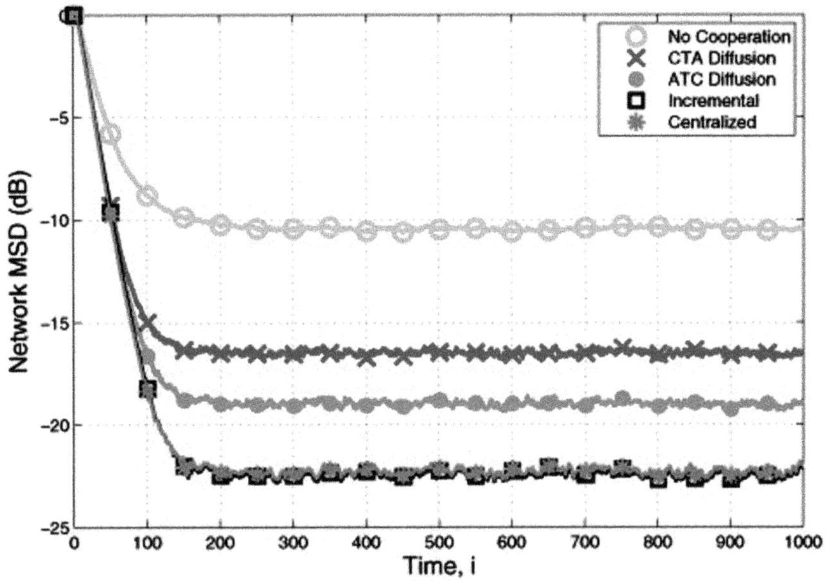

Figure 13. Network EMSE

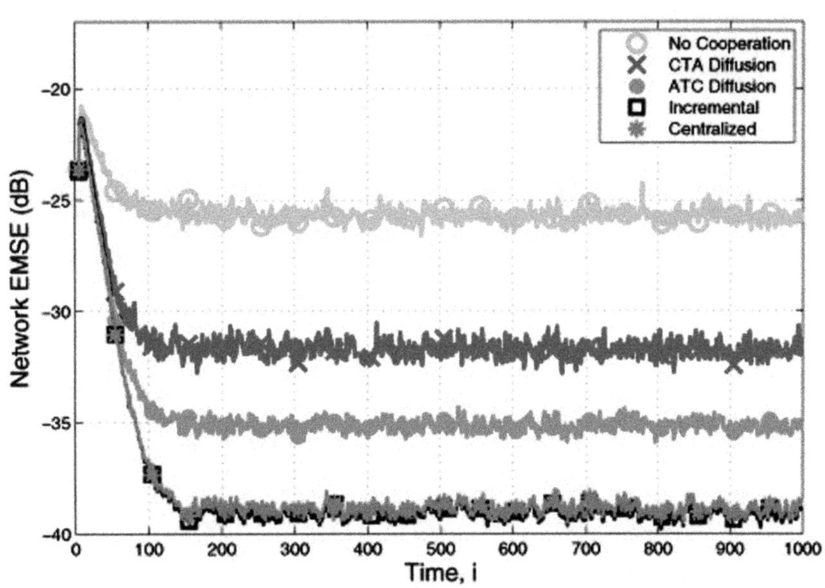

and centralized modes are capable of estimating the most similar PSD, followed by the diffusion, whereas the non-cooperative strategy leads to the least similar PSD estimation.

FUTURE RESEARCH DIRECTIONS

We discussed the methods for adaptive AR parametric modeling and PSD estimation. Real systems, encountered in practice, are always of time-varying nature, which leads to time-varying

Figure 14. Steady-state MSD per node

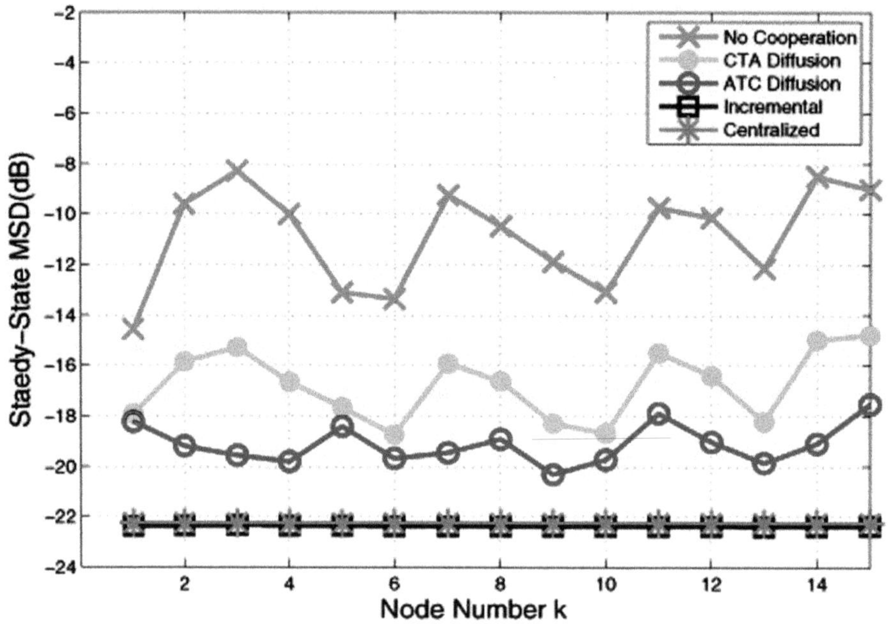

Figure 15. Steady-state EMSE per node

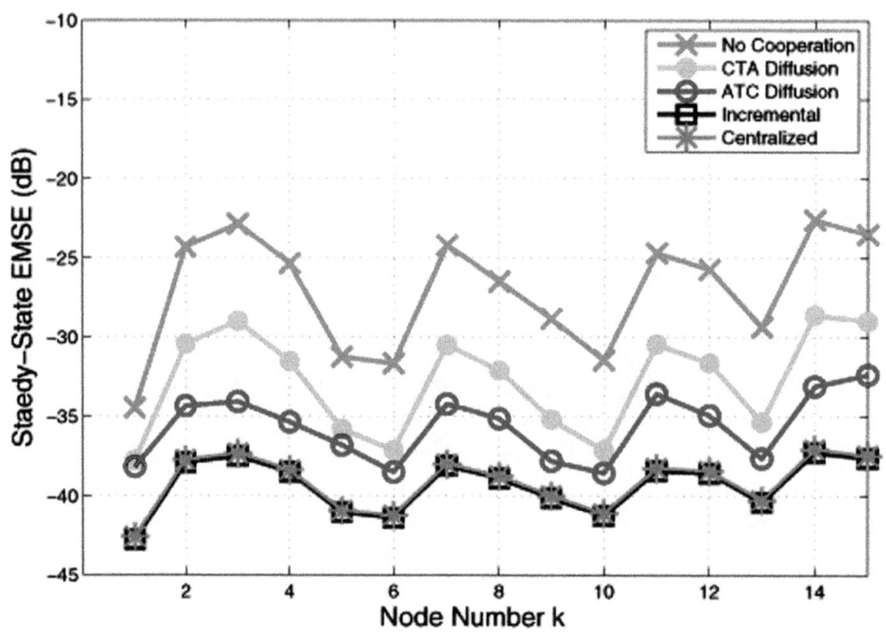

Figure 16. PSD Similarity per node

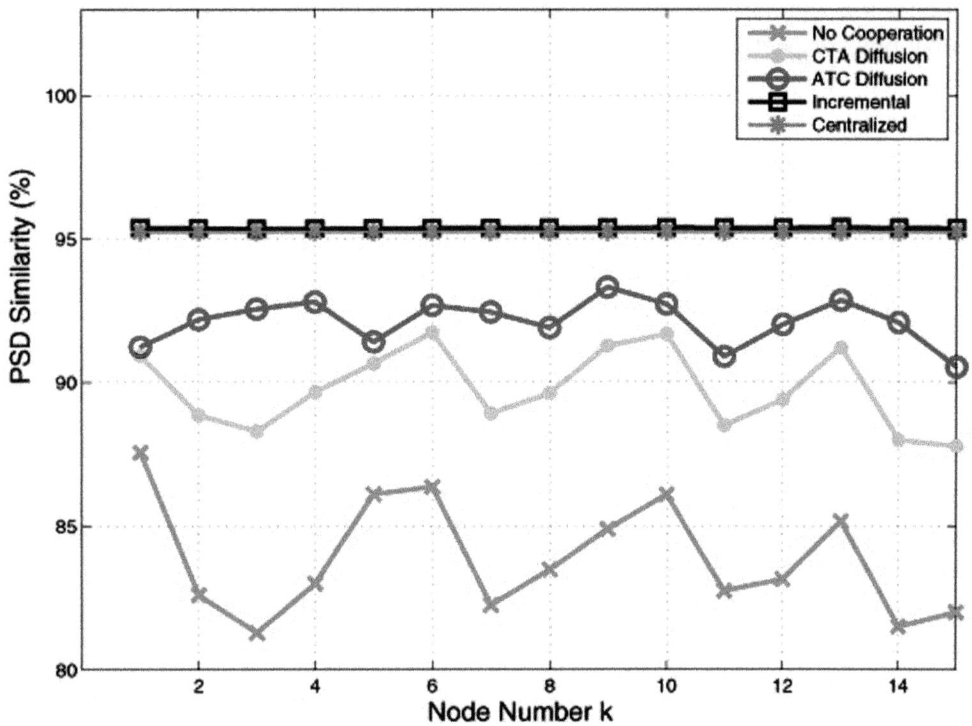

AR systems (TVAR). Studying and analyzing the proposed methods under these conditions helps the problem to be analyzed more realistically. Furthermore, the PSD estimation is generally the first step in signal processing. The other techniques such as detection and classification can make use of the estimated parameters to shape inferences. Studying distributed detection and classification methods, by either using the estimated parameters like the ones in this chapter, or other desired parameter vectors solve the problem in a distributed manner in upper layers of signal processing, and prepares more appropriate results.

CONCLUSION

In this chapter we studied four algorithms for parametric modeling and PSD estimation, based on a set of observations by networked sensors. The adaptive networks were employed in this chapter to estimate the power spectral density. From one hand, these networks are divided into centralized and distributed, and from the other hand into cooperative and non-cooperative. First, the validity of the proposed algorithms was carried out. Theoretical and simulation error criteria were compared in both transient and steady-state conditions, based on the designed simulation scenario. The close relation between the theoretical and simulation results approved the validity of the proposed methods, according to the standard adaptive networks. In the next compartment of the interpretation process, the proposed algorithms were compared with each other both theoretically and numerically. In theoretical comparison, the incremental and centralized algorithms were compared in one division, and the diffusion and non-cooperative strategies in another. Theoretical results demonstrated that among incremental and centralized algorithms, the latter had better or equal performance under steady-state conditions. Furthermore, among diffusion methods ATC, CTA, and non-cooperative modes had more precise estimations, respectively. Additionally, all algorithms were compared numerically in a simulation example, wherein the incremental and centralized modes surpassed the diffusion modes. A considerable distance among cooperative and non-cooperative methods was observed in both transient and steady-state conditions. As a result, the initiated algorithms could estimate the PSD of a WSS random process a great deal more comprehensive with enabling cooperation amongst individual nodes. Moreover, although the centralized method provides fast and accurate PSD estimations in the initial and ideal mode, the incremental algorithm improves the accuracy but at some time expense. As a general conclusion, in essence, the cooperative adaptive parametric modeling and PSD estimation algorithms are the best choices for spatio-temporal processes corrupted by additive noise.

REFERENCES

Agarwal, A., & Jagannatham, A. K. (2014). Distributed Estimation in Homogenous Poisson Wireless Sensor Networks. *IEEE Wireless Communications Letters*, *3*(1), 90–93. doi:10.1109/WCL.2013.112313.130696

Bertsekas, D. P. (1997). A New Class of Incremental Gradient Methods for Least Squares Problems. *SIAM Journal on Optimization*, *7*(4), 913–926. doi:10.1137/S1052623495287022

Blackman, R. B., & Tukey, J. W. (1959). *The measurement of power spectra: from the point of view of communications engineering* (Vol. 1058). Dover Publications New York.

Burg, J. P. (1967). Maximum entropy spectral analysis. In *37th Annual International Meeting Society of Exploration Geophysics.*

Cattivelli, F. S., Lopes, C. G., & Sayed, A. H. (2007). A diffusion RLS scheme for distributed estimation over adaptive networks. In *IEEE 8th Workshop on Signal Processing Advances in Wireless Communications* (pp. 1–5). doi:10.1109/SPAWC.2007.4401393

Cattivelli, F. S., & Sayed, A. H. (2011). Analysis of spatial and incremental LMS processing for distributed estimation. *IEEE Transactions on Signal Processing*, *59*(4), 1465–1480. doi:10.1109/TSP.2010.2100386

Chaudhari, Q. M., & Serpedin, E. (2008). Maximum likelihood estimation of clock parameters for synchronization of wireless sensor networks. In *2008 IEEE International Conference on Acoustics, Speech and Signal Processing* (pp. 2517–2520). doi:10.1109/ICASSP.2008.4518160

Chen, J., & Sayed, A. H. (2011). Diffusion adaptation strategies for distributed optimization and learning over networks. *IEEE Transactions on Signal Processing*, *60*(8), 4289–4305. doi:10.1109/TSP.2012.2198470

Hlinka, O., Hlawatsch, F., & Djuric, P. M. (2013). Distributed particle filtering in agent networks: A survey, classification, and comparison. *IEEE Signal Processing Magazine*, *30*(1), 61–81. doi:10.1109/MSP.2012.2219652

Hou, L., & Bergmann, N. W. (2012). Novel industrial wireless sensor networks for machine condition monitoring and fault diagnosis. *IEEE Transactions on Instrumentation and Measurement*, *61*(10), 2787–2798. doi:10.1109/TIM.2012.2200817

Huang, P., Xiao, L., Soltani, S., Mutka, M. W., & Xi, N. (2013). The evolution of MAC protocols in wireless sensor networks: A survey. *IEEE Communications Surveys and Tutorials*, *15*(1), 101–120. doi:10.1109/SURV.2012.040412.00105

Kay, S. M. (1988). *Modern Spectral Estimation: Theory and Application*. Englewood Cliffs, NJ: Prentice Hall.

Koutsopoulos, I., & Halkidi, M. (2014). Distributed energy-efficient estimation in spatially correlated wireless sensor networks. *Computer Communications*, *45*, 47–58. doi:10.1016/j.comcom.2014.03.012

Liu, X., Gong, X., & Zheng, Y. (2014). Reliable cooperative communications based on random network coding in multi-hop relay WSNs. *IEEE Sensors Journal*, *14*(8), 2514–2523. doi:10.1109/JSEN.2014.2310899

Lopes, C. G., & Sayed, A. H. (2007). Incremental adaptive strategies over distributed networks. *IEEE Transactions on Signal Processing*, *55*(8), 4064–4077. doi:10.1109/TSP.2007.896034

Lopes, C. G., & Sayed, A. H. (2008). Diffusion least-mean squares over adaptive networks: Formulation and performance analysis. *IEEE Transactions on Signal Processing*, *56*(7), 3122–3136. doi:10.1109/TSP.2008.917383

Ma, T., Xu, Z., Hempel, M., Peng, D., & Sharif, H. (2014). Performance analysis of a novel low-complexity high-precision timing synchronization method for wireless sensor networks. *IEEE Transactions on Wireless Communications*, *13*(9), 4758–4765. doi:10.1109/TWC.2014.2331286

Marple, S. L. Jr. (1987). *Digital spectral analysis with applications*. London: Prentice Hall Inc.

Msechu, E. J., & Giannakis, G. B. (2012). Sensor-centric data reduction for estimation with WSNs via censoring and quantization. *IEEE Transactions on Signal Processing*, *60*(1), 400–414. doi:10.1109/TSP.2011.2171686

Naderi, M. Y., Nintanavongsa, P., & Chowdhury, K. R. (2014). RF-MAC: A medium access control protocol for re-chargeable sensor networks powered by wireless energy harvesting. *IEEE Transactions on Wireless Communications*, *13*(7), 3926–3937. doi:10.1109/TWC.2014.2315211

Nedic, A., & Ozdaglar, A. (2009). Distributed subgradient methods for multi-agent optimization. *IEEE Transactions on Automatic Control*, *54*(1), 48–61. doi:10.1109/TAC.2008.2009515

Neuzil, J., Kreibich, O., & Smid, R. (2014). A distributed fault detection system based on iwsn for machine condition monitoring. *IEEE Transactions on Industrial Informatics*, *10*(2), 1118–1123. doi:10.1109/TII.2013.2290432

Papotto, G., Carrara, F., Finocchiaro, A., & Palmisano, G. (2014). A 90-nm CMOS 5-Mbps crystal-less RF-powered transceiver for wireless sensor network nodes. *IEEE Journal of Solid-State Circuits*, *49*(2), 335–346.

Rabbat, M. G., & Nowak, R. D. (2005). Quantized incremental algorithms for distributed optimization. *IEEE Journal on Selected Areas in Communications*, *23*(4), 798–808. doi:10.1109/JSAC.2005.843546

Sayed, A. H. (2013to appear). Diffusion adaptation over networks. In R. Chellappa & S. Theodoridis (Eds.), *E-Reference Signal Processing*. Elsevier.

Sayed, A. H., & Cattivelli, F. S. (2009). Distributed adaptive learning mechanisms. In S. Haykin & K. J. Ray Liu (Eds.), *Handbook on Array Processing and Sensor Networks* (pp. 693–722).

Sayed, A. H., & Lopes, C. G. (2006). Distributed recursive least-squares strategies over adaptive networks. In *Asilomar Conference on Signals Systems and Computers* (pp. 233–237). IEEE. doi:10.1109/ACSSC.2006.356622

Sizov, V., Gashinova, M., Zakaria, N., & Cherniakov, M. (2013). VHF communication channel characterisations over complex wooded propagation paths with applications to ground wireless sensor networks. *IET Microwaves. Antennas & Propagation*, *7*(3), 166–174.

Soltanmohammadi, E., Orooji, M., & Naraghi-Pour, M. (2013). Decentralized hypothesis testing in wireless sensor networks in the presence of misbehaving nodes. *IEEE Transactions on Information Forensics and Security*, *8*(1), 205–215. doi:10.1109/TIFS.2012.2229274

Somov, A., Baranov, A., Spirjakin, D., & Passerone, R. (2014). Circuit design and power consumption analysis of wireless gas sensor nodes: One-sensor versus two-sensor approach. *IEEE Sensors Journal*, *14*(6), 2056–2063. doi:10.1109/JSEN.2014.2309001

Stojcev, M. K., Kosanovic, M. R., & Golubovic, L. R. (2009). Power management and energy harvesting techniques for wireless sensor nodes. In *2009 9th International Conference on Telecommunication in Modern Satellite, Cable, and Broadcasting Services* (Vol. 9, pp. 65–72). doi:10.1109/TELSKS.2009.5339410

Sung, W.-T., & Hsu, Y.-C. (2011). Designing an industrial real-time measurement and monitoring system based on embedded system and ZigBee. *Expert Systems with Applications*, *38*(4), 4522–4529. doi:10.1016/j.eswa.2010.09.126

Tolentino, I. M., Juson, R. S., Tan, B. U., & Talampas, M. R. (2010). Design, development, and evaluation of a simple wireless sensor network for indoor microclimate monitoring. In TENCON 2010 - 2010 IEEE Region 10 Conference (pp. 2018–2023). doi:10.1109/TENCON.2010.5686565

Tsitsiklis, J., Bertsekas, D., & Athans, M. (1986). Distributed asynchronous deterministic and stochastic gradient optimization algorithms. *IEEE Transactions on Automatic Control*, *31*(9), 803–812. doi:10.1109/TAC.1986.1104412

Wang, W.-T., & Ssu, K.-F. (2013). Obstacle detection and estimation in wireless sensor networks. *Computer Networks*, *57*(4), 858–868. doi:10.1016/j.comnet.2012.11.004

Welch, P. (1967). The use of fast Fourier transform for the estimation of power spectra: A method based on time averaging over short, modified periodograms. *IEEE Transactions on Audio and Electroacoustics*, *15*(2), 70–73. doi:10.1109/TAU.1967.1161901

Zheng, W. X. (1998). Adaptive parameter estimation of autoregressive signals from noisy observations. In *ICSP '98. 1998 Fourth International Conference on Signal Processing* (pp. 449–452).

Zhong, X., & Premkumar, A. B. (2014). Multiple wideband source detection and tracking using a distributed acoustic vector sensor array: A random finite set approach. *Signal Processing*, *94*, 583–594. doi:10.1016/j.sigpro.2013.07.008

KEY TERMS AND DEFINITIONS

Adaptive Networks: An Adaptive network contains some adaptive filters which are linked together and cooperatively estimate an unknown parameter.

Autoregressive: An autoregressive (AR) model is a representation of a type of random process; as such, it describes certain time-varying processes in nature, economics, etc. The autoregressive model specifies that the output variable depends linearly on its own previous values.

Centralized: A centralized adaptive algorithm in a distributed sensor network, aggregates all the measurements in a central node and employ an adaptive filter to estimate the unknown parameter.

Diffusion: In adaptive networks with diffusion strategy, each node uses the information of all the adjacent nodes instead of just the previous one. For this purpose, each node makes a local combination of data from all the neighboring nodes and uses it to update its estimation.

Incremental: An incremental adaptive algorithm cycles throughout the adaptive network sequentially. The incremental strategy uses local estimations including the information about the preceding nodes. Each transition leads to a knowledge increment, hence all measurements about the estimated parameter is observed over the last node.

MSD: MSD is one of several measures for evaluating estimation accuracy. It is calculated by squaring the deviation between the desired and obtained parameter and finding the expected value, respectively.

Power Spectral Density (PSD): Power spectral density function (PSD) shows the strength of the variations (energy) as a function of frequency. In other words, it shows at which frequencies variations are strong and at which frequencies variations are weak.

Wireless Sensor Networks (WSN): A wireless sensor network includes numerous spatially distributed autonomous low price sensors to monitor physical or environmental conditions and cooperatively pass their data through the network to a main location.

APPENDIX: PROOF OF MATRIX INVERSION LEMMA

We want to prove the following equality:

$$\bar{\lambda}^T (I - \bar{F}_{\mathrm{cnt}})^{-1} = \frac{\bar{\lambda}^T (2\mu N\Lambda - \mu^2 N^2 \Lambda^2)^{-1}}{1 - \mu^2 N \bar{\lambda}^T (2\mu N\Lambda - \mu^2 N^2 \Lambda^2)^{-1} \bar{\lambda}}, \tag{104}$$

considering $\bar{F}_{cnt}$ to be:

$$\bar{F}_{cnt} = I - 2\mu N\Lambda + \mu^2 N^2 \Lambda^2 + \mu^2 N \bar{\lambda}\bar{\lambda}^T.$$

By inverting both sides of the equation:

$$\frac{1}{2\mu N\Lambda - \mu^2 N^2 \Lambda^2 - \mu^2 N \bar{\lambda} \left(\bar{\lambda}\right)^T} = \left(I - \bar{F}_{cnt}\right)^{-1}, \tag{105}$$

the nominator and denominator of the left hand side (LHS) of the above equation are divided by

$$\left(2\mu N\Lambda - \mu^2 N^2 \Lambda^2\right),$$

hence:

$$\frac{1/\left(2\mu N\Lambda - \mu^2 N^2 \Lambda^2\right)}{\left(2\mu N\Lambda - \mu^2 N^2 \Lambda^2 - \mu^2 N \bar{\lambda} \left(\bar{\lambda}\right)^T\right) / \left(2\mu N\Lambda - \mu^2 N^2 \Lambda^2\right)} = \left(I - \bar{F}_{cnt}\right)^{-1}. \tag{106}$$

Simplifying the nominator and denominator, we get:

$$\frac{\left(2\mu N\Lambda - \mu^2 N^2 \Lambda^2\right)^{-1}}{1 - \mu^2 N \bar{\lambda} \left(\bar{\lambda}\right)^T \left(2\mu N\Lambda - \mu^2 N^2 \Lambda^2\right)^{-1}} = \left(I - \bar{F}_{cnt}\right)^{-1}. \tag{107}$$

In the next step, the nominator and denominator of the LHS are multiplied by $\left(\bar{\lambda}\right)^T$ from left side, leading to the following equation:

$$\frac{\left(\bar{\lambda}\right)^T \left(2\mu N\Lambda - \mu^2 N^2 \Lambda^2\right)^{-1}}{\left(\bar{\lambda}\right)^T \left[1 - \mu^2 N \bar{\lambda} \left(\bar{\lambda}\right)^T \left(2\mu N\Lambda - \mu^2 N^2 \Lambda^2\right)^{-1}\right]} = \left(I - \bar{F}_{cnt}\right)^{-1}. \tag{108}$$

Λ is a diagonal matrix and $\bar{\lambda}\left(\bar{\lambda}\right)^T$ is a square matrix. We know that in case of a diagonal matrix being multiplied by a square matrix, the replacement of the two is allowed, thus:

$$\left(I-\bar{F}_{cnt}\right)^{-1}=\frac{\left(\bar{\lambda}\right)^T\left(2\mu N\Lambda-\mu^2N^2\Lambda^2\right)^{-1}}{\left(\bar{\lambda}\right)^T\left[I-\mu^2N\left(2\mu N\Lambda-\mu^2N^2\Lambda^2\right)^{-1}\bar{\lambda}\left(\bar{\lambda}\right)^T\right]}. \tag{109}$$

Factorizing $\left(\bar{\lambda}\right)^T$ from the right hand side we will have:

$$\left(I-\bar{F}_{cnt}\right)^{-1}=\frac{\left(\bar{\lambda}\right)^T\left(2\mu N\Lambda-\mu^2N^2\Lambda^2\right)^{-1}}{\left[1-\left(\bar{\lambda}\right)^T\mu^2N\left(2\mu N\Lambda-\mu^2N^2\Lambda^2\right)^{-1}\bar{\lambda}\right]\left(\bar{\lambda}\right)^T}.$$

If the whole equation is multiplied by the denominator from the right side, we get:

$$\left[I-\left(\bar{\lambda}\right)^T\mu^2N\left(2\mu N\Lambda-\mu^2N^2\Lambda^2\right)^{-1}\bar{\lambda}\right]\left(\bar{\lambda}\right)^T\left(I-\bar{F}_{cnt}\right)^{-1}=$$
$$\left(\bar{\lambda}\right)^T\left(2\mu N\Lambda-\mu^2N^2\Lambda^2\right)^{-1}.$$

Since the bracketed term is a number, we can divide both sides by that, thus:

$$\left(\bar{\lambda}\right)^T\left(I-\bar{F}_{cnt}\right)^{-1}=\frac{\left(\bar{\lambda}\right)^T\left(2\mu N\Lambda-\mu^2N^2\Lambda^2\right)^{-1}}{\left(1-\left(\bar{\lambda}\right)^T\mu^2N\left(2\mu N\Lambda-\mu^2N^2\Lambda^2\right)^{-1}\bar{\lambda}\right)}, \tag{110}$$

Consequently, the following relation can easily be concluded:

$$\left(\bar{\lambda}\right)^T\left(I-\bar{F}_{cnt}\right)^{-1}=\frac{\left(\bar{\lambda}\right)^T\left(2\mu N\Lambda-\mu^2N^2\Lambda^2\right)^{-1}}{1-\mu^2N\left(\bar{\lambda}\right)^T\left(2\mu N\Lambda-\mu^2N^2\Lambda^2\right)^{-1}\bar{\lambda}}.$$

Chapter 14
Covert Communication in Wireless Sensor Networks for E-Health Applications

Fahimeh Rezaei
University of Nebraska-Lincoln, USA

Michael Hempel
University of Nebraska-Lincoln, USA

Pradhumna Lal Shrestha
University of Nebraska-Lincoln, USA

Sushanta Mohan Rakshit
University of Nebraska-Lincoln, USA

Hamid Sharif
University of Nebraska-Lincoln, USA

ABSTRACT

In this chapter, the authors introduce a threat in Wireless Sensor Network (WSN) E-Health applications that can be used for leaking patients' medical information, identity, personal data, and medical history. This threat is called covert communication, which establishes a secret path hidden within legitimate network traffic to transfer information. Based on the exploit used for this type of communication, covert channels have different categories and functionalities. The authors introduce various covert communication methods and their possible applications in transmitting e-health data secretly in WSNs. In addition, the authors discuss different methods and approaches to detect this type of communication in WSNs to prevent this form of malicious e-health information transfer.

INTRODUCTION

One of the most important applications of Wireless Sensor Networks (WSN) is the area of e-health, where several medical sensors monitor and collect the health information of a patient's body. This information can be collected in a highly localized network type called Body Area Network (BAN), gathered on a local device or monitoring station connected to the Local Area Network (LAN), eventually to be transferred to the hospital or healthcare center via Wide Area Network (WAN) communication. The collected medical information can then be accessed among various groups of doctors, researchers and hospital staff members, but otherwise should be highly access-controlled in accordance with corresponding regulations such as the Health Insurance Portability and Ac-

DOI: 10.4018/978-1-4666-8251-1.ch014

countability Act (HIPAA). Due to the fact that the healthcare information for each patient is confidential and needs to remain protected, the security of WSNs for this specific application is a significant concern. Hence, several research areas and activities are dedicated to e-health data protection and WSN system security to keep medical and personal identity information confidential and prevent malicious or accidental disclosure. Beyond the more traditional attacks and security solutions that are discussed in the literature we would like to introduce the reader to a different potential threat to WSNs that can be used by malicious parties to illegally access patients' identity information, medical history, health status records and more. Although this threat, which is called covert communication, can be employed in other areas of WSNs, the existence of this secret path for leaking e-health information is significantly more dangerous.

This book chapter is organized as follows: In the background section, firstly, e-health WSN architecture and technologies are briefly introduced. This section is continued by current security concerns and areas that are previously discussed in literature. Then we introduce a possible threat that was not considered in literature for e-health applications and outline different methods and techniques that can be used for implementing this threat to leak medical data. In section V, we introduce some methods and techniques to detect this new threat and prevent establishing it for e-health WSN applications. Finally in section VI, we provide a summary and conclusion and the future research track in this area.

BACKGROUND

In this section we first provide a background regarding telemedicine and e-health using WSN devices and technologies. In addition, we review various security topics and concepts that are discussed in previous literature in the area of e-health.

Wireless Sensor Networks in E-Health

Employing recent technologies and smart devices for monitoring and recording biological changes in human body, mostly for medical applications is referred to as e-health. Evidently, taking advantage of e-health technologies and devices reduces dependencies on healthcare staff and specialists as well as frequently health center visits that are required for the patients to record their health condition. Moreover, due to the new and precise hardware and embedded system devices, measured parameters by these devices and new technologies are accurate and reliable. Using the new and high speed wireless technologies also provides an easy way to transfer medical data to a given database in a hospital or healthcare center. This database can be accessible by various groups such as doctors, nurses, researchers, insurance companies and other related groups readily and in a specific format that is required. Hence, the patient does not need to provide the same information to different group multiple times. Therefore, using e-health devices and technologies is reliable, cost effective and convenient for the patients and the parties that need to use patient's medical data and history.

Generally monitoring devices are at least equipped by several sensors to record the biological signals, storage unit to saves the measured parameters and a communication module that can send collected data to a local database. A collection of different medical data or biological signals that can be monitored and measured by BAN devices are shown in Table 1 (Rezaei, Hempel, & Sharif, 2015), as well as their typical required data rate. These sensors are interfaced with a given wireless communication module that can support their required data rate and the redundancy of sending data within a specific time slot. For example if the temperature of a patient's body with 120 bps of data rate needs to be recorded 100 times every second, then a wireless communication technology that provides at least 120 kbps speed of data transmission is required to successfully send data to a destination.

Besides data rate, the wireless communication module is required to have a suitable coverage range to be able to transfer data reliably within a given distance. Therefore, for devices that are connected in a small scale network such as BAN, a wireless module with the coverage of less than couple of meters is sufficient for data transmission. In this case, wireless technologies such as Zarlink or Near Field Communications (NFC) can be employed for a reliable communication. However, for larger scale networks that require to send medical data over 20 meters, wireless communication technologies that are usually used for LAN networks, such as Bluetooth or Zigbee are used. For example, consider a nursing house that monitors different medical and biological status of elderly people in each room and transmits each persons' medical information to a local healthcare database center located in the monitoring room at the same building. This data collected in the nursing center is required to be transferred to a medical clinic or a hospital. Therefore, different tiny devices are attached to specific places of patients' bodies to form a BAN network to collect different information such as Blood Saturation, motion and more. The BAN network can have different topologies such as mesh, star, etc. In case of the mesh topology, tiny devices or sensors are usually connected to a central node and send their collected countermeasures and parameters to this node. The central node is connected to a LAN to transmit the collected data from BAN to the local nursing center database. The collected medical data from this nursing center, and other nursing centers will be transferred to a hospital via WAN, which uses wireless technologies such as WiMAX and LTE to provide high throughput of data transmission and wide coverage area. In Table 2 we summarized the most prevailing wireless communication technologies that are used in e-health systems and their data rate as well as coverage (Rezaei, Hempel, & Sharif, 2015). Figure 1 illustrates a simple and general e-health architecture (Rezaei, Hempel, & Sharif, 2015). As we can see, an end-to-end e-health system usually consists of three stages of data transmission.

In each stage of data transmission ranging from BAN to WAN, attackers or eavesdroppers might break into e-health databases and systems, or listen to the communication link to access medical data and identity of the patients. The risk of information theft, especially in this area, may decrease the interest of using e-health systems and wearable sensors and devices. This arises from

Table 1. Required data rate for Bio1-signals and Medical Data transmission (Rezaei, Hempel, & Sharif, 2015)

Application	Required Data Rate
ECG (Electrocardiogram)	15kbps to 288 kbps
EMG (Electromyography)	320 kbps
EEG(Electroencephalography)	10kbps to 43.2kbps
Blood Saturation	16 bps
Glucose Monitoring	1600 bps
Temperature	120 bps
Motion Sensor	35 kbps
Cochlear Implant	100 kbps
Artificial Retina	50-700 kbps
Audio	4kbps -1Mbps
Voice	50-100 kbps
Heart Sound	120 kbps
Ultrasound, Cardiology, Radiology	2048 kbps
Magnetic Resonance Image	3072 kbps
Scanned X-Ray	14.4 Mbps
Digital Radiography	48 Mbps
Mammogram	192 Mbps
Compressed and Full Motion Video	348 kbps to 1.544 Mbps

Table 2. Different network technologies used in e-2health applications (Rezaei, Hempel, & Sharif, 2015)

Network Scale	Technology	Max. Distance	Max. Data Rate
WAN/MAN	GSM	35 km	270 kbps
	HSDPA and HSUPA	Over 50 km	14 Mbps
	HSPA	Over 50 km	28 Mbps
	IEEE 802.16e (Mobile WiMAX)	Over 80 km	250 Mbps
	LTE	Over 100 km	300 Mbps
	4G and beyond (LTE-Advanced and IEEE 802.16m)	Over 100 km	1 Gbps
LAN	ANT and ANT+	30 m	1000 kbps
	ZigBee	100 m	20-250 kbps
	Bluetooth (Typical Range: Class 2 & 3	1-10 m	3000-25000 kbps
	Bluetooth (Extended Range: Class 1)	100 m	3000-25000 kbps
	Dash7	1000 m	28-200 kbps
	802.11	8000 m	1000-1000000 kbps
BAN	NFC	<0.2 m	106-424 kbps
	Zarlink	<1 m	200-800 kbps
	WUSB	3-10 m	53000-480000 kbps
	Sensium	5-25 m	50 kbps

Figure 1. End-to-end view of a typical e-health architecture 1 (Rezaei, Hempel, and Sharif, 2014)

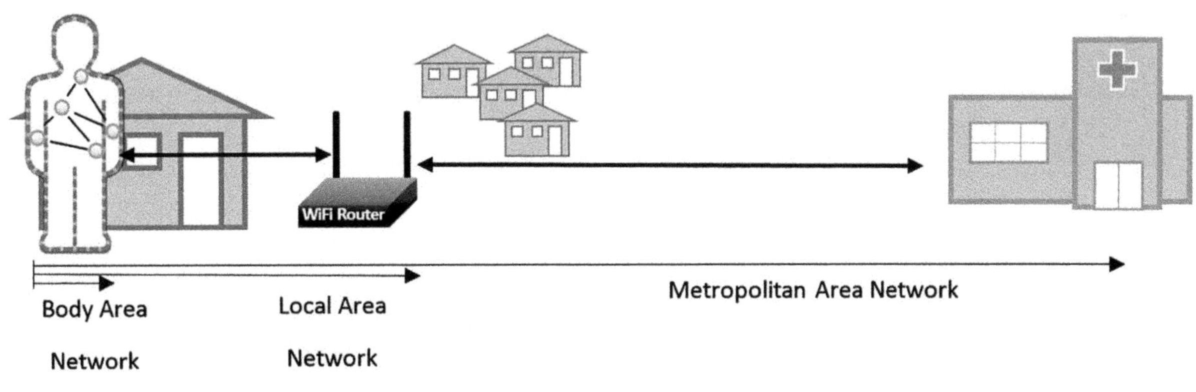

the fact that the patients' may not be willing to share their medical records and history with others for personal and privacy reasons. Moreover, in case if the insurance companies are able to access these data, they can increase the insurance rate or simply do not provide health insurance for the patients who have a risk of getting a specific disease. Hence, both medical data and e-health systems need to be protected to prevent exploiting patients' identity and health information.

In the next section we review various attacks and risks that can impact e-health systems and different solutions for e-health security improvement.

Security and Data Protection in E-Health

In general, security in e-health WSN can be considered as two components: system security, and data security. We discuss each component separately in this section.

1. **System Security:** System security is referred to inaccessible medical data and patients' information and identities that are transferred via wireless communication link or stored in a database. System security consists of different levels according to the article by Al Ameen, Liu and Kwak (2012).
   a. **Administrative Level Security:** The first level which is called administrative level security, is referred to the healthcare staff, doctors and researchers that are using the data, or the engineering staffs that are responsible for maintenance and trouble shooting of the e-health devices. To improve this level of security, a reliable and precise access mechanism to e-health systems must be implemented. A proper access mechanism should establish a hierarchical access to the authorized users. For example, consider the previously discussed nursing house, where there are several people in different positions working and taking care of the elderly people. The maintenance and engineering staffs that install the e-health systems and set up the equipment should have a very limited access to the identity of the elderly people who use and wear the BAN devices. In addition, they should not be able to access any medical data. The nurses and healthcare staffs that are monitoring very basic biological signals such as heart beat or motions of the elderly people, can access some of these rudimentary data, but not their medical history and full information. However, the some physicians and doctors in the hospital that are required to know the comprehensive health status of the patients living in the nursing house, should be able to access all their identity information and medical history. Moreover, a robust and reliable hierarchical access mechanism for e-health systems, should be able to track the identity of the individuals who use and access medical and identity information of patients and provides a specific alarm or indicator for unauthorized accesses.
   b. **Physical Layer Security:** The other level of security that is defined by Al Ameen, Liu and Kwak (2012), is the physical layer security, which is referred to accessing the wearable e-health devices and systems by non-authorized people to extract the medical information. Therefore, at the hardware level, the e-health devices should be designed to stop operating if these devices are handled by un-authorized people.
   c. **Technical Level Security:** Technical level security also is another category, which refers to data transmission and protecting communication links so that the eavesdroppers and attackers cannot access data packets crossing BAN, LAN or WAN. One of the prevalent approaches of accessing network packets in WSNs is to change the routing destination of wireless packets and directing those to a data center other than the host healthcare database.
2. **Information Security:** Beside system security, information security is also another important factor, which is usually referred to data encryption. Using cryptography, transferred data via communication links will be

confidential and cannot be disclosed easily by eavesdroppers. In addition to data encryption, data integrity and data authentication also are a part of information security. Data integrity provides the recipient to realize that the data is not modified or changed. Data authentication ensures that the information is coming from a given expected sender. In addition, Al Ameen, Liu and Kwak (2012) introduced another information security component, which is called freshness protection that prevents the transferred data from retransmission and replaying by the attackers.

The New Threat for WSN E-Health Applications: Covert Communication

Protecting the confidential health care information and patient's identity using the cryptographic algorithms and boosting system security are necessary but not adequate for having a secure and reliable network. These days, another illegal threat for shared environments and network platforms is widely growing, which establishes a hidden communication path over the legitimate communication link. This hidden path can be used by eavesdroppers to transmit patient's healthcare information and identity over this illegal channel. This type of secret communication path is called covert channel and does not fall in any of the discussed security levels.

Using covert channels, one can leak patients' medical information by exploiting the network resources such as the timing between communication packets and unused communication protocol header fields. We define and describe this type of threat in the next section.

COVERT COMMUNICATION

Covert communication is a steganography method that exploits communication resources to establish hidden pathways for transmitting specific information secretly. Covert data is inserted into the communication link within the packet inter-arrival times (Covert Timing Channels – CTC), within packet header or payload fields (Covert Storage Channels – CSC) or a combination thereof (Covert Hybrid Channels – CHC). This type of information hiding strategy can be used to leak sensitive information in WSN e-health applications such as patients' identity, medical history, biological signals and more to coordinate malicious activities.

Figure 2 shows a general perspective of this hidden channel over a communication link. Based on this figure, we introduce the following terminologies that are used towards the end of this chapter:

- **Covert Data:** Refers to the secret information that is going to be hidden within the legitimate packets using resources of the network.
- **Covert Sender:** The Sender that transfers the covert data over the network is called the covert sender.
- **Covert Receiver:** The receiver that is aware of the covert channel functionality and can extract the covert data is called covert receiver.

In addition, the senders and receivers of the legitimate communication link that are not aware of the covert communication and do not participate in such activities are called overt senders and overt receivers, respectively.

Covert Channel Implementation in Different Layers

Based on literature, covert communication can be established within different layers of communication protocols ranging from the MAC layer, such as the work by Radhakrishnan et al. (2013), Holloway et al. (2011), and Wang et al. (2007), to the application layer such as the work

Figure 2. Covert communication for e-health application

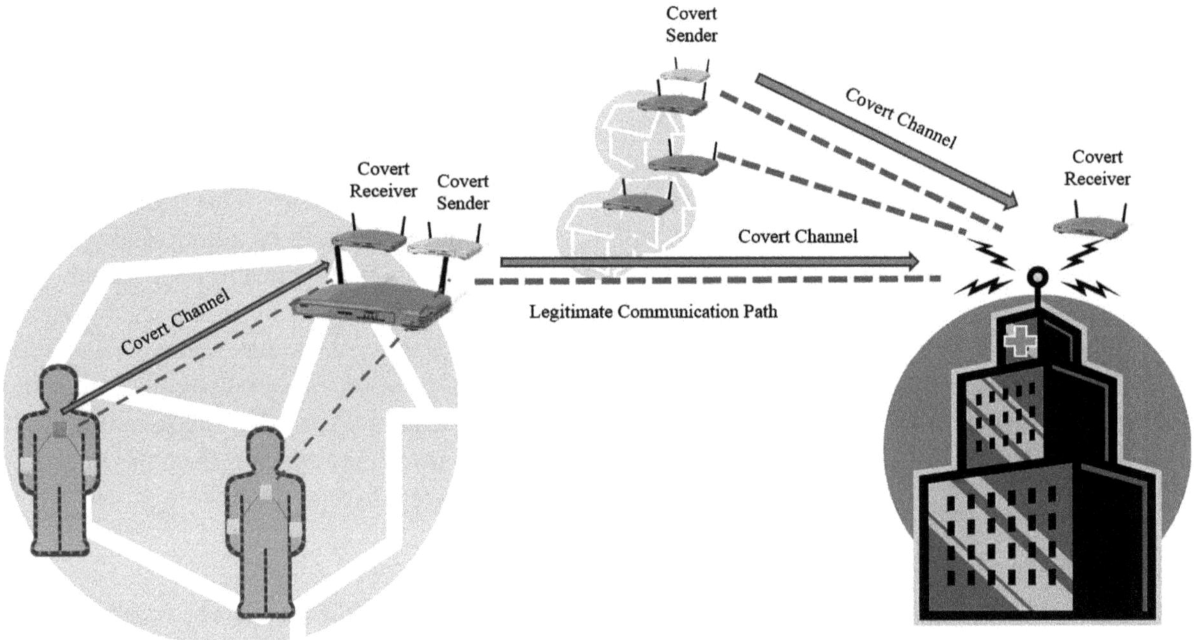

published by Murdoch et al. (2005), Cabuk et al. (2004), and Sellke et al. (2009). For example, in papers by Radhakrishnan et al. (2013), Holloway et al. (2011), and Wang et al. (2007), the authors propose new techniques to implement covert channels in MAC protocol of the 802.11 which is one of the most prevailing wireless technologies that is used for e-health applications such as Pappas et al. (2002) and Monton et al. (2008).

Wang, Deng, and Lee (2007) proposed a covert channel that operates based on the splitting algorithms of colliding nodes in the network. In this covert channel, a node in the wireless network can send a sequence of covert bits by choosing a given path in the splitting tree in a collision resolution period. Covert bits are encoded based on choosing the left or the right node in the splitting tree. The most recent novel approach of establishing a covert channel in MAC layer of 802.11 is proposed by Radhakrishnan, Selcuk and Beyah (2013), as well as Holloway and Beyah (2011) in which the authors utilize Distributed Coordinated Function that is used to avoid collision for covert communication. In this scheme, a sequence of covert bits are encoded to a given back-off time that is shared as a code book with the covert sender and the covert receiver. The covert sender sends a given covert bit sequence by waiting for a pre-defined period of time. The receiver interprets the transmitted covert sequence by waiting for a given time and matching that time to the share code book.

Besides different layers of communication, covert channels can be developed according to different structures. For example, the covert sender and covert receiver can solely be an application or a piece of software, which are installed on the overt sender and overt receiver, and do not have to be a separate hardware or system. Another possible structure is to have a separate systems for covert sender and receiver so that covert sender receives the overt packets from the overt sender and embeds the covert data, and directs the manipulated traffic to the covert receiver. The covert data will be extracted by the covert receiver and the traffic will be flowed towards the destination that is an overt receiver.

Different Types of Covert Channels Based on Network Resources

Covert channels are generally categorized based on the resources that are exploited for embedding secret data. In the following sections we introduce and define the most important categories of covert channels and explain their implementation and functionality.

Covert Storage Channels (CSC)

This type of covert channel takes advantage of the unused communication protocol header fields and possibly a part of the payload that by modifying and embedding covert data into those, the communication will not be impacted. A general architecture of this type of covert communication is shown in Figure 3. As we can see in this figure, the covert data is substituted or inserted into specific fields of legitimate packets that are transferred to a receiver. The covert receiver that is aware of this information hiding technique, is able to extract the covert data, while the legitimate receiver ignores the inserted data and discards that as invalid information. Several CSCs are introduced in the literature such as the work by Krätzer et al. (2006) and Calhoun et al. (2012). Krätzer et al. (2006) evaluate embedding covert data in the MAC protocol layer of 802.11 WLAN. Different fields of the MAC frame are discussed for inserting covert data. Figure 4 shows the MAC frame and the Frame Control Field (FCF) of WLAN, where the covert data can be inserted. As indicated in this article, not all the fields from the MAC frame or frame control field are suitable for embedding covert data. The only fields that can be used for this purpose in MAC frame are: Duration/ID, FCF and CRC. Further in FCF, More Frag, Retry, PwrMgt, and MoreData fields can be used for establishing a covert storage channel. It has been investigated in this article that for the highest capacity of establishing this type of covert channel is 2.1 bytes/sec. Calhoun et al. (2012) improved the capacity of covert channel embedded in the MAC layer of 802.11 up to 60 bps. Although, the reported capacity in these papers might not seem to be very high compared to the actual communication channel, but transmission of covert data for most of the applications does not need to be fast, especially if a password or identification number is transferred via covert channel. Generally, this capacity can be increased depending on the communication link and the type of the wireless communication protocol and packet format.

Establishing a CSC is not solely limited to WLAN communication. The CSCs can be established also in WAN/MAN communication. For example, based on the work by Rezaei et al. (2013), the possibility of embedding covert data in different protocol fields of LTE-Advanced is evaluated and discussed. In addition to the mentioned implementations, one of the most popular implementations of CSC is over the TCP/IP packets that can be used to leak the healthcare information over internet protocol suites. Murdoch and Lewis (2005), evaluate different header fields of TCP/IP packet that can be used for inserting covert data and the ones that can be used for this purpose are highlighted in Figure 4-b. as we can see in this figure, the maximum bits that can be used for embedding covert data belongs to the TCP sequence number. Therefore, using this field, if we assume having the minimum size of a TCP/IP header (i.e. 320 bits), the maximum legitimate communication traffic speed should be at least 160 bps to transfer Blood Saturation of a patient that requires 16 bps of data rate according to Table 1. In Table 3, we present the minimum speed of legitimate communication link that is required to transfer medical information indicated in Table 1, using CSC over TCP/IP header.

3. **Covert Timing Channels (CTC):** This type of covert communication takes advantage of the gaps between inter-arrival times of the communication packets and modulates

Figure 3. Covert storage channel (CSC) general architecture

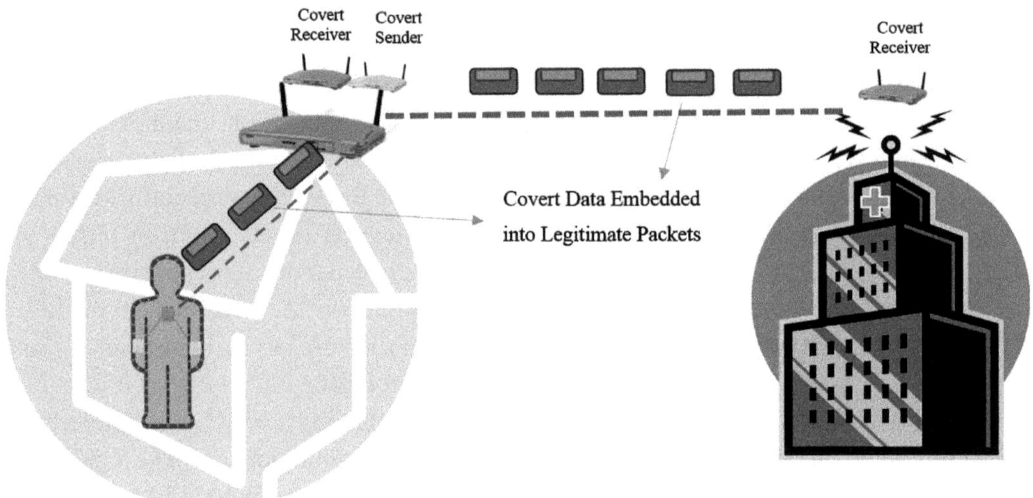

Figure 4. Examples of possible protocol fields for CSC

Bytes 2	**2**	**6**	**6**	**6**	**2**	**6**	**0-2312**	**4**
Frame	Duration	Address	Address	Address	Sequence	Address	Data	CRC

Bits 2	**2**	**4**	**1**	**1**	**1**	**1**	**1**	**1**	**1**	**1**
Protocol Version	Type	Subtype	To DS	From DS	More Frag	Retry	Pwr Mgt	More Data	WEP	Order

a) WLAN and FEC Packet Fields (Krätzer et al. 2006)

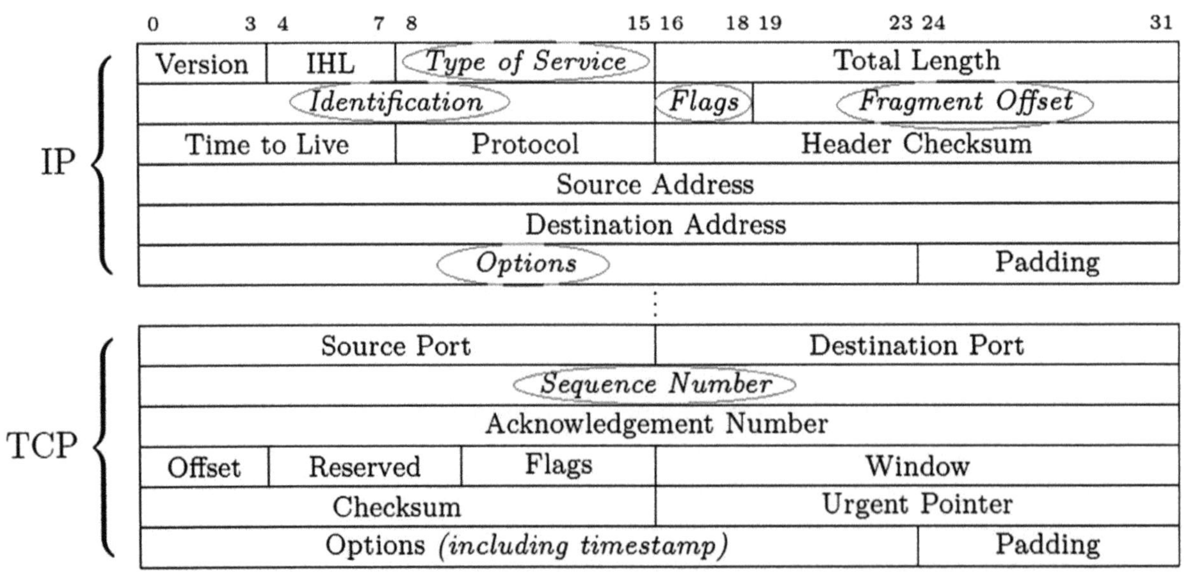

b) IP and TCP Packed Header Fields

Table 3. Communication link speed for implementing CSC to transfer medical data3

Medical Information	Min. Overt Communication Link Speed Required for CSC using SN field in TCP Header (kbps)	Min. Overt Communication Link Speed Required for CSC Using Flag Field in IP Header (kbps)
EMG (Electromyography)	165	1760
EEG(Electroencephalography)	3520	37546.67
Blood Saturation	110	1173.333
Glucose Monitoring	0.172	1.834
Temperature	17.19	183.34
Motion Sensor	1.29	13.75
Cochlear Implant	385	4106.67
Artificial Retina	1100	11733.33
Audio	550	5866.67
Voice	44	469.34
Heart Sound	550	5866.67
Ultrasound, Cardiology, Radiology	1320	14080
Magnetic Resonance Image	22528	240298.7
Scanned X-Ray	33792	360448
Digital Radiography	162201.6	3520
Mammogram	540672	5767168

covert data into a given inter-arrival time pattern over the legitimate traffic. A general architecture of this type of a covert channels is shown in Figure 5. Several techniques and algorithms for embedding covert data in CTC is proposed in literature. For example, based on the work by Cabuk, Brodley and Shields (2004), the covert sender and covert receiver agree on a time slot, called covert time interval, and change the timing of the legitimate packets to be transferred within this time slot or not. This algorithms is known as on-off CTC and is one of the well-known CTC algorithms. A general functionality of this algorithm is shown in Figure 6-a. As illustrated in this figure, if the covert sender wants to send the bit string 1010, it changes the timing of the legitimate packet to be transferred during covert time interval, indicating covert bit' 1'. For covert bit zero, the sender remains silent during covert time interval. On the receiver side, the covert receiver listens to same covert time interval and in the case of receiving any packet in this time slot, the covert bit '1' is interpreted, otherwise covert bit '0' is received. Another reputed and highly cited CTC algorithm is called N-bit to L-packet (Sellke et al., 2009), where a covert bit string of length L is encoded into N inter-packet delays. Figure 6-b presents an example of this CTC algorithm. As we can see in this figure, the encoding of the covert bit string is based on two time slots. For example, in 2 bits to 1 packet scheme, where covert patterns '11', '10', '01' and '00' are intended to be sent covertly, the sender assigns the reference time slot of D to one of the covert bit patterns, for example 50 to '00', and adds the other constant time slot of d for other covert bit patterns. As Figure 6-b presents the covert sender in this example changes the inter-packet time of the legitimate packets respectively to 50, 60, 70 and 80 milliseconds. Then the covert receiver on the other hand, measures the inter-arrival time of the received legitimate packets and interprets the covert bits based on a given measured delay. Figure 6-b represents a general functionality of this algorithm.

One of the main advantages of covert timing channels compared to the covert storage channels is that it cannot be easily detected and disrupted. However, CTC algorithms provide less covert data capacity. In addition, these types of covert channels are more vulnerable to noise and jitter of the legitimate communication. The capacity of these algorithms is bounded to the amount of inter-packet delay that is considered for encoding covert data. As this time decreases, covert data capacity increases proportionally. However, as shown by Shrestha et al. (2014), in the presence of significant network delay jitter and noise over the communication channel, the covert receiver might not be able to interpret the inter-packet delays properly, if the inter-packet delay is too small. This may lead to decoding errors at the receiver. Therefore, the covert interval or the inter-packet delays should be chosen based on the channel conditions and the required speed of sending covert data.

Table 4 shows some scenarios of the two mentioned CTC algorithms and presents the theoretical capacity of covert data that can be transferred using these algorithms. This capacity can be calculated based on the inter-packet delays and covert time interval. In on-off algorithm each covert bit is sent within the given covert time interval, then the capacity can be calculated as 1/ (covert time interval). Therefore, if we assume the covert time interval as 100, then covert data capacity is 10bps. For N-bits to L-packet algorithm, if we assume that the covert bit patterns are uniformly distributed, then the capacity can be calculated as N/ (Average inter-packet delays). Hence, for the example that is illustrated in Figure 6-b, covert data capacity can be calculated as: 2/ ((50+60+70+80)/4) = 30.76 bps. Based on this table and the required data rates for sending biological signals, we can see that sending some of the medical data over covert communication, takes a long time. However, to improve the capacity, decrease the inter-packet time or in case of N-bits to L-packets, choose a large number for L to encode a longer covert bit string to a specific inter-packet delay. Although, the capacity of these two CTC algorithms seems to be low, still can transfer some medical data based on Table 1 such as temperature, and blood saturation.

4. **Covert Hybrid Channels (CHC):** A covert Hybrid channel is referred to a combinational method of storage and timing channels to establish covert communication. Mazurczyk and Lubacz (2010) propose a CHC that inserts covert data into a part of the payload of the legitimate packets and intentionally delays them so that these packets are discarded by the receiver as late packets based on the communication protocol. However, the covert receiver that knows the functionality of the CHC, will keep the intentionally delayed packet and extracts covert date. Figure 7 shows a general structure of this type of covert channel. By using CHC algorithms, some of the disadvantages of the CSC and CTC algorithms such as easy detectability and low capacity can be improved. The capacity of CHC algorithms depend on the implementation and the speed of the communication link as well as the inserted inter-packet delays.

COVERT CHANNEL DETECTION

As we mentioned before, the covert communication can be established to leak patient's personal

Figure 5. Covert timing channel (CTC) general architecture

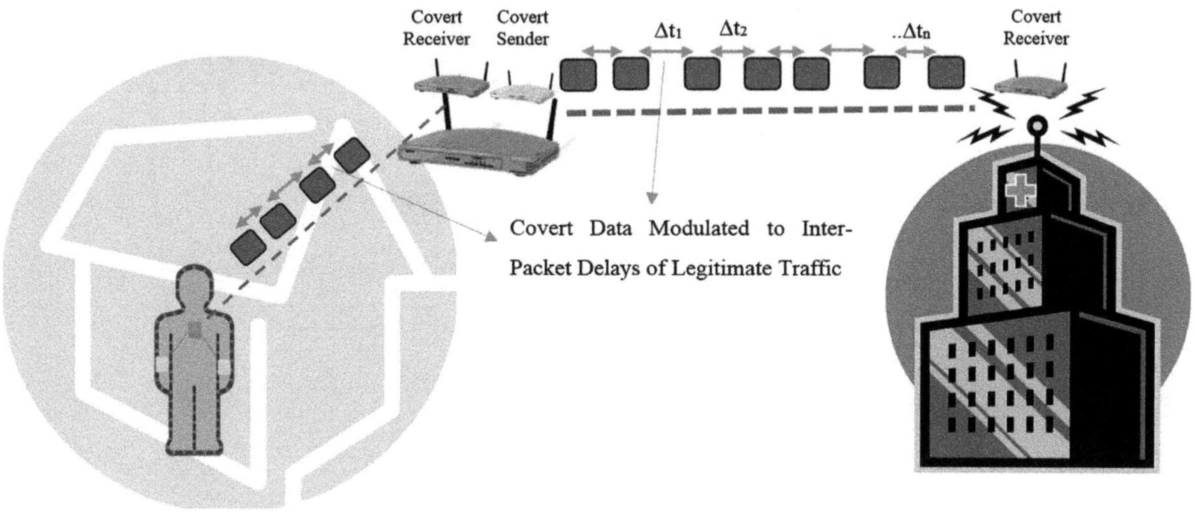

information, identity and medical history and information for malicious purposes. Therefore, covert channels are not desirable in WSN especially in e-health applications and preventing such activities to improve the security of WSN in e-health applications is necessary. According to the covert channel type, there are different detection methods that are reported in literature. Generally, the CSCs are easily detectable and the detection approaches for different covert channels vary in accordance with the modified protocol fields of the packets that carry covert data. For example, Garateguy, Arce and Pelaez (2011) utilized Random Projection (RP) calculation to categorize network packets based on their contents. In this way, the network packets that contain covert data can be recognized by different RP values. While detecting CSC is usually possible by investigating the content of the packets and recognizing unexpected data, detecting CTCs is more complicated. Therefore, in this section we mainly focus on the detection methods that are introduced in literature for CTC algorithms.

Browne (1994), suggested that by calculating the entropy of the system's output and comparing it with the expected entropy, an inference about the presence of a covert channel imposing the difference in entropy could be made. However, he states accurate calculating the entropy of a system is almost impossible. Cabuk, Brodley and Shields (2004) studied the possibility of detecting covert communications by building metrics based on the distribution of packet arrival times. The detector was built on the premise that CTCs would create regular patterns guided by a timing parameters agreed upon by the covert communication parties. Borders and Prakash (2004) considered the detection of a binary channel that uses two distinct delays to transmit covert bit '0' or '1' by using the histogram of the delays. They show that the histogram will have two peaks that correspond to the two associated delay values and the mean of the delay will be somewhere between the two peaks. Based on a similar principle, Berk et al. (2005) proposed a simple equation of calculating the probability that a channel is covert as

$$P = 1 - C_{\mu} / C_{\max},$$

where C_{μ} is the number of packets with mean delay time and $C_{\max}$ is the total number of packets. The possibility of reducing the capacity of timing channels by introducing noise in the system

Figure 6. Examples of covert timing channel (CTC) algorithms (Rezaei et al. IWCMC 2014 and Rezaei et al. MILCOM 2014)

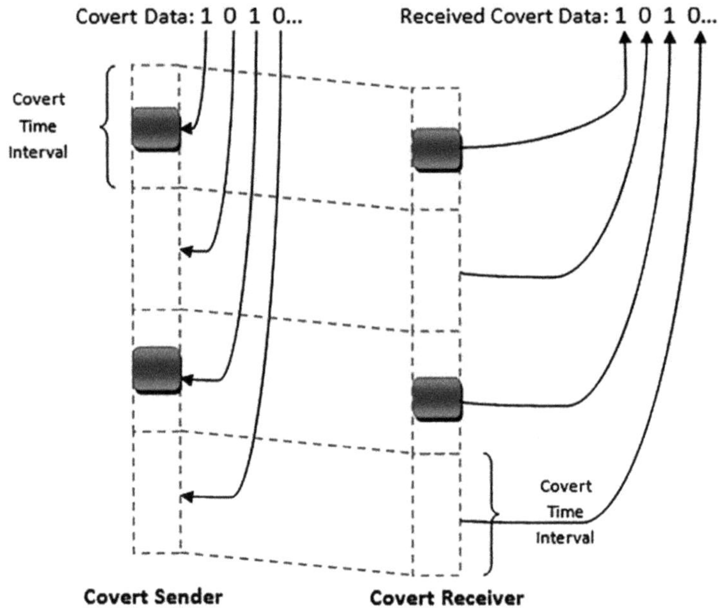

A) On-off CTC

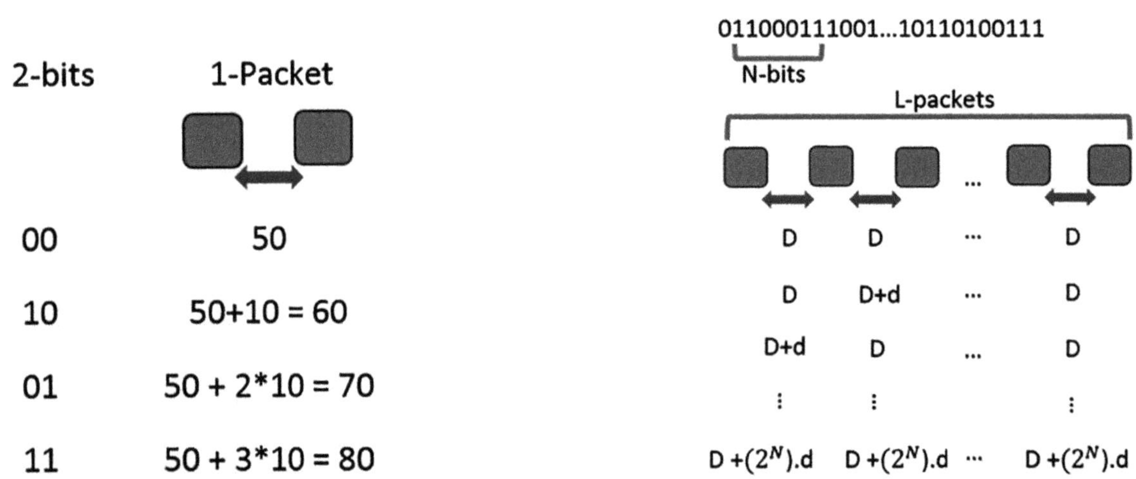

B) N-bits to L-Packet CTC

and jamming have been investigated by Hu et al. (1992), Giles et al. (2000), and Gray et al. (1993).

Gianvecchio and Wang (2011) studied four fingerprints of covert traffic which have been derived from the inter-packet arrival times — Kolmogorov-Smirnov (K-S) test score, regularity score, entropy and corrected conditional entropy (CCE). The first two statistics are the same ones discusses by Cabuk, Brodley and Shields (2004).

Table 4. Examples of CTC covert data capacity 4

	On–Off CTC				N-Bits to L-Packets CTC			
	Covert time interval (ms)				2-bit to 1-packet:		8-bits to 1-packets	
	5	10	25	50	D=10 d=5	D=50 d=10	D=15 d=10	D=45 d=15
Capacity (bps)	200	100	40	20	114.28	30.76	266.67	118.51

Figure 7. Covert hybrid channel general architecture

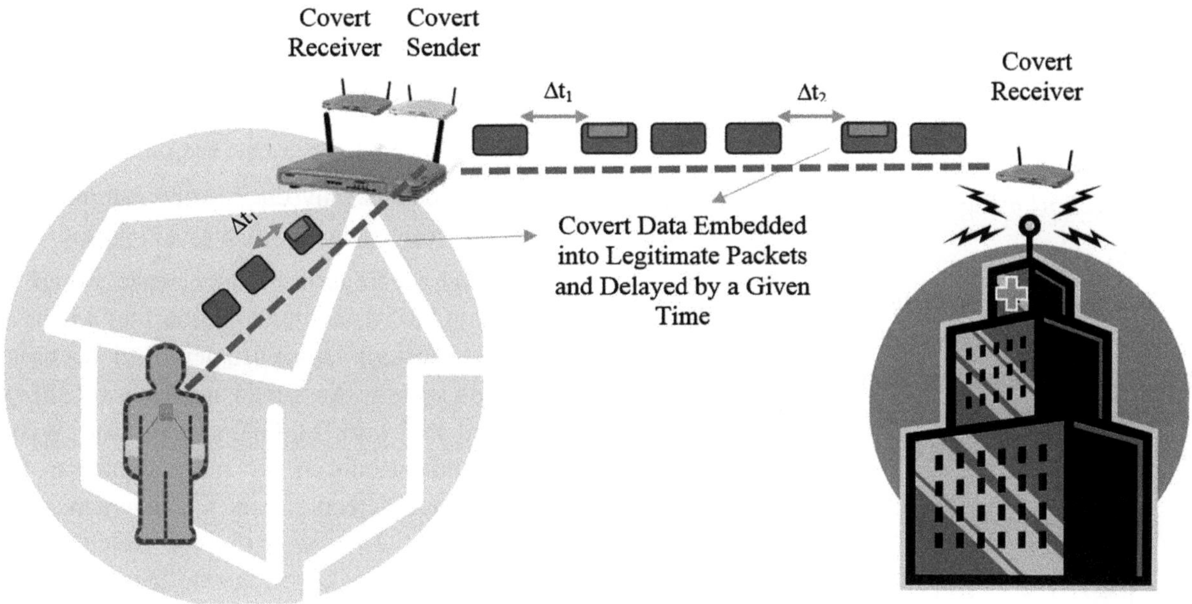

The K-S test score evaluates the hypothesis of whether two random samples are drawn from the same population distribution. The two-sided test utilizes the maximum absolute difference between the cumulative distribution functions of the distributions of the two samples. The test statistic is

$$D = \max_{x}\left(\left|S_1(x) - S_2(x)\right|\right) \quad (1)$$

where $S_1(x)$ and $S_2(x)$ represents the proportion of sample points in the two samples less than or equal to x. The one sided test uses the actual value of the difference rather than the absolute value.

The regularity score is a measures the variation within sequence of samples. Cabuk et al. (2006) have defined the regularity score based on the standard deviation of the continuous block of samples. The sample under test is divided into multiple blocks and the standard deviation of each individual block is computed. Then, the normalized change in standard deviations of the blocks is calculated. The process is repeated multiple times and the standard deviation of the normalized changes are calculated. This metric is defined as the regularity score. If $\sigma_i; i = 1, 2, \ldots, n$ represents the standard deviation of the sample block and n is the total number of samples, the regularity score can be represented by

$$Regularity\,Score = Std.Dev.\left(\frac{\left|\sigma_i - \sigma_j\right|}{\sigma_i}\right); \forall i, j\, and\, i < j \quad (2)$$

From information theoretic point of view, entropy represents the uncertainty in the value of a random variable. Since the distribution of the traffic is unknown, the entropy has to be calculated empirically. The sample points are binned into a fixed number of adjacent bins and am empirical probability distribution is calculated. The range of the bins are chosen so as to make the bins equi-probable. Then, the entropy is calculated as:

$$H(x) = \Sigma_x p(x) \log_2 (x) \quad (3)$$

To capture the correlation between successive sample points, higher level entropy. Successive n-tuples of sample points are distributed across the bins and the entropy evaluated. The size of the tuple defines the level of entropy. Mathematically:

$$H(X_1, \ldots, X_m) = £_x p(x_1, \ldots, x_m) \log_2 (x_1, \ldots, x_m) \quad (4)$$

Similarly, the conditional entropy of a random variable conditioned upon the knowledge of previous values of the random variable is defined as:

$$H(X_m | X_1, \ldots, X_{m-1}) = H(X_1, X_2, \ldots, X_m) - H(X_1, X_2, \ldots, X_{m-1}) \quad (5)$$

It measures the reduction in uncertainty of a random variable when the knowledge of some other values of the random process.

Porta at al. (1998) showed that use of finite length samples to represent a population distribution causes inaccurate estimation of entropy and conditional entropy. This is because a small sample may not be able to properly represent the distribution across the bins that a real population would. The inaccuracy escalates while evaluating higher level entropy. To correct this error, they calculate the number of bins that has a single element in it and this metric after being scaled by the first level entropy represents the correction factor to correct the entropy scores i.e.

$$CCE(X_m | X_1, \ldots, X_{m-1}) = CE(X_m | X_1, \ldots, X_{m-1}) + perc(X_m) EN(X_1) \quad (6)$$

where $perc(X_m)$ represents the number of patterns that occurs only once in the samples and $EN(X_1)$ represents the first level entropy.

Though the fingerprints are generic enough to detect the CTCs under investigation, they will have practical shortcomings when deployed in a blind detection system. Archibald and Ghosal (2014), extended the work of Gianvecchio and Wang (2011) and introduced two more signatures for covert channel detection—the K-L divergence and Welch's t-test. They show that the Welch's t-test detects the Jitterbug covert channel (Shah, Molina, & Blaze, 2006), relatively well compared to other statistics as it looks for the difference in average delay between covert and overt traffic. They also present an SVM based machine learning approach using regularity metrics as the feature points. But this classifier has been tested against only against a single CTC algorithm. Shrestha et al. (2014), discuss the shortcomings of currently available detection mechanisms based on fingerprints derived from inter-packet delays. They show that such mechanisms will perform poorly when the size of the covert message is reduced and it is not possible to detect the presence of such covert message by proportionally scaling the size of traffic under investigation. Few of the results published in the paper are shown in Figures 8 and 9. Figure 8 compares the histogram of the CCE values of covert and overt traffic by embedding 100 bits into a 2000 packets long block. The overlap between the histogram indicates that the measure is not reliable to detect covert communications. This is because the impact of embedding 100 bits into 2000 packets does not change the statistics of the traffic by a lot. Figure 9 compares the histogram of the CCE values when 100 bits are embedded into 100 packets long blocks. However, the histograms still overlap which shows that the detection reliability cannot be improved by scaling the size of the traffic to the covert message size. This is because at small sample sizes, the sample does not represent the population very well and hence the statistics derived from small sample sizes are not reliable. The authors hence also argue for the need of machine learning techniques for generalized detection of CTCs.

FUTURE RESEARCH DIRECTIONS

Based on what we discussed in this chapter, there are various approaches that can be used for leaking medical data and patients' identification and personal information in e-health and telemedicine. Distribution this type of sensitive data that can be abused by beneficial groups is not desirable. Therefore, as the embedding methods and information leakage techniques are growing and new approaches might be introduced in the future, the future research activities can develop corresponding detection methods to verify these vulnerabilities and risks. One of the interesting areas to explore is the real-time covert channel detection that is not widely covered by the existing literature. Due to the fact that the existing detection methods can detect the covert communication activities based on the history of network traffic and by collecting some statistics from the ongoing one, some hidden information can be transferred right before detecting the covert channel and taking a suitable action.

Another tempting areas of research is to be able to detect covert timing channels when the legitimate channel is not ideal. Although establishing these types of communications reliably over noisy legitimate network traffic is a challenging task, but by taking advantage of different error correction techniques in the receiver, the effect of noise can be negligible. However, in this condition detection covert timing channels based on the overt channel conditions will be difficult and is not considered by the reviewed detection methods.

SUMMARY AND CONCLUSION

The application of Wireless Sensor Networks in the area of e-health is growing very fast recently. Different devices that include several sensors are used for monitoring patients' health status and biological signals to prevent frequently visits to the health center and hospital. This contributes to time and cost efficient health checkups and provides immediate assistance from medical staffs in case of abnormal recognition of biological signals and health status. One of the main concerns of using these devices by patients is the security and privacy of medical data and identifications. Health status data and medical records of the patients can be accessible by unauthorized groups and parties when transmitted over different networks, which is not desirable and needs to be prevented. In this chapter we described a general architecture of wireless sensor networks in e-health applications and discussed different communication technologies that are used for data transmission in these devices. Then, we reviewed different levels of security in this area of WSN and introduced a possible a threat in this area, called covert communication. As we discussed in this material, this type of secret path that exploits network resources to leak information can established in different layers of a legitimate communication path for malicious purposes. We have presented different categories and implementations of covert chan-

Figure 8. Normalized Histogram Plot of CCE of Covert and Overt Traffic by analyzing 2,000 packets at a time, with 100 covert bits embedded in each block (Shrestha et al. 2014)

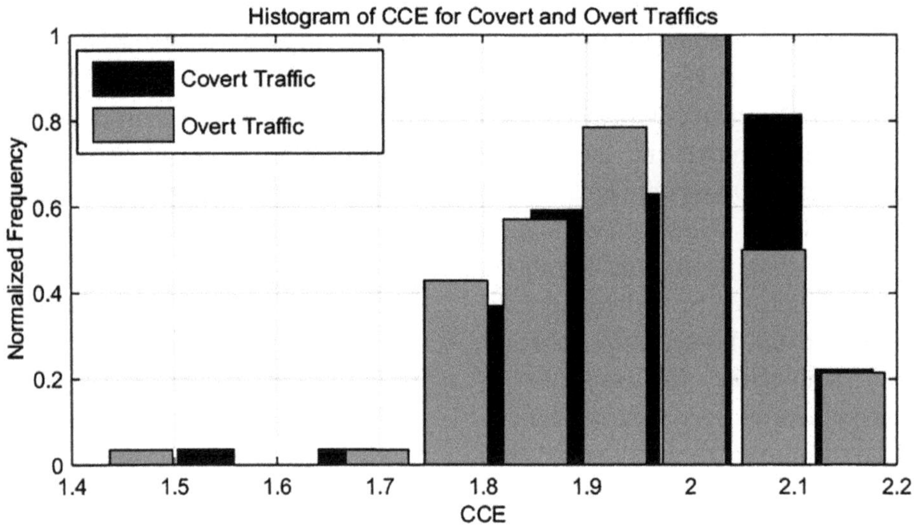

Figure 9. Normalized Histogram Plot of CCE of Covert and Overt Traffic by analyzing 100 packets at a time (Shrestha et al. 2014)

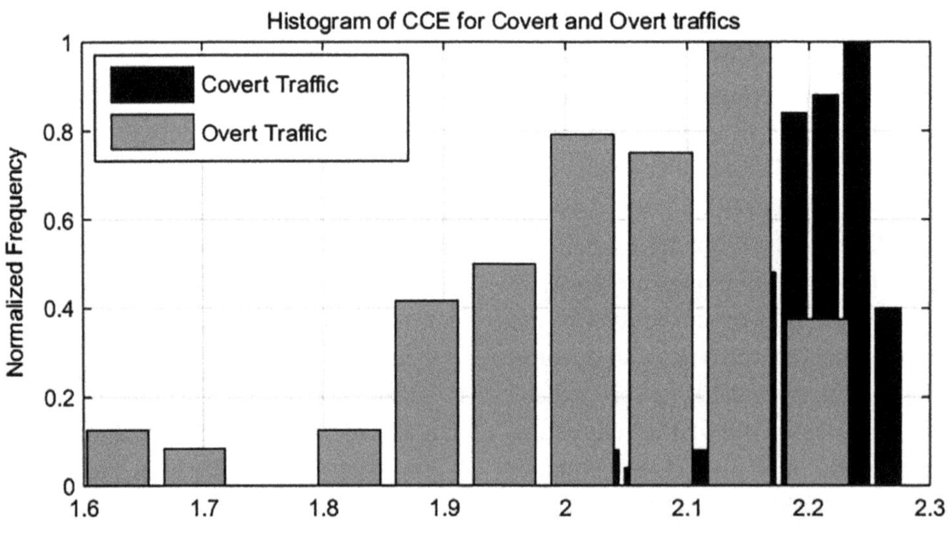

nels that can be used for establishing a hidden path to transfer medical data, patients' identity and their health records to the beneficial group. To prevent and avoid this type of threat in WSN e-health applications we have described several covert communication detection methods to detect exploiting the network resources for malicious use.

REFERENCES

Al Ameen, M., Liu, J., & Kwak, K. (2012). Security and privacy issues in wireless sensor networks for healthcare applications. *Journal of Medical Systems*, *36*(1), 93–101. doi:10.1007/s10916-010-9449-4 PMID:20703745

Archibald, R., & Ghosal, D. (2014). A comparative analysis of detection metrics for covert timing channels. *Computers & Security*, *45*, 284–292. doi:10.1016/j.cose.2014.03.007

Berk, V., Giani, A., Cybenko, G., & Hanover, N. H. (2005). *Detection of covert channel encoding in network packet delays.* (Tech. Rep. TR536). Department of Computer Science, Dartmouth College.

Borders, K., & Prakash, A. (2004, October). Web tap: detecting covert web traffic. In *Proceedings of the 11th ACM conference on Computer and communications security* (pp. 110-120). ACM.

Browne, R. (1994, November). An entropy conservation law for testing the completeness of covert channel analysis. In *Proceedings of the 2nd ACM Conference on Computer and communications security* (pp. 270-281). ACM. doi:10.1145/191177.191244

Cabuk, S., Brodley, C. E., & Shields, C. (2004, October). IP covert timing channels: design and detection. In *Proceedings of the 11th ACM Conference on Computer and Communications Security* (pp. 178-187). ACM.</conf>

Calhoun, T. E. Jr, Cao, X., Li, Y., & Beyah, R. (2012). An 802.11 MAC layer covert channel. *Wireless Communications and Mobile Computing*, *12*(5), 393–405. doi:10.1002/wcm.969

Garateguy, G., Arce, G. R., & Pelaez, J. (2011, March). *Covert Channel detection in VoIP streams.* In Information Sciences and Systems (CISS), 2011 45th Annual Conference on (pp. 1-6). IEEE. doi:10.1109/CISS.2011.5766220

Gianvecchio, S., & Wang, H. (2011). *An entropy-based approach to detecting covert timing channels. IEEE Transactions on Dependable and Secure Computing*, *8*(6), 785–797.

Giles, J., & Hajek, B. (2000). *The jamming game for packet timing channels.* In Proceedings of IEEE International Symposium on Information Theory, 2000. (p. 51). IEEE. doi:10.1109/ISIT.2000.866341

Gray, J. W., III. (1993, May). *On introducing noise into the bus-contention channel.* In Proceedings of IEEE Computer Society Symposium on Research in Security and Privacy, 1993. (pp. 90-98). IEEE. doi:10.1109/RISP.1993.287640

Holloway, R., & Beyah, R. (2011, October). *Covert DCF: A DCF-Based Covert Timing Channel in 802.11 Networks.* In Mobile Ad hoc and Sensor Systems (MASS), 2011 IEEE 8th International Conference on (pp. 570-579). IEEE.

Hu, W. M. (1992). Reducing timing channels with fuzzy time. *Journal of Computer Security*, *1*(3), 233–254.

Krätzer, C., Dittmann, J., Lang, A., & Kühne, T. (2006, September). *WLAN steganography: a first practical review.* In *Proceedings of the 8th workshop on Multimedia and security* (pp. 17-22). ACM.

Latré, B., Braem, B., Moerman, I., Blondia, C., & Demeester, P. (2011). A survey on wireless body area networks. *Wireless Networks*, *17*(1), 1–18. doi:10.1007/s11276-010-0252-4

Mazurczyk, W., & Lubacz, J. (2010). LACK—a VoIP steganographic method. *Telecommunication Systems*, *45*(2-3), 153–163. doi:10.1007/s11235-009-9245-y

Monton, E., Hernandez, J. F., Blasco, J. M., Hervé, T., Micallef, J., Grech, I., & Traver, V. (2008). Body area network for wireless patient monitoring. *IET Communications*, *2*(2), 215–222. doi:10.1049/iet-com:20070046

Murdoch, S. J., & Lewis, S. (2005, January). Embedding covert channels into TCP/IP. In *Information Hiding* (pp. 247–261). Springer Berlin Heidelberg.

Pappas, C., Coscia, E., Dodero, G., Gianuzzi, V., & Earney, M. (2002). *A mobile e-health system based on workflow automation tools.* In Proceedings of the 15th IEEE Symposium on Computer-Based Medical Systems, 2002. (CBMS 2002) (pp. 271-276). IEEE. doi:10.1109/CBMS.2002.1011388

Porta, A., Baselli, G., Liberati, D., Montano, N., Cogliati, C., Gnecchi-Ruscone, T., & Cerutti, S. (1998). Measuring regularity by means of a corrected conditional entropy in sympathetic outflow. *Biological Cybernetics*, *78*(1), 71–78. doi:10.1007/s004220050414 PMID:9485587

Radhakrishnan, S. V., Selcuk Uluagac, A., & Beyah, R. (2013, December). Realizing an 802.11-based covert timing channel using off-the-shelf wireless cards. In *Global Communications Conference (GLOBECOM)*, 2013 IEEE (pp. 722-728). IEEE. doi:10.1109/GLOCOM.2013.6831158

Rezaei, F., Hempel, M., Peng, D., Qian, Y., & Sharif, H. (2013, April). *Analysis and evaluation of covert channels over LTE advanced.* In *Wireless Communications and Networking Conference (WCNC)*, 2013 IEEE (pp. 1903-1908). IEEE. doi:10.1109/WCNC.2013.6554855

Rezaei, F., Hempel, M., Rakshit, S. M., & Sharif, H. (2014). Automated Covert Channel Modeling over Real Network Environment. In *International Wireless Communications and Mobile Computing Conference (IWCMC)*. IEEE. doi:10.1109/IWCMC.2014.6906417

Rezaei, F., Hempel, M., & Sharif, H. (2015 Forthcoming). A Survey of Recent Trends in Wireless Communication Standards, Routing Protocols, and Energy Harvesting Techniques in E-Health Applications. *International Journal of E-Health and Medical Communication.*

Rezaei, F., Hempel, M., Shrestha, P. L., & Sharif, H. (2014). Evaluation and Analysis of Automated Covert Channel Modeling over Real Network Environment. In Proceedings of Military Communication Conference (MILCOM). IEEE.

Sellke, S. H., Wang, C. C., Bagchi, S., & Shroff, N. (2009, April). TCP/IP timing channels: Theory to implementation. In INFOCOM 2009, IEEE (pp. 2204-2212). IEEE.

Shah, G., Molina, A., & Blaze, M. (2006, July). Keyboards and Covert Channels. In *Proceedings of the 15th USENIX Security Symposium*. https://www.usenix.org/legacy/event/sec06/tech/shah/shah.pdf

Shrestha, P., Hempel, M., Rezaei, F., & Sharif, H. (2014). *Leveraging statistical feature points for generalized detection of covert timing channels.* In Proceedings of Military Communication Conference (MILCOM), IEEE. doi:10.1109/MILCOM.2014.10

Shrestha, P. L., Hempel, M., Sharif, H., & Chen, H. H. (2014). *An Event-Based Unified System Model to Characterize and Evaluate Timing Covert Channels*. IEEE Systems Journal.

Vouyioukas, D., Maglogiannis, I., & Komnakos, D. (2007). Emergency M-Health services through high-speed 3G systems: Simulation and performance evaluation. *Simulation*, *83*(4), 329–345. doi:10.1177/0037549707083113

Wang, Z., Deng, J., & Lee, R. B. (2007, May). *Mutual anonymous communications: a new covert channel based on splitting tree MAC.* In INFOCOM 2007. 26th IEEE International Conference on Computer Communications (pp. 2531-2535). IEEE. doi:10.1109/INFCOM.2007.315

Zhao, H., & Shi, Y. Q. (2010, December). *A phase-space reconstruction approach to detect covert channels in TCP/IP protocols.* In Information Forensics and Security (WIFS), 2010 IEEE International Workshop on (pp. 1-6). IEEE.

KEY TERMS AND DEFINITIONS

Covert Communication: Establishing a hidden communication path over the legitimate one to transfer secret data covertly by using different resources of network such as given protocol header fields or inter- packet delay of network traffic. This type of communication sometimes is used for distribution of malicious data and information leakage.

Covert Data: Secret data that is transmitted using covert communication.

Covert Hybrid Channel (CHC): A type of covert communication that uses a combinational method of covert timing channel and covert storage channel for embedding covert data.

Covert Receiver: A designated receiver that is aware of the covert communication protocol and can extract the covert data.

Covert Sender: A designated sender that embeds covert data into the network to establish a covert channel.

Covert Storage Channel (CSC): A type of covert communication that takes advantage of specific packet header fields or a part of the packet payload to embed the covert data.

Covert Timing Channel (CTC): A type of covert communication that embeds covert data into the inter-packet delays of legitimate network traffic.

E-Health: Referred to healthcare practice that utilizes Information Technology, Telecommunication, and recent technologies to provide more reliable and faster services to patients.

Compilation of References

Abdulla, R., & Ismail, W. (2013). Survey of WSN technology based reliable and efficient active RFID. Paper presented at the 2013 IEEE Malaysia International Conference on Communications (MICC) In IEEE, pp. 116-121. doi:10.1109/MICC.2013.6805810

Abolhasan, M., Wysocki, T., & Dutkiewicz, E. (2004). A Review of Routing Protocols for Mobile Ad Hoc Networks. *Ad Hoc Networks*, *2*(1), 1–22. doi:10.1016/S1570-8705(03)00043-X

Abuaitah, G. R., & Wang, B. (2014). Online Data-Centric Anomaly Detection Framework For Sensor Network Deployments. In *The 2014 IEEE International Conference on Computing, Networking and Communications (ICNC)*, (pp. 599-604). doi:10.1109/ICCNC.2014.6785404

Adamchuk, V. I., Hummel, J. W., Morgan, M. T., & Upadhyaya, S. K. (2004). On-the-go soil sensors for precision agriculture. *Computers and Electronics in Agriculture*, *44*(1), 71–91. doi:10.1016/j.compag.2004.03.002

Adams, J. C., & Ramarao, K. V. S. (1989). Distributed diagnosis of Byzantine processors and links. *Proceedings. The 9th International Conference on Distributed Computing Systems*. doi:10.1109/ICDCS.1989.37989

Agarwal, A., & Jagannatham, A. K. (2014). Distributed Estimation in Homogenous Poisson Wireless Sensor Networks. *IEEE Wireless Communications Letters*, *3*(1), 90–93. doi:10.1109/WCL.2013.112313.130696

Ahlswede, R., Cai, N., Li, S.-Y. R., & Yeung, R. W. (2000, July). Network Information Flow. *IEEE Transactions on Information Theory*, *46*(4), 1204–1216. doi:10.1109/18.850663

Ahonen, T., Virrankoski, R., & Elmusrati, M. (2008). Greenhouse monitoring with wireless sensor network. In *Proceedings of the IEEE/ASME International Conference on Mechtronic and Embedded Systems and Applications* (pp. 403–408). doi:10.1109/MESA.2008.4735744

Ainslie, M. A. (2010). *Principles of Sonar Performance Modelling*. Heidelberg, Germany: Springer. doi:10.1007/978-3-540-87662-5

Ainslie, M. A., & McColm, J. G. (1998). A simplified formula for viscous and chemical absorption in seawater. *The Journal of the Acoustical Society of America*, *103*(3), 1671–1672. doi:10.1121/1.421258

Akan, O. B., & Akyildiz, I. F. (2005). ESRT: Event-to-sink reliable transport in wireless sensor networks. *IEEE/ACM Transactions on Networking*, *13*(5), 1003–1016. doi:10.1109/TNET.2005.857076

Akyildiz, I. F., Pompili, D., & Melodia, T. (2004). Challenges for efficient communication in underwater acoustic sensor networks. *ACM SIGBED Review*, *1*(2), 3–8. doi:10.1145/1121776.1121779

Akyildiz, I. F., Pompili, D., & Melodia, T. (2005). Underwater Acoustic Sensor Networks: Research Challenges. *Ad Hoc Networks*, *3*(3), 257–279. doi:10.1016/j.adhoc.2005.01.004

Akyildiz, I. F., Pompili, D., & Melodia, T. (2007). State-of-the-art in protocol research for underwater acoustic sensor networks. *SIGMOBILE Mobile Computing and Communications Review*, *11*(4), 11–22. doi:10.1145/1347364.1347371

Akyildiz, I. F., Su, W., Sankarasubramaniam, Y., & Cayirci, E. (2002). Wireless sensor networks: A survey. *Computer Networks*, *38*(4), 393–422. doi:10.1016/S1389-1286(01)00302-4

Akyildiz, I. F., & Vuran, M. C. (2011). *Wireless Sensor Networks*. New York: Wiley.

Al Ameen, M., Liu, J., & Kwak, K. (2012). Security and privacy issues in wireless sensor networks for healthcare applications. *Journal of Medical Systems*, *36*(1), 93–101. doi:10.1007/s10916-010-9449-4 PMID:20703745

Alekeish, K., & Ezhilchelvan, P. (2012). Consensus in Sparse, Mobile Ad Hoc Networks. *IEEE Transactions on Parallel and Distributed Systems*, *23*(3), 467–474. doi:10.1109/TPDS.2011.182

Ali, A., Rashid, R. A., & Fisal, N. (2008). *Cognitive Forwarding for Real-time Routing in Wireless Sensor Network*.

Ali, A., Rashid, R. A., Arriffian, S. H. F., & Fisal, N. (2007). Optimal forwarding probability for real-time routing in wireless sensor network. Paper presented at the *2007 IEEE International Conference on Telecommunications and Malaysia International Conference on Communications, 2007. ICT-MICC* In IEEE, pp. 419-424. doi:10.1109/ICTMICC.2007.4448673

Ali, S., Fakoorian, A., & Taheri, H. (2007). Optimum Reed-Solomon erasure coding in fault tolerant sensor networks, in *Proceedings of the 4th International Symposium on Wireless Communication Systems (ISWCS)* (pp. 6-10). Trondheim, Norway.

Al-Jemeli, M., & Hussin, F. (2014). An Energy Efficient Cross-Layer Network Operation Model for IEEE 802.15.4-Based Mobile Wireless Sensor Networks. *IEEE Sensors Journal*, *99*, 1–9.

Allred, J., Hasan, A. B., Panichsakul, S., Pisano, W., Gray, P., & Huang, J. (2007, November). *Sensor Flock: An Airborne Wireless Sensor Network of Micro-Air Vehicles*. Paper Presented at the 5th ACM International Conference on Embedded Networked Sensor Systems. Sydney, Australia.

Alnawaiseh, A., & Abdelghany, K. (2012, August). *Team Sense: Energy-Efficient Autonomous Mobile Wireless Sensor Networks for Object Tracking*. Paper presented at the 8th International Wireless Communications and Mobile Computing Conference (IWCMC), Limassol.

Al-Suhail, G. A., Louis, K. W., & Abdallah, T. Y. (2012). Energy Efficiency Analysis of Adaptive Error Correction in Wireless Sensor Networks. *International Journal of Computer Science Issues*, *9*(4), 79–84.

Altamiranda, E., Kiaer, L., & Hu, X. (2009). Condition monitoring and diagnosis for subsea control systems: A subsystem prototype. In *Proceedings of OCEANS 09 – Europe*.

Ameer, P. M., Kumar, A., Manjunath, D., & Boyina, R. (2006, November). *Analysis of network architectures for Zigbee sensor clusters*. Paper presented at the *12th International Telecommunications Network Strategy and Planning Symposium, 2006. Networks*, *2006*, 1–6.

Amundson, I., & Koutsoukos, X. D. (2009, September). *A Survey on Localization for Mobile Wireless Sensor Networks*. Paper Presented at the 15th Annual International Conference on Mobile Computing and Networking. Beijing, China.

Amundson, I., Koutsoukos, X., & Sallai, J. (2008, September). *Mobile Sensor Localization and Navigation Using RF Doppler Shifts*. Paper Presented at the 14th Annual International Conference on Mobile Computing and Networking. San Francisco, CA. doi:10.1145/1410012.1410034

Anandhakumar, P., Priyadarshini, J., Rajeswari, L., Srividhya, S., & Niveditha, C. K. (2011). *Occlusion detection and object tracking using filter banks*. In IEEE International Conference on Recent Trends in Information Technology (pp. 992-998). doi:10.1109/ICRTIT.2011.5972465

Ansari, J., Pankin, D., & Mahonen, P. (2009). Radio-triggered wake-ups with addressing capabilities for extremely low power sensor network applications. *International Journal of Wireless Information Networks*, *16*(3), 118–130. doi:10.1007/s10776-009-0100-6

Ansari, M. S., Mahani, A., & Kavian, Y. S. (2012). *Energy-efficient network design via modeling: optimal designing point for energy, reliability, coverage and end-to-end delay* (pp. 1–8). IET Network; doi:10.1049/iet-net.2012.0024

Anthony, D., Bennett, W. P., Vuran, M. C., Dwyer, M. B., Elbaum, S., Lacy, A., & Wehtje, W. (2012). Sensing through the continent: towards monitoring migratory birds using cellular sensor networks. In *Proceedings of the 11th international conference on Information Processing in Sensor Networks* (pp. 329–340).

Archibald, R., & Ghosal, D. (2014). A comparative analysis of detection metrics for covert timing channels. *Computers & Security*, *45*, 284–292. doi:10.1016/j.cose.2014.03.007

Armholt, M., Junnila, S., & Defee, I. (2007). *A Non-beaconing ZigBee Network Implementation and Performance Study*. Paper presented at the 4th *IEEE International Conference on Communications (ICC '07)*, Glasgow, Scotland, pp. 3232-3236. doi:10.1109/ICC.2007.536

Ashrafuzzaman, K., & Kwak, K. S. (2011). On the performance analysis of the contention access period of IEEE 802.15. 4 MAC. *IEEE Communications Letters*, *15*(9), 986–988. doi:10.1109/LCOMM.2011.071211.111271

Association of American Railroads. (n.d.). *Class I Railroad Statistics*. Retrieved from http://www.aar.org/PubCommon/Documents/AboutTheIndustry/Statistics.pdf

Atmel. (2014). *AVR JTAGICE mkII*. Retrieved October 10, 2014, from http://www.atmel.com/tools/AVR-JTAGICEMKII.aspx

Ayaz, M., Baig, I., Abdullah, A., & Faye, I. (2011). A survey on routing techniques in underwater wireless sensor networks. *Journal of Network and Computer Applications*, *34*(6), 1908–1927. doi:10.1016/j.jnca.2011.06.009

Ayeb, B. & Farhat, A. (2004). A flexible formal framework for masking/demasking faults. *Information Sciences*, *159*, 29–52.

Azlina, N., Aziz, B. A., Mohemmed, A. W., & Alias, M. Y. (2009, March). *A Wireless Sensor Network Coverage Optimization Algorithm Based on Particle Swarm Optimization and Voronoi Diagram.* Paper Presented at IEEE International Conference on Networking, Sensing and Control. Okayama, Japan.

Babiker, A. E., & Zakaria, M. N. B. (2011). Energy efficiency analysis of error correction techniques in underwater wireless sensor networks. *Journal of Engineering Science and Technology*, *6*(1), 17–28.

Babiker, A. E., Zakaria, M. N. B., Yosif, H., & Ibrahim, S. B. (2011). An Efficient Energy Adaptive Hybrid Error Correction Technique for Underwater Wireless Sensor Networks. *World Academy of Science, Engineering and Technology*, *5*, 1108–1124.

Bahl, P., & Padmanabhan, V. (2000). RADAR: An In-Building RF-Based User Location and Tracking System. In *Proceedings of the IEEE INFOCOM '00, March 2000.* doi:10.1109/INFCOM.2000.832252

Bahl, P., Balachandran, A., & Padmanabhan, V. (2000). *Enhancements to the RADAR User Location and Tracking System* (Tech. Rep., MSR, 2000).

Balakrishnan, G., Yang, M., Jiang, Y., & Kim, Y. (2007). *Performance Analysis of Error Control Codes for Wireless Sensor Networks*. Paper presented at 4th International Conference on Information Technology, Las Vegas, Nevada. doi:10.1109/ITNG.2007.149

Barboni, L. & Valle, M. (2008). Wireless Sensor Networks Power-Aware Deployment. *SENSORCOMM 2008, August 2008, Cap Esterel, France* (pp. 252-257), DOI: .10.1109/SENSORCOMM.2008.48

Baronti, P., Pillai, P., Chook, V. W. C., Chessa, S., Gotta, A., & Hu, Y. F. (2007). Wireless sensor networks: A survey on the state of the art and the 802.15. 4 and ZigBee standards. *Computer Communications*, *30*(7), 1655–1695. doi:10.1016/j.comcom.2006.12.020

Barrenetxea, G., Ingelrest, F., Schaefer, G., & Vetterli, M. (2008). Wireless sensor networks for environmental monitoring: The sensorscope experience. In *IEEE International Zurich Seminar on Communications* (pp. 98–101).

Barrenetxea, G., Ingelrest, F., Schaefer, G., & Vetterli, M. (2008). The hitchhiker's guide to successful wireless sensor network deployments. In *Proceedings of the 6th ACM conference on Embedded network sensor systems* (pp. 43–56). doi:10.1145/1460412.1460418

Barrios, C. Himberg, H. Motai, Y. Sadek, A. (2006, September). *Multiple Model Framework of Adaptive Extended Kalman Filtering for Predicting Vehicle Location.* In IEEE Intelligent Transportation Systems Conference. Toronto, Canada.

Batalin, M. A., & Sukhatme, G. S. (2004). Coverage, Exploration and Deployment by a Mobile Robot and Communication Network. *Telecommunication Systems, 26*(2-4), 181–196. doi:10.1023/B:TELS.0000029038.31947.d1

Beckwith, R., Teibel, D., & Bowen, P. (2004, October). Unwired wine: Sensor networks in vineyards. In Sensors, 2004. Proceedings of IEEE (pp. 561-564). IEEE.

Behera, S., & Maity, C. (2008). *Active RFID tag in Real Time Location System.* Paper presented at the 5th International Multi-Conference on Systems, Signals and DevicesAmman (IEEE SSD 2008), pp. 1-7. doi:10.1109/SSD.2008.4632779

Bensky, A. (2007). *Wireless positioning technologies and applications.* Artech House.

Bergbreiter, S., & Pister, K. (2003, October). *Cotsbots: An Off-the-Shelf Platform for Distributed Robotics.* Paper Presented at the IEEE/RSJ International Conference on Intelligent Robots and Systems. Las Vegas, Nevada. doi:10.1109/IROS.2003.1248878

Berk, V., Giani, A., Cybenko, G., & Hanover, N. H. (2005). *Detection of covert channel encoding in network packet delays.* (Tech. Rep. TR536). Department of Computer Science, Dartmouth College.

Bertsekas, D. P. (1997). A New Class of Incremental Gradient Methods for Least Squares Problems. *SIAM Journal on Optimization, 7*(4), 913–926. doi:10.1137/S1052623495287022

Bharathidasan, A., & Ponduru, V. A. S. (2003). *Sensor Networks: An Overview. Unpublished overview.* Davis: University of California.

Bhuiyan, M. Z. A., Wang, G.-, Zhang, L., & Peng, Y. (2010). Prediction-based energy-efficient target tracking protocol in wireless sensor networks. *Journal of Central South University of Technology, 17*(2), 340–348. doi:10.1007/s11771-010-0051-1

Bijwaard, D. J. A., van Kleunen, W. A. P., Havinga, P. J. M., Kleiboer, L., & Bijl, M. J. J. (2011). Industry: using dynamic wsns in smart logistics for fruits and pharmacy. In *Proceedings of the 9th ACM Conference on Embedded Networked Sensor Systems* (pp. 218–231). doi:10.1145/2070942.2070965

Blackman, R. B., & Tukey, J. W. (1959). *The measurement of power spectra: from the point of view of communications engineering* (Vol. 1058). Dover Publications New York.

Bolot, J. C. (2005). End-to-end packet delay and loss behavior in the internet, in *Proceedings of the ACM SIGCOMM Conference on Communications Architectures, Protocols and Applications (SIGCOMM)* (pp.13-17). San Francisco, California, USA.

Borders, K., & Prakash, A. (2004, October). Web tap: detecting covert web traffic. In *Proceedings of the 11th ACM conference on Computer and communications security* (pp. 110-120). ACM.

Borms, J., Steenhaut, K., & Lemmens, B. (2010). Low-overhead dynamic multi-channel mac for wireless sensor networks. In *Wireless Sensor Networks* (pp. 81–96). Springer Berlin Heidelberg. doi:10.1007/978-3-642-11917-0_6

Brekhovskikh, L. M., & Lysanov, Y. P. (1982). *Fundamentals of ocean acoustics.* Berlin, Germany: Springer-Verlag. doi:10.1007/978-3-662-02342-6

Brooks, R. R., Griffin, C., & Friedlander, D. S. (2002). Self-Organized Distributed Sensor Network Entity Tracking. *International Journal of High Performance Computing Applications, 16*(3), 207–219. doi:10.1177/10943420020160030201

Browne, R. (1994, November). An entropy conservation law for testing the completeness of covert channel analysis. In *Proceedings of the 2nd ACM Conference on Computer and communications security* (pp. 270-281). ACM. doi:10.1145/191177.191244

Buettner, M., Yee, G. V., Anderson, E., & Han, R. (2006, October). X-MAC: a short preamble MAC protocol for duty-cycled wireless sensor networks. In*Proceedings of the 4th international conference on Embedded networked sensor systems* (pp. 307-320). ACM. doi:10.1145/1182807.1182838

Buonadonna, P., Gay, D., Hellerstein, J., Hong, W., & Madden, S. (2005). Task: sensor network in a box. In *Proceedings of the Second European Workshop on Wireless Sensor Networks*, (pp. 133–144). doi:10.1109/EWSN.2005.1462005

Buratti, C., & Verdone, R. (2009). Performance Analysis of IEEE 802.15.4 Non Beacon-Enabled Mode. *IEEE Transactions on Vehicular Technology*, *58*(7), 3480–3493. doi:10.1109/TVT.2009.2014956

Burg, J. P. (1967). Maximum entropy spectral analysis. In *37th Annual International Meeting Society of Exploration Geophysics*.

Burrell, J., Brooke, T., & Beckwith, R. (2004). Vineyard computing: Sensor networks in agricultural production. *Pervasive Computing, IEEE*, *3*(1), 38–45. doi:10.1109/MPRV.2004.1269130

Burrowes, B., & Khan, J. Y. (2011). Short-Range Underwater Acoustic Communication Networks. In N. A. Cruz (Ed.), *Autonomous Underwater Vehicles* (pp. 174–198). Rijeka, Croatia: Intech Europe. doi:10.5772/24098

Buskens, R. W., & Bianchini, R. P. J. (1993). Distributed on-line diagnosis in the presence of arbitrary faults. *FTCS-23 The Twenty-Third International Symposium on Fault-Tolerant Computing*.

Cabuk, S., Brodley, C. E., & Shields, C. (2004, October). IP covert timing channels: design and detection. In *Proceedings of the 11th ACM Conference on Computer and Communications Security* (pp. 178-187). ACM.</conf>

Cai, J., & Goodman, D. (1997). General packet radio service in GSM. *Communications Magazine, IEEE*, *35*(10), 122–131. doi:10.1109/35.623996

Cai, S., Gao, Z., Yang, D. S., & Yao, N. (2013). A network coding based protocol for reliable data transfer in underwater acoustic sensor. *Ad Hoc Networks*, *11*(5), 1603–1609. doi:10.1016/j.adhoc.2013.02.001

Cai, W., Jin, X., Zhang, Y., Chen, K., & Tang, J. (2006). *Research on Reliability Model of Large-Scale Wireless Sensor Networks, international conference on wireless communication, networking and mobile computing* (pp. 1–4).

Calhoun, T. E. Jr, Cao, X., Li, Y., & Beyah, R. (2012). An 802.11 MAC layer covert channel. *Wireless Communications and Mobile Computing*, *12*(5), 393–405. doi:10.1002/wcm.969

Callaway, E., Gorday, P., Hester, L., Gutierrez, J. A., Naeve, M., Heile, B., & Bahl, V. (2002). Home networking with IEEE 802.15.4: A developing standard for low-rate wireless personal area networks. *IEEE Communications Magazine*, *40*(8), 70–77. doi:10.1109/MCOM.2002.1024418

Cardei, M., Yang, Y., & Wu, J. (2008, June). *Non-Uniform Sensor Deployment in Mobile Wireless Sensor Networks*. Paper Presented at the International Symposium on a World of Wireless, Mobile and Multimedia Networks. Newport Beach, CA. doi:10.1109/WOWMOM.2008.4594823

Cardei, M., & Wu, J. (2006). Energy-efficient coverage problems in wireless ad-hoc sensor networks. *Computer Communications*, *29*(4), 413–420. doi:10.1016/j.comcom.2004.12.025

Castagnetti, A., Pegatoquet, A., Le, T. N., & Auguin, M. (2014). *A Joint Duty-Cycle and Transmission Power Management for Energy Harvesting WSN* (pp. 928–936). IEEE Transactions on Industrial Informatics.

Cattivelli, F. S., Lopes, C. G., & Sayed, A. H. (2007). A diffusion RLS scheme for distributed estimation over adaptive networks. In *IEEE 8th Workshop on Signal Processing Advances in Wireless Communications* (pp. 1–5). doi:10.1109/SPAWC.2007.4401393

Cattivelli, F. S., & Sayed, A. H. (2011). Analysis of spatial and incremental LMS processing for distributed estimation. *IEEE Transactions on Signal Processing*, *59*(4), 1465–1480. doi:10.1109/TSP.2010.2100386

CC2420. (2012). *Cc2420 datasheet*. Retrieved June 16, 2012, from http://inst.eecs.berkeley.edu/ cs150/Documents/CC2420.pdf

Cella, U. M., Johnstone, R., & Shuley, N. (2009). *Electromagnetic wave wireless communication in shallow water coastal environment: theoretical analysis and experimental results*. In *Proceedings of 4th ACM Int. Workshop on Underwater Networks* (pp. 9:1–9:8). doi:10.1145/1654130.1654139

Chan, C. Y. M., & Motani, M. (2007). An integrated energy efficient data retrieval protocol for underwater delay tolerant networks. In Proceedings of MTS/IEEE OCEANS 07 – Europe (pp. 1-6). doi:10.1109/OCEANSE.2007.4302341

Chandrasekaran, S., Cooper, O., Deshpande, A., Franklin, M. J., Hellerstein, J. M., Hong, W., & Shah, M. A. (2003). Telegraphic: Continuous dataflow processing for an uncertain world. In *Proceedings of the CIDR Conference*.

Chang, J.-H., & Tassiulas, L. (2004). Maximum lifetime routing in wireless sensor networks. *IEEE/ACM Transactions on Networking*, *12*(4), 609–619. doi:10.1109/TNET.2004.833122

Chaudhari, Q. M., & Serpedin, E. (2008). Maximum likelihood estimation of clock parameters for synchronization of wireless sensor networks. In *2008 IEEE International Conference on Acoustics, Speech and Signal Processing* (pp. 2517–2520). doi:10.1109/ICASSP.2008.4518160

Chen, F., & Dressler, F. (2007). A Simulation Model of IEEE 802.15. 4 in OMNeT+. 6. Fachgespräch Sensor netzwerke. pp. 35.

Chen, C. (2010). Design of a Child Localization System on RFID and Wireless Sensor Networks. *Journal of Sensors*, *2010*, 1–8. doi:10.1155/2010/450392

Cheng, C.-F., Wang, S.-C., & Liang, T. (2008). Byzantine Agreement & Fault Diagnosis Agreement in Dynamic Ad-Hoc Environment. *Fundamenta Informaticae*, *89*, 161–187.

Chen, J., Li, S., & Sun, Y. (2007). Novel Deployment Schemes for Mobile Sensor Networks. *Sensors (Basel, Switzerland)*, *7*(11), 2907–2919. doi:10.3390/S7112907

Chen, J., & Sayed, A. H. (2011). Diffusion adaptation strategies for distributed optimization and learning over networks. *IEEE Transactions on Signal Processing*, *60*(8), 4289–4305. doi:10.1109/TSP.2012.2198470

Chen, K., Ma, M., Cheng, E., Yuan, F., & Su, W. (2014). A Survey on MAC Protocols for Underwater Wireless Sensor Networks. *IEEE Communications Surveys and Tutorials*, *16*(3), 1433–1447. doi:10.1109/SURV.2014.013014.00032

Chen, Y., Trappe, W., & Martin, R. (2007). Detecting and localizing wireless spoofing attacks. In *Proceedings of the 4th Annual IEEE Communications Society Conference on Sensor, Mesh and Ad Hoc Communications and Networks (SECON)* (pp. 193–202).

Cherkaoui, E. H. Azad S., Casari, P., Toni, L., Agoulmine, N. & Zorzi, M. (2012). Packet Error Recovery via Multipath Routing and Reed-Solomon Codes in Underwater Networks. In *Proceedings of MTS/IEEE OCEANS 12*.

Chiang, M. L., Wang, S. C., & Tseng, L. Y. (2010). The anatomy study of consensus agreement in MANETs. *Computers & Electrical Engineering*, *36*(3), 234–253. doi:10.1016/j.compeleceng.2009.09.001

Chitre, M., Shahabudeen, S., Freitag, L., & Stojanovic, M. (2008). Recent Advances in Underwater Acoustic Communications & Networking. In Proceedings of MTS/IEEE OCEANS 2008 (pp. 1-10). doi:10.1109/OCEANS.2008.5152045

Cho, H., Choi, H., Lee, W., Jung, Y., & Baek, Y. (2006). LITeTag: Design and implementation of an RFID system for IT-based port logistics. *Journal of Communication*, *1*(4), 48–57.

Chung, M. Y., Lee, T. J., & Lee, H. R. (2006, July). MAC throughput limit analysis of slotted CSMA/CA in IEEE 802.15.4 WPAN. *IEEE Communications Letters*, *10*(7), 561–563. doi:10.1109/LCOMM.2006.1673013

Coates, R. (1989). *Underwater Acoustic Systems*. New York, NY: Wiley.

Comfort, L. K., Znati, T., & Voortman, M., Xerandy & Freitag L. E. (2012). Early Detection of Near-Field Tsunamis using Underwater Sensor Networks. *Journal of Tsunami Society International*, *31*(4), 232–243.

Costello, D. J., Hagenauer, J., Imai, H., & Wicker, S. B. (1998). Applications of Error-Control Coding. *IEEE Transactions on Information Theory*, *44*(6), 2531–2560. doi:10.1109/18.720548

Crocker, M. J. (Ed.). (1998). *Handbook of Acoustics*. New York, NY: Wiley.

Crossbow Technology. (2009). *TelosB mote platform*.

Cui, J. H., Kong, J., Gerla, M., & Zhou, S. (2006). The challenges of building mobile underwater wireless networks for aquatic applications. *IEEE Network*, *20*(3), 12–18. doi:10.1109/MNET.2006.1637927

Curtin, T. B., & Bellingham, J. G. (2009). Progress toward autonomous ocean sampling networks. *Deep-sea Research. Part II, Topical Studies in Oceanography*, *56*(3-5), 62–67. doi:10.1016/j.dsr2.2008.09.005

da Silva, A. P. R., Martins, M. H. T., Rocha, B. P. S., Loureiro, A. A. F., Ruiz, L. B., & Wong, H. C. (2005). Decentralized intrusion detection in wireless sensor networks. In *Proceedings of the 1st ACM international workshop on Quality of service & security in wireless and mobile networks (Q2SWinet)* (pp. 16–23).

Dahnil, D. P., Singh, Y. P., & Ho, C. K. (2012). Topology-controlled adaptive clustering for uniformity and increased lifetime in wireless sensor networks. *The Institution of Engineering and Technology*, *2*(4), 318–327.

Dargie, W., & Poellabauer, C. (2010). *Fundamentals of Wireless Sensor Networks: Theory and Practice* (1st edition). Wiley.

Dawson-Haggerty, S., Lanzisera, S., Taneja, J., Brown, R., & Culler, D. (2012). @scale: Insights from a large, long-lived appliance energy wsn. In *Proceedings of the 11th international conference on Information Processing in Sensor Networks*, (pp. 37–48). doi:10.1145/2185677.2185683

De Couto, D. S. J., Aguayo, D., Bicket, J., & Morris, R. (2003). A high-throughput path metric for multi-hop wireless routing. In *Proceedings of the 9th annual international conference on Mobile computing and networking (MobiCom)* (pp. 134–146).

Deldar, F., & Yaghmaee, M. H. (2011). Designing a Prediction-based Clustering Algorithm for Target Tracking in Wireless Sensor Networks. In *International Symposium on Computer Networks and Distributed Systems (CNDS)*. doi:10.1109/CNDS.2011.5764572

Demirkol, I., Ersoy, C., & Alagoz, F. (2006). MAC protocols for wireless sensor networks: A survey. *IEEE Communications Magazine*, *44*(4), 115–121. doi:10.1109/MCOM.2006.1632658

Digi International Inc. (n.d.). *Xbee Module Datasheet*. Retrieved from www.digi.com/pdf/ds_xbeezbmodules.pdf

Digi International. (2008). XBee/XBee-PRO - ZB OEM RF Modules datasheet. Retrrieved March 2 2014, from http://ftp1.digi.com/support/documentation/90000976_a.pdf Dong-Liang, W., Ng, W. W. Y., Yeung, D. S. & Hai-Lan, D. (2009) In *International Conference on Machine Learning and Cybernetics.*, Vol. 4 Baoding, pp. 2330-2335.

Di, M., Joo, E. M., & Beng, L. H. (2008). A Comprehensive Study of Kalman Filter and Extended Kalman Filter for Target Tracking in Wireless Sensor Networks. In *IEEE International Conference on Systems, Man and Cybernetics* (pp. 2792-2797).

Ding, M., Chen, D., Xing, K., & Cheng, X. (2005). Localized fault-tolerant event boundary detection in sensor networks. In *Proceedings of the 24th IEEE International Conference on Computer Communications (INFOCOM)*, (volume 2, pp. 902–913).

Dini, G., Pelagatti, M., & Savino, I. M. (2008). An algorithm for reconnecting wireless sensor network partitions. In *Proceedings of the 5th European conference on Wireless sensor networks (EWSN)* (pp. 253–267). doi:10.1007/978-3-540-77690-1_16

Douceur, J. R. (2002). The sybil attack. In *Revised Papers from the First International Workshop on Peer-to-Peer Systems (IPTPS)* (pp. 251–260).

DSPComm. (2014). *Aquacomm: Underwater Wireless Modem*. Retrieved April 30, 2014, from http://www.dspcomm.com

Duffy, J. E., Amaral-Zettler, L. A., Fautin, D. G., Paulay, G., Rynearson, T. A., Sosik, H. M., & Stachowicz, J. J. (2013). Envisioning a Marine Biodiversity Observation Network. *Bioscience*, *63*(5), 350–361. doi:10.1525/bio.2013.63.5.8

Dyo, V., & Mascolo, C. (2007, November). *A Node Discovery Service for Partially Mobile Sensor Networks*. Paper Presented at the 2nd International Workshop on Middleware for Sensor Networks. Newport Beach, CA. doi:10.1145/1376860.1376863

Dyo, V., & Mascolo, C. (2008, June). *Efficient Node Discovery in Mobile Wireless Sensor Networks*. Paper Presented at the 4th IEEE International Conference on Distributed Computing in Sensor Systems. Santorini Island, Greece. doi:10.1007/978-3-540-69170-9_33

Edwards, M., Donelson, J. I., Zavis, W., Prabhakaran, A., Brabb, D., & Jackson, A. (2005, March 16-18). Improving freight rail safety with on-board monitoring and control systems. In *Proceedings of the 2005 ASME/IEEE Joint Rail Conference* (pp. 117-122).

Eisenman, S. B., Miluzzo, E., Lane, N. D., Peterson, R. A., Ahn, G.-S., & Campbell, A. T. (2007, November). *The BikeNet Mobile Sensing System for Cyclist Experience Mapping*. Paper Presented at the 5th International Conference on Embedded Networked Sensor Systems. New York, NY. doi:10.1145/1322263.1322273

Elbergui, A., Quidu, I., Zerr, B., & Solaiman, B. (2012). *Model Based Classification of Mine-Like Objects in Sidescan Sonar Using the Highlight Information.* In *proceedings of 11th European Conference on Underwater Acoustics ECUA 2012* (pp. 1158-1165).

Elhadef, M., Boukerche, A., & Elkadiki, H. (2007). An adaptive fault identification protocol for an emergency/rescue-based wireless and mobile ad-hoc network. In *Proceedings of the 21st International Parallel and Distributed Processing Symposium, IPDPS 2007.*

El-Hoiydi, A., & Decotignie, J. D. (2004). WiseMAC: An ultra low power MAC protocol for multi-hop wireless sensor networks. In *Algorithmic Aspects of Wireless Sensor Networks* (pp. 18–31). Springer Berlin Heidelberg. doi:10.1007/978-3-540-27820-7_4

Elson, J., Girod, L., & Estrin, D. (2002, December). *Fine-Grained Network Time Synchronization Using Reference Broadcasts.* Paper Presented at the 5th Symposium on Operating Systems Design and Implementation. Boston, MA. doi:10.1145/1060289.1060304

Eriksson, J., Girod, L., Hull, B., Newton, R., Madden, S., & Balakrishnan, H. (2008, June). *The Pothole Patrol: Using a Mobile Sensor Network for Road Surface Monitoring.* Paper Presented at the 6th International Conference on Mobile Systems, Applications, and Services. Breckenridge, CO. doi:10.1145/1378600.1378605

Esteves, J., Carvalho, A., & Couto, C. (2003, August). *Generalized Geometric Triangulation Algorithm for Mobile Robot Absolute Self-Localization.* Paper Presented at the IEEE International Symposium on Industrial Electronics. doi:10.1109/ISIE.2003.1267272

Fagui, L., Yang, Z., & Wenhui, L. (2009) In. *Standard System Framework of RFID Application in Logistics.* Paper presented at the *Third International Symposium on Intelligent Information Technology Application(IITA 2009)*, Vol. 3 Nanchang, pp. 44-47.

Fan, G. J., & Jin, S. Y. (2010). Coverage Problem in Wireless Sensor Network: A Survey. *Journal of Networks*, *5*(9), 1033–1040. doi:10.4304/jnw.5.9.1033-1040

Fanimokun, A., & Frolik, J. (2003). Effects of natural propagation environments on wireless sensor network coverage area. In *Proceedings of the 35th Southeastern Symposium System Theory (SSST03).* 16-18 March 2003.

Farooqi, A. H., & Khan, F. A. (2009). Intrusion detection systems for wireless sensor networks: A survey. In Communication and Networking, (vol. 56, pp. 234–241). doi:10.1007/978-3-642-10844-0_29

Farr, N., Bowen, A., Ware, J., Pontbriand, C. & Tivey, M. (2010). An integrated, underwater optical/acoustic communications system. In *Proceedings of MTS/IEEE OCEANS* 10.

Fischer, M. J., & Lynch, N. A. (1981). A Lower Bound for the Time to Assure Interactive Consistency. *Information Processing Letters*, *14*(4), 183–186. doi:10.1016/0020-0190(82)90033-3

Fisher, F. H., & Simmons, V. P. (1977). Sound absorption in sea water. *The Journal of the Acoustical Society of America*, *62*(3), 558–564. doi:10.1121/1.381574

Flowers, D., & Yang, Y. (n.d.). *The MiWi Wireless Networking Protocol is a simple protocol designed for low data rate, short distance, low-cost networks.* Retrieved from http://ww1.microchip.com/downloads/en/AppNotes/AN1066%20-%20MiWi%20App%20Note.pdf

Fonseca, R., Gnawali, O., Jamieson, K., & Levis, P. (2007). Four-bit wireless link estimation. In Proceedings of HotNets VI.

Fragouli, C., Le Boudec, J.-Y., & Widmer, J. (2006). Network Coding: An instant Primer. *Computer Communication Review*, *36*(1), 63–68. doi:10.1145/1111322.1111337

Franklin, M. J., Jeffery, S. R., Krishnamurthy, S., & Reiss, F. (2005). Design considerations for high fan-in systems: The hifi approach. In *Proceedings of the CIDR Conference* (pp. 290–304).

Fu, Q., Banitalebi, B., Zhang, L., & Beigl, M. (2013, March). *Energy-Efficient Collaborative Data Collection in Mobile Wireless Sensor Networks.* Paper Presented at the 47th Annual Conference on Information Sciences and Systems, Baltimore, MD.

Gandham, S. R., Dawande, M., Prakash, R., & Venkatesan, S. (2003, December). *Energy Efficient Schemes for Wireless Sensor Networks with Multiple Mobile Base Stations.* Paper Presented at the Global Telecommunications Conference, San Francisco, CA. doi:10.1109/GLOCOM.2003.1258265

Ganeriwal, S., Balzano, L. K., & Srivastava, M. B. (2008). Reputation-based framework for high integrity sensor networks. *ACM Transactions on Sensor Networks, 4*(3), 15:1–15:37.

Ganesan, D., Govindan, R., Shenker, S., & Estrin, D. (2001). Highly-resilient, energy efficient multipath routing in wireless sensor networks, *Proceeding ACM MobiHoc'01*, Long Beach, CA, pp. 251-254. doi:10.1145/501416.501452

Gao, M., Soh, W.-S., & Tao, M. (2009). A Transmission Scheme for Continuous ARQ Protocols over Underwater Acoustic Channels. In *Proceedings of IEEE International Conference on Communications* (pp. 1-5). doi:10.1109/ICC.2009.5198684

Garateguy, G., Arce, G. R., & Pelaez, J. (2011, March). *Covert Channel detection in VoIP streams.* In Information Sciences and Systems (CISS), 2011 45th Annual Conference on (pp. 1-6). IEEE. doi:10.1109/CISS.2011.5766220

Garetto, M., Gribaudo, M., Chiasserini, C., & Leonardi, E. (2007, October*). A Distributed Sensor Relocation Scheme for Environmental Control.* Paper Presented at the IEEE International Conference on Mobile Ad Hoc and Sensor Systems, Pisa, Italy.

Gianvecchio, S., & Wang, H. (2011). *An entropy-based approach to detecting covert timing channels. IEEE Transactions on Dependable and Secure Computing, 8*(6), 785–797.

Gilani, M. H. S., Sarrafi, I., & Abbaspour, M. (2011). An adaptive CSMA/TDMA hybrid MAC for energy and throughput improvement of wireless sensor networks. *Ad Hoc Networks, 11*(4), 1290–1304.

Giles, J., & Hajek, B. (2000). *The jamming game for packet timing channels.* In Proceedings of IEEE International Symposium on Information Theory, 2000. (p. 51). IEEE. doi:10.1109/ISIT.2000.866341

Girod, L., Lukac, M., Trifa, V., & Estrin, D. (2006, October). *The Design and Implementation of a Self-Calibrating Acoustic Sensing Platform.* Paper Presented at the 4th International Conference on Embedded Networked Sensor Systems, Boulder, CO. doi:10.1145/1182807.1182815

Gislason, D. (2008). *Zigbee wireless networking.* Newnes.

Global Positioning System. (n.d.). *GPS.gov Web site.* Retrieved from http://www.gps.gov/

Gnawali, O., Fonseca, R., Jamieson, K., Moss, D., & Levis, P. (2009). Collection tree protocol. In *Proceedings of the 7th ACM Conference on Embedded Networked Sensor Systems* (pp. 1–14). doi:10.1145/1644038.1644040

Gobriel, S., Mosse, D., & Cleric, R. (2009, June). TDMA-ASAP: Sensor network TDMA scheduling with adaptive slot-stealing and parallelism. In *29th IEEE International Conference on Distributed Computing Systems, 2009.* (pp. 458-465). IEEE.

Gray, J. W., III. (1993, May). *On introducing noise into the bus-contention channel.* In Proceedings of IEEE Computer Society Symposium on Research in Security and Privacy, 1993. (pp. 90-98). IEEE. doi:10.1109/RISP.1993.287640

Guillemette, M. G., Fontaine, I., & Caron, C. (2008). *Hybrid RFID-GPS Real-Time Location System for Human Resources: Development, Impacts and Perspectives.* Paper presented at the *Proceedings of the 41st Annual Hawaii International Conference on System Sciences*Waikoloa, HI, pp. 406-406. doi:10.1109/HICSS.2008.195

Guo, P., Jiang, T., Zhang, Q., & Zhang, K. (2012). Sleep scheduling for critical event monitoring in wireless sensor networks. *IEEE Transactions on Parallel and Distributed Systems*, *23*(2), 345–352. doi:10.1109/TPDS.2011.165

Guo, S., Zhong, Z., & He, T. (2009). Find: faulty node detection for wireless sensor networks. In *Proceedings of the 7th ACM Conference on Embedded Networked Sensor Systems (SenSys)*, (pp. 253–266). doi:10.1145/1644038.1644064

Guo, Z., Wang, B., Xie, P., Zeng, W., & Cui, J.-H. (2009). Efficient error recovery with network coding in underwater sensor networks. *Ad Hoc Networks*, *7*(4), 791–802. doi:10.1016/j.adhoc.2008.07.011

Guo, Z., Xie, P., Cui, J. H., & Wang, B. (2006). On applying network coding to underwater sensor networks. In *Proceedings of 1st CM International Workshop on Underwater Networks* (pp. 109-112). doi:10.1145/1161039.1161062

Gupchup, J., Sharma, A., Terzis, A., Burns, R., & Szalay, A. (2008). The perils of detecting measurement faults in environmental monitoring networks (Invited Paper). In *Proceedings of DCoSS.*

Hamming, R. W. (1950). Error Detecting and Error Correcting Codes. *The Bell System Technical Journal*, *29*(2), 147–160. doi:10.1002/j.1538-7305.1950.tb00463.x

Hamouda, Y. E. M., & Phillips, C. (2011). Adaptive sampling for energy-efficient collaborative multi-target tracking in wireless sensor networks. *The Institution of Engineering and Technology*, *1*, 15–25.

Harris, A. F., & Zorzi, M. (2007). *Modeling the underwater acoustic channel in ns2*. In *Proceedings of 2nd International Conference on Performance Evaluation Methodologies and Tools*. (pp. 1-8). doi:10.4108/nstools.2007.2024

Hartung, C., Balasalle, J., Han, R., Hartung, C., Balasalle, J., & Han, R. (2005). *Node compromise in sensor networks: The need for secure systems* (Tech. Rep. CU-CS-990-05), University of Colorado, Boulder, CO.

Hashemi, H. (1993, July). The Indoor Radio Propagation Channel. *Institute of Electrical and Electronics Engineers*, *81*(7), 943–968. doi:10.1109/5.231342

Hauer, J.-H., Handziski, V., & Wolisz, A. (2009). Experimental Study of the Impact of WLAN Interference on IEEE 802.15.4 Body Area Networks. In *Proceedings of the 6th European Conference on Wireless Sensor Networks, Cork, Ireland, 2009* (LNCS 5432, pp: 17 – 32), doi:10.1007/978-3-642-00224-3_2

He, T., Huang, C., Blum, B. M., Stankovic, J. A., & Abdelzaher, T. (2003). Range-free localization schemes for large scale sensor networks. In *9th annual international conference on MobiCom 2003, San Diego, CA, USA 2003* (pp. 81–95), doi:10.1145/938985.938995

Heidemann, J. Ye., & Wills, W., J., Syed A. & Li Y. (2006). *Research Challenges and Applications for Underwater Sensor Networking*. IEEE Wireless Communications and Networking Conference, 228-235. doi:10.1109/WCNC.2006.1683469

Heinzelman, W. B., Chandrakasan, A. P., & Balakrishnan, H. (2002). An Application-Specific Protocol Architecture for Wireless Microsensor Networks. *IEEE Transactions on Wireless Communications*, *1*(4), 660–670. doi:10.1109/TWC.2002.804190

He, J., Tang, Z., Chen, H. H., & Wang, S. (2008). An accurate Markov model for slotted CSMA/CA algorithm in IEEE 802.15. 4 networks. *IEEE Communications Letters*, *12*(6), 420–422. doi:10.1109/LCOMM.2008.071460

Herman, T., Pemmaraju, S., Pilard, L., & Mjelde, M. (2007). Temporal partition in sensor networks. *In Proceedings of the 9th International Conference on Stabilization, Safety, and Security of Distributed Systems (SSS)* (pp. 325–339). doi:10.1007/978-3-540-76627-8_25

He, T., Krishnamurthy, S., Luo, L., Yan, T., Gu, L., Stoleru, R., & Krogh, B. (2006). Vigilnet: An integrated sensor network system for energy-efficient surveillance. *ACM Transactions on Sensor Networks*, *2*(1), 1–38. doi:10.1145/1138127.1138128

He, Y., & Xu, T. The research of non-uniform node distribution inwireless sensor networks. *IEEE Int. Symp. Information Engineeringand Electronic Commerce*, July2010

Hlinka, O., Hlawatsch, F., & Djuric, P. M. (2013). Distributed particle filtering in agent networks: A survey, classification, and comparison. *IEEE Signal Processing Magazine*, *30*(1), 61–81. doi:10.1109/MSP.2012.2219652

Ho, T., Karger, D. R., Médard, M., & Koetter, R. (2003). *Network Coding from a Network Flow Perspective*. Paper presented at IEEE International Symposium on Information Theory, 2003, Yokohama, Japan. doi:10.1109/ISIT.2003.1228458

Ho, T., Koetter, R., Médard, M., Karger, D. R., & Effros, M. (2003). *The Benefits of Coding over Routing in a Randomized Setting*. Paper presented at IEEE International Symposium on Information Theory, Yokohama, Japan. doi:10.1109/ISIT.2003.1228459

Holloway, R., & Beyah, R. (2011, October). *Covert DCF: A DCF-Based Covert Timing Channel in 802.11 Networks*. In Mobile Ad hoc and Sensor Systems (MASS), 2011 IEEE 8th International Conference on (pp. 570-579). IEEE.

Hou, L., & Bergmann, N. W. (2012). Novel industrial wireless sensor networks for machine condition monitoring and fault diagnosis. *IEEE Transactions on Instrumentation and Measurement*, *61*(10), 2787–2798. doi:10.1109/TIM.2012.2200817

Howard, A., Matari'c, M. J., & Sukhatme, G. S. (2002). *Efficient Incremental Sensor Network Deployment Algorithm*. Unpublished dissertation, Federal University, Brazil.

Howard, A., Matari'c, M. J., & Sukhatme, G. S. (2002). Mobile Sensor Network Deployment Using Potential Fields: A Distributed, Scalable Solution to the Area Coverage Problem. In A. Howard (Ed.), *Distributed Autonomous Robotic Systems 5* (pp. 299–308). Springer Japan Publishing. doi:10.1007/978-4-431-65941-9_30

Howitt, I., & Gutierrez, J. A. (2003). IEEE 802.15.4 low rate: Wireless personal area network coexistence issues. In 2003 IEEE Wireless Communications and Networking (Vol. 3, pp. 1481-1486). New Orleans, LA, USA.

Hsiao, H.-S., Chin, Y.-H., & Yang, W.-P. (2000). Reaching Fault Diagnosis Agreement Under a Hybrid Fault Model. *IEEE Transactions on Computers*, *49*(9), 980–986.

Hsua, J. M., Chenb, C. C., & Li, C. C. (2012). POOT: An efficient object tracking strategy based on short-term optimistic predictions for face-structured sensor networks. *Computers & Mathematics with Applications (Oxford, England)*, *63*(2), 391–406. doi:10.1016/j.camwa.2011.07.034

Hu, L., & Evans, D. (2004, September). *Localization for Mobile Sensor Networks*. Paper Presented at the 10th Annual International Conference on Mobile Computing and Networking, Philadelphia, PA.

Huang, C.-F., Yang, T. C., Kuo, N.-T., Liu, J.-Y., & Schindall, J. (2012). *Experimental Demonstration of Current Mapping Using Distributed Networked Underwater Sensors*. Presented at Seventh ACM International Conference on Underwater Networks and Systems (WUWNet '12). doi:10.1145/2398936.2398983

Huang, P., Xiao, L., Soltani, S., Mutka, M. W., & Xi, N. (2013). The evolution of MAC protocols in wireless sensor networks: A survey. *IEEE Communications Surveys and Tutorials*, *15*(1), 101–120. doi:10.1109/SURV.2012.040412.00105

Huang, Y. K., Pang, A. C., & Hung, H. N. (2009). A comprehensive analysis of low-power operation for beacon-enabled IEEE 802.15. 4 wireless networks. *IEEE Transactions on Wireless Communications*, *8*(11), 5601–5611. doi:10.1109/TWC.2009.081485

Hu, W. M. (1992). Reducing timing channels with fuzzy time. *Journal of Computer Security*, *1*(3), 233–254.

Hu, Y., Perrig, A., & Johnson, D. B. (2002). *Wormhole detection in wireless ad hoc networks. Technical report*. Department of Computer Science, Rice University.

IEEE (2009). IEEE Standard for Information technology-Telecommunications and information exchange between systems-Local and metropolitan area networks-Specific requirements Part 15.4: Wireless Medium Access Control (MAC) and Physical Layer (PHY) Specifications for Low-Rate Wireless Personal Area Networks (WPANs) Amendment 2: Alternative Physical Layer Extension to support one or more of the Chinese 314-316 MHz, 430-434 MHz, and 779-787 MHz bands. *IEEE Std 802.15.4c-2009 (Amendment to IEEE Std 802.15.4-2006)*, c1-21.

IEEE-SA. (2003). *IEEE 802.15.4, Part 15.4: Wireless Medium Access Control (MAC) and Physical Layer (PHY) Specifications for Low Rate Wireless Personal Area Networks (LR WPANs)*. IEEE SA Standards Board.

IEEE-SA. (2006). *IEEE Standard 802.15.4, Part 15.4: Wireless Medium Access Control and Physical Layer Specification for Low Rate Wireless Personal Area Networks*. IEEE Std. 802.15.4.

IEEE-SA. (2012, April 5). *IEEE 802.11: Wireless LAN Medium Access Control (MAC) and Physical Layer (PHY) Specifications* (2012 revision).

IEEE-SA. (n.d.). *IEEE 802.2 standards*. Retrieved from http://standards.ieee.org/about/get/802/802.2.html

Ingelrest, F., Barrenetxea, G., Schaefer, G., Vetterli, M., Couach, O., & Parlange, M. (2010). Sensorscope: Application-specific sensor network for environmental monitoring. *ACM Transactions on Sensor Networks, 6*(2), 17:1–17:32.

Intel Lab Data. (2012). *Intel lab data*. Retrieved December 2, 2012, from http://db.csail.mit.edu/labdata/labdata.html

ISA. (n.d.). *ISA.org Web site*. Retrieved from http://www.isa.org/

Islam, K., Shen, W., & Wang, X. (2012).Wireless sensor network reliability and security in factory automation. IEEE Transactions on Systems, Man AND Cybernetics, *42*(6).

Iyer, Y., Gandham, S., & Venkatesan, S. (2005). STCP: a generic transport layer protocol for wireless sensor networks, in *Proceedings of 14th International Conference on Computer Communications and Networks (ICCCN)* (pp. 449-454). San Diego, California, USA. doi:10.1109/ICCCN.2005.1523908

Jae-Han, L., & Byung Tae, J. (2008) In. *Dynamic Duty Cycle Adaptation to Real-Time Data in IEEE 802.15.4 Based WSN*. Paper presented at the 5th IEEE Consumer Communications and Networking ConferenceLas Vegas (CCNC 2008), NV, pp. 353-357.

Jawhar, I., Mohamed, N., Al-Jaroodi, J., & Zhang, S. (2013). *An Efficient Framework for Autonomous Underwater Vehicle Extended Sensor Networks for Pipeline Monitoring*. Paper presented at 11th IEEE International Symposium on Robotic and Sensors Environments (ROSE), Washington, DC. doi:10.1109/ROSE.2013.6698430

Jennic (2006). Application Note: JN-AN-1035, Calculating 802.15.4 Data Rates. Retrieved March 2 2014, from: http://www.jennic.com/files/support_files/JN-AN-1035%20Calculating%20802-15-4%20Data%20Rates-1v0.pdf

Jin, S., Yue, W., & Sun, Q. (2012). Performance analysis of the sleep/wakeup protocol in a wireless sensor network. *International Journal of Innovative Computing, Information, & Control, 8*(5), 3833–3844.

Jin-Shyan, L., Yu-Wei, S., & Chung-Chou, S. (2007). A Comparative Study of Wireless Protocols: Bluetooth, UWB, ZigBee, and Wi-Fi. Paper presented at the *33rd Annual Conference of the IEEE Industrial Electronics Society (IECON 2007)*. Taipei, Taiwan.

Julier, S. J., & Uhlmann, J. K. (1997). A new extension of the Kalman filter to nonlinear systems. In *Proceedings of the Society of Photo-Optical Instrumentation Engineers*.

Jung, C. Y., Hwang, H. Y., Sung, D. K., & Hwang, G. U. (2009). Enhanced Markov chain model and throughput analysis of the slotted CSMA/CA for IEEE 802.15. 4 under unsaturated traffic conditions. *IEEE Transactions on Vehicular Technology, 58*(1), 473–478. doi:10.1109/TVT.2008.923669

Jurdak, R., Ruzzelli, A. G., & O'Hare, G. (2008). *Adaptive Radio Modes in Sensor Networks: How Deep to Sleep?* Paper presented at the *Annual IEEE Communications Society Conference on Sensor, Mesh and Ad Hoc Communications and Networks*San Francisco *(SECON '08) 5th*, CA, pp. 386-394. doi:10.1109/SAHCN.2008.54

Jurdak, R., Ruzzelli, A. G., & O'Hare, G. M. P. (2008, August 3-7). Multi-Hop RFID Wake-Up Radio: Design, Evaluation and Energy Tradeoffs. In *Proceedings of 17th International Conference on Computer Communications and Networks, 2008 (ICCCN '08)* (pp. 1-8). doi:10.1109/ICCCN.2008.ECP.124

Jurdak, R., Ruzzelli, A. G., Barbirato, A., & Boivineau, S. (2011). Octopus: Monitoring, visualization, and control of sensor networks. *Wireless Communication & Mobile Computing*, *11*(8), 1073–1091. doi:10.1002/wcm.826

Jurdak, R., Wang, X., Obst, O., & Valencia, P. (2011). Wireless sensor network anomalies: Diagnosis and detection strategies. In Intelligence-Based. *Systems Engineering*, *10*, 309–325.

Kaiser, W. J., Pottie, G. J., Srivastava, M., Sukhatme, G. S., Villasenor, J., & Estrin, D. (2003). Networked infomechanical systems (nims) for ambient intelligence (Technical Report 31). University of California Los Angeles, Networked Embedded Systems Lab, Los Angeles, CA.

Kam, C., Kompella, S., Nguyen, G. D., Ephremides, A., & Zaihan, J. (2014). Frequency Selection and Relay Placement for Energy Efficiency in Underwater Acoustic Networks. *IEEE Journal of Oceanic Engineering*, *39*(2), 331–342. doi:10.1109/JOE.2013.2245057

Kansal, A., Somasundara, A. A., Jea, D. D., Srivastava, M. B., & Estrin, D. (2004, June). *Intelligent Fluid Infrastructure for Embedded Networks.* Paper Presented at the 2nd International Conference on Mobile Systems, Applications, and Services, Boston, MA. doi:10.1145/990064.990080

Karlof, C., Sastry, N., & Wagner, D. (2004). Tinysec: A link layer security architecture for wireless sensor networks. In *Proceedings of the 2nd international conference on Embedded networked sensor systems (SenSys)*, (pp. 162–175). doi:10.1145/1031495.1031515

Karlof, C., & Wagner, D. (2003). Secure routing in wireless sensor networks: Attacks and countermeasures. In *Proceedings of the First IEEE International Workshop on Sensor Network Protocols and Applications*, (pp. 113 – 127). doi:10.1109/SNPA.2003.1203362

Kar, S., & Moura, J. M. F. (2007). Distributed average consensus in sensor networks with random link failures and communication channel noise. In *Conference Record: Asilomar Conference on Signals, Systems and Computers*. 676–680. doi:10.1109/ACSSC.2007.4487299

Karvonen, H., Shelby, Z., & Pomalaza-Raez, C. (2004). *Coding for energy efficient wireless embedded networks*. Paper presented at the International Workshop on Wireless Ad-hoc Networks, Oulu, Finland. doi:10.1109/IWWAN.2004.1525590

Kashani, Z. H., & Shiva, M. (2006). BCH Coding and Multi-hop Communication in Wireless Sensor Networks. Paper presented at the International Conference on Embedded and Ubiquitous Computing, Barcelona, Spain. doi:10.1109/WOCN.2006.1666619

Kay, S. M. (1988). *Modern Spectral Estimation: Theory and Application*. Englewood Cliffs, NJ: Prentice Hall.

Ke, R., Peng, G., & Wang, X. (2008). *Further Research of RFID Applying on Exhibition Logistics*. Paper presented at the *International Conference on Cyberworlds*Hangzhou (pp. 631-634).

Kennedy, J., & Eberhart, R. (1995). Particle Swarm Optimization. In *Proceedings of IEEE International Conference on Neural Networks. IV* (pp. 1942–1948).

Kennedy, J., & Eberhart, R. (2001). *Swarm Intelligence*. Morgan Kaufmann.

Khan, M. M. H., Le, H. K., LeMay, M., Moinzadeh, P., Wang, L., & Yang, Y., … Jin, X. (2010). Diagnostic powertracing for sensor node failure analysis. In *Proceedings of the 9th ACM/IEEE International Conference on Information Processing in Sensor Networks (IPSN)*, (pp. 117–128). doi:10.1145/1791212.1791227

Khedr, A. M., & Osamy, W. (2011). Effective target tracking mechanism in a self-organizing wireless sensor network. *Journal of Parallel and Distributed Computing*, *71*(10), 1318–1326. doi:10.1016/j.jpdc.2011.06.001

Khosravi, A., Mohammadi, K., & Shiroie, M. (2011). Locating malicious links in fully-connected networks using a formal framework. In *Proceedings of theICSEng 2011: International Conference on Systems Engineering*. 247–250. doi:10.1109/ICSEng.2011.51

Kim, H., Kim, E., & Han, K. (2006). An Energy Efficient Tracking method in Wireless Sensor Networks. In Next Generation Teletraffic and Wired/Wireless Advanced Networking (LNCS, 4003, pp. 278-286). doi:10.1007/11759355_27

Kim, S., Fonseca, R., & Culler, D. (2004). Reliable transfer on wireless sensor networks. In *Proceedings of the 1st Annual IEEE Communications Society Conference on Sensor and Ad Hoc Communications and Networks (IEEE SECON)* (pp. 449-459). Santa Clara, California, USA.

Kim, H., & Ayurzana, O. (2009). Improvement of data receive ratio in remote water meter system by upgrading sensor. *International Journal of Control. Automation and Systems*, *7*(1), 145–150. doi:10.1007/s12555-009-0118-7

Kim, H., Chung, J. M., & Kim, C. H. (2009). Secured communication protocol for internetworking ZigBee cluster networks. *Computer Communications*, *32*(13-14), 1531–1540. doi:10.1016/j.comcom.2009.05.014

Kim, K.-H., & Shin, K. G. (2011). Self-reconfigurable wireless mesh networks. *IEEE/ACM Transactions on Networking*, *19*(2), 393–404. doi:10.1109/TNET.2010.2096431

Kim, S., Pakzad, S., Culler, D., Demmel, J., Fenves, G., Glaser, S., & Turon, M. (2006). Wireless sensor networks for structural health monitoring. In *Proceedings of the 4th international conference on Embedded networked sensor systems* (pp. 427–428). doi:10.1145/1182807.1182889

Kinney, P. (2003). *Wireless Control that Simply Works* In *Communications design.* Paper presented at the *Communications design conference*, Vol. 2. Kinney Consulting LLC.

Kloos, G., Guivant, J. E., Nebot, E. M., & Masson, F. (2006). Range Based Localisation Using RF and the Application to Mining Safety. In *Proceedings of International Conference on Intelligent Robots and Systems*, October 2006, Beijing, China (pp: 1304-1310), doi:10.1109/IROS.2006.281894

Koetter, R., & Médard, M. (2003). An Algebraic Approach to Network Coding. *IEEE/ACM Transactions on Networking*, *11*(5), 782–795. doi:10.1109/TNET.2003.818197

Koneru, A. (2006, December). *Comparative Study of RSS-based Collaborative Localization Methods in Wireless Sensor Networks.* Master's Thesis, University of North Texas.

Koutsopoulos, I., & Halkidi, M. (2014). Distributed energy-efficient estimation in spatially correlated wireless sensor networks. *Computer Communications*, *45*, 47–58. doi:10.1016/j.comcom.2014.03.012

Kramer, M., & Geraldy, A. (2006). Energy measurements for micaz node (Tech. Rep. KrGe06). University of Kaiserslautern, Kaiserslautern, Germany.

Krätzer, C., Dittmann, J., Lang, A., & Kühne, T. (2006, September). *WLAN steganography: a first practical review*. In *Proceedings of the 8th workshop on Multimedia and security* (pp. 17-22). ACM.

Krishna, C. R., & Yadav, P. S. (2014). *A Hybrid Localization Scheme for Underwater Wireless Sensor Networks.* In *proceedings of International Conference on Issues and Challenges in Intelligent Computing Techniques (ICICT)*, (pp. 579-582). doi:10.1109/ICICICT.2014.6781345

Krontiris, I., Giannetsos, T., & Dimitriou, T. (2008). Launching a sinkhole attack in wireless sensor networks; the intruder side. In *Proceedings of the 4th IEEE International Conference on Wireless and Mobile Computing, Networking and Communications (WIMOB)*, (pp. 526–531). doi:10.1109/WiMob.2008.83

Kshirsagar, R. V., & Jirapure, B. (2011). A Survey on Fault Detection and Fault Tolerance in Wireless Sensor Networks. *International Journal of Computers and Applications*, *3*(1), 130–138.

Kuk Liu, Z., & Liu, Z. (2011). *A Dynamic Clustering Protocol for Wireless Sensor Networks.* In *IEEE International Conference on Computational and Information Sciences.*

Kulkarni, R. V. & Venayagamoorthy, G. K. (2010). Particle Swarm Optimization in Wireless Sensor Networks: A Brief Survey. *IEEE Transaction on systems, Man, and Cybernetics - part C: Applications and reviews.* doi: .0.1109/TSMCC.2010.2054080

Kumar, P., Gu, X., Nes, M., Mushtaq, Q., & Schiller, J. (2010). *Optimizing Duty-Cycle for Delay and Energy Bound WSN Applications.* Paper presented at the *IEEE 24th International Conference on Advanced Information Networking and Applications Workshops Perth*, (WAINA) (pp. 692-697).

Kumar, A., Namboothiri, P., Deshpande, S., Vidhyadharan, S., Sivalingam, K., & Murty, S. (3-5Feb.2012). Testbed based throughput analysis in a Wireless Sensor Network. In *National Conference on Communications (NCC)* (pp. 1-5). doi:10.1109/NCC.2012.6176804

Kun, H., Zhu, Y., Wang, W., & Wang, H. (2014, October). *Cross-Layer Design of MAC-PHY for Optimal Capacity of Multimedia Wireless Sensor Networks.* Paper Presented at the ACM Research in Adaptive and Convergent Systems (RACS), Towson, MD.

Kusy, B., Ledeczi, A., & Koutsoukos, X. (2007, November). *Tracking Mobile Nodes Using RF Doppler Shifts.* Paper presented at the 5th International Conference on Embedded Networked Sensor Systems, Sydney, Australia. doi:10.1145/1322263.1322267

Kusy, B., Sallai, J., Balogh, G., Ledeczi, A., Protopopescu, V., Tolliver, J., . . . Parang, M. (2007, June). *Radio Interferometric Tracking of Mobile Wireless Nodes.* Paper Presented at the 5th International Conference on Mobile Systems, Applications and Services. San Juan, Puerto Rico.

Kusy, B., Dutta, P., Levis, P., Maroti, M., Ledeczi, A., & Culler, D. (2006). Elapsed Time on Arrival: A Simple and Versatile Primitive for Canonical Time Synchronization Services. *International Journal of Ad Hoc and Ubiquitous Computing*, *1*(4), 239–251. doi:10.1504/IJAHUC.2006.010505

Labrador, Y., Karimi, M., Pan, D., & Miller, J. (2009). Modulation and error correction in the underwater acoustic communication channel. *International Journal of Computer Science and Network Security*, *9*(7), 123–130.

Lal, R., Reicosky, D. C., & Hanson, J. D. (2007). Evolution of the plow over 10,000 years and the rationale for no-till farming. *Soil & Tillage Research*, *93*(1), 1–12. doi:10.1016/j.still.2006.11.004

Lamport, L., Shostak, R., & Pease, M. (1982). The Byzantine Generals Problem. *ACM Transactions on Programming Languages and Systems*, *4*(3), 382–401. doi:10.1145/357172.357176

Langendoen, K., & Reijers, N. (2003). Distributed localization in wireless sensor networks: A quantitative comparison. *Computer Networks*, *43*(4), 499–518. doi:10.1016/S1389-1286(03)00356-6

Lashkari, A., Danesh, M., & Samadi, B. (2009Aug.8-11). A survey on wireless security protocols (WEP, WPA and WPA2/802.11i). In *2nd IEEE International Conference on Computer Science and Information Technology* (pp. 48-52).

Latré, B., Braem, B., Moerman, I., Blondia, C., & Demeester, P. (2011). A survey on wireless body area networks. *Wireless Networks*, *17*(1), 1–18. doi:10.1007/s11276-010-0252-4

Lee J., Kim J., Lee J., Jang, Y. & Dho, K. (2008). An improved ARQ scheme in underwater acoustic sensor networks, In *Proceedings of MTS/IEEE OCEANS 08.*

Lee, E. F., Wang, C., & Xiao, L. (2006). *A Study of Radio Signal Behaviors in Complex Environments* (Tech. Rep. MSU-CSE-06-6). Computer Science Department. Michigan State University.

Lee, J., Dharne, A., & Jayasuria, S. (2007, July). *Potential Field Based Hierarchical Structure for Mobile Sensor Network Deployment.* Paper Presented at the American Control Conference. New York, NY. doi:10.1109/ACC.2007.4282565

Lee, B. H., & Wu, H. K. (2009). Study on a delayed backoff algorithm for IEEE 802.15.4 low-rate wireless personal area networks. *IET Communications*, *3*(7), 1089–1096. doi:10.1049/iet-com.2007.0517

Lee, K. W., & Ewe, H. T. (2007). Performance study of Byzantine Agreement Protocol with artificial neural network. *Information Sciences*, *177*(21), 4785–4798. doi:10.1016/j.ins.2007.04.011

Lee, S. H., Lee, S., Song, H., & Lee, H. S. (2011). Gradual Cluster Head Election for High Network Connectivity in Large-Scale Sensor Networks. In *IEEE International Conference on Advanced Communication Technology*, pp. 168-172.

Le, H. C., Guyennet, H., Felea, V., & Zerhouni, N. (2007). *A low latency MAC scheme for event-driven wireless sensor networks* (pp. 291–301). Mobile Ad-Hoc and Sensor Networks.

Lei, Z., & Zhi, W. (2006). *Integration of RFID into Wireless Sensor Networks: Architectures, Opportunities and Challenging Problems.* Paper presented at the Fifth International Conference on Grid and Cooperative Computing WorkshopsHunan (GCCW '06), pp. 463-469.

Lenders, V., Wagner, J., & May, M. (2006). Analyzing the impact of mobility in ad hoc networks. In *Proceedings of the 2nd international workshop on Multi-hop ad hoc networks: from theory to reality* (REALMAN) (pp. 39–46). doi:10.1145/1132983.1132991

Le, T., Hu, W., Corke, P., & Jha, S. (2009). ERTP: Energy-efficient and reliable transport protocol for data streaming in wireless sensor networks. *Computer Communications, 32*(7-10), 1154–1171. doi:10.1016/j.comcom.2008.12.045

Levis, P., Madden, S., Polastre, J., Szewczyk, R., Whitehouse, K., Woo, A., . . . Culler, D. (2005). TinyOS: An operating system for sensor networks. In Ambient intelligence (pp. 115-148). Springer Berlin Heidelberg.

Levis, P., Lee, N., Welsh, M., & Culler, D. (2003, November). TOSSIM: Accurate and scalable simulation of entire TinyOS applications. In *Proceedings of the 1st international conference on Embedded networked sensor systems*(pp. 126-137). ACM. doi:10.1145/958491.958506

Levis, P., Patel, N., Culler, D., & Shenker, S. (2004). Trickle: A self-regulating algorithm for code propagation and maintenance in wireless sensor networks. In *Proceedings of the First USENIX/ACM Symposium on Networked Systems Design and Implementation (NSDI)*, (pp. 15–28).

Liang, Z., Ma, S., & Zhang, Q. (2011).Analysis of wireless sensor sleep mechanism with group arrival queuing model.*The Eighth IEEE Consumer Communications and Networking Conf.*, (pp. 570–574.) doi:10.1109/CCNC.2011.5766545

Libelium. (n.d.). *Libelium Waspmote Wi-Fi exapansion board.* Retrieved from http://www.libelium.com/expansion_radio_board/

Libelium. (n.d.). *Libelium Waspmote.* Retrieved from http://www.libelium.com/products/waspmote

Li, J., Zhang, B., & Cui, L. (2010). A Restricted Delaunay Tringulation Graph Based Algorithm for Self-Deployment in Mobile Sensor Networks. *Journal of Computer Information Systems, 6*(10), 3155–3162.

Lim, S., Ji, Y., Cho, J., & An, S. (2006). An ultra low power medium access control protocol with the divided preamble sampling. In *Ubiquitous Computing Systems* (pp. 210–224). Springer Berlin Heidelberg. doi:10.1007/11890348_17

Lim, S., Kim, S., Cho, J., & An, S. (2006, January). Medium access control with an energy-efficient algorithm for wireless sensor networks. In *Personal Wireless Communications* (pp. 334–343). Springer Berlin Heidelberg. doi:10.1007/11872153_29

Lin, E. Y., Rabaey, J. M., & Wolisz, A. (2004, June). Power-efficient rendez-vous schemes for dense wireless sensor networks. In *IEEE International Conference on Communications* (Vol. 7, pp. 3769-3776). IEEE. doi:10.1109/ICC.2004.1313259

Lin, Y. B., Chang, K. C., Chen, J. C., Wang, P. H., & Li, L. S. (2006). *ZigBee Sensor Network for Structural Health Monitoring.* Paper presented at the 4th International Conference on Earthquake Engineering.Taipei, Taiwan

Lindhorst, T., Lukas, G., & Nett, E. (2010). Modeling fast link failure detection for dependable wireless mesh networks. In *Proceedings of the 2010 Ninth IEEE International Symposium on Network Computing and Applications (NCA)*, (pp. 44–51). doi:10.1109/NCA.2010.13

LinkQuest. (2014). *Underwater Acoustic Modem Models.* Retrieved April 30, 2014, from http://www.linkquest.com

Li, Q. (2012). *Digital Sonar Design in Underwater Acoustics: Principles and Applications.* Berlin, Germany: Springer. doi:10.1007/978-3-642-18290-7

Li, S.-Y. R., Yeung, R. W., & Cai, N. (2003). Linear Network Coding. *IEEE Transactions on Information Theory, 49*(2), 371–381. doi:10.1109/TIT.2002.807285

Liu, C.-H., & Ssu, K.-F. (2008, September). *A Moving Algorithm for Non-Uniform Deployment in Mobile Sensor Networks.* Paper Presented at the 6th International Conference on Mobile Systems, Applications, and Services. Breckenridge, CO. doi:10.1145/1506270.1506316

Liu, T., Sadler, C. M., Zhang, P., & Martonosi, M. R. (2004, June). *Implementing Software on Resource-Constrained Mobile Sensors: Experimences with Impala and ZebraNet*. Paper Presented at the 2nd International Conference on Mobile Systems, Applications, and Services. Boston, MA. doi:10.1145/990064.990095

Liu, B., Garcin, F., Ren, F., & Lin, C. (2008) *A study of forward error correction schemes for reliable transport in underwater sensor networks*. In *Proceedings of the 5th Annual IEEE Communications Society Conference on Sensor, Mesh and Ad Hoc Communications and Networks*, 197–205. doi:10.1109/SAHCN.2008.33

Liu, J.-Y., & Huang, C.-F. (1999). Surface-Generated Noise in an Ocean Waveguide with a Transition Layer of Continuously Varying Density and Sound Speed. *Journal of Computational Acoustics*, *7*(4), 253–268. doi:10.1142/S0218396X99000175

Liu, K., Ma, Q., Zhao, X., & Liu, Y. (2011). Self-diagnosis for large scale wireless sensor networks. In *Proceedings of IEEE INFOCOM* (pp. 1539 –1547). doi:10.1109/INFCOM.2011.5934944

Liu, X., Gong, X., & Zheng, Y. (2014). Reliable cooperative communications based on random network coding in multi-hop relay WSNs. *IEEE Sensors Journal*, *14*(8), 2514–2523. doi:10.1109/JSEN.2014.2310899

Liu, Y., He, Y., Li, M., Wang, J., Liu, K., Mo, L., & Li, X.-Y. (2011). Does wireless sensor network scale? A measurement study on greenorbs. In *Proceedings of IEEE INFOCOM* (pp. 873 –881).

Li, X. (2007, November). Collaborative localization with received-signal strength in wireless sensor networks. *IEEE Transactions on Vehicular Technology*, *56*(6), 3807–3817. doi:10.1109/TVT.2007.904535

Li, Y., & Parker, L. (2008). Detecting and monitoring time-related abnormal events using a wireless sensor network and mobile robot. In *Proceedings of the IEEE/RSJ International Conference on Intelligent Robots and Systems (IROS)*, (pp. 3292–3298).

Lonare, S., & Wahane, G. (2012). A Survey on Energy Efficient Routing Protocols in Wireless Sensor Network. *IEEE Communications Survey and Tutorial, VOL, 15*(2), 551–591.

Loo, C. E., Ng, M. Y., Leckie, C., & Palaniswami, M. (2006). Intrusion detection for routing attacks in sensor networks. *International Journal of Distributed Sensor Networks*, *2*(4), 313–332. doi:10.1080/15501320600692044

Lopes, C. G., & Sayed, A. H. (2007). Incremental adaptive strategies over distributed networks. *IEEE Transactions on Signal Processing*, *55*(8), 4064–4077. doi:10.1109/TSP.2007.896034

Lopes, C. G., & Sayed, A. H. (2008). Diffusion least-mean squares over adaptive networks: Formulation and performance analysis. *IEEE Transactions on Signal Processing*, *56*(7), 3122–3136. doi:10.1109/TSP.2008.917383

Lopez, M., Sabater, J., Daemitabalvandani, M., Sabater, J., Gómez, J. M., Carmona, M., & Herms, A. (2013). Software management of power consumption in WSN based on duty cycle algorithms. Paper presented at the EUROCON, 2013 IEEE, and pp. 399-406. doi:10.1109/EUROCON.2013.6625014

Lorincz, K., & Welsh, M.,(2006). MoteTrack: A Robust, Decentralized Approach to RF-Based Location Tracking. In *Proceedings of Personal and Ubiquitous Computing*.

Lou, W. (2005). "Am efficient N-to-1 multipath routing protocol in wireless sensor networks", Proceeding IEEE MASS'05 (pp. 1-8). Washington, DC.

Luo, J., Wang, D., & Zhang, Q. (2009, January). *Double Mobility: Coverage of the Sea Surface with Mobile Sensor Networks*. Paper Presented at ACM SIGMOBILE Mobile Computing and Communications Review. New York, NY. doi:10.1145/1558590.1558601

Lurton, X. (2010). *An Introduction to Underwater Acoustics: Principles and Applications*. Berlin, Germany: Springer. doi:10.1007/978-3-642-13835-5

Lymberopoulos, D., Lindsey, Q., & Savvides, A. (2006). *An Empirical Analysis of Radio Signal Strength Variability in IEEE 802.15.4 Networks Using Monopole Antennas*. (ENALAB Tech. Rep. 050501).

Ma, J., Chen, Q., Zhang, D., & Ni, L. M. (2006). *An Empirical Study of Radio Signal Strength in Sensor Networks Using MICA2 Nodes* (HKUST, Tech. Rep.).

Ma, J., Zhou, X., Li, S., & Li, Z. (2011, October). Connecting agriculture to the Internet of Things through sensor networks. In *Internet of Things (iThings/CPSCom), 2011 International Conference on and 4th International Conference on Cyber, Physical and Social Computing* (pp. 184-187). IEEE. doi:10.1109/iThings/CPSCom.2011.32

Ma, T., Hempel, M., Hua, K., Peng, D., & Sharif, H. (2010, October). *A Novel Cooperative Image Transmission Scheme in Wireless Sensor Networks.* Paper Presented at the 35th IEEE Conference on Local Computer Networks. Denver, CO. doi:10.1109/LCN.2010.5735711

Madden, S. R., Franklin, M. J., Hellerstein, J. M., & Hong, W. (2005). Tinydb: An acquisitional query processing system for sensor networks. *ACM Transactions on Database Systems*, *30*(1), 122–173. doi:10.1145/1061318.1061322

Madden, S., Franklin, M. J., Hellerstein, J. M., & Hong, W. (2002). Tag: a tiny aggregation service for ad-hoc sensor networks. In *Proceedings of the 5th symposium on operating systems design and implementation (OSDI)*. doi:10.1145/1060289.1060303

Maggs, M. K., O'Keefe, S. G., & Thiel, D. V. (2012). Consensus clock synchronization for wireless sensor networks. *IEEE Sensors Journal*, *12*(6), 2269–2277. doi:10.1109/JSEN.2011.2182045

Mahani, A., Tahmasebi, F., Nezamabadi-pour, H. (2011). *Two tiers wireless mesh networks: optimal configuration, ICCKE11* (pp. 235–239).

Mahani, A., Ansari, M. S., & Kavian, Y. S. (2012). Reliability or performance: a tradeoff in wireless sensor networks, 8th IEEE, *IET International Symposium on Communication Systems, Networks and Digital Signal Processing*.

Mahani, A., Askari-Hemmat, M. H., & Kavian, Y. S. (2012). Duplication Avoidance for Energy Efficient Wireless Sensor Networks. In *8th IEEE, IET International Symposium on Communication Systems, Networks and Digital Signal Processing*.

Mahani, A., Kargar, A., Kavian, Y. S., & Rashvand, H. F. (2012). Non-uniform distribution of multi-hop sensor networks: Performance improvement and energy hole mitigation. *IET Wireless Sensor Systems*, *2*(4), 302–308. doi:10.1049/iet-wss.2012.0032

Mahasukhon, P., Sharif, H., Hempel, M., Zhou, T., Ma, T., & Shrestha, P. L. (4-8July2011). A study on energy efficient multi-tier multi-hop wireless sensor networks for freight-train monitoring. In *7th International Wireless Communications and Mobile Computing Conference (IWCMC)*, (pp. 297-301). doi:10.1109/IWCMC.2011.5982549

Mahasukhon, P., Sharif, H., Hempel, M., Zhou, T., Wang, W., & Ma, T. (2010). Multi-Tier Multi-Hop Routing in Large-Scale Wireless Sensor Networks for Freight Train Monitoring. In *ITS World Congress*.

Mahmood, M. A., & Seah, W. (2011). Event reliability in wireless sensor networks. In International Conference on Intelligent Sensors, Sensor networks and Information Processing (pp. 377-382).

Mahmood, M. A., & Seah, W. K. G. (2011). Event Reliability in Wireless Sensor Networks. In *Proceedings of the seventh International Conference on Intelligent Sensors, Sensor Networks and Information Processing (ISSNIP)* (pp. 1-6). Adelaide, Australia.

Majstor, F. (20-24Oct.2003). WLAN security threats & solutions. In *28th Annual IEEE International Conference on Local Computer Networks*, (p. 650).

Mallela, S., & Masson, G. M. (1978). Diagnosable Systems for Intermittent Faults. *IEEE Transactions on Computers*, *27*(6), 560–566. doi:10.1109/TC.1978.1675147

Mamdani, E. H., & Assilian, S. (1975). An experiment in linguistic synthesis with a fuzzy logic controller. *International Journal of Man-Machine Studies*, *7*(1), 1–13. doi:10.1016/S0020-7373(75)80002-2

Manolakis, D. (1996). Efficient Solution and Performance Analysis of 3-D Position Estimation by Trilateration. *IEEE Transactions on Aerospace and Electronic Systems*, *32*(4), 1239–1248. doi:10.1109/7.543845

Marchi, B., Grilo, A., & Nunes, M. (2007). DTSN: Distributed transport for sensor networks, in *Proceedings of the 12th IEEE Symposium on Computers and Communications (ISCC)*, Aveiro, Portugal, pp. 165-172.

Marks, M. (2010). A Survey of multi-objective deployment in wireless sensor networks. *Journal of Telecommunications & Information Technology*, *36*(3).

Maroti, M., Kusy, B., Balogh, G., Volgyesi, P., Nadas, A., Molnar, K., ... Ledeczi, A. (2005, November). *Radio Interferometric Geolocation*. Paper presented at the 3rd International Conference on Embedded Networked Sensor Systems, Sydney, Australia. doi:10.1145/1098918.1098920

Maróti, M., Kusy, B., Simon, G., & Lédeczi, A. (2004). The flooding time synchronization protocol. In *Proceedings of the 2nd international conference on Embedded networked sensor systems*, (pp. 39–49). doi:10.1145/1031495.1031501

Marple, S. L. Jr. (1987). *Digital spectral analysis with applications*. London: Prentice Hall Inc.

Martins, J., Giulietti, A., & Strum, M. (2002). *Performance comparison of convolutional and block turbo codes for WLAN applications*. In *Proceedings of the Fourth IEEE International Caracas Conference on Devicces, Circuits and Systems*, (pp. T032-1–T032-5). doi:10.1109/ICCDCS.2002.1004115

Martirosyan, A., Boukerche, A., Richard, W., & Pazzi, N. (2008). *Energy-aware and quality of service-based routing in wireless sensor networks and vehicular ad hoc networks* (pp. 669–681). Institut TELECOM and Springer-Verlag France.

Ma, T., Xu, Z., Hempel, M., Peng, D., & Sharif, H. (2014). Performance analysis of a novel low-complexity high-precision timing synchronization method for wireless sensor networks. *IEEE Transactions on Wireless Communications*, *13*(9), 4758–4765. doi:10.1109/TWC.2014.2331286

Mazurczyk, W., & Lubacz, J. (2010). LACK—a VoIP steganographic method. *Telecommunication Systems*, *45*(2-3), 153–163. doi:10.1007/s11235-009-9245-y

McGillem, C., & Rappaport, T. (1989). A Beacon Navigation Method for Autonomous Vehicles. *IEEE Transactions on Vehicular Technology*, *38*(3), 132–139. doi:10.1109/25.45466

Médard, M., & Sprintson, A. (2012). *Network Coding*. Oxford, England: Academic Press.

Medeiros, H., Park, J., & Kak, A. C. (2008). Distributed Object Tracking Using a Cluster-Based Kalman Filter in Wireless Camera Networks. *IEEE Journal of Selected Topics in Signal Processing*, *2*(4), 448–463. doi:10.1109/JSTSP.2008.2001310

MEMSIC. (2014). *Memsic wireless sensor networks*. Retrieved October 10, 2014, http://www.memsic.com/wireless-sensor-networks/

MEMSIC. (n.d.). *MICAz Datasheet*. Retrieved from http://www.memsic.com/support/documentation/wireless-sensor-networks/category/7-datasheets.html?download=148%3Amicaz

Meyer, F. J., & Pradhan, D. K. (1991). Consensus with dual failure modes. *IEEE Transactions on Parallel and Distributed Systems*, *2*(2), 214–222. doi:10.1109/71.89066

Miao, X., Liu, K., He, Y., Liu, Y., & Papadias, D. (2011). Agnostic diagnosis: Discovering silent failures in wireless sensor networks. In *Proceedings of IEEE INFOCOM* (pp. 1548–1556). doi:10.1109/INFCOM.2011.5934945

Mirsadeghi, M. (2014). *Energy Efficient Mobile Target Tracking by Wireless Sensor Networks*. A Thesis Submitted as a Partial Fulfillment of the Requirements for the Degree of Master of science in Electronic (M. Sc.).

Mirsadeghi, M. Mahani, A. (2014). Energy efficient fast predictor for WSNbased target tracking. *Springer Annals of Telecommunications*. doi:.10.1007/s12243-014-0430-y

Mirsadeghi, M., & Mahani, A. (2013). *A novel distributed clustering protocol using fuzzy logic*. In Conference on Electronics, Telecommunications and Computers (CETC).

Mirsadeghi, M., & Mahani, A. (2013). *Low power prediction mechanism for WSN-based object tracking*. In *Conference on Electronics, Telecommunications and Computers (CETC)*.

Mirshahi, S., Uysal, S., & Akbari, A. (2013). Integration of RFID and WSN for supply chain intelligence system. Paper presented at the 2013 International Conference on Electronics, Computers and Artificial Intelligence (ECAI), pp.1, 6, 27-29. doi:10.1109/ECAI.2013.6636160

Mishra, N., Chebrolu, K., Raman, B., & Pathak, A. (2006). *Wake-on-WLAN*. Paper presented at the International World Wide Web Conference Committee Edinburgh (IW3C2), Scotland (pp. 761-769).

Mishra, S., Jena, L., & Pradhan, A. (2012). Fault Tolerance in Wireless Sensor Networks. *International Journal of Advanced Research in Computer Science and Software Engineering*, *2*(10), 146–153.

Misic, J., Shafi, S., & Misic, V. B. (2005). Avoiding the Bottlenecks in the MAC Layer in 802.15.4 Low-Rate WPAN. In *Proceedings of 11th International Conference on Parallel and Distributed Systems Workshops* (pp. 363-367).

Mo, L., He, Y., Liu, Y., Zhao, J., Tang, S.-J., Li, X.-Y., & Dai, G. (2009). Canopy closure estimates with greenorbs: sustainable sensing in the forest. In *Proceedings of the 7th ACM Conference on Embedded Networked Sensor Systems* (pp. 99–112). doi:10.1145/1644038.1644049

Monowar, M. M., & Alam, M. M. Obaidur Rahman, Md., Hong, C. S., & Lee, S. (2010). A load-aware energy-efficient and throughput-maximized asynchronous duty cycle MAC for wireless sensor networks (pp. 777–794). Institut Télécom and Springer-Verlag.

Monton, E., Hernandez, J. F., Blasco, J. M., Hervé, T., Micallef, J., Grech, I., & Traver, V. (2008). Body area network for wireless patient monitoring. *IET Communications*, *2*(2), 215–222. doi:10.1049/iet-com:20070046

Moon, T. K. (2005). *Error Correction Coding*. Hoboken, NJ: Wiley. doi:10.1002/0471739219

Moore, D., Leonard, J., Rus, D., & Teller, S. (2004, November). *Robust Distributed Network Localization with Noisy Range Measurements*. Paper presented at the 2nd International Conference on Embedded Networked Sensor Systems. Baltimore, MD. doi:10.1145/1031495.1031502

Morelos-Zaragoza, R. H. (2006). *The Art of Error Correcting Coding*. Chichester, England: Wiley. doi:10.1002/0470035706

Moteiv Corporation Inc (2006). *Ultra low power IEEE 802.15.4 compliant wireless sensor module* (Tmote Sky datasheet).

Mpitziopoulos, A., Gavalas, D., Konstantopoulos, C., & Pantziou, G. (2009). A survey on jamming attacks and countermeasures in wsns. *IEEE Communications Surveys and Tutorials*, *11*(4), 42–56. doi:10.1109/SURV.2009.090404

Msechu, E. J., & Giannakis, G. B. (2012). Sensor-centric data reduction for estimation with WSNs via censoring and quantization. *IEEE Transactions on Signal Processing*, *60*(1), 400–414. doi:10.1109/TSP.2011.2171686

Mulligan, G. (2007). The 6LoWPAN architecture. In *Proceedings of the 4th workshop on Embedded networked sensors*, ACM. doi:10.1145/1278972.1278992

Munoz, D., Lara, F., Vargas, C., & Enriquez-Caldera, R. (2009). *Position Location Techniques and Applications*. Academic Press.

Murdoch, S. J., & Lewis, S. (2005, January). Embedding covert channels into TCP/IP. In *Information Hiding* (pp. 247–261). Springer Berlin Heidelberg.

Murphy, K. (1998). *Kalman filter toolbox for Matlab*. Retrieved from http://www.cs.ubc.ca/~murphyk/Software/Kalman/kalman.html

Naderi, M. Y., Nintanavongsa, P., & Chowdhury, K. R. (2014). RF-MAC: A medium access control protocol for re-chargeable sensor networks powered by wireless energy harvesting. *IEEE Transactions on Wireless Communications*, *13*(7), 3926–3937. doi:10.1109/TWC.2014.2315211

Naik, L. N., & Agnew, G. (2006). Data capacity improvement of wireless sensornetworks using non-uniform sensor distribution,Int. J. Distrib. *SensorNetw.*, *2*(2), 121–145.

Namos (2012). *Namos deployment -james reserve (lake fulmor) august 28th ~ september 1st 2006*. Retrieved June 3, 2012, from http://robotics.usc.edu/~namos/data/jr aug 06/

nationalatlas.gov. (n.d.). *Overview of Freight Railroad Industry*. Retrieved from http://nationalatlas.gov/articles/transportation/a_freightrr.html

Navarro-Serment, L. E., Grabowski, R., Paredis, C. J., & Khosla, P. K. (2002). Millibots: The Development of a Framework and Algorithms for a Distributed Heterogeneous Robot Team. *IEEE Robotics and Automation Magazine*, *9*(4), 31–40. doi:10.1109/MRA.2002.1160069

Nedic, A., & Ozdaglar, A. (2009). Distributed subgradient methods for multi-agent optimization. *IEEE Transactions on Automatic Control*, *54*(1), 48–61. doi:10.1109/TAC.2008.2009515

Neugebauer, M., Plonnigs, J., & Kabitzsch, K. (2005January). A new beacon order adaptation algorithm for IEEE 802.15.4 networks. In *Proceedings of European Workshop on Wireless Sensor Networks (EWSN)* (pp. 302-311).

Neuzil, J., Kreibich, O., & Smid, R. (2014). A distributed fault detection system based on iwsn for machine condition monitoring. *IEEE Transactions on Industrial Informatics*, *10*(2), 1118–1123. doi:10.1109/TII.2013.2290432

Newsome, J., Shi, E., Song, D., & Perrig, A. (2004). The sybil attack in sensor networks: analysis & defenses. In *Proceedings of the 3rd international symposium on Information processing in sensor networks (IPSN)* (pp. 259–268). doi:10.1145/984622.984660

Ngai, E., Liu, J., & Lyu, M. (2006). On the intruder detection for sinkhole attack in wireless sensor networks. In *Proceedings of the IEEE International Conference on Communications (ICC)* (volume 8, pp. 3383–3389).

Ni, K., Ramanathan, N., Chehade, M. N. H., Balzano, L., Nair, S., Zahedi, S., ... Srivastava, M. (2009). Sensor network data fault types. *ACM Transactions of Sensor Networks*, *5*(3), 25:1–25:29.

Nicolaou, N., See, A., & Xie, P. Cui, J.-H., & Maggiorini, D. (2007). Improving the Robustness of Location based Routing for Underwater Sensor Networks. In Proceedings of MTS/IEEE OCEANS Europe.

Niculescu, D., & Nath, B. (2003, January). DV Based Positioning in Ad hoc Networks. *Journal of Telecommunication Systems*, *22*(1), 267–280. doi:10.1023/A:1023403323460

Ni, L. M., Liu, Y., Lau, Y. C., & Patil, A. P. (2004). LANDMARC: Indoor Location Sensing Using Active RFID. *Wireless Networks*, 701–710.

Ni, S.-Y., Tseng, Y.-C., Chen, Y.-S., & Sheu, J.-P. (1999). The broadcast storm problem in a mobile ad hoc network. In *Proceedings of the 5th annual ACM/IEEE international conference on Mobile computing and networking (MobiCom)* (pp. 151– 162). doi:10.1145/313451.313525

Okun, M. (2009). On the round complexity of Byzantine agreement without initial set-up. *Information and Computation*, *207*(12), 1351–1368. doi:10.1016/j.ic.2009.07.002

Olfati-Saber, R., Fax, J. A., & Murray, R. M. (2007). Consensus and Cooperation in Networked Multi-Agent Systems. *Proceedings of the IEEE*, *95*(1), 215–233. doi:10.1109/JPROC.2006.887293

Ondrej, S., Zdenek, B., Petr, F., & Ondrej, H. (2006). *ZigBee Technology and Device Design*. Paper presented at the International Conference on Networking, International Conference on Systems and International Conference on Mobile Communications and Learning Technologies *(ICN/ICONS/MCL 2006)* (pp. 129-129).

Osterlind, F., & Dunkels, A. (2008). *Approaching the Maximum 802.15.4 Multi-hop Throughput* (SICS Tech. Rep. T2008:05).

Otero, C. E., Kostanic, I., & Otero, L. D. (2009). Development of a simulator for stochastic deployment of wireless sensor networks. *Journal of Networks*, *4*(8), 754–762. doi:10.4304/jnw.4.8.754-762

Ouyang, R. W., Wong, A. K. S., & Lea, C.-T. (2010, March). Received signal strength based wireless localization via semidefinite programming: Noncooperative and cooperative schemes. *IEEE Transactions on Vehicular Technology*, *59*(3), 1307–1318. doi:10.1109/TVT.2010.2040096

Øvsthus, K., & Kristensen, L. M. (2014). *An industerial perspective on wireless sensor networks – a survey of requirements, protocols and challenges* (pp. 1391–1412). IEEE Communication Surveys & Tutorials.

Pala, Z., & Inanc, N. (2007). *Smart Parking Applications Using RFID Technology*. Paper presented at the *1st Annual RFID Eurasia* Istanbul (pp. 1-3).

Pantazis, N. A., Nikolidakis, S. A., & Vergados, D. D. (2013). Energy efficient routing protocols in wireless sensor networks: A survey. *IEEE Communications Surveys and Tutorials*, *15*(2), 551–591. doi:10.1109/SURV.2012.062612.00084

Pant, B., & Alkin, O. (2012, October). Correlated Movement Mobility Model and Constant Acceleration Model for EKF-Based Tracking Applications. In *IEEE Wireless and Mobile Computing* (pp. 869–874). Networking and Communications.

Papotto, G., Carrara, F., Finocchiaro, A., & Palmisano, G. (2014). A 90-nm CMOS 5-Mbps crystal-less RF-powered transceiver for wireless sensor network nodes. *IEEE Journal of Solid-State Circuits*, *49*(2), 335–346.

Pappas, C., Coscia, E., Dodero, G., Gianuzzi, V., & Earney, M. (2002). *A mobile e-health system based on workflow automation tools.* In Proceedings of the 15th IEEE Symposium on Computer-Based Medical Systems, 2002. (CBMS 2002) (pp. 271-276). IEEE. doi:10.1109/CBMS.2002.1011388

Paradis, L., & Han, Q. (2007). A Survey of Fault Management in Wireless Sensor Networks. *Journal of Network and Systems Management*, *15*(2), 171–190. doi:10.1007/s10922-007-9062-0

Parameshwaraiah, R., Ramesh, R., & Gurumurthy, N. K. (2013). Real-Time Underwater Communication Technique for Energy Efficient Ocean Monitoring. Presented at *Seventh International Conference on Sensor Technologies and Applications (SENSORCOMM 2013)*, Barcelona, Spain.

Park, J., Mackay, S., & Wright, E. (2003). *Practical Data Communications for Instrumentation and Control*. Oxford, England: Newnes.

Park, S. J., Sivakumar, R., Akyildiz, I., & Vedantham, R. (2008). GARUDA: Achieving effective reliability for downstream communication in wireless sensor networks. *IEEE Transactions on Mobile Computing*, *07*(02), 214–230. doi:10.1109/TMC.2007.70707

Park, T. R., Kim, T. H., Choi, J. Y., Choi, S., & Kwon, W. H. (2005). Throughput and energy consumption analysis of IEEE 802.15.4 slotted CSMA/CA. *Electronics Letters*, *41*(18), 1017–1019. doi:10.1049/el:20051662

Partan, J., Kurose, J., & Levine, B. N. (2006). A survey of practical issues in underwater networks. In *Proceedings of ACM International Conference on Underwater Sensor Networks and Systems* (pp.17-24). doi:10.1145/1161039.1161045

Pasqualetti, F., Bicchi, A., & Bullo, F. (2012). Consensus computation in unreliable networks: A system theoretic approach. *IEEE Transactions on Automatic Control*, *57*(1), 90–104. doi:10.1109/TAC.2011.2158130

Pathan, A. K., & Hong, C. S. (2008). *SERP: secure energy-efficient routing protocol for densely deployed wireless sensor networks* (pp. 529–541). Institut TELECOM and Springer-Verlag France.

Patheja, P. S., Waoo, A., & Shrivastava, P. (2012). Fault Tolerant, Energy Saving Method for Reliable Information Propagation in Sensor Network. *International Journal of Smart Sensors and Ad Hoc Networks*, *1*(4), 34–38.

Patro. Kumar, R. (2007). Analysis and improvement of contention access protocol in IEEE 802.15. 4 star network. *Mobile Adhoc and Sensor Systems, 2007. MASS 2007. IEEE Internatonal Conference on. IEEE.*

Pei, D., Zhao, X., Massey, D., & Zhang, L. (2004). A study of BGP path vector route looping behavior. In *Proceedings of the 24th International Conference on Distributed Computing Systems (ICDCS)*, (pp. 720–729). doi:10.1109/ICDCS.2004.1281640

Peng, C., Guo, J., Pei, L., & Wen, L. (2010). Energy-Efficiency Strategies in Wireless Sensor Networks. *Control Engineering of China, 17*(3).

Peng, M., Chen, H., Xiao, Y., Ozdemir, S., Vasilakos, A. V., & Wu, J. (2011). Impact of sensor node distributions on coverage in sensor networks. *Elsevier Int. J. Parallel and Distrib. Comput.*, *71*(12), 1578–1591. doi:10.1016/j.jpdc.2011.04.011

Pereora, P. R., Grilo, A., Rocha, F., Nunes, M. S., Casaca, A., Chaudet, C., ... Johansson, M. (2007). *End-to-End Reliability in Wireless Sensor Networks: Survey and Research Challenges*. Workshop on IP QoS and Traffic Control, Lisbon, Portugal.

Perrig, A., Szewczyk, R., Tygar, J. D., Wen, V., & Culler, D. E. (2002). Spins: Security protocols for sensor networks. *Wireless Networks*, *8*(5), 521–534. doi:10.1023/A:1016598314198

Petrut, T., Ioana, C., Mauuary, D., Mallet, J., & Phillipe, O. (2014). Distributed Data Classification in Underwater Acoustic Sensors Based on Local Time-Frequency Coherence Analysis. In proceedings of MTS/IEEE OCEANS 14 Taipei, 6 pages doi:10.1109/OCEANS-TAIPEI.2014.6964531

Pires, W. R., J., de Paula Figueiredo, T., Wong, H., & Loureiro, A. (2004). Malicious node detection in wireless sensor networks. In *Proceedings of the 18th International Parallel and Distributed Processing Symposium (IPDPS)*. doi:10.1109/IPDPS.2004.1302934

Polastre, J. Hill, J. and Culler, D (2004). *Versatile low power media access for wireless sensor networks* (pp. 95-107).

Polastre, J., Hill, J., & Culler, D. (2004). *Versatile low power media access for wireless sensor networks.* Paper presented at the Conference on Embedded Networked Sensor Systems (SenSys; ACM) (pp. 3-5). Baltimore, Maryland, USA.

Polastre, J., Szewczyk, R., & Culler, D. (2005, April). Telos: enabling ultra-low power wireless research. In *IPSN 2005. Fourth International Symposium on Information Processing in Sensor Networks* (pp. 364-369). IEEE. doi:10.1109/IPSN.2005.1440950

Polastre, J., Hill, J., & Culler, D. (2004, November). Versatile low power media access for wireless sensor networks. In *Proceedings of the 2nd international conference on Embedded networked sensor systems* (pp. 95-107). ACM. doi:10.1145/1031495.1031508

Polastre, J., Szewczyk, R., & Culler, D. (2005). Telos: enabling ultra-low power wireless research. In *Proceedings of the 4th international symposium on Information processing in sensor networks.*

Pollin, S., Ergen, M., Ergen, S., Bougard, B., Der Perre, L., Moerman, I., & Catthoor, F. (2008). Performance analysis of slotted carrier sense IEEE 802.15. 4 medium access layer. *IEEE Transactions on Wireless Communications*, *7*(9), 3359–3371. doi:10.1109/TWC.2008.060057

Porta, A., Baselli, G., Liberati, D., Montano, N., Cogliati, C., Gnecchi-Ruscone, T., & Cerutti, S. (1998). Measuring regularity by means of a corrected conditional entropy in sympathetic outflow. *Biological Cybernetics*, *78*(1), 71–78. doi:10.1007/s004220050414 PMID:9485587

Postel. J. (1981). *Transmission Control Protocol* (IETF RFC 793).

Preisig, J. (2007). Acoustic propagation considerations for underwater acoustic communications network development. *Mobile Computing and Communications Review*, *11*(44), 2–10. doi:10.1145/1347364.1347370

Proakis, J. G., Sozer, E. M., Rice, J. A., & Stojanovic, M. (2001). Shallow water acoustic networks. *IEEE Communications Magazine*, *39*(11), 114–119. doi:10.1109/35.965368

Proctor, A. A., Kennedy, J., Gamroth, E., & Bradley, C. (2010). The Ocean Technology Test Bed: From Concept to Operation. In Proceedings of OCEANS 10.

Purvis, M., Sambells, J., & Turner, C. (2006). *Beginning Google maps applications with PHP and Ajax.* Michael Purvis, Jeffrey Sambells, and Cameron Turner.

Rabbat, M. G., & Nowak, R. D. (2005). Quantized incremental algorithms for distributed optimization. *IEEE Journal on Selected Areas in Communications*, *23*(4), 798–808. doi:10.1109/JSAC.2005.843546

Radhakrishnan, S. V., Selcuk Uluagac, A., & Beyah, R. (2013, December). Realizing an 802.11-based covert timing channel using off-the-shelf wireless cards. In *Global Communications Conference (GLOBECOM)*, 2013IEEE (pp. 722-728). IEEE. doi:10.1109/GLOCOM.2013.6831158

Radio-Electronics.com. (n.d.). *Networking Topologies.* Retrieved from http://www.radio-electronics.com/info/wireless/ieee-802-15-4/mesh-networking-topology-topologies.php

Radio-Electronics.com. (n.d.). *Wireless Standards Technology.* Retrieved from http://www.radio-electronics.com/info/wireless/ieee-802-15-4/wireless-standard-technology.php

Rajasegarar, S., Bezdek, J. C., Leckie, C., & Palaniswami, M. (2007). Analysis of anomalies in ibrl data from a wireless sensor network deployment. In *Proceedings of the 2007 International Conference on Sensor Technologies and Applications (SENSORCOMM)*, (pp. 158–163). Washington, DC, USA. doi:10.1109/SENSORCOMM.2007.4394914

Rajendran, V., Obraczka, K., & Garcia-Luna-Aceves, J. J. (2006). Energy-efficient, collision-free medium access control for wireless sensor networks. *Wireless Networks*, *12*(1), 63–78. doi:10.1007/s11276-006-6151-z

Rajesh, S. M. (2013). *Integration of Active RFID and WSN for real time low-cost data monitoring of patients in hospitals*. Paper presented at the 2013 International Conference on Control, Automation, Robotics and Embedded Systems (CARE) (pp. 16-18). doi:10.1109/CARE.2013.6733768

Rakshit, S. M. (2013). *Design and Implementation of a Hybrid Technology Networking System*. Dissertations & Student Research in Computer Electronics & Engineering.

Rakshit, S. M., Hempel, M., Sharif, H., Punwani, J., & Stewart, M. (2012). Hybrid Technology Networking: A Novel Wireless Networking Approach for Real-Time Railcar Status Monitoring. In *ASME Rail Transportation Division Fall Technical Conference*, (pp. 1-7). doi:10.1115/RTDF2012-9402

Rakshit, S. M., Hempel, M., Sharif, H., Punwani, J., Stewart, M., & Mehrvarzi, S. (2012). Challenges in current Wireless Sensor Technology for Railcar Status Monitoring for North America's Freight Railroad Industry. In *ASME/ASCE/IEEE Joint Rail Conference*, (pp. 1-9). doi:10.1115/JRC2012-74114

Ram, M., & Kumar, S. (2014). *Analytical energy consumption model for MAC protocols in wireless sensor networks*. Paper presented at the In 2014 International Conference on IEEE Signal Processing and Integrated Networks (SPIN) (pp. 444-447). doi:10.1109/SPIN.2014.6776994

Ramachandran, I., Das, A. K., & Roy, S. (2007). Analysis of the contention access period of IEEE 802.15. 4 MAC. *ACM Transactions on Sensor Networks*, *3*(1). doi:10.1145/1210669.1210673

Ramanathan, N., Balzano, L., Burt, M., Estrin, D., Harmon, T., Harvey, C., ... Srivastava, M. (2006). *Rapid deployment with confidence: Calibration and fault detection in environmental sensor networks* (Tech. Rep.). Center for Embedded Networked Sensing, UCLA and Department of Civil and Environmental Engineering, MIT.

Ramanathan, N., Balzano, L., Burt, M., Estrin, D., Harmon, T., Harvey, C., ... Srivastava, M. (2006). *Rapid deployment with confidence: Calibration and fault detection in environmental sensor networks* (Tech. Rep.). Center for Embedded Networked Sensing, UCLA and Department of Civil and Environmental Engineering, MIT.

Ramanathan, N., Chang, K., Kapur, R., Girod, L., Kohler, E., & Estrin, D. (2005). Sympathy for the sensor network debugger. In *Proceedings of the 3rd international conference on Embedded networked sensor systems (SenSys)*, (pp. 255–267). doi:10.1145/1098918.1098946

Rashid, R. A., Sarijari, M. A. B., Rahim, M. R. A., & Skudai, J. (2008). *Wireless Home Security and Automation System Utilizing ZigBee based Multi-hop Communication*. Paper presented at the National Conference on Telecommunication Technology (NCTT). Putrajaya, Malaysia.

Reichenbach, F., Blumenthal, J., & Timmermann, D. (2006). Improved Precision of Coarse Grained Localization in Wireless Sensor Networks.*9th Conference on DSD 2006, Dubrovnik, Croatia, 2006* (pp. 630-637), doi:10.1109/DSD.2006.61

Rezaei, F., Hempel, M., & Sharif, H. (2015 Forthcoming). A Survey of Recent Trends in Wireless Communication Standards, Routing Protocols, and Energy Harvesting Techniques in E-Health Applications. *International Journal of E-Health and Medical Communication*.

Rezaei, F., Hempel, M., Shrestha, P. L., & Sharif, H. (2014). Evaluation and Analysis of Automated Covert Channel Modeling over Real Network Environment. In Proceedings of Military Communication Conference (MILCOM). IEEE.

Rezaei, F., Hempel, M., Peng, D., Qian, Y., & Sharif, H. (2013, April). *Analysis and evaluation of covert channels over LTE advanced*. In *Wireless Communications and Networking Conference (WCNC)*, 2013IEEE (pp. 1903-1908). IEEE. doi:10.1109/WCNC.2013.6554855

Rezaei, F., Hempel, M., Rakshit, S. M., & Sharif, H. (2014). Automated Covert Channel Modeling over Real Network Environment. In *International Wireless Communications and Mobile Computing Conference (IWCMC)*. IEEE. doi:10.1109/IWCMC.2014.6906417

Rezaii, T. Y., & Tinati, M. A. (2011). *Distributed multi-target tracking using joint probabilistic data association and average consensus filter* (pp. 553–566). Institut Télécom and Springer-Verlag.

Rhee, I., Warrier, A., Aia, M., Min, J., & Sichitiu, M. L. (2008). Z-MAC: A hybrid MAC for wireless sensor networks.[TON]. *IEEE/ACM Transactions on Networking, 16*(3), 511–524. doi:10.1109/TNET.2007.900704

Ribeiro, A., Schizas, I. D., Roumeliotis, S. I., & Giannakis, G. B. (2010). Kalman filtering in wireless sensor network. *IEEE Control Systems Magazine, 30*(2), 66–86. doi:10.1109/MCS.2009.935569

Rizzo, L. (1997). Effective erasure codes for reliable computer communication protocols. *ACM Computer Communication Review, 27*(02), 24–36. doi:10.1145/263876.263881

Rodrigues, A., Camilo, T., Silva, J., & Boavida, F. (2012). Diagnostic tools for wireless sensor networks: A comparative survey. *Journal of Network and Systems Management*, 1–45.

Rohm, D., Goyal, M., Xie, W., Polepalli, B., Hosseini, H., Divjak, A., & Bashir, Y. (2009). *Dynamic Reconfiguration in Beaconless IEEE 802.15.4 Networks Under Varying Traffic Loads.* Paper presented at the Global Telecommunications Conference, (GLOBECOM 2009). IEEE, Honolulu, Hawaii.

Romer, K., & Mattern, F. (2004). The Design Space of Wireless Sensor Networks. *IEEE Wireless Communications, 11*(6), 54–61. doi:10.1109/MWC.2004.1368897

Roy, S., Jandhyala, V., Smith, J., Wetherall, D., Otis, B., Chakraborty, R., & Sample, A. (2010). RFID: From supply chains to sensor nets. *Proceedings of the IEEE, 98*(9), 1583–1592. doi:10.1109/JPROC.2010.2053690

Ruzzelli, A. G., Jurdak, R., & O'Hare, G. M. (2007). *On the RFID wake-up impulse for multi-hop sensor networks.*

Safa, H. El-Hajj, W., & Zoubian. (2013). A robust topology control solution for the sink placement problem in WSNs. *Journal of Network and Computer Applications.*

Sahinoglu, Z., Gezici, S., & Guvenc, I. (2008). *Ultra-wideband positioning systems: Theoretical limits, ranging algorithms, and protocols.* Cambridge university press. doi:10.1017/CBO9780511541056

Sahota, H., Kumar, R., Kamal, A., & Huang, J. (2010, June). An energy-efficient wireless sensor network for precision agriculture. In *IEEE Symposium on Computers and Communications (ISCC)* (pp. 347-350). IEEE. doi:10.1109/ISCC.2010.5546508

Sahota, H., Kumar, R., & Kamal, A. (2011). A wireless sensor network for precision agriculture and its performance. *Wireless Communications and Mobile Computing, 11*(12), 1628–1645. doi:10.1002/wcm.1229

Salvatore, C., Bocchino, S., Petracca, M., Pelliccia, R., Ghibaudi, M., & Pagano, P. (2012). WSN and RFID integrated solution for advanced safety systems in industrial plants. Paper presented at the *20th International Conference on Telecommunications and Computer Networks (SoftCOM)* (pp. 11-13).

Sandirigama, M., & Idamekorala, R. (2009, Sept. 26-27). Security weaknesses of WEP protocol IEEE 802.11b and enhancing the security with dynamic keys. In *2009 IEEE Toronto International Conference-Science and Technology for Humanity* (pp. 433-438).

Sankarasubramaniam, Y., Akyildiz, I. F., & McLaughlin, S. W. (2003). Energy efficiency based packet size optimization in wireless sensor networks. In *Proceedings of the First IEEE International Workshop on Sensor Network Protocols and Applications*, (pp. 1-8). doi:10.1109/SNPA.2003.1203351

Savarese, C., Rabaey, J., & Langendoen, K. (2002). Robust positioning algorithms for distributed ad-hoc wireless sensor networks. In *USENIX technical annual conference, vol. 2. Monterey, CA, 2002.*

Savvides, A., Han, C., & Srivastava, M. B. (2001). Dynamic fine-grained localization in ad-hoc networks of sensors. In *7th annual international conference on MobiCom 2001, Rome, Italy* (pp. 166–179), doi:10.1145/381677.381693

Savvides, A., Park, H., & Srivastava, M. (2002). The Bits and Flops of the N-Hop Multilateration Primitive for Node Localization Problems. *First ACM International Workshop on Wireless Sensor Networks and Application*, Atlanta, GA, September 2002. doi:10.1145/570738.570755

Sayed, A. H. (2013to appear). Diffusion adaptation over networks. In R. Chellappa & S. Theodoridis (Eds.), *E-Reference Signal Processing*. Elsevier.

Sayed, A. H., & Cattivelli, F. S. (2009). Distributed adaptive learning mechanisms. In S. Haykin & K. J. Ray Liu (Eds.), *Handbook on Array Processing and Sensor Networks* (pp. 693–722).

Sayed, A. H., & Lopes, C. G. (2006). Distributed recursive least-squares strategies over adaptive networks. In *Asilomar Conference on Signals Systems and Computers* (pp. 233–237). IEEE. doi:10.1109/ACSSC.2006.356622

Schaar, M. V. D., & Shankar, N. D. S. (2005). Cross-Layer Wireless Multimedia Transmission: Challenges, Principles, and New Paradigms. *IEEE Wireless Communications, 12*(4), 50–58. doi:10.1109/MWC.2005.1497858

Schultz, J. J., Healy, C. A., Parker, K., & Lowers, B. (2013). Detecting submerged objects: The application of side scan sonar to forensic contexts. *Forensic Science International, 231*(1–3), 306–316. doi:10.1016/j.forsciint.2013.05.032 PMID:23890654

Schurgers, C., Tsiatsis, V., Ganeriwal, S., & Srivastava, M. (2002). Optimizing sensor networks in the energy-latency-density design space. *Mobile Computing. IEEE Transactions on, 1*(1), 70–80.

SD Card Association. (n.d.). *SD Association Web site.* Retrieved from https://www.sdcard.org/

Seah, W. K. G., & Xian-Tan, H. (2006). Multipath Virtual Sink Architecture for Underwater Sensor Networks. In *Proceedings of MTS/IEEE OCEANS Asia Pacific 2006,* (pp. 1-6). doi:10.1109/OCEANSAP.2006.4393933

Sellke, S. H., Wang, C. C., Bagchi, S., & Shroff, N. (2009, April). TCP/IP timing channels: Theory to implementation. In INFOCOM 2009, IEEE (pp. 2204-2212). IEEE.

Setton, E., Yoo, T., Zhu, X., Goldsmith, A., & Girod, B. (2005). Cross-layer Design of Ad Hoc Networks for Real-time Video Streaming. *IEEE Wireless Communications, 12*(4), 59–65. doi:10.1109/MWC.2005.1497859

Shah, G., Molina, A., & Blaze, M. (2006, July). Keyboards and Covert Channels. In *Proceedings of the 15th USENIX Security Symposium.* https://www.usenix.org/legacy/event/sec06/tech/shah/shah.pdf

Shah, K., & Kumar, M. (2007). *Distributed Independent Reinforcement Learning (DIRL) Approach to Resource Management in Wireless Sensor Networks.* Paper presented at the IEEE Internatonal Conference on Mobile Adhoc and Sensor SystemsPisa (MASS 2007), Italy (pp. 1-9). doi:10.1109/MOBHOC.2007.4428658

Shaikh, F. K., Khelil, A., & Suri, N. (2007). On Modeling the Reliability of Data Transport in Wireless Sensor Networks, *Proceedings of 15th EUROMICRO International Conference on Parallel, Distributed and Network-Based Processing,* pp. 395-402. doi:10.1109/PDP.2007.55

Shaikh, F., Khelil, A., Ali, A., & Suri, V. (2010). Trccit: Tunable reliability with congestion control for information transport in wireless sensor networks, in *5th Annual International ICST Wireless Internet Conference (WICON),* Singapore, pp. 1-9. doi:10.4108/ICST.WICON2010.8623

Shakkottai, S., Srikant, R., & Shroff, N. (2005). Unreliable sensor grids: Coverage, connectivity and diameter. *Ad Hoc Networks, 3*(6), 702–716. doi:10.1016/j.adhoc.2004.02.001

Shang, Y., Ruml, W., Zhang, Y., & Fromherz, M. (2004). Localization from Connectivity in Sensor Networks. *IEEE Transactions on Parallel and Distributed Systems, 15*(11), 961–974. doi:10.1109/TPDS.2004.67

Sharif, B. S., Neasham, J., Hinton, O. R., & Adams, A. E. (2000). A Computationally Efficient Doppler Compensation System for Underwater Acoustic Communications. *IEEE Journal of Oceanic Engineering, 25*(1), 52–61. doi:10.1109/48.820736

Sharma, A. B., Golubchik, L., & Govindan, R. (2010). Sensor faults: Detection methods and prevalence in real-world datasets. *ACM Transactions on Sensor Networks, 6*(3), 23:1–23:39.

Sharma, A., Golubchik, L., & Govindan, R. (2007). On the prevalence of sensor faults in real-world deployments. In *Proceedings of the 4th Annual IEEE Communications Society Conference on Sensor, Mesh and Ad Hoc Communications and Networks,* (pp. 213–222). doi:10.1109/SAHCN.2007.4292833

Shelby, Z., Pomalaza-Raez, C., & Haapola, C. (2004). *Energy optimization in multihop wireless embedded and sensor networks*. Paper presented at the 15th IEEE International Symposium on Personal, Indoor and Mobile Radio, Barcelona, Spain. doi:10.1109/PIMRC.2004.1370868

Sheng, X., Changfeng, X., & Zhangsong, S. (2011). An Adaptive Target Tracking Scheme for Binary Wireless Sensor Networks. In *IEEE International Conference on Electronics, Communications and Control* (pp. 121-124). doi:10.1109/ICECC.2011.6067926

Sheu, S. T., Shih, Y. Y., & Chen, L. W. (2005). An adaptive interleaving access scheme (IAS) for IEEE 802.15.4 WPANs. In *Proceedings of IEEE Vehicular Technology Conference (VTC)*, 3, 1523-1527.

Shih, K., Wang, S. S., Chen, H. C., & Yang, P. H. (2008). *CollECT: Collaborative event detection and tracking in wireless heterogeneous sensor networks*. Elsevier B.V.

Shi, L., Liao, Q., He, Y., Li, R., Striegel, A., & Su, Z. (2011). SAVE: Sensor anomaly visualization engine. In *Proceedings of the IEEE Conference on Visual Analytics Science and Technology*, (pp. 201–210). doi:10.1109/VAST.2011.6102458

Shrestha, P., Hempel, M., Rezaei, F., & Sharif, H. (2014). *Leveraging statistical feature points for generalized detection of covert timing channels*. In Proceedings of Military Communication Conference (MILCOM), IEEE. doi:10.1109/MILCOM.2014.10

Shrestha, P. L., Hempel, M., Sharif, H., & Chen, H. H. (2014). *An Event-Based Unified System Model to Characterize and Evaluate Timing Covert Channels*. IEEE Systems Journal.

Shrivastava, N., Suri, S., & T'oth, C. D. (2005). Detecting cuts in sensor networks. In *Proceedings of the 4th international symposium on Information processing in sensor networks (IPSN).*

Shu, F., Sakurai, T., Vu, H. L., & Zukerman, M. (2006). Optimizing the IEEE 802.15. 4 MAC. In TENCON 2006. 2006 IEEE Region 10 Conference (pp. 1-4).

Siadat, S. H., & Selamat, A. (2008). *Location-based system for mobile devices using RFID*. Paper presented at the Second Asia International Conference on Modelling & SimulationKuala Lumpur, Malaysia (pp. 291-296). doi:10.1109/AMS.2008.44

Sibley, G. T., Rahimi, M. H., & Sukhatme, G. S. (2002, May). *Robomote: A Tiny Mobile Robot Platform for Large-Scale Sensor Networks*. Paper Presented at the IEEE International Conference on Robotics and Automation. Washington, DC. doi:10.1109/ROBOT.2002.1014697

Sichitiu, M. L., Ramadurai, V., & Peddabachagari, P. (2003). Simple algorithm for outdoor localization of wireless sensor networks with inaccurate range measurements.*International Conference on Wireless Networks*, 2003 (pp. 300-305).

Singh, S., Woo, M., & Raghavendra, C. S. (1998). Power-aware routing in mobile ad hoc networks. In *Proceedings of the 4th annual ACM/IEEE international conference on Mobile computing and networking (MobiCom)* (pp. 181–190). doi:10.1145/288235.288286

Siu, H. S., Chin, Y. H., & Yang, W. P. (1996). A note on consensus on dual failure modes. *IEEE Transactions on Parallel and Distributed Systems*, *7*(3), 225–230. doi:10.1109/71.491575

Siu, H. S., Chin, Y. H., & Yang, W. P. (1998). Byzantine agreement in the presence of mixed faults on processors and links. *IEEE Transactions on Parallel and Distributed Systems*, *9*(4), 335–345. doi:10.1109/71.667895

Sizov, V., Gashinova, M., Zakaria, N., & Cherniakov, M. (2013). VHF communication channel characterisations over complex wooded propagation paths with applications to ground wireless sensor networks. *IET Microwaves. Antennas & Propagation*, *7*(3), 166–174.

Sizun, H. (2004). *Radio Wave Propagation for Telecommunication Applications*. Berlin, Heidelberg: Springer.

Sohraby, K., Minoli, D., & Znati, T. (2007). *Wireless Sensor Networks: Technology, Protocols, and Applications*. Wiley-Interscience. doi:10.1002/047011276X

Soltanmohammadi, E., Orooji, M., & Naraghi-Pour, M. (2013). Decentralized hypothesis testing in wireless sensor networks in the presence of misbehaving nodes. *IEEE Transactions on Information Forensics and Security*, *8*(1), 205–215. doi:10.1109/TIFS.2012.2229274

Somov, A., Baranov, A., Spirjakin, D., & Passerone, R. (2014). Circuit design and power consumption analysis of wireless gas sensor nodes: One-sensor versus two-sensor approach. *IEEE Sensors Journal*, *14*(6), 2056–2063. doi:10.1109/JSEN.2014.2309001

Song, G., Zhuang, W., Wei, Z., & Song, A. (2006, October*).Racemote: A Mobile Node for Wireless Sensor Networks.* Paper presented at the 5th IEEE Conference on Sensors. Daegu.

Sorber, J., Kostadinov, A., Garber, M., Brennan, M., Corner, M. D., & Berger, E. D. (2007). Eon: a language and runtime system for perpetual systems. In *Proceedings of the 5th international conference on Embedded networked sensor systems* (pp. 161–174). New York, NY, USA. doi:10.1145/1322263.1322279

Sozer, E. M., Stojanovic, M., & Proakis, J. G. (2000). Underwater Acoustic Networks. *IEEE Journal of Oceanic Engineering*, *25*(1), 72–83. doi:10.1109/48.820738

Srinivasan, K., & Levis, P. (2006). RSSI is Under Appreciated.*Third Workshop on EmNets, 2006.*

Srouji, M. S., Wang, Z., & Henkel, J. (2011). RDTS: A Reliable Erasure-Coding Based Data Transfer Scheme for Wireless Sensor Networks, in *Proceedings of the 17th International Conference on Parallel and Distributed Systems (ICPADS)*, Tainan, Taiwan, pp. 481-488. doi:10.1109/ICPADS.2011.104

Staddon, J., Balfanz, D., & Durfee, G. (2002). Efficient tracing of failed nodes in sensor networks. In *Proceedings of the 1st ACM international workshop on Wireless sensor networks and applications (WSNA)*, (pp. 122–130). doi:10.1145/570738.570756

Stann, F., & Heidemann, J. (2003). RMST: reliable data transport in sensor networks, in *Proceedings of the First IEEE International Workshop on Sensor Network Protocols and Applications (SNPA)*, Anchorage, Alaska, USA, pp. 102- 112. doi:10.1109/SNPA.2003.1203361

Stargate (2014). *Stargate: Distinguishing capabilities*. Retrieved October 10, 2014, from http://platformx.sourceforge.net/home.html

Stargate Wi-Fi. (n.d.). *WiFi FAQ*. Retrieved from http://platformx.sourceforge.net/Documents/nuts/WiFiFAQ.html

Stargate. (n.d.). *Stargate Gateway*. Retrieved from www.willow.co.uk/html/spb400-_stargate_gateway.html

Stoep, J. V. (2009). *Design and Implementation of Reliable Localization Algorithms using Received Signal Strength*. Ms Thesis report, University of Washington, 2009.

Stojanovic, M. (2008). *Underwater Acoustic Communications: Design Considerations on the Physical Layer*. Paper presented at 5th Annual Conference on Wireless on Demand Network Systems and Services, Garmisch-Partenkirchen, Germany. doi:10.1109/WONS.2008.4459349

Stojanovic, M. (2006). On the relationship between capacity and distance in an underwater acoustic communication channel. In *Proceedings of ACM International Conference on Underwater Sensor Networks and Systems* (pp. 41–47). doi:10.1145/1161039.1161049

Stojanovic, M., Catipovic, J. A., & Proakis, J. G. (1991). Phase-coherent digital communications for underwater acoustic channels. *IEEE Journal of Oceanic Engineering*, *16*(1), 12–20.

Stojcev, M. K., Kosanovic, M. R., & Golubovic, L. R. (2009). Power management and energy harvesting techniques for wireless sensor nodes. In *2009 9th International Conference on Telecommunication in Modern Satellite, Cable, and Broadcasting Services* (Vol. 9, pp. 65–72). doi:10.1109/TELSKS.2009.5339410

Stoleru, R., & Stankovic, J. A. (2004). Probability Grid: A Location Estimation Scheme for Wireless Sensor Networks.*IEEE International Conference on Sensor and Ad-Hoc Communications and Networks (SECON)*, Santa Clara, CA, 2004. doi:10.1109/SAHCN.2004.1381945

Stoyanova, T., Kerasiotis, F., Prayati, A., & Papadopoulos, G. (2009). A Practical RF Propagation Model for Wireless Network Sensors. SENSORCOMM 2009, Athens-Vouliagmeni, Greece (pp. 194-199).

Stoyanova, Ts., Kerasiotis, K., Antonopoulos, C., & Papadopoulos, G. (2014). RSS-based Localization for Wireless Sensor Networks in Practice. 9th IEEE/IET CSNDSP 2014, Manchester, UK, 23-25/July 2014

Stoyanova, T., Kerasiotis, F., Efstathiou, K., & Papadopoulos, G. (2010). *Modeling of the RSS Uncertainty for RSS-Based Outdoor Localization and Tracking Applications in Wireless Sensor Networks. SENSORCOMM 2010, July, 2010* (pp. 45–50). Italy: Venice/Mestre.

Stoyanova, T., Kerasiotis, F., Prayati, A., & Papadopoulos, G. (2009, December). Evaluation of Impact Factors on RSS Accuracy for Localization and Tracking Applications in Sensor Networks. *Telecommunications Systems Journal*, *42*, 235–248.

Stoyanova, T., Kerasiotis, K., & Papadopoulos, G. (2010). Deployment Considerations for Reliable Communication in Wireless Sensor Networks. *International Journal on Advances in Networks and Services*, *3*(1&2), 151–162.

Su, W., & Lim, T. L. (2006, June). *Cross-Layer Design and Optimization for Wireless Sensor Networks.* Paper presented at the IEEE 7th Software Engineering, Artificial Intelligence, Networking, and Parallel/Distributed Computing. Las Vegas, NV.

Suh, C., Joung, J. E., & Ko, Y. B. (2007). New RF Models of TinyOS Simulator for IEEE 802.15.4 Standard. *Proc. of WCNC 2007* (pp: 2238-2242). doi:10.1109/WCNC.2007.418

Suh, C., Mir, Z. H., & Ko, Y. B. (2008). Design and implementation of enhanced IEEE 802.15. 4 for supporting multimedia service in Wireless Sensor Networks. *Computer Networks*, *52*(13), 2568–2581. doi:10.1016/j.comnet.2008.03.011

Sung, W.-T., & Hsu, Y.-C. (2011). Designing an industrial real-time measurement and monitoring system based on embedded system and ZigBee. *Expert Systems with Applications*, *38*(4), 4522–4529. doi:10.1016/j.eswa.2010.09.126

Suzuki, T., Kato, K., Makihara, E., Kobayashi, T., Kono, H., Sawai, K., . . . Yamashiro, H. (2014). Development of Underwater Monitoring Wireless Sensor Network to Support Coral Reef Observation.International Journal of Distributed Sensor Networks, Article 189643.

Svensson, D. (2010). *Target Tracking in Complex Scenarios*. Göteborg: Department of Signals and Systems, Chalmers University of Technology.

Szewczyk, R., Mainwaring, A., Polastre, J., Anderson, J., & Culler, D. (2004). An analysis of a large scale habitat monitoring application. In *Proceedings of the 2nd international conference on Embedded networked sensor systems*, (pp. 214–226). doi:10.1145/1031495.1031521

Szewczyk, R., Polastre, J., Mainwaring, A. M., & Culler, D. E. (2004). Lessons from a sensor network expedition. In *Proceedings of the First European Workshop on Wireless Sensor Networks* (pp. 307–322).

Tadayon, N., Khoshroo, S., Askari, E., Wang, H., & Michel, H. (2013). Power management in SMAC-based energy-harvesting wireless sensor networks using queuing analysis. *Journal of Network and Computer Applications*, *36*(3), 1008–1017. doi:10.1016/j.jnca.2012.01.009

Taheri, H., Neamatollahi, P., Younis, O. M., Naghibzadeh, S., & Yaghmaee, M. H. (2012). An energy-aware distributed clustering protocol in wireless sensor networks using fuzzy logic. *Elsevier Ad Hoc Networks*, *10*(7), 1469–1481. doi:10.1016/j.adhoc.2012.04.004

Tam, V., Cheng, K.-Y., & Lui, K.-S. (2006, July). Using Micro-Genetic lgorithms to Improve Localization in Wireless Sensor Networks. *Journal of Communication*, *1*(4), 1–10.

Tan, G., Stephen, A., Jarvis, Member, & Kermarrec, A. M. (2009). Diffusion Based self-deployment algorithm for mobile sensor networks deployment schemes for mobile sensor networks. *IEEE Transactions on Mobile Computing*, *8*(6).

Tan, H., Seah, W. K. G., & Doyle, L. (2007). A multi-hop ARQ protocol for underwater acoustic networks. In Proceedings of OCEANS 07 Europe (pp. 1-6). doi:10.1109/OCEANSE.2007.4302195

Tao, Z., Panwar, S., Gu, D., & Zhang, J. (2006). Performance analysis and a propose improvement for the IEEE 802.15.4 contention access period. *Wireless Communications and Networking Conference*. 4, 1811-1818.

Tariq, M., Zhou, Z., Park, Y. J., & Sato, T. (2010, September). *Diffusion Based Self-Deployment Algorithm for Mobile Sensor Networks*. Paper presented at the 72nd IEEE Vehicular Technology Conference, Ottawa, ON. doi:10.1109/VETECF.2010.5594514

Teng, J., Snoussi, H., & Richard, C. (2007). Prediction-based Proactive Cluster Target Tracking Protocol for Binary Sensor Network. In *IEEE International Symposium on Signal Processing and Information Technology* (pp. 234-239). doi:10.1109/ISSPIT.2007.4458178

Texas Instruments. (2014, September). *Low-Power SoC (System-on-Chip) with MCU, Memory, Sub-1 GHz RF Transceiver, and USB Controller*. http://www.ti.com/lit/gpn/cc1110f32

Tezcan, N., & Wang, W. (2007). ART: An asymmetric and reliable transport mechanism for wireless sensor networks. *International Journal of Sensor Networks*, *02*(3-4), 188–200. doi:10.1504/IJSNET.2007.013199

The MultihopLQI Protocol. (2012). *The MultihopLQI Protocol*. Retrieved June 16, 2012, from http://www.tinyos.net/tinyos-2.x/tos/lib/net/lqi

Thorp, W. H. (1965). Deep Ocean Sound Attenuation in the Sub- and Low-Kilocycle-per-Second Region. *The Journal of the Acoustical Society of America*, *38*(4), 648. doi:10.1121/1.1909768

Thorp, W. H. (1967). Analytic Description of the Low-Frequency Attenuation Coefficient. *The Journal of the Acoustical Society of America*, *42*(1), 270. doi:10.1121/1.1910566

Tian, Z., Yuan, D., & Liang, Q. (2008). Energy efficiency analysis of error control schemes in wireless sensor networks. In *proceedings of the International Wireless Communications and Mobile Computing Conference*, (pp. 401-405). doi:10.1109/IWCMC.2008.70

Tiny, O. S. (2014). *Tinyos home page*. Retrieved October 10, 2014, from http://www.tinyos.net/

TinyNode. (2014). *Tinynode: Real life wireless sensor networks*. Retrieved October 10, 2014, from http://www.tinynode.com/

Tolentino, I. M., Juson, R. S., Tan, B. U., & Talampas, M. R. (2010). Design, development, and evaluation of a simple wireless sensor network for indoor microclimate monitoring. In TENCON 2010 - 2010 IEEE Region 10 Conference (pp. 2018–2023). doi:10.1109/TENCON.2010.5686565

Tolle, G., & Culler, D. (2005). Design of an application-cooperative management system for wireless sensor networks. In *Proceedings of the Second European Workshop on Wireless Sensor Networks*, (pp. 121–132). doi:10.1109/EWSN.2005.1462004

Tolle, G., Polastre, J., Szewczyk, R., Culler, D., Turner, N., Tu, K., & Hong, W. (2005). A macroscope in the redwoods. In *Proceedings of the 3rd international conference on Embedded networked sensor systems (SenSys)* (pp. 51–63). doi:10.1145/1098918.1098925

Tran, S. P. M., & Yang, T. A. (2006). OCO: Optimized Communication and Organization for Target Tracking in Wireless Sensor Networks. In *Proceedings of the IEEEE International Conference on Sensor Networks*.

Tripathy, A. K., & Chinara, S. (2012). Staggered Clustering Protocol: SCP an Efficient Clustering Approach for Wireless Sensor Network. In *IEEE World Congress on Information and Communication Technologies*. doi:10.1109/WICT.2012.6409209

Tsai, H. W., Chu, C. P., & Chen, T. S. (2007). Mobile object tracking in wireless sensor Networks. *Computer Communications*, *30*(8), 1811–1825. doi:10.1016/j.comcom.2007.02.018

Tsai, M. F., Shieh, C.-K., Ke, C.-H., & Deng, D.-J. (2010). Sub-packet forward error correction mechanism for video streaming over wireless networks. *Multimedia Tools and Applications*, *47*(1), 49–69. doi:10.1007/s11042-009-0406-5

Tsiftes, N., & Dunkels, A. (2011). A database in every sensor. In *Proceedings of the 9th ACM Conference on Embedded Networked Sensor Systems (SenSys)* (pp. 316–332).

Tsitsiklis, J., Bertsekas, D., & Athans, M. (1986). Distributed asynchronous deterministic and stochastic gradient optimization algorithms. *IEEE Transactions on Automatic Control*, *31*(9), 803–812. doi:10.1109/TAC.1986.1104412

Tyagi, S., & Kumar, N. (2013). A systematic review on clustering and routing techniques based upon LEACH protocol for wireless sensor network. *Journal of Network and Computer Applications*, *36*(2), 623–645. doi:10.1016/j.jnca.2012.12.001

Urban, H. G. (2002). *Handbook of Underwater Acoustic Engineering*. Bremen, Germany: ATLAS Elektronik GmbH.

Urick, R. J. (1983). *Principles of Underwater Sound*. Los Altos, CA: Peninsula Publishing.

US Department of Transportation, & the Federal Railroad Administration. (n.d.). *Freight Rail Today*. Retrieved from http://www.fra.dot.gov/Page/P0362

US Department of Transportation; Federal Railroad Administration. (n.d.). *Freight Rail Overview*. Retrieved from http://www.fra.dot.gov/Page/P0362

USB Standard. (n.d.). *USB 3.1 Specification*. Retrieved from http://www.usb.org/developers/docs/

Vaghefi, R. M., Gholami, M. R., & Strom, E. G. (2010, November). Bearing-only target localization with uncertainties in observer position. *IEEE PSPB Operations Manual, amended 19*.

Vaidehi, V., Sandhya, M., Karthika, J., & Azhagarasan, S. (2011). Power optimization for object detection and tracking in wireless sensor networks. In *Recent Trends in Information Technology (ICRTIT)*, (pp. 555-560). doi:10.1109/ICRTIT.2011.5972433

Vaidya, N. H., & Pradhan, D. K. (1994). Safe system level diagnosis. *IEEE Transactions on Computers*, *43*(3), 367–370. doi:10.1109/12.272443

Vaidya, N., & Das, S. R. (2008). Rfid-based networks: Exploiting diversity and redundancy. *Mobile Computing and Communications Review*, *12*(1), 2–14. doi:10.1145/1374512.1374514

Van Dam, T., & Langendoen, K. (2003, November). An adaptive energy-efficient MAC protocol for wireless sensor networks. In *Proceedings of the 1st international conference on Embedded networked sensor systems* (pp. 171-180). ACM. doi:10.1145/958491.958512

Van Phuong, T., Hung, L. X., Cho, S. J., Lee, Y.-K., & Lee, S. (2006). An anomaly detection algorithm for detecting attacks in wireless sensor networks. In *Proceedings of the 4th IEEE international conference on Intelligence and Security Informatics (ISI)* (pp. 735–736). doi:10.1007/11760146_111

Vasilescu, G. (2005). *Electronic Noise and Interfering Signals: Principles and Applications*. Springer-Verlag.

Venkatesan. L., Shanmugavel. S., Subramaniam. C. (2013).A survey on modeling and enhancing reliability of wireless sensor network, pp. 41-51, doi.org/.<ALIGNMENT.qj></ALIGNMENT>10.4236/wsn.2013.53006

Verma, S., Chug, N., & Gadre, D. V. (2010, March). Wireless sensor network for crop field monitoring. In *2010 International Conference on Recent Trends in Information, Telecommunication and Computing (ITC)* (pp. 207-211). IEEE. doi:10.1109/ITC.2010.21

Vouyioukas, D., Maglogiannis, I., & Komnakos, D. (2007). Emergency M-Health services through high-speed 3G systems: Simulation and performance evaluation. *Simulation*, *83*(4), 329–345. doi:10.1177/0037549707083113

W¨alchli, M., & Braun, T. (2009). Efficient signal processing and anomaly detection in wireless sensor networks. In *Proceedings of the EvoWorkshops on Applications of Evolutionary Computing* (pp. 81–86).

Wan, C. Y., Campbell, A., & Krishnamurthy, L. (2005). Pump-slowly, fetchquickly (PSFQ), a reliable transport protocol for sensor networks. *IEEE Journal on Selected Areas in Communications*, *23*(04), 862–872. doi:10.1109/JSAC.2005.843554

Wang, C., Chen, W., & Li, W. (2014, April). *Energy Efficient Distributed Kalman Filter for Wireless Sensor Networks*. Paper presented at the 11th IEEE International Conference on Networking, Sensing and Control, Miami, FL. doi:10.1109/ICNSC.2014.6819618

Wang, G., Cao, G., & LaPorta, T. (2003, November*).A Bidding Protocol for Deploying Mobile Sensors*. Paper presented at the 11th IEEE International Conference on Network Protocols, Atlanta, GA.

Wang, G., Cao, G., Porta, T. L., & Zhang, W. (2005, March). *Sensor Relocation in Mobile Sensor Networks.* Paper presented at the 24th Annual Joint Conference of the IEEE Computer and Communications Societies, Miami, FL.

Wang, Q., Hempstead, M., & Yang, W. (2006, September). *A Realistic Power Consumption Model for Wireless Sensor Network Devices.* Paper presented at the 3rd Annual IEEE Communications Society on Sensor and Ad Hoc Communications and Networks. Reston, VA. doi:10.1109/SAHCN.2006.288433

Wang, Y., Zhao, X., Wu, Y., & Xu, P. (2008) In. *The research of RFID middleware's data management model.* Paper presented at the IEEE International Conference on Automation and Logistics Qingdao (ICAL 2008) (pp. 2565-2568).

Wang, Z. Lou, W. Wang, Z. Ma, J., & Chen, H. (2013). A Hybrid Cluster-Based Target Tracking Protocol for Wireless Sensor Networks. *International Journal of Distributed Sensor Networks*, Article ID 494863.

Wang, Z., Deng, J., & Lee, R. B. (2007, May). *Mutual anonymous communications: a new covert channel based on splitting tree MAC.* In INFOCOM 2007. 26th IEEE International Conference on Computer Communications (pp. 2531-2535). IEEE. doi:10.1109/INFCOM.2007.315

Wang, G., Cao, G., & Porta, T. L. (2004). Movement-Assisted Sensor Deployment. *IEEE Transactions on Mobile Computing*, *5*(6), 640–652. doi:10.1109/TMC.2006.80

Wang, G., & Yang, K. (2011, May). A New Approach to Sensor Node Localization Using RSS Measurements in Wireless Sensor Networks. *IEEE Transactions on Wireless Communications*, *10*(5), 1389–1395. doi:10.1109/TWC.2011.031611.101585

Wang, S. C., & Yan, K. Q. (2000). Reaching fault diagnosis agreement on dual link failure mode. *Proceedings Seventh International Conference on Parallel and Distributed Systems.* doi:10.1109/ICPADS.2000.857710

Wang, S.-S., Yan, K.-Q., & Wang, S.-C. (2010). An optimal solution for byzantine agreement under a hierarchical cluster-oriented mobile ad hoc network. *Computers & Electrical Engineering*, *36*(1), 100–113. doi:10.1016/j.compeleceng.2009.06.002

Wang, W., Peng, D., Wang, H., Sharif, H., & Chen, H. H. (2008). Energy-Constrained Distortion Reduction Optimization for Wavelet-Based Coded Image Transmission in Wireless Sensor Networks. *IEEE Transactions on Multimedia*, *10*(6), 1169–1180. doi:10.1109/TMM.2008.2001354

Wang, W.-T., & Ssu, K.-F. (2013). Obstacle detection and estimation in wireless sensor networks. *Computer Networks*, *57*(4), 858–868. doi:10.1016/j.comnet.2012.11.004

Wang, X., Ma, J. J., Wang, S., & Bi, D. W. (2007). Cluster-based Dynamic Energy Management for Collaborative Target Tracking in Wireless Sensor Networks. *Sensors (Basel, Switzerland)*, *7*(7), 1193–1215. doi:10.3390/s7071193

Wang, X., Ma, J. J., Wang, S., & Bi, D. W. (2007). Prediction-based Dynamic Energy Management in Wireless Sensor Networks Sensor Networks. *Sensors (Basel, Switzerland)*, 7.

Wang, X., Xing, G., & Zhang, Y. et al.. (2003). Integrated Coverage and Connectivity Configuration in Wireless Sensor Networks, *1st ACM Conference on Embedded Networked Sensor Systems.* doi:10.1145/958491.958496

Wang, Y., Attebury, G., & Ramamurthy, B. (2006). A survey of security issues in wireless sensor networks. *IEEE Communications Surveys and Tutorials*, *8*(2), 2–23. doi:10.1109/COMST.2006.315852

Wang, Z., Lou, W., Wang, Z., Ma, J., & Chen, H. (2010). A Novel Mobility Management Scheme for Target Tracking in Cluster-Based Sensor Networks. *Springer-Verlag Distributed Computing in Sensor Systems*, *6131*, 172–186. doi:10.1007/978-3-642-13651-1_13

Want, R., Hopper, A., Falcao, V., & Gibbons, J. (1992). The Active Badge Location System. *ACM Transactions on Information Systems*, *10*(1), 91–102. doi:10.1145/128756.128759

Wei, Y., Heidemann, J., & Estrin, D. (2002) In *INFOCOM 2002. An energy-efficient MAC protocol for wireless sensor networks.* Paper presented at the *Twenty-First Annual Joint Conference of the IEEE Computer and Communications Societies*, (Vol. 3, pp. 1567-1576) New York, USA.

Wei, Y., Heidemann, J., & Estrin, D. (2004). Medium access control with coordinated adaptive sleeping for wireless sensor networks. *IEEE/ACM Transactions on Networking*, *12*(3), 493–506. doi:10.1109/TNET.2004.828953

Welch, G., & Bishop, G. (2006). *An Introduction to the Kalman Filter* (Tech. Rep. 95-041), UNC-Chapel Hill. Retrieved July 24, 2006, from: http://www.cs.unc.edu/~welch/kalman/

Welch, P. (1967). The use of fast Fourier transform for the estimation of power spectra: A method based on time averaging over short, modified periodograms. *IEEE Transactions on Audio and Electroacoustics*, *15*(2), 70–73. doi:10.1109/TAU.1967.1161901

Wen, H., Lin, C., Ren, F., Yue, Y., & Huang, X. (2007). Retransmission or Redundancy: transmission reliability in wireless sensor networks. In *IEEE 4th International Conference on Mobile Adhoc and Sensor Systems (MASS)* (pp. 1-7). Pisa, Italy.

Wenz, G. M. (1962). Acoustic Ambient Noise in the Ocean: Spectra and Sources. *The Journal of the Acoustical Society of America*, *34*(12), 1936–1956. doi:10.1121/1.1909155

Werner-Allen, G., Lorincz, K., Johnson, J., Lees, J., & Welsh, M. (2006). Fidelity and yield in a volcano monitoring sensor network. In *Proceedings of the 7th symposium on Operating systems design and implementation (OSDI)* (pp. 381–396).

Wheeler, A. (2007). Commercial Applications of Wireless Sensor Networks Using ZigBee. *IEEE Communications Magazine*, *45*(4), 70–77. doi:10.1109/MCOM.2007.343615

Whitehouse, K., Karlof, C., & Culler, D. (2007, January). A Practical Evaluation of Radio Signal Strength for Ranging-based Localization. *Mobile Computing and Communications Review*, *11*(1), 41–52. doi:10.1145/1234822.1234829

Wi-Fi Alliance. (n.d.). *Wi Fi Alliance Web site*. Retrieved from http://www.wi-fi.org

Wikipedia. (n.d.). *Automatic Equipment Identification*. Retrieved from http://en.wikipedia.org/wiki/Automatic_equipment_identification</eref>

Wikipedia. (n.d.). *Devices in Ad-Hoc Mode*. Retrieved from http://en.wikipedia.org/wiki/Peer-to-peer

Wikipedia. (n.d.). *IEEE 802.11 Standard*. Retrieved from http://en.wikipedia.org/wiki/IEEE_802.11

Wikipedia. (n.d.). *IEEE 802.15.4*. Retrieved from https://en.wikipedia.org/wiki/IEEE_802.15.4

Wikipedia. (n.d.). *Wayside Monitoring Equipment*. Retrieved from http://en.wikipedia.org/wiki/Hot_box

Wikipedia. (n.d.). *Wi-Fi Architecture*. Retrieved from http://en.wikipedia.org/wiki/IEEE_802.11

Wikipedia. (n.d.). *ZigBee Network Topologies*. Retrieved from http://en.wikipedia.org/wiki/IEEE_802.15.4

Willig, A., & Karl, H. (2005). Data transport reliability in wireless sensor networks: A survey of issues and solutions. *Praxis der Informationsverarbeitung und Kommunikation*, *28*(2), 86–92.

Winston, J. J., & Paramasivan, B. (2011). A Survey on Connectivity Maintenance and Preserving Coverage for Wireless Sensor Networks. *International Journal of Research and Review in Wireless Sensor Networks*, *1*(2), 11–18.

Wittenburg, G., Dziengel, N., Wartenburger, C., & Schiller, J. (2010). A system for distributed event detection in wireless sensor networks. In *Proceedings of the 9th ACM/IEEE International Conference on Information Processing in Sensor Networks (IPSN)*, (pp. 94–104). doi:10.1145/1791212.1791225

Wong, K. J., & Arvind, D. K. (2006, May). SpeckMAC: low-power decentralised MAC protocols for low data rate transmissions in specknets. In *Proceedings of the 2nd international workshop on Multi-hop ad hoc networks: from theory to reality* (pp. 71-78). ACM. doi:10.1145/1132983.1132996

Woo, A., & Culler, D. C. (2004). A transmission control scheme for media access in sensor *annual international conference on mobile computing and networking* networks, in proceeding of 2001. ACM (mobicom'01), Rome, Italy, pp.221-235.

Woo, A., Tong, T., & Culler, D. (2003). Taming the underlying challenges of reliable multihop routing in sensor networks. In *Proceedings of the 1st international conference on Embedded networked sensor systems (SenSys)*, (pp. 14–27). doi:10.1145/958491.958494

Wood, T. (2008). *Mathematical Modelling Of Single Target SONAR and RADAR Contact Tracking*. Thesis submitted for transfer to DPhil status Michaelmas Term.

Wood, A., & Stankovic, J. (2002). Denial of service in sensor networks. *Computer*, *35*(10), 54–62. doi:10.1109/MC.2002.1039518

Wotton, P. (2006). Providing reliable sensing and control using ZigBee wireless networks. *RF DESIGN magazine*, *29*(7), 18.

Wu, C., Lee, K., & Chung, Y. (2006, July). *A Delaunay Triangulation Based Method for Wireless Sensor Network Deployment.* Paper Presented at the 12th International Conference on Parallel and Distributed Systems. Minneapolis, MN.

Wu, X., Chen, G. H., & Das, S. K. (2008). Avoiding energy holes in wirelesssensor networks with non-uniform node distribution. *IEEE Transactions on Parallel and Distributed Systems*, *19*(5), 1–11.

Xia, J., Tang, Z., Shi, X., Fan, L., & Li, H. (2011, December). An environment monitoring system for precise agriculture based on wireless sensor networks. In *Seventh International Conference on Mobile Ad-hoc and Sensor Networks (MSN)* (pp. 28-35). IEEE. doi:10.1109/MSN.2011.16

Xiao, W., Tham, C. K., & Das, S. K. (2010). Collaborative sensing to improve information quality for target tracking in wireless sensor networks. In 8th IEEE International Conference on Pervasive Computing and Communications Workshops (PERCOM Workshops) (pp. 99-104). doi:10.1109/PERCOMW.2010.5470610

Xiao, Z., He, C., & Jiang, L. (2010). Slot-based model for IEEE 802.15. 4 MAC with sleep mechanism. *IEEE Communications Letters*, *14*(2), 154–156. doi:10.1109/LCOMM.2010.02.091981

Xie, P., Cui, J. H., & Lao, L. (2006). VBF:vector-based forwarding protocol for underwater sensor networks. In *Proceedings of International Federation for Information Processing Networking Conference* (pp. 228-235). doi:10.1007/11753810_111

Xie, M., Han, S., Tian, B., & Parvin, S. (2011). Anomaly detection in wireless sensor networks: A survey. *Journal of Network and Computer Applications*, *34*(4), 1302–1325. doi:10.1016/j.jnca.2011.03.004

Xie, P., & Cui, J. H. (2007). An FEC-based Reliable data transport protocol for underwater sensor networks. In *Proceedings of the 16th International Conference on Computer Communications and Networks* (pp. 747-753). doi:10.1109/ICCCN.2007.4317907

Xing, K., Srinivasan, S. S. R., Rivera, M. J. M., Li, J., & Cheng, X. (2010). Network Security, chapter Attacks and Countermeasures in Sensor Networks. *Survey (London, England)*, 251–272.

Xiong, Z., Sottile, F., Spirito, M. A., & Garello, R. (2011). Hybrid indoor positioning approaches based on WSN and RFID. Paper presented at the *2011 4th IFIP International Conference on* IEEE in *New Technologies, Mobility and Security (NTMS)* (pp. 1-5).

Xue, F., & Kumar, P. R. (2004). The number of neighbours needed for connectivity of wireless networks. *Wireless Networks*, *10*(2), 169–181. doi:10.1023/B:WINE.0000013081.09837.c0

Xu, J., Li, K., & Min, G. (2011). Reliable and Energy-Efficient Multipath Communications in Underwater Sensor Networks. *IEEE Transactions on Parallel and Distributed Systems*, *23*(7), 1326–1335. doi:10.1109/TPDS.2011.266

Xu, Y., Winter, J., & Lee, W. C. (2004). Prediction-based strategies for energy saving in object tracking sensor networks. In *Proceedings of the Fifth IEEE International Conference on Mobile Data Management (MDM'04)* (pp. 346-357).

Yadav, A., Naik, N., Ananthasayanam, M. R., Gaur, A., & Singh, Y. N. (2012). A Constant Gain Kalman Filter Approach to target tracking in Wireless Sensor Networks. In *IEEE International Conference on Industrial and Information Systems* (pp. 1-7). doi:10.1109/ICIInfS.2012.6304803

Yang, O., & Heinzelman, W. (2011). Sleeping multipath routing: a trade-off between reliability and lifetime in wireless sensor networks. In *Global Communication Conf., GlobeCom* (pp. 1–5).

Yang, H., & Sikdar, B. A. (2003). Protocol for Tracking Mobile Targets using Sensor Networks. In *Proceedings of IEEE Workshop*.

Yang, J., Soffa, M. L., Selavo, L., & Whitehouse, K. (2007). Clairvoyant: a comprehensive source-level debugger for wireless sensor networks. In *Proceedings of the 5th International Conference on Embedded Networked Sensor Systems (SenSys)* (pp. 189–203).

Yang, L., Feng, C., Rozenblit, J. W., & Qiao, H. (2006). Adaptive Tracking in Distributed Wireless Sensor Networks.*Proceedings of the 13th Annual IEEE International Symposium and Workshop on Engineering of Computer Based Systems*. doi:10.1109/ECBS.2006.20

Yang, W. C., Fu, Z., Kim, J. H., & Park, M. S. (2007). *An Adaptive Dynamic Cluster-Based Protocol for Target Tracking in Wireless Sensor Networks* (pp. 157–167). Advances in Data and Web Management. doi:10.1007/978-3-540-72524-4_19

Yang, X., Ong, K. G., Dreschel, W. R., Zeng, K., Mungle, C. S., & Grimes, C. A. (2002). Design of a Wireless Sensor Network for Long-term, In-Situ Monitoring of an Aqueous Environment. *Sensors (Basel, Switzerland)*, *2*(11), 455–472. doi:10.3390/s21100455

Yang, Y., Zong, C., Sun, Y., & Yang, J. (2010). Network Coding based reliable disjoint and braided multipath routing for sensor networks. *Journal of Network and Computer Applications*, *33*(4), 422–432. doi:10.1016/j.jnca.2010.02.003

Yan, K. Q., & Wang, S. C. (2005). Reaching fault diagnosis agreement on an unreliable general network. *Information Sciences*, *170*(2-4), 397–407. doi:10.1016/j.ins.2004.03.011

Yarvis. M., Kushalnagar. N., Singh. H., Rangarajan. A., Liu. Y., & Singh. S. (2005). Exploiting heterogeneity in sensor networks. In *Proceeding IEEE INFOCOM'05*, (Vol. 2, pp. 878-890).

Ye, W., Heidemann, J., & Estrin, D. (2004). Medium access control with coordinated adaptive sleeping for wireless sensor networks. *IEEE/ACM Transactions on Networking*, *12*(3), 493-506.

Yedavalli, K., Krishnamachari, B., Ravula, S., & Srinivasan, B. (2005). Ecolocation: A Technique for RF Based Localization in Wireless Sensor Networks. IPSN 2005, Los Angeles, CA, April 2005.

Yeh, W. C., Lin, Y. C., & Chung, Y. Y. (2010). Performance analysis of cellular automata Monte Carlo simulation for estimating network reliability. *International Journal of Expert Systems with Application*, *37*(5), 3537–3544. doi:10.1016/j.eswa.2009.09.070

Ye, W., Heidemann, J., & Estrin, D. (2002). An energy-efficient MAC protocol for wireless sensor networks. In *INFOCOM 2002. Twenty-First Annual Joint Conference of the IEEE Computer and Communications Societies. Proceedings of the IEEE*, *3*, 1567–1576.

Yick, J., Mukherjee, B., & Ghosal, D. (2008). Wireless sensor network survey. *Computer Networks*, *52*(12), 2292–2330. doi:10.1016/j.comnet.2008.04.002

Yu, C. W., Wang, C. H., Hsu, L. C., & Cheng, K. J. (2008, September). *Coverage Algorithms in GPS-Less Wireless Mobile Sensor Networks*. Paper Presented at the 5th International Conference on Mobile Technology, Applications, and Systems, Mobility Conference. Yilan, Taiwan. doi:10.1145/1506270.1506368

Yu, B., & Xiao, B. (2006). Detecting selective forwarding attacks in wireless sensor networks. In *Proceedings of the 20th International Parallel and Distributed Processing Symposium (IPDPS)*.

Yu, Z., & Tsai, J. (2008). A framework of machine learning based intrusion detection for wireless sensor networks. In *IEEE International Conference on Sensor Networks, Ubiquitous and Trustworthy Computing (SUTC)* (pp. 272–279). doi:10.1109/SUTC.2008.39

Zhang, B., Sukhatme, G. S., & Requicha, A. A. G. (2004). Adaptive sampling for marine microorganism monitoring. In *Proceedings of IEEE/RSJ International Conference on Intelligent Robots and Systems*, 2, (pp. 1115-1122).

Zhang, H., Arora, A., Choi, Y. R., & Gouda, M. G. (2005). RBC: Reliable bursty convergecast in wireless sensor networks, in *Proceedings of the 6th ACM international symposium on Mobile ad hoc networking and computing (MobiHoc)*, Urbana-Champaign, Illinois, USA, pp. 266-276.

Zhang, H., & Hou, J. C. (2005, March). Maintaining Sensing Coverage and Connectivity in Large Sensor Networks. *Ad Hoc & Sensor Wireless Networks, 1*, 89–124.

Zhang, W., & Cao, G. (2004). *DCTC: Dynamic Convoy Tree-Based Collaboration for Target Tracking in Sensor Networks*. Published in the IEEE Transactions on Wireless Communications.

Zhang, Y., Meratnia, N., & Havinga, P. (2010). Outlier detection techniques for wireless sensor networks: A survey. *IEEE Communications Surveys and Tutorials, 12*(2), 159–170. doi:10.1109/SURV.2010.021510.00088

Zhao, H., & Shi, Y. Q. (2010, December). *A phase-space reconstruction approach to detect covert channels in TCP/IP protocols*. In Information Forensics and Security (WIFS), 2010 IEEE International Workshop on (pp. 1-6). IEEE.

Zhao, Z., & Guo, S. (2010). An FHMA acoustic communication system for multiple underwater robots. In *Proceedings of IEEE International Conference on Information and Automation (ICIA)*, (pp. 1223 – 1228). doi:10.1109/ICINFA.2010.5512294

Zheng, J., & Jamalipour, A. (2009). *Wireless sensor networks A networking perspective*. Hoboken, New Jersey: John Wiley.

Zheng, J., & Lee, M. J. (2004). Will IEEE 802.15.4 make ubiquitous networking a reality? A discussion on a potential low power low bit rate standard. *IEEE Communications Magazine, 42*(6), 140–146. doi:10.1109/MCOM.2004.1304251

Zheng, W. X. (1998). Adaptive parameter estimation of autoregressive signals from noisy observations. In *ICSP '98.1998 Fourth International Conference on Signal Processing* (pp. 449–452).

Zhong, X., & Premkumar, A. B. (2014). Multiple wideband source detection and tracking using a distributed acoustic vector sensor array: A random finite set approach. *Signal Processing, 94*, 583–594. doi:10.1016/j.sigpro.2013.07.008

Zhou, Y., & Yu, L. (2007). The Current Status of Energy-Efficiency for Wireless Sensor Networks. *Network Communications and Security,* 128-130.

Zhou, H., Guan, X., & Wu, C. (2008). RTMC: Reliable transport with memory consideration in wireless sensor networks, in *IEEE International Conference on Communications (ICC)*, Beijing, China, pp. 2819-2824.

Zhoul, H., Chen, X., Liu, X. & Yang, J. (2009). Applications of Zigbee wireless technology to measurement system in grain storage. *Computer and Computing Technologies in Agriculture II,* 3, 2021-2029.

Zhu, C., Shu, L., Hara, T., Wang, L., & Nishio, S. (2010, August). *Research Issues on Mobile Sensor Networks.* Paper presented at the 5th International Conference on Software Testing on Communications and Networking. Beijing, China.

Zigbee Alliance. (2006). *Zigbee-2006 specification.* Retrieved March 2 2014, from http://www.zigbee.org/

ZigBee Alliance. (2006, Dec 1). *ZigBee Specification* (ZigBee document 053474r13). Retrieved from https://openwsn.googlecode.com/files/ZigBee-Specification-2006.pdf

Zigbee, (2003). *CC2420: 2.4 GHz IEEE 802.15.4 / Zigbee RF Transceiver.* Chipcon AS SmartRF® CC2420 Preliminary Datasheet (rev 1.0), 2003-11-17.

ZigBee. (n.d.). *ZigBee Primer.* Retrieved from http://www.daintree.net/downloads/whitepapers/ZigBee_primer.pdf

ZigBee. (n.d.). *ZigBee.* Retrieved from http://www.sensor-networks.org/index.php?page=0903503549

Zou, Y., & Chakrabarty, K. (2003, April). *Sensor Deployment and Target Localization Based on Virtual Forces.* Paper presented at the 23nd Annual Joint Conference of the IEEE Computer and Communications. San Francisco, CA. doi:10.1109/INFCOM.2003.1208965

About the Contributors

Hamid Sharif is the Charles J. Vranek Distinguished Professor in the Department of Computer and Electronics Engineering at the University of Nebraska-Lincoln. He is also the Director of Advanced Telecommunications Engineering Laboratory (TEL) at University of Nebraska. Professor Sharif has published a large number of research articles in international journals and conferences and has been the recipient of a number of awards and best papers. Dr. Sharif has been serving on many IEEE and other international journal editorial boards and currently is the co-editor-in-chief for the Wiley Journal of Security and Communication Networks. He has contributed to the IEEE in many roles including the elected Chair of the IEEE Nebraska Section, elected Chair of the Nebraska Computer Chapter, elected Chair of the Nebraska Communication Chapter, and the Chapter Coordinator for the IEEE Region 4 in US. He received his BS, MS, and PhD all in Electrical Engineering from University of Iowa (1982), University of Missouri-Columbia (1984), and University of Nebraska-Lincoln (1996) respectively.

Yousef S. Kavian received the B.Sc. (Hons.) degree in electronic engineering from Shahid Beheshti University, Tehran, Iran, in 2001, the M.Sc. degree in control engineering from the Amirkabir University of Technology, Tehran, in 2003, and the Ph.D. degree in electronic engineering from the Iran University of Science and Technology, Tehran, in 2008. After one year lecturing at Shahid Beheshti University in 2008, he joined Shahid Chamran University, Ahvaz, Iran, as an Assistant Professor. His research interests include digital circuits and systems design, distributed systems, and networking. He has published three books, five book chapters, and around 90 technical papers in journals and international conferences. He has over 13 years of industrial experience and collaborations.

***

Raed Abdulla received the Bachelor's degree in Electrical engineering from AL-Mustansiria University, Baghdad, Iraq in 1997, the Master degree in Electronic Systems Design Engineering from Universiti Sains Malaysia, Malaysia in 2008 and he received his PhD in Wireless and Mobile Systems from Universiti Sains Malaysia, Malaysia in 2012. He served Universiti Sains Malaysia, Malaysia from 2008 until 2011 as Postgraduate Teaching Assistant and in the year 2012 he works at Asia Pacific University of Technology and innovation (APU), Malaysia as a Senior lecturer until now. His main areas of research are active RFID (Radio Frequency Identification) and Wireless Sensor Networks. His research output was translated to numbers of awards and publications. He is also a member in the institution of engineering and technology (IET).

Giovani Rimon Abuaitah received his Ph.D. degree in computer science and engineering from Wright State University in 2013. He joined the Business Analytics team at IBM in 2013 after his graduation. His dissertation research focused on leveraging machine learning and data mining techniques to detect anomalies in real-world sensor network deployments. Giovani's current research interests span the areas of sensor networks and the Internet of Things (IoT), sensor data stream mining, data analytics and predictive analytics, behavioral modeling, network and information security, and anomaly detection.

Mohammad Saeed Ansari, received the B.Sc degree with honors in electrical and electronics engineering from Iran University of Science and Technology (IUST), Tehran, Iran, in 2013. He is currently working toward the M.Sc degree in the Electrical Engineering Department of IUST. His research interests include wireless sensor networks, Fault tolerant and testable design, multi-core and reconfigurable architectures.

Michael Hempel received his Ph.D. in computer engineering at the University of Nebraska-Lincoln, USA. He is currently working as a Research Assistant Professor at the Advanced Telecommunication Engineering Laboratory (TEL) at the University of Nebraska-Lincoln. His research interests include wireless communication protocol design and performance analysis, wireless multimedia services, and distributed computing. For his research in networking he has also been developing various network simulation and analysis solutions for streaming media and WiFi/WiMAX technologies. He has authored or co-authored over 80 publications in major international journals and conferences, and served as a TPC member on numerous international conferences. He also serves as an Associate Editor for Wiley's Security and Communication Networks (SCN).

Kun Hua is an Assistant Professor of Electrical and Computer Engineering Department, Lawrence Technological University, Southfield, MI, USA since 2010. He received the B.Sc. (with First Class Honors) and M.Sc. degrees from Electrical and Computer Engineering, Xi'an Jiaotong University, China, in 1999 and 2004 respectively. He earned his PhD degree in computer and electronic engineering from the University of Nebraska Lincoln, Nebraska, USA in 2008. He continued his research in the University of Nebraska Lincoln as a Post-doctorial researcher in 2009. His current research interests are in the areas of wireless communication and multimedia signal processing. Prof. Hua is a senior member of IEEE. He served as an Associate Editor of Wiley Security and Communication Networks Journal.

Widad Ismail graduated from University of Huddersfield, UK in 1999 and earned First Class Honors in Electronics and Communications Engineering and she received her PhD in Electronics Engineering from University of Birmingham, UK in 2004. She is currently an associate professor at the School of Electrical and Electronics Engineering, USM in Nibong Tebal, Penang, Malaysia. She has contributed extensively in research and in the areas of Radio Frequency Identification (RFID), Active Integrated Antennas (AIA), RF systems and Wireless Systems Design. She has initiated Auto-ID Laboratory (AIDL), Malaysia in 2008 as a research and commercialize oriented centre where the main objective is to become a hub for research and commercialization activities. These research works have produced 8 filed patents, 4 international awards, 4 commercialized research products & more than 100 publications including international journal papers, conference/seminars and other publications. She is also a member of IEEE and Wireless World Research Forum (WWRF). Email: eewidad@usm.my.

Ahmed E. Kamal is a professor of Electrical and Computer Engineering at Iowa State University in the USA. He received a B.Sc. (distinction with honors) and an M.Sc. both from Cairo University, Egypt, and an M.A.Sc. and a Ph.D. both from the University of Toronto, Canada, all in Electrical Engineering. He is a Fellow of the IEEE and a senior member of the Association of Computing Machinery. He is also an IEEE Communications Society Distinguished Lecturer for 2013 and 2014. His research interests include cognitive radio networks, optical networks, wireless sensor networks, and performance evaluation. He received the 1993 IEE Hartree Premium for papers published in Computers and Control in IEE Proceedings, and the best paper award of the IEEE Globecom 2008 Symposium on Ad Hoc and Sensors Networks Symposium. He was the chair or co-chair of the Technical Program Committees of a number of IEEE sponsored conferences, and serves on the editorial boards of the IEEE Communications Surveys and Tutorials, the Computer Networks journal, and the Optical Switching and Networking journal.

Fotis Kerasiotis is a researcher. He received his Diploma in Electrical and Computer Engineering from the University of Patras, Greece, in 2006. From 2006 until today, he has been working as a Research Assistant at the University of Patras, Greece, towards his Ph.D, and as a Researcher at ISI (www.isi.gr), Patras, Greece, working on EU research Projects. His research interests are in the area of real-time distributed-embedded systems and wireless sensor networks, focusing on middleware, network protocols design, methodologies for distributed devices, control applications design and embedded systems development, development of device architecture, RF-signal propagation modeling, target localization and tracking techniques.

Adel Khosravi received the B.Sc. degree in electronic engineering from Shahrekord University in 2008, and the M.Sc. degree in electronic engineering from Iran University of Science and Technology (IUST) in 2010. He is currently a Ph.D. candidate in electrical engineering department of Shahid Chamran University of Ahvaz. His research interests are in the area of fault tolerant networks, reliability computing in wireless networks, distributed computing and stochastic approximations in sensor networks.

Ratnesh Kumar is a Professor of Electrical & Computer Engineering at the Iowa State University, and prior to which he was with the ECE Dept. at the Univ. of Kentucky. He received B.Tech. in Electrical Eng. from Indian Institute of Technology, Kanpur (IITK) in 1987, and M.S. and Ph.D. in Electrical & Computer Engineering from the Univ. of Texas, Austin (UTAustin) in 1989 and 1991, respectively. Ratnesh's research interest spans sensors, networks, controls and software with application domains of cyberphysical (hybrid) systems, embedded and real-time systems, model-based software and web-services, power systems, energy harvesting, and sustainable agriculture. Ratnesh received Gold Medals from IITK, MCD Fellowship and Dissertation Award from UTAustin, Fellowships from NASA-Ames, Applied Research Lab-Penn State Univ, Idaho National Lab, and United Technologies Research Center, and several awards from NSF (including Research Initiation Award), DoE, ONR, General Motors, and Adobe. Ratnesh is a Fellow of the IEEE for contributions to discrete event system modeling, control, diagnosis and applications . Ratnesh has served on a number of editorial boards for IEEE and ACM, has been General/Program Chair and also given keynote talks at IEEE and ACM conferences.

Mark Leeson received the degrees of BSc and BEng with First Class Honors in Electrical and Electronic Engineering from the University of Nottingham, UK, in 1986 followed by a PhD in Engineering from the University of Cambridge, UK, in 1990. From 1990 to 1992 he worked as a Network Analyst for National Westminster Bank in London. After holding academic posts in London and Manchester, he joined the School of Engineering at Warwick, where he is now a Reader. His major research interests are coding and modulation, ad hoc networking and optical communication systems with over 220 publications to date. He is a Senior Member of the IEEE and a Fellow the UK Institute of Physics.

Ali Mahani, received the B.Sc degree in Electronic Engineering from the Shahid Bahonar University of Kerman, Kerman, Iran in 2001, the M.Sc degree and PhD both in Electronic Engineering from Iran University of Science and Technology (IUST) in 2003 and 2009 respectively. Since then he has been with the Department of Electrical and Electronic Engineering of Shahid Bahonar University of Kerman, where he is currently an assistant professor. His interests focus on Fault tolerant & Reliable digital circuits, Test and Testable Design and Communication Networks. Dr Mahani has more than 60 publications in these fields.

Maryam sadat Mirsadeghi, received the B.Sc and M.Sc degree both in electronic engineering from Shahid Bahonar University of Kerman, Kerman, Iran, in 2011 and 2014 respectively. Her research interests include wireless sensor networks, Digital Circuit design.

Hamed Nosrati received his B.Sc. degree in electronics engineering from Shahrood University of Technology, Shahrood, Iran, in 2010. Since October 2011, he is an M.Sc. Student at Sahand University of Technology, Tabriz, IRAN. In 2013, he received his M.Sc. degree in Communication Systems from Sahand University of Technology, Tabriz, Iran. His research activities include statistical signal processing, model based signal processing, array processing, adaptive networks, and distributed sensor networks.

George Papadopoulos received his Master's and Ph.D. degrees at M.I.T. in 1964 and 1970 respectively, all in Electrical Engineering. H was Professor in the Department of Electrical and Computer Engineering at the University of Patras and Director of the Applied Electronics Laboratory (1974-2007), and Professor Emeritus afterwards. Dr. Papadopoulos was Director of the Industrial Systems Institute (1998-2003) and Chairman of the Scientific Board of the Greek National Research Center DEMOKRITOS (1997-2001). He was Visiting Professor in the ECE Department of the University of MASS in Amherst (1984-1985) and served as Chairman of the ECE Department and Head of the Electronics and Computers Division. Recent research interests cover real time industrial networks, wireless and networked embedded systems, wireless sensor networks and RF communications. He has participated as coordinator, partner or subcontractor in many EU funded projects.

Sahil Patel graduated from Visvesvaraya Technological University (V.T.U.), India, with a Bachelor of Engineering in Telecommunications, achieving First Class with Distinction in 2012. He recently received his MSc in Electronic Systems with Communications with Distinction from the University of Warwick, UK. His major skills and interests cover a wide range of software languages and tools, leading to his employment in technology and finance going forward.

Sushanta Mohan Rakshit received his Bachelor of Technology degree from St. Thomas' College of Engineering and Technology, Kolkata, India in Electronics and Communication Engineering. He obtained his M.S. in Telecommunications Engineering from the University of Nebraska-Lincoln. He is currently pursuing his Ph.D. in Computer Engineering from the Department of Computer and Electronics Engineering at the University of Nebraska-Lincoln. His research interests include wireless sensor networks, sensor hardware design, and protocol design among others.

Hadi Rasouli received the B.Sc. degree in electrical engineering from Shahid Rajaee University, Tehran, Iran, in 2009, and the M.Sc. (Hons.) degree from Shahid Chamran University, Ahvaz, Iran, in 2014. His research interests are wireless sensor networks and digital circuits and systems.

Fahimeh Rezaei is a Ph.D. candidate in the Department of Computer and Electronics Engineering at the University of Nebraska-Lincoln, where she received her M.S. degree in Telecommunications Engineering in 2010. She is a Research Assistant at the Advanced Telecommunication Engineering Laboratory (TEL) and her research interests include covert channels and anonymizing networks, multimedia security and content protection, wireless communications and network protocols.

Herman S Sahota received his Ph.D in Computer Engineering from Iowa State University in 2013, with a thesis on the design of wireless sensor networks for precision agriculture applications. Herman received his B.Tech. in Electrical Engineering and M.Tech in Information and Communication Technology from the Indian Institute of Technology, Delhi (IITD) in 2006. Currently, he is working in the area of large scale data analytics at IBM, San Jose, California. Herman's research interests include design of wireless sensor networks, and machine learning on large datasets.

Mohammad Hossein Sedaaghi was born in Tehran, Iran. He received the B.Sc. and M.Sc. degrees from the Sharif University of Technology, Tehran, in 1986 and 1987, respectively. In 1998, he received the Ph.D. degree from Liver-pool University. From 2006 to 2007 he was a research fellow with the AIIA lab, Aristotle University of Thessaloniki, Greece. He is now a professor at Sahand University of Technology, Tabriz, Iran. He is the author of the textbook *DSP with simulation of analytical propositions and applied examples in Matlab* (Sahand University of Technology Press, 2012). Also, he is a co-author of over 20 journal papers. Prof. Sedaaghi has been the supervisor of several research projects in the field of digital signal processing and pattern recognition. His research interests include signal/image processing, pattern recognition, machine learning and biometrics.

Mousa Shamsi received his B.Sc. degree in Electrical Engineering (major: Electronics) from Tabriz University, in 1995. In 1996, he joined the University of Tehran, Tehran, Iran. He received his M.Sc. degree in Electrical Engineering (major: Biomedical Engineering) from this university in 1999. From 1999 to 2002, he taught as a lecturer at the Sahand University of Technology, Tabriz, Iran. In 2002, he entered the University of Tehran as a Ph.D. candidate. From 2002 to 2008, he was a Ph.D. student at the University of Tehran in Bioelectrical Engineering. In 2006, he was granted with the Iranian government scholarship as a visiting researcher at the Ryukyus University, Okinawa, Japan. From December 2006 to May 2008, he was a visiting researcher at this University. He received his PhD degree in Electrical Engineering (major: Biomedical Engineering) from University of Tehran in December 2008. From December 2008 to April 2013, he was an assistant professor at Faculty of Electrical Engineering, Sahand University of technology, Tabriz, Iran. From April 2013, he is an associate professor at Faculty of Electrical Engineering, Sahand University of technology, Tabriz, Iran. His research interests include biomedical image and signal processing, genomic signal processing, pattern recognition and adaptive networks.

Pradhumna Lal Shrestha received the B.E. in electronics and communication from Tribhuvan University, Kathmandu, Nepal, in 2007, and the M.S. degree in telecommunications engineering from the University of Nebraska-Lincoln, Lincoln, in 2011, where he is currently working towards the Ph.D. degree in the Department of Computer and Electronics Engineering, University of Nebraska-Lincoln, Lincoln. His research interests include theoretical and mathematical modeling, wireless communications, signal processing, high-speed networks, network security, among others.

Tsenka Stoyanova received her Diploma in Electrical Engineering from the Technical University of Varna, Bulgaria, in 1994 and Ph.D degree in Electrical Engineering from the University of Patras, Greece, in 2011. She worked as a Researcher in the marine and radar system development group at "Tcherno More"LTD, Varna, Bulgaria (1995-1997) and as a Researcher and Teaching Assistant at the Technical University of Varna, Bulgaria (1998-2001). From 2002 until today, she has been working as a Research Assistant at the University of Patras, Greece, and as a Researcher at ISI, Patras, Greece working on EU research Projects. Her research interests include wireless communication networks, RF signals propagation and modeling, deployment algorithms, target localization and tracking techniques.

Sayed Mostafa Taheri received his B.Sc. (Hons) in Electrical Engineering and Communications from Ferdowsi University of Mashhad (FUM), IR. He is currently a postgraduate student in Signal Processing and Communications at School of Electronics and Engineering of the College of Science and Engineering, The University of Edinburgh, United Kingdom. He has also been a research affiliate at Department of Communications and Electronics, Barthawa Institute of Technology, IR, since 2012. Mostafa is the co-author of two books, several peer reviewed research papers, and author of two IR patents. His research interests include wireless communications and signal processing. He is a member of Edinburgh University Young Scientific Researcher Association (EUYSRA).

Bin Wang received the Ph.D. degree in electrical engineering from the Ohio State University in 2000. He joined the Wright State University in September 2000, where he is currently full professor of computer science and engineering. His research interests include optical networks, real-time computing, mobile and wireless networks, cognitive radio networks, trust and information security, and semantic web. He was a recipient of the US Department of Energy Career Award and the AFOSR senior research fellowship. His research has been supported by US Department of Energy, National Science Foundation, Air Force Office of Scientific Research, Air Force Research Laboratories, Ohio Supercomputer Center, and the State of Ohio. He is a senior member of the IEEE.

Yuenong Zhu comes from Beijing, China. She lives in Southfield, MI, USA. She was born on December 17th, 1991. Her current research interests are in the areas of wireless communication and multimedia signal processing. She is a graduate student at Lawrence Technological University's Department of Electrical and Computer Engineering from 2013. She studied in Liaoning Technological University for her Bachelor degree of Telecommunications Engineering from 2010 to 2013.

Index

L

M

N

O

P

R

S

T

U

V

W

Z

www.ingramcontent.com/pod-product-compliance
Ingram Content Group UK Ltd.
Pitfield, Milton Keynes, MK11 3LW, UK
UKHW050952290926
15470UKWH00005B/1778